# Current Developments in Cell Biology

# Current Developments in Cell Biology

Editor: Samantha Granger

R CALLISTO
REFERENCE

www.callistoreference.com

**Callisto Reference,**
118-35 Queens Blvd., Suite 400,
Forest Hills, NY 11375, USA

Visit us on the World Wide Web at:
www.callistoreference.com

ISBN: 978-1-64116-147-3 (Hardback)

**Cataloging-in-Publication Data**

Current developments in cell biology / edited by Samantha Granger.
    p. cm.
Includes bibliographical references and index.
ISBN 978-1-64116-147-3
1. Cytology. 2. Cells. 3. Biology. I. Granger, Samantha.
QH581.2 .C87 2019
571.6--dc23

# Table of Contents

**Permissions**

**List of Contributors**

**Index**

# Preface

This book aims to highlight the current researches and provides a platform to further the scope of innovations in this area. This book is a product of the combined efforts of many researchers and scientists, after going through thorough studies and analysis from different parts of the world. The objective of this book is to provide the readers with the latest information of the field.

The study of the different structures and functions of the cell falls under the discipline of cell biology. It explains the organization of cell organelles, their physiological properties, signaling pathways, metabolic processes and interactions with the environment. The studies are done on a microscopic and molecular level. It encompasses the study of both prokaryotic and eukaryotic cells. Various technological implements are employed, such as optical microscopy, electron microscopy, scanning electron microscopy and fluorescence microscopy, besides many others. Cell culture, immunostaining, computational genomics, etc. are some of the different methods used in cell biology studies. This book contains some path-breaking studies in the field of cell biology. It traces the progress of this field and highlights some of its key concepts and applications. It will provide comprehensive knowledge to the readers.

I would like to express my sincere thanks to the authors for their dedicated efforts in the completion of this book. I acknowledge the efforts of the publisher for providing constant support. Lastly, I would like to thank my family for their support in all academic endeavors.

**Editor**

# The role of neural connexins in HeLa cell mobility and intercellular communication through tunneling tubes

Lina Rimkutė, Vaidas Jotautis, Alina Marandykina, Renata Sveikatienė, Ieva Antanavičiūtė
and Vytenis Arvydas Skeberdis[*]

**Abstract**

**Background:** Membranous tunneling tubes (TTs) are a recently discovered new form of communication between remote cells allowing their electrical synchronization, migration, and transfer of cellular materials. TTs have been identified in the brain and share similarities with neuronal processes. TTs can be open-ended, close-ended or contain functional gap junctions at the membrane interface. Gap junctions are formed of two unapposed hemichannels composed of six connexin (Cx) subunits. There are evidences that Cxs also play channel-independent role in cell adhesion, migration, division, differentiation, formation of neuronal networks and tumorigenicity. These properties of Cxs and TTs may synergetically determine the cellular and intercellular processes. Therefore, we examined the impact of Cxs expressed in the nervous system (Cx36, Cx40, Cx43, Cx45, and Cx47) on: 1) cell mobility; 2) formation and properties of TTs; and 3) transfer of siRNA between remote cells through TTs.

**Results:** We have identified two types of TTs between HeLa cells: F-actin rich only and containing F-actin and α-tubulin. The morphology of TTs was not influenced by expression of examined connexins; however, Cx36-EGFP-expressing cells formed more TTs while cells expressing Cx43-EGFP, Cx45, and Cx47 formed fewer TTs between each other compared with wt and Cx40-CFP-expressing cells. Also, Cx36-EGFP and Cx40-CFP-expressing HeLa cells were more mobile compared with wt and other Cxs-expressing cells. TTs containing Cx40-CFP, Cx43-EGFP, or Cx47 gap junctions were capable of transmitting double-stranded small interfering RNA; however, Cx36-EGFP and Cx45 were not permeable to it. In addition, we show that Cx43-EGFP-expressing HeLa cells and laryngeal squamous cell carcinoma cells can couple to the mesenchymal stem cells through TTs.

**Conclusions:** Different Cxs may modulate the mobility of cells and formation of TTs in an opposite manner; siRNA transfer through the GJ-containing TTs is Cx isoform-dependent.

**Keywords:** Tunneling tubes, Connexins, Gap junction channels, Cell mobility, siRNA transport

## Background

Directed cell migration is a pivotal process for normal development and morphogenesis of most animals, wound healing, tissue renewal, immune responses, angiogenesis, and tumor metastasis [1, 2]. During these processes, cells are subjected to stress and increased energy demands. A growing body of evidence suggests that a newly discovered form of intercellular communication referred to as "intercellular bridges" or "tunneling nanotubes" and "tunneling tubes" (TTs) contributes to cell movement [3–5] and provides means for energy supply to the remote cells by transporting ATP and even mitochondria [5–7]. Basically, TTs form when filopodial or lamellipodial protrusions from one cell attach to the target cell or during dislodgement of abutted cells [8]. In these ways, remote cells can establish open-ended, close-ended, or gap junction (GJ)-based communication. TTs have been shown to be implicated in the intercellular


* Correspondence: arvydas.skeberdis@lsmuni.lt
Institute of Cardiology, Lithuanian University of Health Sciences, 17 Sukilėlių Ave., 50009 Kaunas, Lithuania

electrical coupling and $Ca^{2+}$ flux; transfer of organelles or proteins; virus, pathogenic prion, and protein transmission; cell migration; and bacteria capture (reviewed in refs. [8–11]).

Recently, it has been proposed that connexins (Cxs), in addition to their canonical function of composing GJs, play a channel-independent role in cell adhesion, migration, division, differentiation, and tumorigenicity (reviewed in refs. [12–14]). Among 21 isoforms of Cxs found in the human genome, the role of only Cx26, Cx31.1, Cx32, and Cx43 in these processes has been described in the scientific literature so far [14]. It has been shown that Cx26 inhibits cell migration by altering the distribution of actin filaments; Cx31.1 decreases cell proliferation, delays the cell cycle at the G1 phase, and decreases migration and invasion of lung cancer cells; Cx32 increases cell proliferation, migration, and invasion; Cx43 increases cell migration, induces actin cytoskeleton reorganization, and reduces cell proliferation. Cxs interacting with cytoskeletal and tight junction proteins [12, 15] contribute to the regulation of cell migration, directed outgrowing of filopodial and lamellipodial protrusions [12, 16–18], and intercellular communication through TTs [5].

Eleven isoforms of Cxs have been identified in the nervous system where they can play an important role in the directed migration of cells, formation of neural processes, and progression of brain tumors [19]. Astrocytes express high levels of Cxs and can couple to neurons and oligodendrocytes. Astrocyte dysfunction may cause neuroautoimmune diseases, neoplasms, and epilepsy [20]. Neuronal processes share structural and functional similarities with TTs, and the directed formation of TTs between developing neurons and astrocytes has been demonstrated [21]. Cx43 accumulation at the tips of filopodium-like structures of astrocytes [22] may cause more frequent filopodium formation [23], stabilization of the leading edge protrusions in neuronal cells [24], and biological molecule transmission via TT-like structures [25]. Thus, TTs and GJs in the brain are likely to facilitate the intercellular exchange of materials and possibly genetic information. The last may be of particular importance in determining the stem cell differentiation, cancer invasion, and metastasis. The role of Cxs in cancer is controversial as well as tissue- and cancer stage-specific. Reduced Cx expression, or redistribution from the membrane to the cytoplasm, has been documented in a variety of cancers, including colon, lung, ovarian, breast, endometrial, and renal cell carcinomas and sarcomas, gliomas as well as in pre-cancerous tissues such as that of cervix. Up-regulated Cx expression has also been frequently described, and examples include breast cancer, skin cancers and various squamous cell carcinomas, colon cancer, and pancreatic cancer. Even within the same tumor type,

both increased and decreased Cx expression can be found (reviewed in ref. [26]). In gliomas, a decrease in Cx43 expression is associated with increasing proliferation and a higher tumor grade, but low-grade gliomas (for example Grade II) show increased levels of Cx43 (reviewed in ref. [27]). GJ intercellular communication between glioma cells and endothelial cells is also thought to play a critical role in glioma invasion. Up-regulation of Cx43 in micrometastases of breast cancer appears to facilitate their attachment to the pulmonary endothelium [27]. Thus, it looks like that Cx down-regulation facilitates cancer cell escape from solid tumors, while up-regulation promotes the formation of metastasis.

Valiūnas and colleagues [28] have demonstrated that small RNAs may be delivered through GJs composed of Cx43 but not of Cx26 or Cx32 in HeLa, Mβ16tsA (*wt*), and human mesenchymal stem cells. Also, GJ-dependent transfer of si/miRNAs has been shown to occur between primary cardiac myocytes [29]; human cardiac stem cells and postmitotic myocytes [30]; bone marrow stromal and breast cancer cells [31]; glioma cells [32]; glioma stem cells and MSCs [33]. It is assumed that transfer of small RNAs with high molecular weight via GJs is possible due to rod-shaped morphology of siRNAs, a diameter of which allows their passage through GJs with larger pores [34, 35]. However, the transfer of siRNAs between abutted cells through GJs is under debate so far due to experimental difficulties to reject the pinocytotic pathway of transfer [34]. Our previous study was the first that demonstrated the transfer of double stranded siRNA between remote human laryngeal squamous cell carcinoma (LSCC) cells through open-ended and even through Cx43 GJ-containing TTs [5].

In the present study, we used the HeLa cell model to examine the impact of neural Cxs (Cx36, Cx40, Cx43, Cx45, and Cx47) on the following: 1) cell mobility; 2) formation and properties of TTs; and 3) transfer of siRNA between remote cells through TTs.

## Results

### General properties of TTs between HeLa cells

To examine the impact of different Cx expression on TT morphology, HeLa cells were stably transfected with Cx36-EGFP, Cx40-CFP, Cx43-EGFP, Cx45, or Cx47. Non-transfected HeLa *wt* cells were used as control. We found that HeLa cells, either *wt* or expressing different Cxs, in the culture formed intercellular TTs of various width (ranging from < 200 nm to > 2 μm) and length (up to 70 μm; only TTs longer than 10 μm were taken into account). Time-lapse imaging revealed highly dynamic formation of filopodium-like TTs that were identified as not touching the substratum (Fig. 1a-c). The diameter of the thinnest TTs (<200 nm) could not be measured precisely by conventional optical microscopy as well as

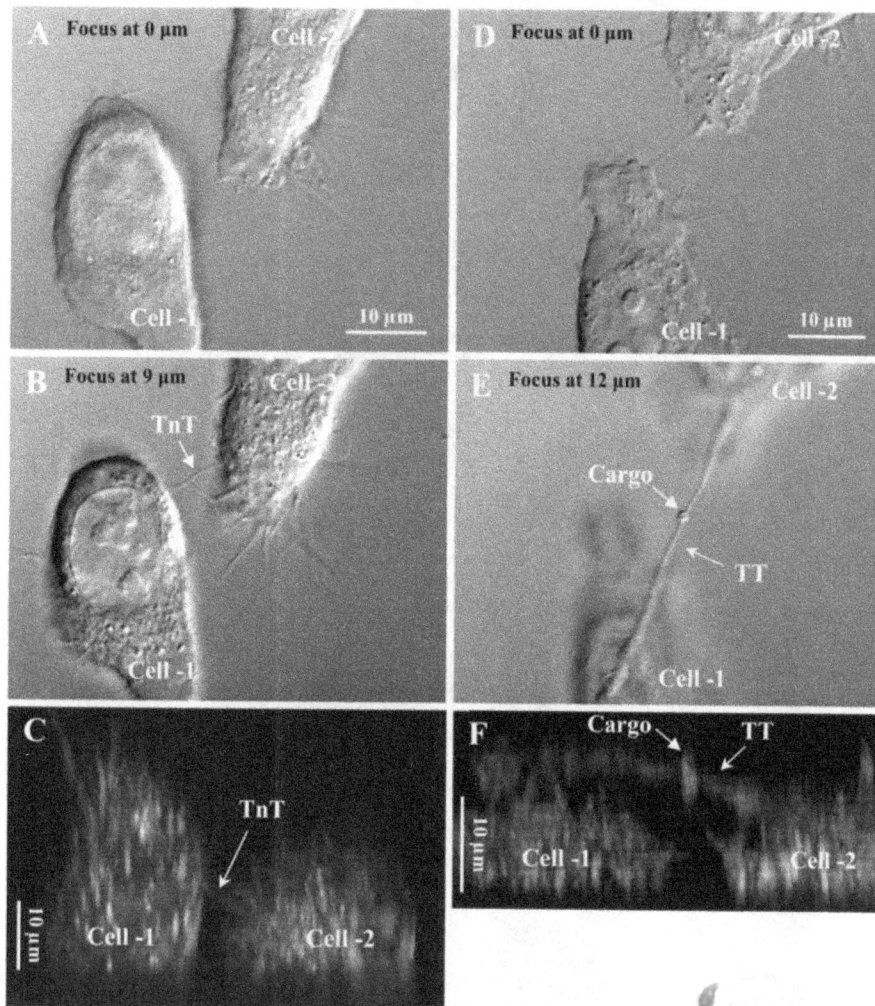

**Fig. 1** Formation of TTs between HeLa cells. **a-c** TTs formed by the filopodium outgrowth mechanism. **d-f** TTs formed in the process of cell division and successive dislodgment or by the lamellipodium outgrowth mechanism. In both the cases, the pictures represent the top view of cells at a different focus **a** and **b**; **d** and **e**) and Z-X reconstruction showing TTs raised above the substratum (**c** and **f**)

their electrical and permeability properties could not be examined due to a short lifetime (tens of seconds).

Much thicker TTs (>300 μm) formed during cell division and subsequent dislodgment or by the lamellipodium outgrowth mechanism. These TTs also were found raised above the substratum (Fig. 1d-f) and were involved in cargo transport either inside the TTs or along their outer surface (indicated by arrows in Fig. 1e and f). However, the leading edges of lamellipodium extensions were usually attached to the substratum and participated in cell motility and TT formation. The lifetime of these TTs lasted tens of minutes and even hours and allowed to use the dual whole-cell patch-clamp technique and fluorescence microscopy for characterization of their formation and properties.

HeLa cells grown to confluence on the glass coverslips formed numerous GJ plaques that can be visible due to chimeric fluorescent proteins (Fig. 2a and b). As it was demonstrated before, abutted HeLa cells expressing Cxs used in the current study formed functional GJs permeable to fluorescent dyes of different molecular weight and net charge [36–38]. In contrast, abutted HeLa *wt* cells did not exhibit any electrical coupling or dye transfer between cells.

However, in this study, the cells were grown at relatively low density and fluorescently tagged proteins helped us confirm the presence and site of GJ plaques in the TT in addition to electrical measurements. We identified two types of TTs between *wt* or different Cx-expressing HeLa cells: TTs containing only F-actin (F-TTs) (Fig. 2c and d) and those containing F-actin and α-tubulin (Fα-TTs) (Fig. 2e and f). The cells were labeled with phalloidin and anti-α-tubulin to visualize the actin network and microtubules, respectively. HeLa cells on average formed 17 and

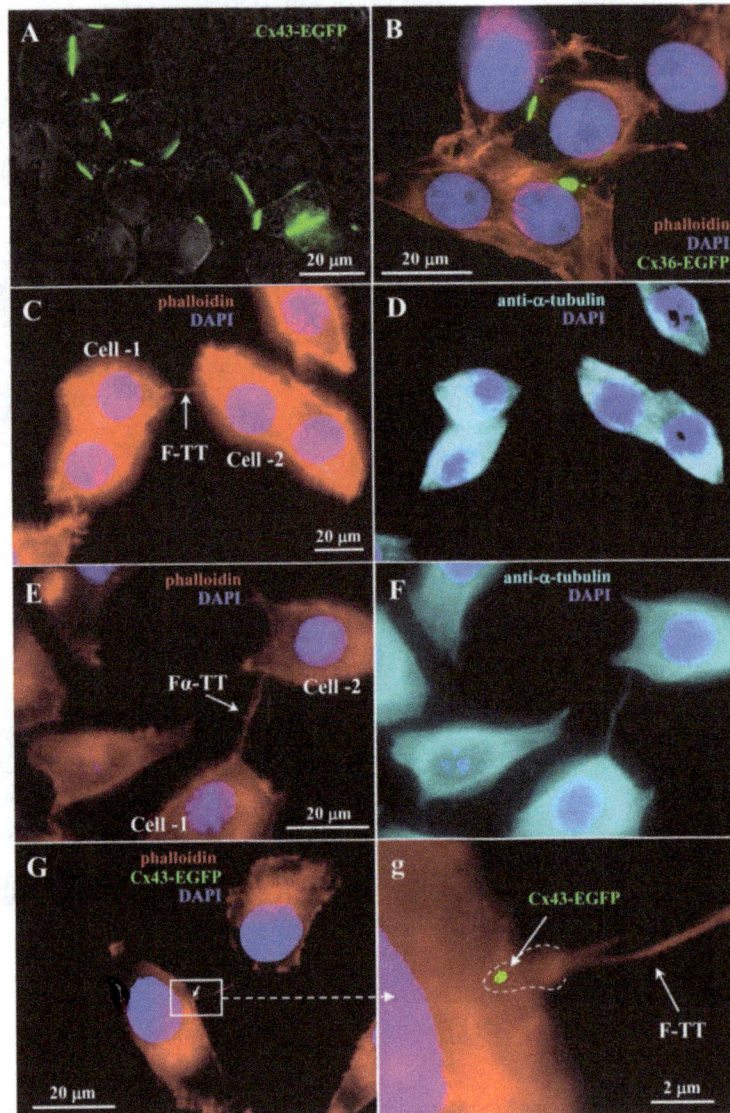

**Fig. 2** Types of TTs formed between HeLa cells. **a** and **b** A typical view of Cx43-EGFP- and Cx36-EGFP-expressing HeLa cells, respectively, exhibiting multiple fluorescent GJ plaques. **c** and **d** Only F-actin-containing F-TTs. **e** and **f** F-actin- and α-tubulin-containing Fα-TTs. (**G** and **g**) TTs formed GJs at the Cx-expressing cell border (see green dots and a white arrow indicating Cx43-EGFP cluster)

83 % of F-TTs and Fα-TTs, respectively, and this proportion was not affected by Cxs expression. Cells expressing different Cxs formed TTs with clusters of respective Cxs at the membrane interface with the remote cell (Fig. 2G and g).

### The impact of different Cxs on formation and electrical properties of TTs

The properties of TTs between *wt* and Cx-expressing HeLa cells are presented in Table 1. None of these connexins affected the geometry of TTs; however, the number of TTs calculated per 1 mm$^2$ or per 100 cells was significantly higher between cells expressing Cx36-EGFP and lower between cells expressing Cx43-EGFP, Cx45, and Cx47 compared with HeLa *wt* cells (Fig. 3).

The electrical properties of TTs were examined by the dual whole-cell patch-clamp technique (Fig. 4a). HeLa cells formed functional GJ-containing TTs independent on the isoform of expressed Cx as confirmed by the measurement of voltage gating typical of GJs. Electrical coupling and voltage gating were estimated by applying 30-s voltage ramps from 0 to −120 mV in the cell-1 (Fig. 4b, upper panel) and measuring the current response in the cell-2 (Fig. 4b, middle panel demonstrates the typical $I_T$ response of open-ended TTs, and lower panel, of GJ-containing TTs). $g_T$-$V_T$ dependences (Fig. 4c) were calculated from $I_T$ responses to the $V_T$ ramps shown in Fig. 4b. Fig. 4d and Table 1 show that $g_T$ strongly depended on the single channel conductance

**Table 1** Summary of properties of open-ended and GJ-containing TTs

| Cx isoform | L (μm) | Number of TTs per mm²/per 100 cells[c] | $g_T$ (nS) | siRNA/AF488permeability (×10⁻¹⁴ cm³/s) |
|---|---|---|---|---|
| TT/(−)GJ(HeLa *wt*) | $16.3 \pm 0.4 (n = 214)$ | $51 \pm 8/9.5 \pm 1.3$ | $6.0 \pm 1.1^{[a]}(n = 17)$ | $65 \pm 38 (n = 5)$ |
| TT/Cx36-EGFP | $17.0 \pm 0.4 (n = 330)$ | $79 \pm 9^*/30 \pm 7.1^*$ | $0.8 \pm 0.2 (n = 16)$ | $-^{[b]}(n = 5)$ |
| TT/Cx40-CFP | $15.9 \pm 0.5 (n = 126)$ | $46 \pm 6/9.7 \pm 1.8$ | $16.5 \pm 2.8 (n = 20)$ | $4.4 \pm 2.0 (n = 6)$ |
| TT/Cx43-EGFP | $17.0 \pm 1.1 (n = 96)$ | $23 \pm 4^*/4.7 \pm 0.7^*$ | $8.8 \pm 3.4 (n = 7)$ | $32 \pm 17 (n = 4)$ |
| TT/Cx45 | $14.3 \pm 1.0 (n = 68)$ | $5.7 \pm 1.6^*/1.4 \pm 0.3^*$ | $4.1 \pm 0.5 (n = 41)$ | $-^{[b]}(n = 8)$ |
| TT/Cx47 | $15.1 \pm 0.5 (n = 112)$ | $27 \pm 3^*/5.0 \pm 0.9^*$ | $3.0 \pm 0.6 (n = 15)$ | $4.3 \pm 1.9 (n = 6)$ |

"−" nonpermeable; the number of experiments is indicated in parentheses; the diameter of TTs did not depend on the isoform of expressed Cxs and it was 0.9 μm on average ($n = 210$; varied from 0.4 to 2.4 μm)
*$p < 0.05$ compared with HeLa *wt* cells
[a]the majority of junctions were closed-ended (53 out of 70); [b] Cx36-EGFP and Cx45 GJs were impermeable to siRNA/AF488; however, they were permeable to AF488 ($6.3 \pm 1.9 \times 10^{-14}$ cm³/s ($n = 5$) and $6.4 \pm 4.9 \times 10^{-14}$ cm³/s ($n = 4$), respectively); [c]Cell densities measured 36 h after seeding were $81 \pm 5$; $75 \pm 6$; $78 \pm 3$; $73 \pm 7$; $70 \pm 3$; and $79 \pm 5$ cells per region for *wt*; Cx36-EGFP-; Cx40-CFP-; Cx43-EGFP-; Cx45-; and Cx47-expressing HeLa cells, respectively

of the expressed Cx (presented in the Discussion). Interestingly, $g_T$ of open-ended TTs between HeLa *wt* cells was smaller than that between particular Cx-expressing cells. These TTs did not distinguish as having the highest conductance presumably because during cell dislodgment in the process of cytokinesis, open-ended TTs rapidly turn into close-ended ones (~76 % of TTs between HeLa *wt* cells were close-ended) before they rupture. This observation supports the significance of GJs in determining the strength of communication between cells. In Cx-transfected cells, the ratio of close-ended, open-ended, and GJ-containing TTs did not depend on the Cx type

and was ~1:2:4 ($n = 94$). In rare cases, TTs between Cx-expressing HeLa cells did not couple the cells electrically for 3 possible reasons: 1) TTs were close-ended; 2) GJ-dependent electrical coupling was not established yet in the process of *de novo* formation of TT; and 3) GJ-dependent electrical coupling was already lost due to cell separation.

In general, $g_T$ should at least in part depend directly on TT width ($d_T$) and inversely on TT length ($L_T$); however, in our experiments $g_T$ only moderately correlated with TT geometry suggesting that the total conductance of TTs is more complex. For instance, $g_T$

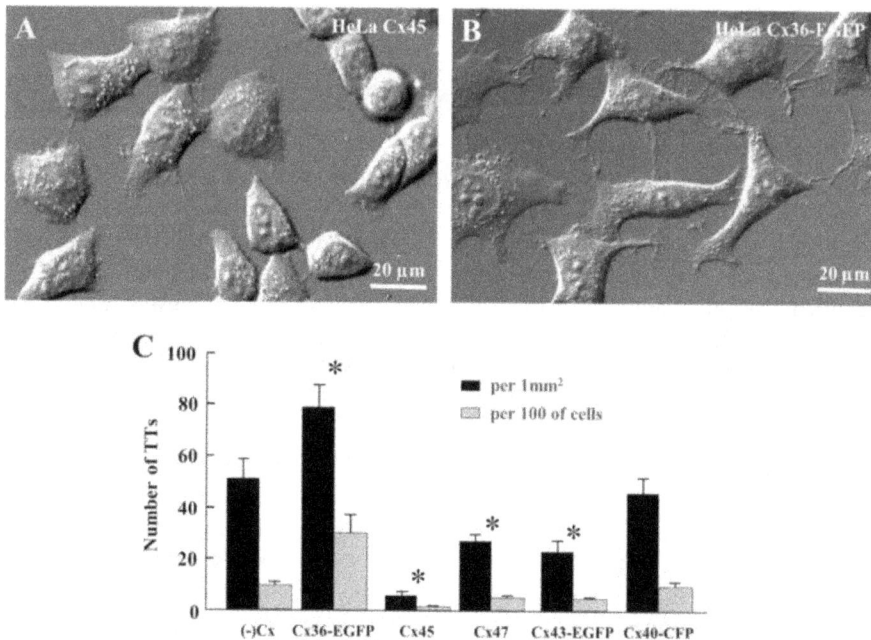

**Fig. 3** Comparison of TT formation capabilities of HeLa cells expressing different **Cxs**. A typical view of TT formation between HeLa Cx45 (**a**) and Cx36-EGFP cells (**b**). **c** *Wt* and each Cx-expressing HeLa cells were seeded in 24-well plates with glass coverslips on the bottom at equal densities ($3 \times 10^4$ in each well) and 36 h later were examined using differential interference contrast microscopy with ×20 lens. The experiments were repeated in 2 passages. The number of TTs was counted in 15 randomly selected regions per coverslip and presented as the average values per 1 mm² or per 100 cells ($n = 2*15 = 30$) (see Table 1). *$p < 0.05$

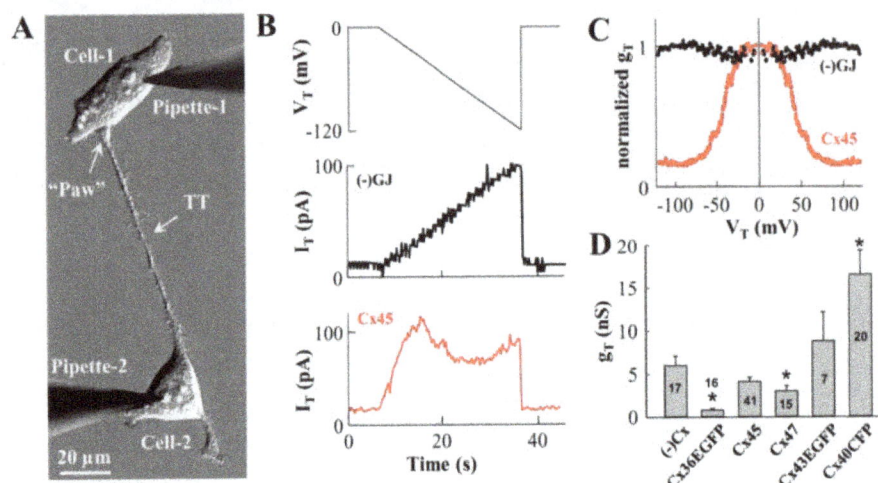

**Fig. 4** Characterization of electrical properties of TTs formed between HeLa *wt* and Cx36-EGFP-, Cx40-CFP-, Cx43-EGFP-, Cx45-, and Cx47-expressing cells. **a** The TT connecting a pair of HeLa Cx45 cells. **b** Electrical properties of TTs were evaluated by applying a voltage ramp of negative polarity from 0 to −120 mV (**b**, upper panel) to the cell-1 and measuring junctional current in the cell-2 (**b**, middle panel demonstrates a typical $I_T$ response of open-ended TTs, and lower panel, of GJ containing TTs); **c** Typical $g_T$-$V_T$ dependences of open-ended (black) and GJ-containing (red) TTs calculated from $I_T$ responses to $V_T$ ramps shown in (**b**) with their symmetric counterparts. **d** Summary of conductances of TTs between HeLa *wt* and different Cx-expressing cells (the number of experiments is indicated on the bars; exact numbers are presented in **Table** 1). *$p < 0.05$

strongly depends on the number of functional GJ channels that are not related to TT geometry and size of GJ plaques [37, 39, 40]. Also, not the external diameter of the TT, but the internal one, is a $g_T$-limiting factor, and unfortunately, there are no means for estimation of its dimensions.

### The impact of different connexins on the mobility of HeLa cells

The mobility properties of different Cx-expressing HeLa cells were examined by the wound healing assay (Fig. 5). Occupation of the scraped area by Cx36-EGFP- and Cx40-CFP-expressing cells was faster compared with *wt* or Cx43-EGFP-, Cx45-, and Cx47-expressing cells; however, there was no statistically significant difference between *wt* and Cx43-EGFP-, Cx45-, or Cx47-expressing cells. The percentage of the occupied scraped area after 12 h was as follows: in HeLa *wt*, 42.8 ± 6.0 %; HeLa Cx36-EGFP, 69.6 ± 5.4 %; HeLa Cx40-CFP, 68.3 ± 6.7 %; HeLa Cx43-EGFP, 29.7 ± 1.1 %; HeLa Cx45, 29.0 ± 2.4 %; and in HeLa Cx47 cells, 35.3 ± 3.3 %.

### Cx isoform-specific permeability of TTs to siRNA

Previously, we have reported that small RNAs (siRNA/AF488, negative control double stranded siRNA conjugated with AF488) were capable of transiting between LSCC cells through open-ended and Cx43 GJ-containing TTs. In this study, we examined whether TTs containing GJs composed of different neural Cxs were permeable to siRNA using the same approach as described previously

[5]. To measure TT permeability, the pipette-1 containing siRNA/AF488 (2 μM) was attached to the cell-1 (Fig. 6a) and after opening the patch, siRNA diffused to the cell-1 followed by its transfer or not through the TT to the cell-2 (Fig. 6b). Typically, accumulation of siRNA/AF488 in the cell-2 started after ~10 min delay compared with cell-1. The total permeability of TT, $P_T$, was evaluated using equation 3, which accounted for changes in fluorescence intensity in the cell-1 ($FI_1$, Fig. 6c) and the cell-2 ($FI_2$, Fig. 6d). At the end of siRNA transfer measurement, the patch in the cell-2 was opened to measure $g_T$ and $g_T$-$V_T$.

As it is demonstrated in Figs. 6a-d, TTs between HeLa cells expressing Cx45 and Cx36-EGFP were impermeable to siRNA/AF488 even though they exhibited substantial electrical coupling and permeability to AF488 (Figs. 6e and f). However, TTs containing Cx40-CFP, Cx43-EGFP, and Cx47 GJs as well as open-ended TTs between non-transfected HeLa *wt* cells were permeable to siRNA/AF488 (Figs. 6g-j, Table 1).

Additional file 1: Figure S1 demonstrates that in the monolayer of HeLa *wt* cells, AF488 injected into the single cell does not spread to the adjacent cells (A-E), and there is no electrical coupling between the abutted HeLa *wt* cells (F and G).

### Discussion

Agnati and his colleagues in their elegant review divided the intercellular communication in the brain into two main modes: wiring transmission (neuronal processes

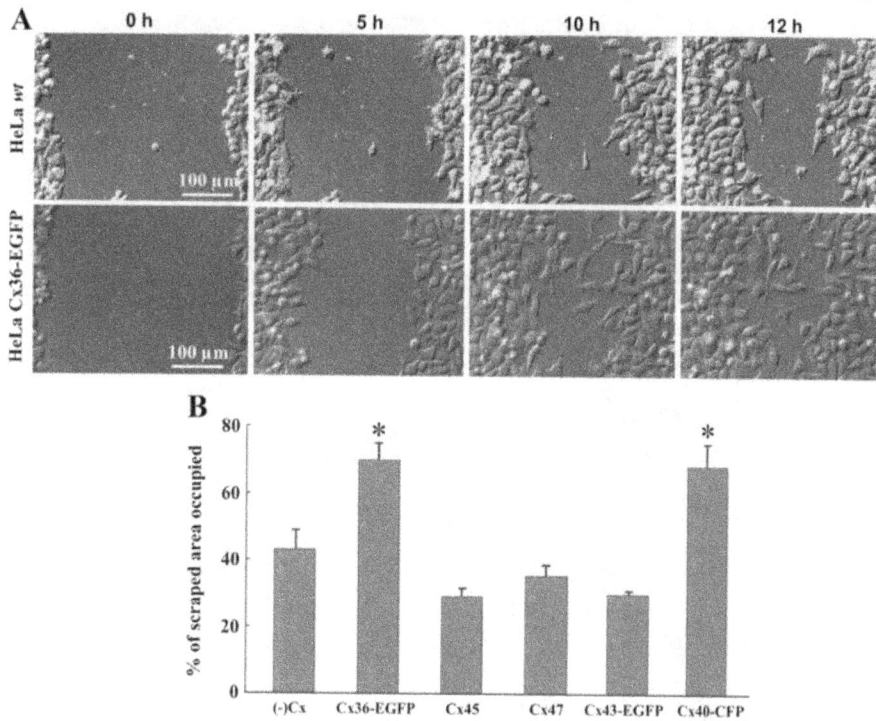

**Fig. 5** Evaluation of mobility properties of different Cx-expressing HeLa cells by the wound healing assay. **a** The representative pictures of wound healing in HeLa *wt* and HeLa Cx36-EGFP cell monolayers at different time points (5, 10, and 12 h after wound formation). **b** The percentage of the scraped area occupied 12 h after wound formation in different Cx-expressing HeLa cell monolayers ($n = 3$). *$p < 0.05$ compared with HeLa *wt* cells

and TTs with their electrical and chemical synapses) and volume transmission (extracellular vesicles) [41]. However, both these modes can be encompassed by TTs that connect cells over long distances by establishing open-ended (direct cell-to-cell channels), close-ended (synaptic transmission and/or active transport), or GJ-based connections. In such a way, TTs are implicated in intercellular electrical and metabolic coupling as well as transfer of vesicles, proteins, organelles, and even genetic material. All these processes are involved not only in the normal functioning of the brain but also in the development and progression of neurodegenerative diseases and brain tumors. Our recent study has demonstrated that TTs containing Cx43 GJs were capable of transferring siRNA/AF488 [5] and raised a question whether TTs containing GJs composed of other neural Cxs are permeable to it. In parallel, we evaluated a GJ channel- and hemichannel-independent impact of the same Cxs on HeLa cell migration and development of TTs.

The novelty and main findings of the current study are the following: 1) Cx36-EGFP promotes while Cx43-EGFP, Cx47, and especially Cx45 inhibit the formation of TTs between HeLa cells; 2) Cx36-EGFP- and Cx40-CFP-expressing HeLa cells demonstrate better mobility properties; 3) TTs containing Cx40-CFP, Cx43-EGFP, and Cx47 are permeable while containing Cx36-EGFP and Cx45 are not permeable to siRNAs.

As it is seen from Table 1, the conductance of GJ-containing TTs depended on the type of the expressed Cx, i.e. Cxs with higher single channel conductance (single channel conductances of Cx36, Cx45, Cx47, Cx43, and Cx40 are $\sim 10$, 30, 55, 100, and 170 pS, respectively (reviewed in ref. [19])) determined the higher conductance of the TT assuming that the number of channels in the GJ plaque-containing TTs was alike in all cases since TT geometry was not affected by the type of the expressed Cx. Open-ended TTs of HeLa *wt* cells did not distinguish by the highest conductance, presumably because open-ended TTs rapidly turn into close-ended during cell dislodgment in the process of cytokinesis ($\sim 76$ % of the TTs between HeLa *wt* cells were close-ended). This observation supports the role of GJs in determining the strength of communication between remote cells connected through TTs.

One of the steps in the cell motility cycle is integrin-dependent adhesion to the substrate [42]. Our observation that Cxs localize on the tips of lamellipodium-like protrusions and at their contact with the remote cell suggests that Cxs may interact with cellular adhesion and tight junctional proteins of other cells. For instance, Cx43 has been shown to exert effects on migration by interfering with receptor signaling, cytoskeleton remodeling, and tubulin dynamics [15]; in the developing brain, Cx43

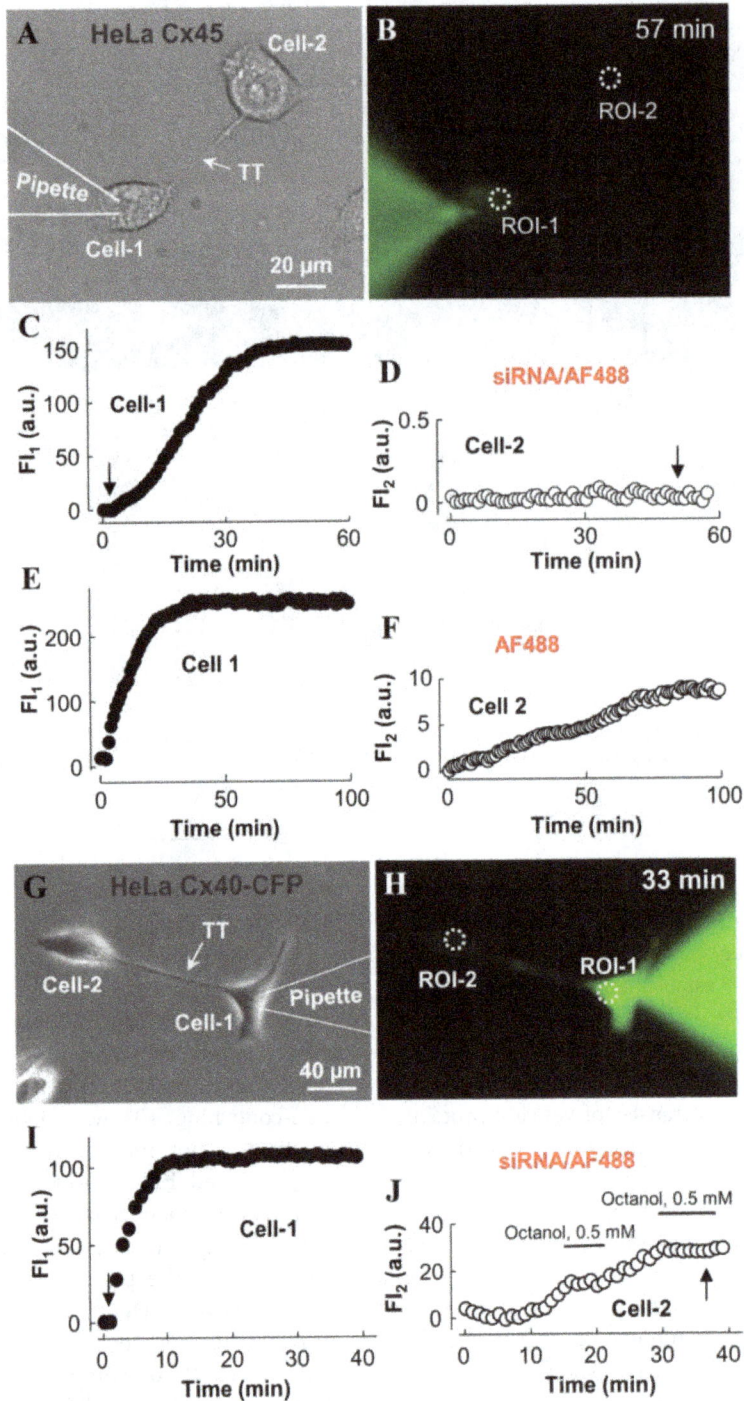

**Fig. 6** SiRNA transfer through TTs between different Cx-expressing HeLa cells. **a** A pair of Cx45-expressing HeLa cells connected through the TT. **b** SiRNA/AF488 (2 μM) introduced into the patch pipette entered the cell-1 after patch opening, spread along the TT but did not enter the remote cell-2. **c** and **d** Kinetics of siRNA/AF488 accumulation in the cell-1 and the cell-2, respectively. $g_T$ measured at the end of the experiment was 2.6 nS. Arrows point to the moments of patch opening in the cell-1 and the cell-2. **e** and **f** Kinetics of AF488 accumulation in the cell-1 and the cell-2, respectively. **g** A pair of Cx40-CFP-expressing HeLa cells connected through the TT. **h** After the patch was opened in the cell-1, siRNA/AF488 entered the cell-1, spread along the TT, and accumulated in the cell-2. **i** and **j** Kinetics of siRNA/AF488 accumulation in the cell-1 and the cell-2, respectively. siRNA/AF488 accumulation in the cell-2 was arrested by octanol (0.5 mM), a GJ blocker. $g_T$ measured at the end of the experiment was 10 nS

adhesion-promoting properties facilitate the migration of pyramidal cell precursors from the ventricular zone toward the ventricular plate [43, 44]. Also, the expression of Cx43 inversely correlated with the migration rate in the culture of canine brain tumor cells [45]. Similarly, in our study, Cx43-EGFP- as well as Cx45- and Cx47-expressing HeLa cells demonstrated tendency to reduced mobility compared with HeLa *wt* cells (Fig. 5b). Interestingly, cells expressing the same Cxs exhibited reduced formation of TTs while Cx36-EGFP-expressing cells formed the largest number of TTs and demonstrated the highest mobility (Fig. 3c); Cx40-CFP increased cell mobility but TT formation did not differ from that in HeLa *wt* cells. These observations suggest that cell mobility and TT formation may be regulated through different mechanisms that may be controlled by different Cxs in an opposite manner.

The Cx isoform-specific permeability of TTs to siRNA/ AF488 suggests that GJs may play an important role in forming the limiting barrier of genetic material transfer through homotypic and heterotypic homocellular and heterocellular connections. TTs containing GJs composed of Cxs with the lowest single channel conductances (Cx36-EGFP and Cx45) were impermeable to siRNA/ AF488; however, they both were permeable to AF488 (Fig. 6). The differences in permeability could be determined by different diameters of GJ pores and by distribution of fixed charge sites in the pore. The main factor that can facilitate the effective permeability of genetic material is rod-shaped morphology of a siRNA molecule [34, 35]. Also, it is worth noting that siRNA transfer is not a simple process of diffusion but rather energy-dependent, motor protein-involving transportation. For instance, kinesin and dynein have been shown to be motor proteins associated with microtubule transport [46] that can be blocked by azide, an inhibitor of ATP synthesis [47, 48]. This assumption stays in line with our observation that $g_T$ and $P_T$ of siRNA did not correlate either among Cxs with different single channel conductances (see Table 1) or in the single Cx series of experiments.

Epithelial-to-mesenchymal and mesenchymal-to-epithelial transitions play crucial roles in cancer metastasis, and these processes can be controlled and reversed by miRNAs [49]. The known pathways of miRNA transfer between cells are the following: extracellular vesicles (exosomes, ectosomes, and apoptotic bodies); circulating RNA in a vesicle-independent form; synapses; GJs and TTs [50]. The last pathway would be the most swift and efficient; however, it still is not definitely proven. A general view that Cxs are down-regulated in cancer cells needs to be revised because many studies including our own [5] have demonstrate the presence of Cxs and GJs in tumor tissues. Even though epithelial-to-mesenchymal transition is associated with Cx down-regulation or loss of communication through GJs between cells that communicate normally (this process is related to the onset of neoplasia and tumorigenesis), mesenchymal-to-epithelial transition is related to the up-regulation of Cx expression and acquisition of novel GJ-based communication between cell types that do not communicate in healthy tissue (these processes are related to angiogenesis, invasion, and metastasis) (reviewed in refs. [12, 51]). For instance, in glioma cell populations, the over-expression of Cxs and GJs between tumor and non-tumor glia cells facilitates the invasion of glioma cells [52]. The expression of Cx43 in the glioma core is very heterogeneous: Cx expression is restricted to minor populations of cells endowed with invasive and cancer stem cell-like properties and able to migrate, but other cells non-expressing Cx43 are able to proliferate [27]. Migrating glioma cells expressing Cx43 may then be able to induce the development of secondary or recurrent gliomas with GJs [13]. Cx43 expression is increased in breast cancer cells predestined to spread to the brain [53]. Disseminating breast cancer or melanoma cells migrate along the luminal surface searching for suitable sites to extravasate and form functional GJs (Cx43, Cx26) with brain endothelial and/or glial cells to initiate brain metastasis, and first lesions develop in Cx-rich vasculature and stroma of the brain [54]. Recent studies have suggested the involvement of miRNA in coordination of the gene expression program determining tumor metastasis [55]. Delivery of miRNA to remote cells can be facilitated by open-ended or GJ-containing TTs. The first evidence of the effectiveness of GJ-dependent pathway was provided by Valiunas and colleagues, who demonstrated that siRNA delivered through GJs down-regulated a reporter gene in the recipient cell [35].

## Conclusions

Our data demonstrate a new modulatory effect of different neural Cxs on cell migration, TT formation, and permeability to siRNA. These results may contribute to the knowledge about mechanisms of cancer invasion and metastasis.

## Methods

### Cell lines and culture conditions

Experiments were performed on HeLa (human cervix carcinoma, ATCC CCL-2, Manassas, VA, USA) cells stably transfected with Cxs tagged with green or cyan fluorescent proteins (Cx36-EGFP, Cx43-EGFP and Cx40-CFP) or untagged Cx45 and Cx47. Stable HeLa cell lines expressing Cxs used in this study were obtained in collaboration with the laboratory of Dr. F. Bukauskas (Albert Einstein College of Medicine, New York, USA). Briefly, vectors were transfected into HeLa cells using Lipofectamine 2000 (Invitrogen, USA) and following the transfection protocol of manufacturer. Cell lines expressing Cx36-EGFP and

Cx43-EGFP were selected using 500 µg/ml G418/geneticin (Sigma-Aldrich Co.), whereas 1 µg/ml puromycin (Invitrogen, USA) was used for selection of HeLa Cx40-CFP, Cx45, and Cx47 cell lines. The construction protocols of vectors are described elsewhere [56, 57]. Cells were grown in DMEM medium containing 10 % fetal bovine serum (FBS), penicillin/streptomycin mix (100 U/ml penicillin and 100 µg/ml streptomycin; Gibco Laboratories). Typically, the cells were analyzed on the second day after passage. LSCC cells were prepared as described elsewhere [5].

### Time lapse imaging

Time lapse imaging of HeLa cell mobility and TT formation in the culture medium was performed at 37 °C in the humidified atmosphere of 5 % $CO_2$ using an incubation system INUBG2E-ONICS (Tokai Hit, Shizuoka-ken, Japan) with an incubator mounted on the stage of motorized Olympus IX81 microscope (Olympus Europe holding Gmbh, Hamburg, Germany) with Orca-$R^2$ cooled digital camera (Hamamatsu Photonics K.K., Japan), fluorescence excitation system MT10 (Olympus Life Science Europa Gmbh, Hamburg, Germany), and XCELLENCE software (Olympus Soft Imaging Solutions Gmbh, München, Germany).

### Wound healing assay

Cell migration analysis was performed by the wound healing assay. Cells were grown to confluence on the glass coverslips. Then the monolayer was scraped with a sterile surgical blade. The cells were washed with fresh growth medium to remove cell debris. Wound healing was evaluated by measuring the cell-free area remaining in the wound [58] after 5, 10, or 12 h.

### Electrophysiological measurements

For simultaneous electrophysiological and fluorescence recording, cells grown onto glass coverslips were transferred to an experimental chamber with constant flow-through perfusion mounted on the stage of the inverted microscope Olympus IX8. Junctional conductance $g_T$ between the cells connected by the TT was measured using the dual whole-cell patch-clamp technique. Cell-1 and cell-2 of a cell pair were voltage clamped independently with a patch-clamp amplifier MultiClamp 700B (Molecular Devices, Inc., USA) at the same holding potential, $V_1 = V_2$. Voltages and currents were digitized using a Digidata 1440A data acquisition system (Molecular Devices, Inc., USA) and acquired and analyzed using pClamp 10 software (Molecular Devices, Inc., USA). By stepping the voltage in the cell-1 ($\Delta V_1$) and keeping the other constant, junctional current was measured as the change in current in the unstepped cell-2, $I_T = \Delta I_2$. Thus, $g_T$ was obtained from the ratio $-I_T/\Delta V_1$, where $\Delta V_1$ is equal to

transjunctional voltage ($V_T$) and negative sign indicates that the junctional current measured in cell-2 is oppositely oriented to the one measured in cell-1. To minimize the effect of series resistance on the measurements of $g_T$ [59], we maintained pipette resistances below 3 MOhms. Patch pipettes were pulled from borosilicate glass capillary tubes with filaments. Experiments were performed at room temperature in modified Krebs-Ringer solution (in mM): NaCl, 140; KCl, 4; $CaCl_2$, 2; $MgCl_2$, 1; glucose, 5; pyruvate, 2; HEPES, 5 (pH 7.4). Patch pipettes were filled with saline containing (in mM): KCl, 130; Na aspartate, 10; MgATP, 3; $MgCl_2$, 1; $CaCl_2$, 0.2; EGTA, 2; HEPES, 5 (pH = 7.3).

### Fluorescence Imaging and siRNA Transfer Studies

Fluorescence signals were acquired using the Olympus IX81 microscope with Orca-$R^2$ digital camera, fluorescence excitation system MT10, and XCELLENCE software. For siRNA transfer studies, siRNA conjugated with Alexa Fluor-488 fluorescent dye (siRNA/AF488, QIAGEN, Venlo, Netherlands) or Alexa Fluor-488 hydrazide (AF488, Life Technologies) was introduced into cell-1 of a pair through a patch pipette in whole-cell voltage-clamp mode. Typically, this resulted in loading of the cell-1, followed by siRNA/AF488 or AF488 transfer via the TT to the neighboring cell-2. At the end of siRNA/AF488 or AF488 transfer measurement, the patch in the cell-2 was opened to measure $g_T$ in dual whole-cell patch-clamp mode. The presence or absence of GJ in the TT was checked by measuring $g_T$-$V_T$. Evaluation of GJ permeability to fluorescent dyes from changes in fluorescence intensity in both cells was previously described elsewhere [36, 39, 40]. In brief, the cell-to-cell flux ($J_T$) of the dye in the absence of transjunctional voltage ($V_T = 0$ mV) can be determined from changes of dye concentration in the cell-2 ($\Delta C_2$) over the time interval ($\Delta t$) as follows:

$$J_T = \frac{vol_2 \cdot \Delta C_2}{\Delta t} \qquad (1)$$

where $vol_2$ is the volume of cell-2. Then, according to the modified [60] Goldman-Hodgkin-Katz (GHK) equation [61], the total junctional permeability ($P_T$) can be described in consequence:

$$P_T = \frac{J_T}{C_1 - C_2} = \frac{vol_2 \cdot \Delta C_2}{\Delta t \cdot (C_1 - C_2)} \qquad (2)$$

where $C_1$ and $C_2$ are dye concentrations in the cell-1 (dye donor) and the cell-2 (dye recipient), respectively. Cell volume was approximated as a hemisphere. The diameter of a hemisphere was determined by averaging the longest and the shortest diameters of the cell; the volume of examined HeLa cells was ~1800 µm$^3$ on average. Assuming that the dye concentration is directly proportional to fluorescence intensity (C = k • FI), equation 2 can be modified as follows:

$$P_T = \frac{vol_2 \cdot \Delta FI_2}{\Delta t \cdot (FI_1 - FI_2)} \tag{3}$$

where $\Delta FI_2 = FI_{2,n+1} - FI_{2,n}$ is the change in FI in cell-2 over time, $\Delta t = (t_{n+1} - t_n)$; n is the nth time point in the recording. To minimize siRNA/AF488 bleaching, studies were performed using time-lapse imaging, which exposed cells to low-intensity light for ~0.5 s every 1 min.

## Immunocytochemistry of cells
Cells were grown in 24-well plates with glass coverslips on the bottom, fixed with 4 % paraformaldehyde for 15 min, and permeabilized with 0.2 % Triton X-100/PBS for 3 min. Coverslips were incubated for 1 h with mouse anti-α-tubulin (Sigma-Aldrich, Steinheim, Germany) primary antibody, then rinsed with 1 % BSA/PBS and incubated for 30 min with goat anti-mouse IgG H&L (Cy5) (Abcam Cambridge, UK) secondary antibody. The F-actin network was visualized using Alexa Fluor 594 phalloidin (Invitrogen, USA), coverslips were incubated with the dye for 30 min at 37 °C. Analysis was performed with the Olympus IX81 microscope equipped with Orca-R$^2$ digital camera, fluorescence excitation system MT10, and XCELLENCE software.

## Data analysis and statistics
The analysis was performed using SigmaPlot software (Systat, Richmond, CA, USA), and averaged data are reported as means ± SEM. For statistical evaluation, the Student's $t$ test was used, and a difference was considered statistically significant when p was < 0.05.

## Additional file

**Additional file 1: Figure S1.** Dye permeability and electrical coupling between HeLa *wt* cells. (A-C) AF488 was introduced into a single cell (ROI-1) of the HeLa *wt* cell monolayer, and the time-lapse imaging of fluorescence intensity was monitored in the surrounding cells (ROI-2 – ROI-6). (D and E) Kinetics of AF488 accumulation in the injected cell and neighboring cells, respectively (background fluorescence subtracted) (*n* = 6). The arrow indicates the moment of patch opening. Circles filled with different colors in E represent FI from ROI-2 – ROI-6. (F-G) Electrical coupling of abutted HeLa *wt* cells was evaluated by applying voltage ramp (V$_j$) of negative polarity from 0 to −120 mV (G, upper panel) to the cell-1 and measuring junctional current (I$_j$) (G, lower panel) in the cell-2 (*n* = 17). (TIF 1708 kb)

## Abbreviations
TT: tunneling tube; F-TT: tunneling tube containing only F-actin; Fα-TT: tunneling tube containing F-actin and α-tubulin; C: concentration of the dye; Cx: connexin; FI: fluorescence intensity of the dye; GJ: gap junction; g$_j$: junctional conductance of abutted cells; g$_T$: conductance of the TT; I$_T$: current through the TT; J$_T$: cell-to-cell flux of the dye; P$_T$: permeability of the TT; siRNA: small interfering RNA; V$_j$: transjunctional voltage of abutted cells; V$_T$: voltage across the TT; LSCC: human laryngeal squamous cell carcinoma.

## Competing interests
The authors declare no financial and non-financial competing interests.

## Authors' contributions
V.A.S. designed research; V.A.S., L.R., A.M., V.J., and R.S. performed patch-clamp, dye transfer and time-lapse experiments; I.A. prepared Cxs-transfected HeLa cells and performed immunocytochemistry experiments; V.A.S., I.A., L.R., and A.M. analyzed data and wrote the paper. All authors read and approved the final manuscript.

## Acknowledgments
This research was funded by a grant (No. LIG-13/2012) from the Research Council of Lithuania.

## References
1. Bravo-Cordero JJ, Hodgson L, Condeelis J. Directed cell invasion and migration during metastasis. Curr Opin Cell Biol. 2012;24(2):277–83.
2. Friedl P, Gilmour D. Collective cell migration in morphogenesis, regeneration and cancer. Nature Reviews. 2009;10(7):445–57.
3. Chauveau A, Aucher A, Eissmann P, Vivier E, Davis DM. Membrane nanotubes facilitate long-distance interactions between natural killer cells and target cells. Proc Natl Acad Sci U S A. 2010;107(12):5545–50.
4. Zani BG, Edelman ER. Cellular bridges: Routes for intercellular communication and cell migration. Commun Integr Biol. 2010;3(3):215–20.
5. Antanavičiūtė I, Rysevaitė K, Liutkevičius V, Marandykina A, Rimkutė L, Sveikatienė R, et al. Long-distance communication between laryngeal carcinoma cells. PLoS One. 2014;9(6):e99196.
6. Koyanagi M, Brandes RP, Haendeler J, Zeiher AM, Dimmeler S. Cell-to-cell connection of endothelial progenitor cells with cardiac myocytes by nanotubes: a novel mechanism for cell fate changes? Circ Res. 2005;96(10):1039–41.
7. Pasquier J, Guerrouahen BS, Al Thawadi H, Ghiabi P, Maleki M, Abu-Kaoud N, et al. Preferential transfer of mitochondria from endothelial to cancer cells through tunneling nanotubes modulates chemoresistance. J Transl Med. 2013;11:94.
8. Abounit S, Zurzolo C. Wiring through tunneling nanotubes - from electrical signals to organelle transfer. J Cell Sci. 2012;125(Pt 5):1089–98.
9. Kimura S, Hase K, Ohno H. Tunneling nanotubes: emerging view of their molecular components and formation mechanisms. Exp Cell Res. 2012;318(14):1699–706.
10. Austefjord MW, Gerdes HH, Wang X. Tunneling nanotubes: Diversity in morphology and structure. Commun Integr Biol. 2014;7(1):e27934.
11. Hurtig J, Chiu DT, Onfelt B. Intercellular nanotubes: insights from imaging studies and beyond. Wiley Interdiscip Reviews. 2010;2(3):260–76.
12. Defamie N, Chepied A, Mesnil M. Connexins, gap junctions and tissue invasion. FEBS Lett. 2014;588(8):1331–8.
13. Naus CC, Laird DW. Implications and challenges of connexin connections to cancer. Nat Rev Cancer. 2010;10(6):435–41.
14. Zhou JZ, Jiang JX. Gap junction and hemichannel-independent actions of connexins on cell and tissue functions - an update. FEBS Lett. 2014;588(8):1186–92.
15. Kameritsch P, Pogoda K, Pohl U. Channel-independent influence of connexin 43 on cell migration. Biochim Biophys Acta. 2012;1818(8):1993–2001.
16. Xu X, Li WE, Huang GY, Meyer R, Chen T, Luo Y, et al. Modulation of mouse neural crest cell motility by N-cadherin and connexin 43 gap junctions. J Cell Biol. 2001;154(1):217–30.
17. Francis R, Xu X, Park H, Wei CJ, Chang S, Chatterjee B, et al. Connexin43 modulates cell polarity and directional cell migration by regulating microtubule dynamics. PLoS. 2011;6(10):e26379.
18. Machtaler S, Dang-Lawson M, Choi K, Jang C, Naus CC, Matsuuchi L. The gap junction protein Cx43 regulates B-lymphocyte spreading and adhesion. J Cell Sci. 2011;124(Pt 15):2611–21.
19. Rackauskas M, Neverauskas V, Skeberdis VA. Diversity and properties of connexin gap junction channels. Medicina. 2010;46(1):1–12.
20. Moinfar Z, Dambach H, Faustmann PM. Influence of drugs on gap junctions in glioma cell lines and primary astrocytes in vitro. Front Physiol. 2014;5:186.
21. Wang X, Bukoreshtliev NV, Gerdes HH. Developing neurons form transient nanotubes facilitating electrical coupling and calcium signaling with distant astrocytes. PLoS One. 2012;7(10):e47429.

22. Yamane Y, Shiga H, Asou H, Haga H, Kawabata K, Abe K, et al. Dynamics of astrocyte adhesion as analyzed by a combination of atomic force microscopy and immuno-cytochemistry: the involvement of actin filaments and connexin 43 in the early stage of adhesion. Arch Histol Cytol. 1999;62(4):355–61.

23. Crespin S, Bechberger J, Mesnil M, Naus CC, Sin WC. The carboxy-terminal tail of connexin43 gap junction protein is sufficient to mediate cytoskeleton changes in human glioma cells. J Cell Biochem. 2010;110(3):589–97.

24. Elias LA, Wang DD, Kriegstein AR. Gap junction adhesion is necessary for radial migration in the neocortex. Nature. 2007;448(7156):901–7.

25. Wang X, Veruki ML, Bukoreshtliev NV, Hartveit E, Gerdes HH. Animal cells connected by nanotubes can be electrically coupled through interposed gap-junction channels. Proc Natl Acad Sci U S A. 2010;107(40):17194–9.

26. Aasen T. Connexins: junctional and non-junctional modulators of proliferation. Cell Tissue Res. 2014;360(3):685–99.

27. Sin WC, Crespin S, Mesnil M. Opposing roles of connexin43 in glioma progression. Biochim Biophys Acta. 2012;1818(8):2058–67.

28. Valiunas V, Polosina YY, Miller H, Potapova IA, Valiuniene L, Doronin S, et al. Connexin-specific cell-to-cell transfer of short interfering RNA by gap junctions. J Physiol. 2005;568:459–68.

29. Kizana E, Cingolani E, Marban E. Non-cell-autonomous effects of vector-expressed regulatory RNAs in mammalian heart cells. Gene Ther. 2009;16(9):1163–8.

30. Hosoda T, Zheng H, Cabral-da-Silva M, Sanada F, Ide-Iwata N, Ogorek B, et al. Human cardiac stem cell differentiation is regulated by a mircrine mechanism. Circulation. 2011;123(12):1287–96.

31. Lim PK, Bliss SA, Patel SA, Taborga M, Dave MA, Gregory LA, et al. Gap junction-mediated import of microRNA from bone marrow stromal cells can elicit cell cycle quiescence in breast cancer cells. Cancer Res. 2011;71(5):1550–60.

32. Katakowski M, Buller B, Wang X, Rogers T, Chopp M. Functional microRNA is transferred between glioma cells. Cancer Res. 2010;70(21):8259–63.

33. Lee HK, Finniss S, Cazacu S, Bucris E, Ziv-Av A, Xiang C, et al. Mesenchymal stem cells deliver synthetic microRNA mimics to glioma cells and glioma stem cells and inhibit their cell migration and self-renewal. Oncotarget. 2013;4(2):346–61.

34. Brink PR, Valiunas V, Gordon C, Rosen MR, Cohen IS. Can gap junctions deliver? Biochim Biophys Acta. 2012;1818(8):2076–81.

35. Valiunas V, Wang HZ, Li L, Gordon C, Valiuniene L, Cohen IS, et al. A comparison of two cellular delivery mechanisms for small interfering RNA. Physiol Rep. 2015;3(2):e12286.

36. Rackauskas M, Verselis VK, Bukauskas FF. Permeability of homotypic and heterotypic gap junction channels formed of cardiac connexins mCx30.2, Cx40, Cx43, and Cx45. Am J Physiol Heart Circ Physiol. 2007;293(3):H1729–36.

37. Marandykina A, Palacios-Prado N, Rimkute L, Skeberdis VA, Bukauskas FF. Regulation of connexin36 gap junction channels by n-alkanols and arachidonic acid. J Physiol. 2013;591(Pt 8):2087–101.

38. Skeberdis VA, Rimkute L, Skeberdyte A, Paulauskas N, Bukauskas FF. pH-dependent modulation of connexin-based gap junctional uncouplers. J Physiol. 2011;589(Pt 14):3495–506.

39. Palacios-Prado N, Briggs SW, Skeberdis VA, Pranevicius M, Bennett MV, Bukauskas FF. pH-dependent modulation of voltage gating in connexin45 homotypic and connexin45/connexin43 heterotypic gap junctions. Proc Natl Acad Sci U S A. 2010;107(21):9897–902.

40. Palacios-Prado N, Sonntag S, Skeberdis VA, Willecke K, Bukauskas F. Gating, permselectivity and pH-dependent modulation of channels formed by connexin57, a major connexin of horizontal cells in the mouse retina. J Physiol. 2009;587:3251–69.

41. Agnati LF, Guidolin D, Maura G, Marcoli M, Leo G, Carone C, et al. Information handling by the brain: proposal of a new "paradigm" involving the roamer type of volume transmission and the tunneling nanotube type of wiring transmission. J Neural Transm. 2014;121(12):1431–49.

42. Petrie RJ, Doyle AD, Yamada KM. Random versus directionally persistent cell migration. Nature Reviews. 2009;10(8):538–49.

43. Cina C, Maass K, Theis M, Willecke K, Bechberger JF, Naus CC. Involvement of the cytoplasmic C-terminal domain of connexin43 in neuronal migration. J Neurosci. 2009;29(7):2009–21.

44. Fushiki S, Perez Velazquez JL, Zhang L, Bechberger JF, Carlen PL, Naus CC. Changes in neuronal migration in neocortex of connexin43 null mutant mice. J Neuropathol Exp Neurol. 2003;62(3):304–14.

45. McDonough WS, Johansson A, Joffee H, Giese A, Berens ME. Gap junction intercellular communication in gliomas is inversely related to cell motility. Int J Dev Neurosci. 1999;17(5–6):601–11.

46. Mi L, Xiong R, Zhang Y, Li Z, Yang W, Chen J-Y, et al. Microscopic observation of the intercellular transport of CdTe quantum dot aggregates through tunneling-nanotubes. J Biomat Nanobiotech. 2011;2(2):173–80.

47. Onfelt B, Nedvetzki S, Benninger RK, Purbhoo MA, Sowinski S, Hume AN, et al. Structurally distinct membrane nanotubes between human macrophages support long-distance vesicular traffic or surfing of bacteria. J Immunol. 2006;177(12):8476–83.

48. Wang ZG, Liu SL, Tian ZQ, Zhang ZL, Tang HW, Pang DW. Myosin-driven intercellular transportation of wheat germ agglutinin mediated by membrane nanotubes between human lung cancer cells. ACS Nano. 2012;6(11):10033–41.

49. Lu M, Jolly MK, Levine H, Onuchic JN, Ben-Jacob E. MicroRNA-based regulation of epithelial-hybrid-mesenchymal fate determination. Proc Natl Acad Sci U S A. 2013;110(45):18144–9.

50. Mittelbrunn M, Sanchez-Madrid F. Intercellular communication: diverse structures for exchange of genetic information. Nature Reviews. 2012;13(5):328–35.

51. Kotini M, Mayor R. Connexins in migration during development and cancer. Dev Biol. 2015;401(1):143–51.

52. Zhang YW, Nakayama K, Nakayama K, Morita I. A novel route for connexin 43 to inhibit cell proliferation: negative regulation of S-phase kinase-associated protein (Skp 2). Cancer Res. 2003;63(7):1623–30.

53. Bos PD, Zhang XH, Nadal C, Shu W, Gomis RR, Nguyen DX, et al. Genes that mediate breast cancer metastasis to the brain. Nature. 2009;459(7249):1005–9.

54. Stoletov K, Strnadel J, Zardouzian E, Momiyama M, Park FD, Kelber JA, et al. Role of connexins in metastatic breast cancer and melanoma brain colonization. J Cell Sci. 2013;126(Pt 4):904–13.

55. Ma L, Teruya-Feldstein J, Weinberg RA. Tumour invasion and metastasis initiated by microRNA-10b in breast cancer. Nature. 2007;449(7163):682–8.

56. Bukauskas FF, Jordan K, Bukauskiene A, Bennett MV, Lampe PD, Laird DW, et al. Clustering of connexin 43-enhanced green fluorescent protein gap junction channels and functional coupling in living cells. Proc Natl Acad Sci U S A. 2000;97(6):2556–61.

57. Teubner B, Degen J, Sohl G, Guldenagel M, Bukauskas FF, Trexler EB, et al. Functional expression of the murine connexin 36 gene coding for a neuron-specific gap junctional protein. J Membr Biol. 2000;176(3):249–62.

58. Goyal P, Behring A, Kumar A, Siess W. STK35L1 associates with nuclear actin and regulates cell cycle and migration of endothelial cells. PLoS One. 2011;6(1):e16249.

59. Wilders R, Jongsma HJ. Limitations of the dual voltage clamp method in assaying conductance and kinetics of gap junction channels. Biophys J. 1992;63:942–53.

60. Verselis V, White RL, Spray DC, Bennett MV. Gap junctional conductance and permeability are linearly related. Science. 1986;234:461–4.

61. Hille B. Ion channels of excitable membranes. 3rd ed. Sunderland: Sinauer Associates; 2001.

# 2

# The absence of dysferlin induces the expression of functional connexin-based hemichannels in human myotubes

Luis A. Cea[1*], Jorge A. Bevilacqua[1,2], Christian Arriagada[1], Ana María Cárdenas[3], Anne Bigot[4], Vincent Mouly[4], Juan C. Sáez[3,5] and Pablo Caviedes[6]

## Abstract

**Background:** Mutations in the gene encoding for dysferlin cause recessive autosomal muscular dystrophies called dysferlinopathies. These mutations induce several alterations in skeletal muscles, including, inflammation, increased membrane permeability and cell death. Despite the fact that the etiology of dysferlinopathies is known, the mechanism that explains the aforementioned alterations is still elusive. Therefore, we have now evaluated the potential involvement of connexin based hemichannels in the pathophysiology of dysferlinopathies.

**Results:** Human deltoid muscle biopsies of 5 Chilean dysferlinopathy patients exhibited the presence of muscular connexins (Cx40.1, Cx43 and Cx45). The presence of these connexins was also observed in human myotubes derived from immortalized myoblasts derived from other patients with mutated forms of dysferlin. In addition to the aforementioned connexins, these myotubes expressed functional connexin based hemichannels, evaluated by ethidium uptake assays, as opposed to myotubes obtained from a normal human muscle cell line, RCMH. This response was reproduced in a knock-down model of dysferlin, by treating RCMH cell line with small hairpin RNA specific for dysferlin (RCMH-sh Dysferlin). Also, the presence of $P2X_7$ receptor and the transient receptor potential channel, TRPV2, another $Ca^{2+}$ permeable channels, was detected in the myotubes expressing mutated dysferlin, and an elevated resting intracellular $Ca^{2+}$ level was found in the latter myotubes, which was in turn reduced to control levels in the presence of the molecule D4, a selective Cx HCs inhibitor.

**Conclusions:** The data suggests that dysferlin deficiency, caused by mutation or downregulation of dysferlin, promotes the expression of Cx HCs. Then, the *de novo* expression Cx HC causes a dysregulation of intracellular free $Ca^{2+}$ levels, which could underlie muscular damage associated to dysferlin mutations. This mechanism could constitute a potential therapeutical target in dysferlinopathies.

**Keywords:** Dysferlinopathy, Membrane permeability, Calcium

* Correspondence: luiscea@med.uchile.cl
[1]Program of Anatomy and Developmental Biology, Faculty of Medicine, Institute of Biomedical Sciences, University of Chile, Av. Independencia #1027, Independencia, Santiago, Chile
Full list of author information is available at the end of the article

## Background

Dysferlinopathies are muscular dystrophies caused by mutations in dysferlin, a 230 kDa membrane protein, mainly localized in the sarcolemma [1] and the T-tubule system [2, 3]. It is accepted that dysferlin participates in membrane resealing after damage [4, 5]. This protein is weakly expressed in myoblasts (myogenic precursor cells), but highly expressed in adult skeletal muscle [6]. Clinically, dysferlinopathies manifest between the second and third decade of life in previously asymptomatic patients. At onset, most patients refer weakness in the lower extremities, difficulty in running or climbing stairs, sometimes accompanied with pain [7]. Regarding the alterations produced by dysferlin mutations in muscle, prior reports have detected the presence of inflammation [8–10]; disruption of the T-tubule structure, which in turn was ameliorated by reduction of external $[Ca^{2+}]$ or blocking of L-type $Ca^{2+}$ channels with diltiazem [3], suggesting that a deregulated entry of external $Ca^{2+}$ may underlie damage in dysferlin null myofibers. In addition, an altered permeability to dyes such as Evans blue has been reported in skeletal muscles from $Dysf^{-/-}$ mice (an animal model of dysferlinopathy) [4], and previous reports have demonstrated that Evans blue crosses the cell membrane through Cx HC [11]. Hence, these results strongly suggest the presence of connexin based hemichannels (Cx HC) in the sarcolemma of myofibers from the animal model. As we previously reported, the *de novo* expression of Cx HCs has been observed in similar pathologies, where they mediate myofiber atrophy induced by denervation [11]. Interestingly, only a mild muscular atrophy was observed after denervation in Cx43 and Cx45 KO mice [11]. Since Cx HC are nonselective channels permeable to ions (e.g. $Ca^{2+}$ and $Na^+$) and small compounds, including signaling molecules such as ATP and $NAD^+$ and dyes including ethidium ($Etd^+$) and Evans blue [12, 13], the altered membrane permeability caused by the Cx HC expression could contribute to the development of the muscular atrophy. Indeed, the *de novo* expression of Cx HCs promotes the increase of oxidative stress in pathological conditions such as muscle denervation [14] and they constitute a mechanism of ATP release in several muscle pathologies [11, 12, 14].

To date there is no effective treatment to arrest or even reduce the symptomatology of the patients affected with dysferlinopathies. Nevertheless, the introduction of a mini-dysferlin in animal models of the disease ($Dysf^{-/-}$ mice) results in the recovery of membrane resealing function. However, the progressive degeneration, ascertained from muscle histology studies, remains unabated [15]. The aforementioned evidence points to the existence of an additional pathological mechanism, triggered by the absence of dysferlin. In the present work we evaluated whether myotubes of patients suffering from dysferlinopathies, as well as in other in vitro models of dysferlin deficiency, express Cx HCs and whether the expression of these types of channels alters the sarcolemma permeability, and increases intracellular free $Ca^{2+}$ in these cells.

## Results

### Human muscles bearing dysferlin mutations express connexins 40.1, 43 and 45

We analyzed the presence of connexin proteins by immnunofluorescent microscopy in human muscles biopsies from patients bearing dysferlin mutations (see methods for dysferlin mutations), the absence of dysferlin was confirmed by immunohistochemistry assays (data not shown). As shown in Fig. 1, connexins 40.1, 43 and 45 (green signal, Fig. 1) were detected in biopsies from patients with dysferlinopathy but not in biopsies of control subjects (control). These proteins colocalized with the plasma membrane protein spectrin (Fig. 2) [16], indicating that all three connexins are present in the sarcolemma. Using immunofluoresence, we next evaluated the presence of the purinergic receptor P2X$_7$ and the transient receptor potential cation channel subfamily V member 2 (TRPV2), which have been previously associated with muscular atrophy [11]. P2X$_7$ receptors were detected in one of the two patients evaluated, whereas TRPV2 was found in the biopsies of both patients (Fig. 3). Conversely, in control patients (patients without a muscular pathology) both receptors were absent (Fig. 3).

Because $Ca^{2+}$ influx is reportedly increased in human muscles bearing dysferlin mutations [17], we evaluated the presence, in human dysferlin-mutated myotubes (HDMM) (using immunofluorescent miscroscopy), of different $Ca^{2+}$ channels; connexin-based hemichannels, transient potential receptor TRPV2 and P2X$_7$, all of which have been previously linked in muscular pathologies [11, 12, 18]. We observed the presence of Cxs 40.1, 43 and 45 in all HDMM, although they have different dysferlin mutations (Fig. 4, green signal), they were absent in normal myotubes (Control, RCMH cells). On the other hand, TRPV2 channels and P2X$_7$ receptors were only present in 107 and 379 (dysferlin-mutated cell lines, see methods) derived myotubes (Fig. 4). TRPV2 channels were functional in 379 myotubes, as evidenced by their response to stimulation with 2-aminoethoxydiphenyl borate (2-APB), a selective TRPV2 agonist [19], which induces a sustained increase in $Ca^{2+}$ levels (Additional file 1: Figure S1).

### Absence of dysferlin in control RCMH myotubes mimics the dysferlin mutation effect

To demonstrate that the absence of dysferlin, regardless of the type of mutation, can induce the expression of

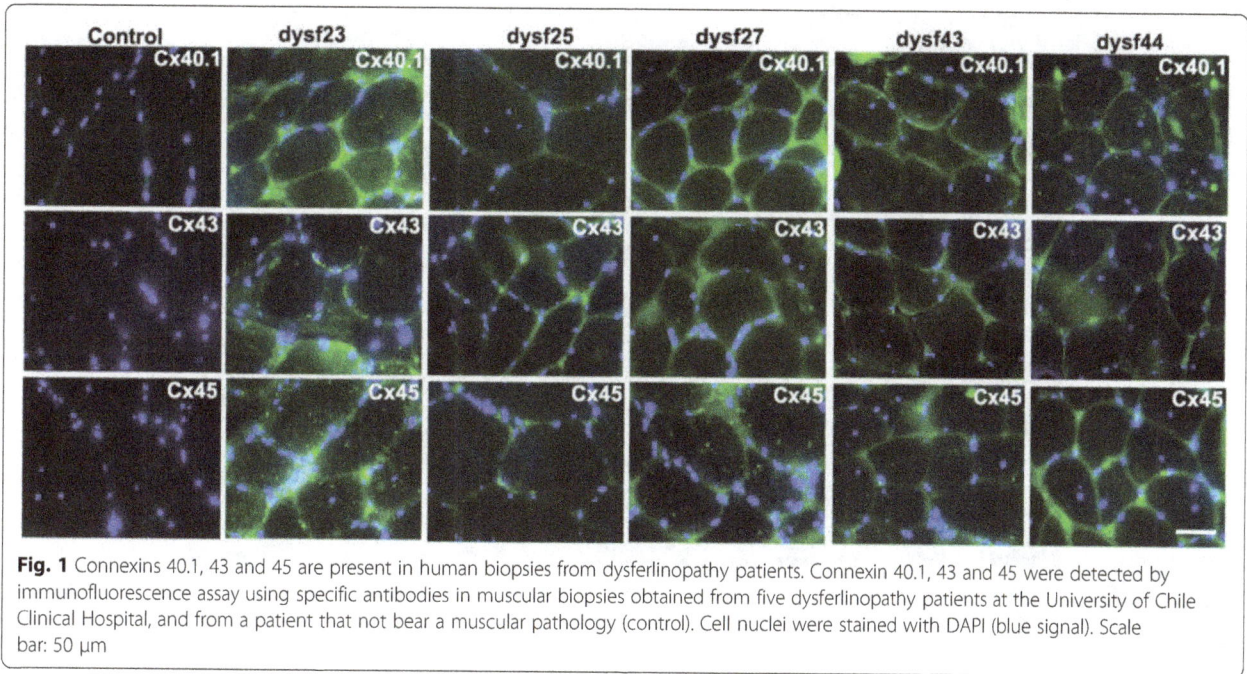

**Fig. 1** Connexins 40.1, 43 and 45 are present in human biopsies from dysferlinopathy patients. Connexin 40.1, 43 and 45 were detected by immunofluorescence assay using specific antibodies in muscular biopsies obtained from five dysferlinopathy patients at the University of Chile Clinical Hospital, and from a patient that not bear a muscular pathology (control). Cell nuclei were stained with DAPI (blue signal). Scale bar: 50 μm

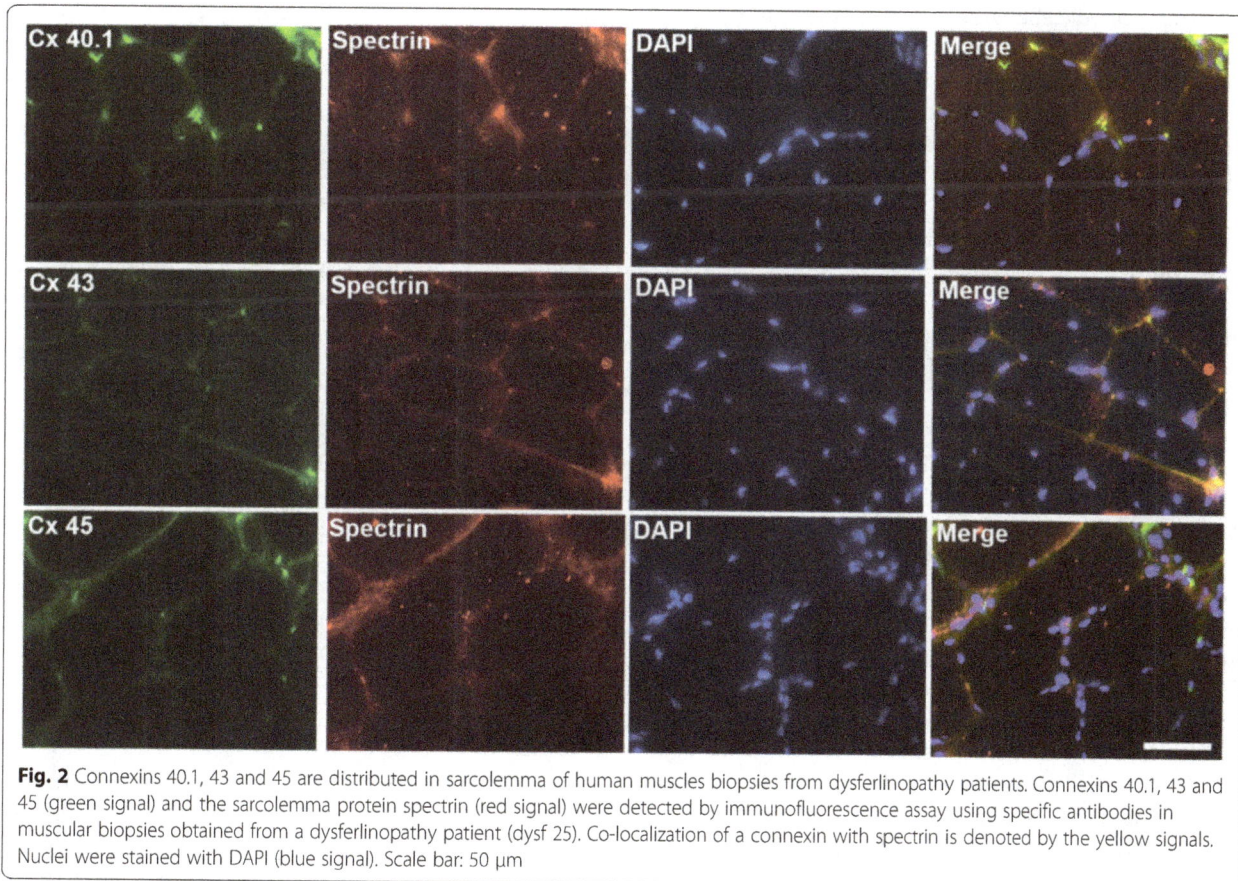

**Fig. 2** Connexins 40.1, 43 and 45 are distributed in sarcolemma of human muscles biopsies from dysferlinopathy patients. Connexins 40.1, 43 and 45 (green signal) and the sarcolemma protein spectrin (red signal) were detected by immunofluorescence assay using specific antibodies in muscular biopsies obtained from a dysferlinopathy patient (dysf 25). Co-localization of a connexin with spectrin is denoted by the yellow signals. Nuclei were stained with DAPI (blue signal). Scale bar: 50 μm

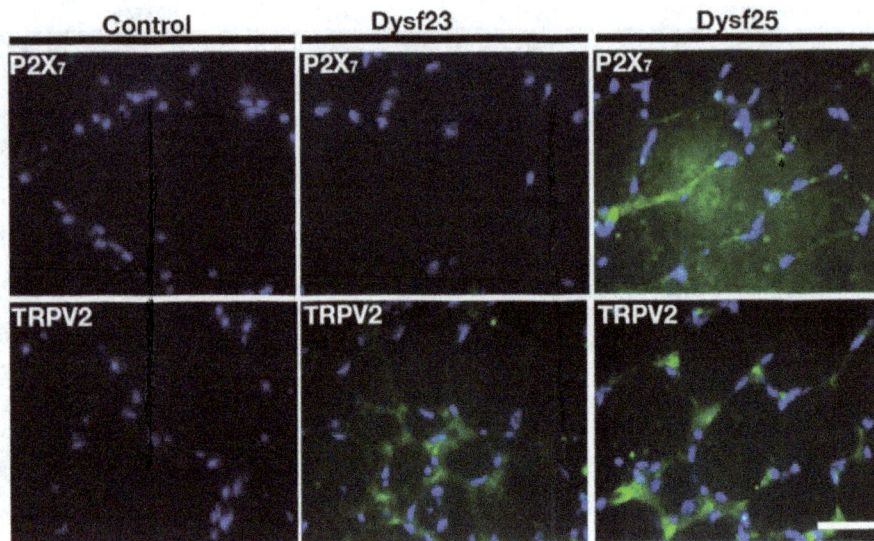

**Fig. 3** Presence of P2X$_7$ receptor and TRPV2 channel in muscular biopsies from dysferlinopathy patients. The purinergic receptor P2X$_7$ and the transient receptor potential vanilloid type 2 (TRPV2) channel were detected by immunofluorescence studies using specific antibodies in cross sections of human muscular biopsies from two dysferlinopathy patients and a control patient (patient without a muscular pathology). Nuclei were stained with DAPI (blue signal). Scale bar: 50 μm

connexins in normal human myotubes, we knocked-down expression of dysferlin in RCMH cells by transfecting a small hairpin RNA specific for dysferlin, and transfected cells were subsequently differentiated to myotubes (RCMH-sh Dysf). In the absence of dysferlin, RCMH cells express connexins 40.1, 43 and 45 (green signal), which were in control myotubes (RCMH) that express dysferlin (red signal, Fig. 5).

### Human dysferlin-mutated myotubes express functional connexin based hemichannels

Because Cx proteins are present in dysferlin-mutated myotubes (Fig. 4), we analyzed whether these proteins were indeed forming functional hemichannels (Cx HCs) in these cells. Hemichannel activity was evaluated by ethidium (Etd$^+$) uptake assays [20], which revealed that HDMM demonstrated elevated Etd$^+$ uptake compared with control myotubes (Fig. 6). We detected a 4-fold increment in the 107 and 379 cell lines, compared with RCMH, whereas AB320 and ER (human dysferlin-mutated cell lines, see methods) showed an increment of 2-fold compared to RCMH. Etd$^+$ uptake was successfully inhibited with external application of 50 μM carbenoxolone (Fig. 5a), a Cx HC inhibitor [21]. In addition, we observed that different dysferlin mutations exhibit different Cx HC activity. Indeed, cell lines 107 and 379 presented the highest Cx HC activity (Fig. 6b). In addition, RCMH myotubes where dysferlin was silenced via specific small hairpin RNA resulted in the expression of functional Cx HCs to levels comparable to 107 and 379 myotubes (Fig. 6b).

### Inhibition of connexin-based hemichannels reduces elevated intracellular basal Ca$^{2+}$ signals

Considering that Ca$^{2+}$ influx is reportedly increased in human muscles bearing dysferlin mutations [17], and that human dysferlin-mutated myotubes express functional Cx HCs (Fig. 6), which in turn are non-selective permeable Ca$^{2+}$ channels [22], we analyzed basal cytosolic Ca$^{2+}$ levels in the dysferlin-deficient and control lines using FURA 2-AM assays. Dysferlin-mutated myotubes showed significantly elevated Ca$^{2+}$ levels compared to control myotubes (Fig. 7), a finding consistent with the increased Cx HC activity observed in the four cell lines evaluated herein. Since Cx HCs were present and the basal intracellular Ca$^{2+}$ level was elevated in dysferlin-mutated myotubes, we then investigated whether the inhibition of Cx HCs could prevent this phenomenon. Therefore, human dysferlin-mutated myotubes were treated daily for 7 days with 100 nM D4, a selective Cx HC blocker that does not inhibit gap junction channels, thus allowing the fusion of myoblasts to form myotubes where gap junction channels are relevant [23]. Incubation with D4 significantly reverted the intracellular basal Ca$^{2+}$ signals to values comparable to those of control cells (Fig. 7), suggesting that these channels are responsible of this response.

### Discussion

In the present report, we have demonstrated that muscles from patients diagnosed with dysferlinopathy express Cxs 40.1, 43 and 45 at the plasma membrane. Additionally, four immortalized muscle cell lines

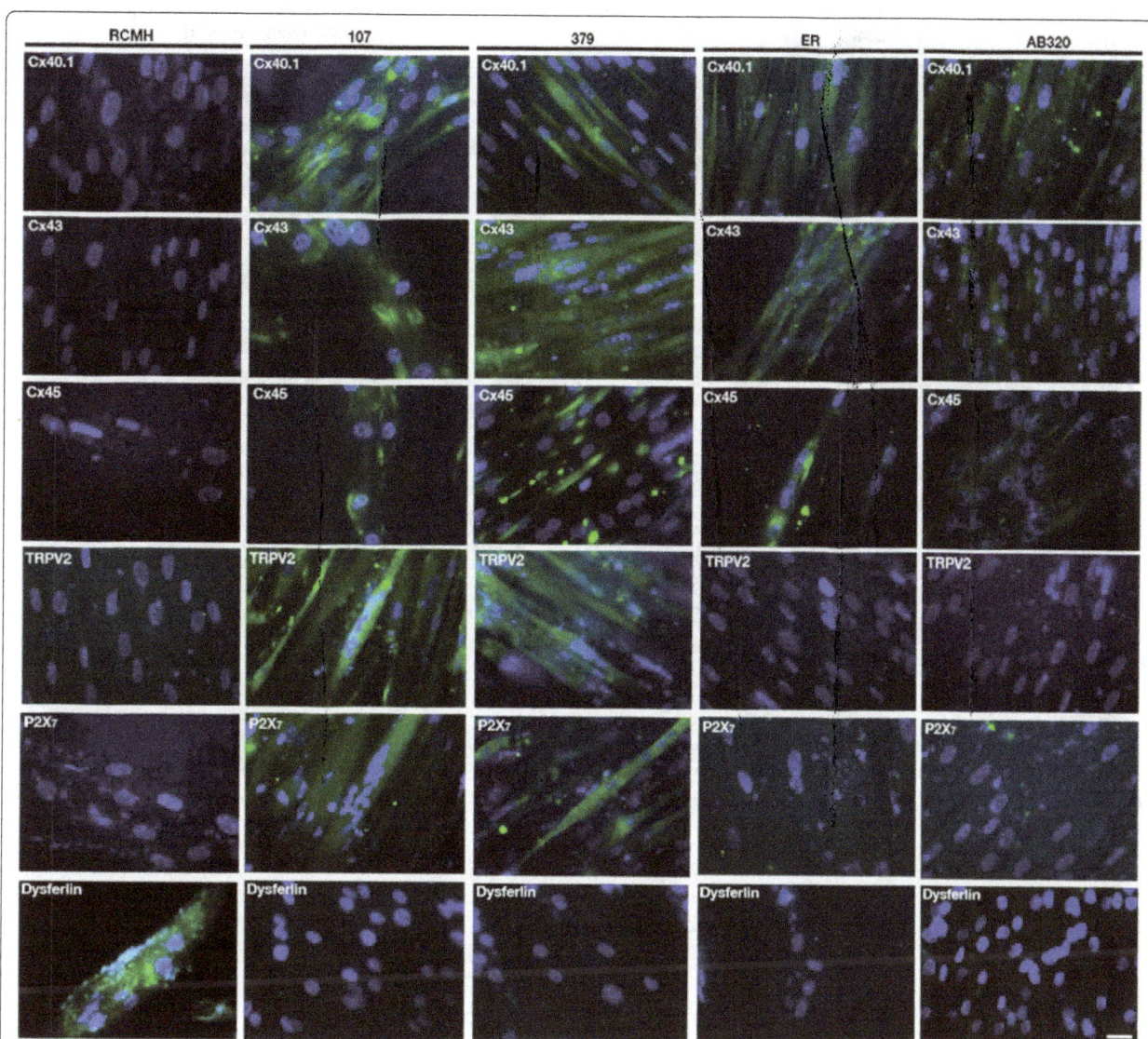

**Fig. 4** Human dysferlin-mutated myotubes present connexins 40.1, 43 and 45. Human dysferlin-mutated myotubes were obtained by differentiation of dysferlin-mutated myoblast lines named 107, 379, AB320 and ER, each one bearing different dysferlin mutations. Connexins 40.1, 43 and 45 were detected by immunofluorescence studies using specific antibodies. All cell lines showed positive reactivity to Cxs (green signal) compared with control myotubes (RCMH), where Cxs reactivity was not detected. $P2X_7$ receptor and TRPV2 channel were also detected by immunofluorescence studies using specific antibodies. Only 107 and 379 myotubes expressed $P2X_7$ receptors and TRPV2 channels. Nuclei were stained with DAPI (blue signal). Scale bar: 30 μm. $n = 4$ cell cultures for each cell line

generated from muscle samples of dysferlinopathy patients also expressed these Cxs, which in turn, formed functional Cx-based hemichannels. This phenomenon was also induced in control myotubes (RCMH, which normally do not express Cxs), after decreasing dysferlin expression through the transfection (in myoblast stage) of a specific small hairpin RNA against dysferlin. Moreover, human dysferlin-mutated 107 and 379 myotubes expressed two additionals $Ca^{2+}$ permeable channels, the TRPV2 channel and $P2X_7$ receptor. These myotubes also exhibited greater Cx HC activity, and such an elevated Cx HC

activity could explain the presence of these channels in the myotubes, which in turn could also contribute to the elevated intracellular $Ca^{2+}$ signal previously reported in dysferlin KO skeletal muscle fibers [3] and confirmed here in our model. Anyway, it seems that the presence of Cx HCs is sufficient to explain the elevated basal intracellular $Ca^{2+}$ levels in dysferlin-deficient myocytes, especially if we consider that the specific inhibition of these channels with a selective blocker (molecule D4), was enough to reduce the elevated basal $Ca^{2+}$ concentration to levels similar to those of control cells. The mechanisms regulating $Ca^{2+}$ in

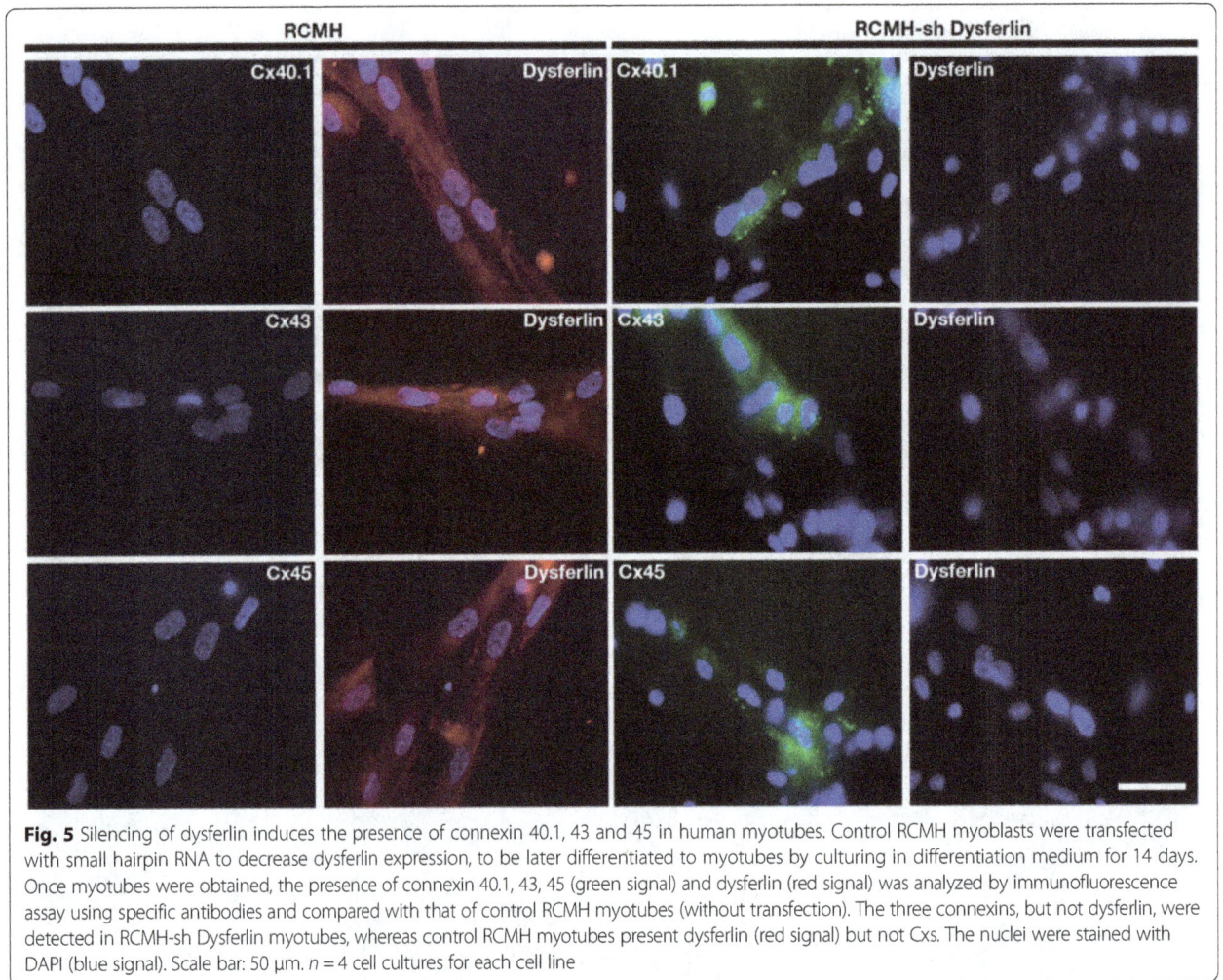

**Fig. 5** Silencing of dysferlin induces the presence of connexin 40.1, 43 and 45 in human myotubes. Control RCMH myoblasts were transfected with small hairpin RNA to decrease dysferlin expression, to be later differentiated to myotubes by culturing in differentiation medium for 14 days. Once myotubes were obtained, the presence of connexin 40.1, 43, 45 (green signal) and dysferlin (red signal) was analyzed by immunofluorescence assay using specific antibodies and compared with that of control RCMH myotubes (without transfection). The three connexins, but not dysferlin, were detected in RCMH-sh Dysferlin myotubes, whereas control RCMH myotubes present dysferlin (red signal) but not Cxs. The nuclei were stained with DAPI (blue signal). Scale bar: 50 μm. $n = 4$ cell cultures for each cell line

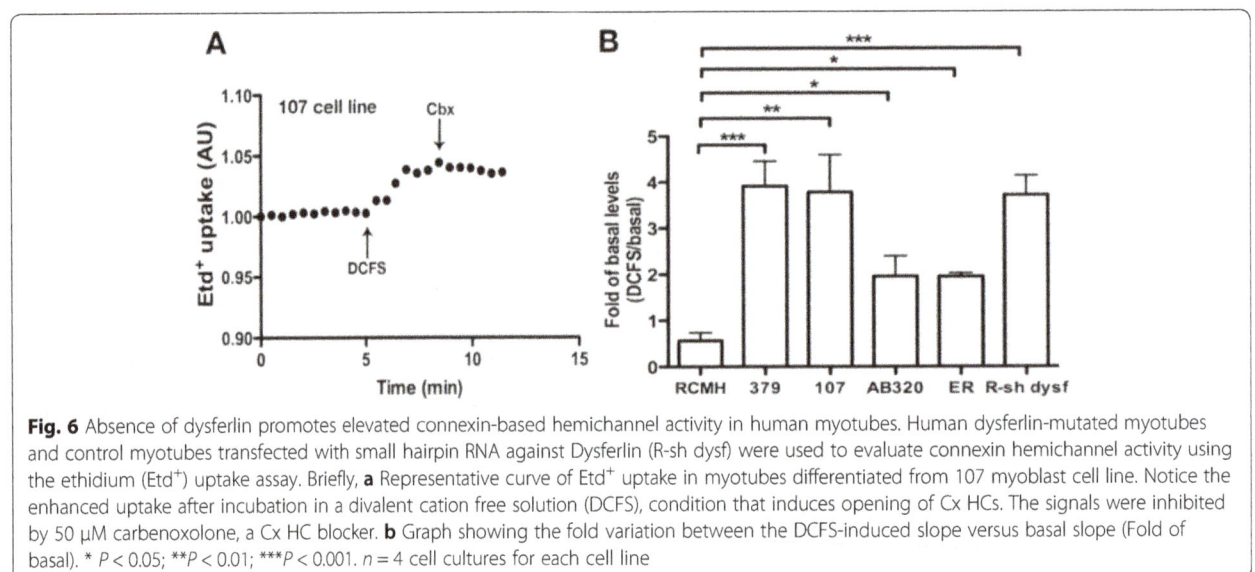

**Fig. 6** Absence of dysferlin promotes elevated connexin-based hemichannel activity in human myotubes. Human dysferlin-mutated myotubes and control myotubes transfected with small hairpin RNA against Dysferlin (R-sh dysf) were used to evaluate connexin hemichannel activity using the ethidium (Etd$^+$) uptake assay. Briefly, **a** Representative curve of Etd$^+$ uptake in myotubes differentiated from 107 myoblast cell line. Notice the enhanced uptake after incubation in a divalent cation free solution (DCFS), condition that induces opening of Cx HCs. The signals were inhibited by 50 μM carbenoxolone, a Cx HC blocker. **b** Graph showing the fold variation between the DCFS-induced slope versus basal slope (Fold of basal). * $P < 0.05$; ** $P < 0.01$; *** $P < 0.001$. $n = 4$ cell cultures for each cell line

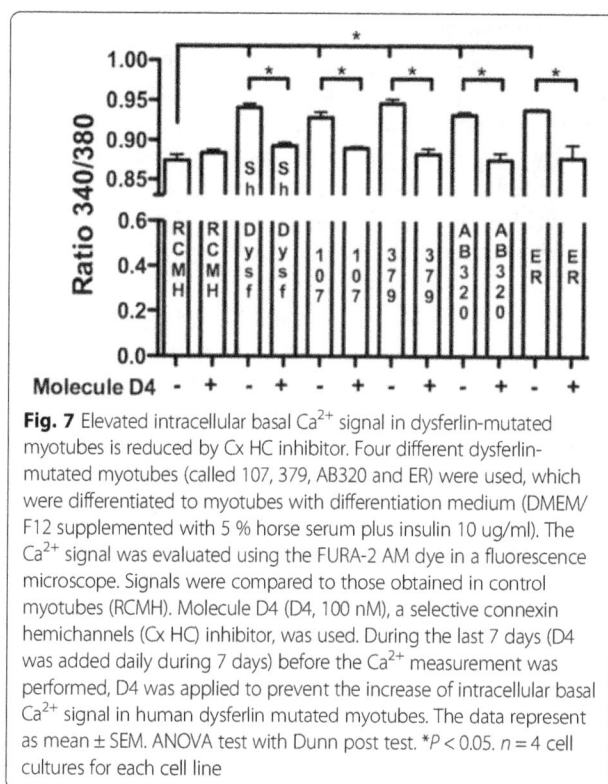

**Fig. 7** Elevated intracellular basal $Ca^{2+}$ signal in dysferlin-mutated myotubes is reduced by Cx HC inhibitor. Four different dysferlin-mutated myotubes (called 107, 379, AB320 and ER) were used, which were differentiated to myotubes with differentiation medium (DMEM/F12 supplemented with 5 % horse serum plus insulin 10 ug/ml). The $Ca^{2+}$ signal was evaluated using the FURA-2 AM dye in a fluorescence microscope. Signals were compared to those obtained in control myotubes (RCMH). Molecule D4 (D4, 100 nM), a selective connexin hemichannels (Cx HC) inhibitor, was used. During the last 7 days (D4 was added daily during 7 days) before the $Ca^{2+}$ measurement was performed, D4 was applied to prevent the increase of intracellular basal $Ca^{2+}$ signal in human dysferlin mutated myotubes. The data represent as mean ± SEM. ANOVA test with Dunn post test. *$P < 0.05$. $n = 4$ cell cultures for each cell line

myotubes, such as the $Na^+/Ca^{2+}$ exchanger, $Ca^{2+}$ binding proteins and most importantly the sarcoplasmic/endoplasmic reticulum $Ca^{2+}$ ATPases (SERCA), are extremely efficient in controlling the cytosolic levels of this cation [24]. Even so, integration of Cx hemichannels into the plasma membrane increased the basal $Ca^{2+}$ concentration by ~8 %, which may be enough to alter cellular homeostasis of myocytes. This increment in $Ca^{2+}$ levels could be enough to alter the cellular homeostasis of the myocytes. Indeed, an altered cytosolic $Ca^{2+}$ level could induce several changes in the myotubes, including an increased proteolytic activity [25], which occurs in muscles from patients bearing dysferlin mutations [26]. The latter could then drive atrophy in affected muscles, and also to cause the degeneration of muscles by apoptosis or necrosis [18]. These findings have been previously observed in Duchenne muscular dystrophy were resting $Ca^{2+}$ levels in human DMD myotubes were increased compared with normal myotubes [27].

Therefore, in addition to the deficit in membrane repair mechanism found in dysferlin-mutated myofibers [4], we propose herein a second pathological mechanism as a consequence of dysferlin mutations, which involves Cx HCs. These non-selective channels could explain several of the reported features in dysferlin-deficient muscles, including the increased membrane permeability [4], higher proteasome activity [26, 28], activation of inflammation signals [8–10] and muscle degeneration [29]. The latter interpretation is supported by prior reports showing the

role of Cx HC in altered membrane permeability and atrophy of skeletal muscles induced by denervation [11, 12]. Although the mechanism responsible of the induction of connexins by the absence of dysferlin still remains to be elucidated, the fact that the knock-down of dysferlin induces the *de novo* expression of functional connexin-based hemichannels suggests a direct relation between the absence of dysferlin and the presence of connexin. However, it has been demonstrated that some conditions such as inflammation induces the expression of functional connexin-based hemichannels in skeletal muscles through proinflammatory cytokines and the transcription factor NFκB [11, 12]. Inflammation has been described as a key feature in dysferlin-mutated skeletal muscles [9]. Whether the induction of Cx HC involves an activation of NFκB still needs to be investigated.

## Conclusions
In this study, we have demonstrated the presence of connexins 40.1, 43 and 45 in muscular biopsies of patients with different dysferlin mutations. This was confirmed in myotubes derived from cell lines of human dysferlinopathy patients, as well as in normal myotubes where dysferlin was knocked down. The Cxs expressed formed functional connexin based hemichannels, that are responsible of elevated basal intracellular $Ca^{2+}$ levels in human myotubes. The presence of these channels may represent a novel pathological mechanism in dysferlinopathies, as well as a most attractive therapeutical target, considering their localization at the membrane level.

## Methods
### Reagents
Anti-rabbit or anti-mouse IgG antibodies-conjugated to Cy2 (green) or Cy3 (red) were purchased from Jackson immunoresearch laboratories (West Grove, PA, USA). Ethidium (Etd⁺) bromide was from GIBCO/BRL (*Grand Island*, NY, USA), fluoromount-G was from Electron Microscopy Science (Hatfield, PA, USA), Previously described polyclonal anti-Cx40.1, -43, and -Cx45 antibodies were used [11]. Synthesis and characterization of D4 molecule will be published elsewhere.

### Generation of dysferlin-mutated myoblast cell lines
The protocol used was obtained from Philippi et al. [30]. Briefly, primary myoblasts isolated by protease digestion were obtained from fresh muscle biopsies and expanded at 37 °C in skeletal muscle growth medium (PromoCell, Heidelberg, Germany) supplemented with 10 % FCS (Gibco, Paisely, UK). The cultures were enriched in myoblasts by immuno-magnetic cell sorting using anti-CD56/NCAM antibody coated magnetic beads (Miltenyi Biotech, Bergisch Gladbach, Germany). Purity of the myoblast preparation was verified by staining with an

anti-desmin antibody (DAKO) revealing more than 95 % desmin-positive cells.

## Immortalization of primary human myoblasts and their differentiation into myotubes

Primary human dysferlin-deficient myoblast lines were transduced with pBABE retroviral vectors carrying Cdk4 and hTERT. Puromycin and neomycin were used as selection markers, respectively and isolation of individual myogenic clones was carried out as described by Mamchaoui et al. [31]. The immortalized dysferlin-deficient human myoblast lines were cultured in growth medium consisting of 1 vol 199 Medium (Invitrogen, Carlsbad, CA)/4 vol DMEM (Invitrogen) supplemented with 20 % foetal calf serum (Invitrogen), 2.5 ng/ml HGF (Invitrogen), 0.1 μM Dexamethasone (Sigma-Aldrich, St. Louis, MO) and 50 μg/ml Gentamycin (Invitrogen). Differentiation into myotubes was initiated at approximately 90 % confluence by cultivation in differentiation medium (DMEM/F12, 5 % horse serum, insulin 10 μg/ml) for 14 days.

Four different dysferlin-mutated myoblast lines were generated and called 107, 379, AB320 and ER, each one bearing a different dysferlin mutation as follow: **107**, c.855 + 1delG c.895G > A; **379**, c.1448C > A c.107 T > A; **AB320**, c.342-1G > A HTZ c.3516_3517delTT (p.Ser1173X) HTZ; **ER**, G1628R(c.4882G > A)HMZ.

## Intracellular Ca$^{2+}$ Signals

Basal intracellular Ca$^{2+}$ signals were evaluated in myotubes by using the ratiometric dye FURA 2-AM. The myotubes were incubated in Krebs-Ringer buffer (in mM: 145 NaCl, 5 KCl, 3 CaCl$_2$, 1 MgCl$_2$, 5.6 glucose, 10 HEPES-Na, pH7.4) containing FURA2-AM dye (2 μM) during 55 min at room temperature. Then, the Ca$^{2+}$ signal was evaluated in a Nikon Eclipse *Ti* microscope equipped with epifluorescence illumination, and images were obtained by using a Clara camera (Andor), at 2 wavelength (λ) 340 and 380 nm, calculating the ratio 340 vs 380.

## Membrane permeability to dyes

The uptake of ethidium (Etd$^+$) was evaluated by using time lapse measurements. Human myotubes plated onto glass cover slips were washed twice with Krebs-buffer solution. For time lapse measurements myotubes were incubated in Krebs buffer solution containing 5 μM Etd$^+$ bromide. Etd$^+$ fluorescence was recorded in regions of interest located in nuclei of the myofibers. An epifluorescence Nikon *Ti* microscope was used. Images were captured with a Clara camera (Andor) every 30 s. Image processing was performed off-line with ImageJ software (NIH, Bethesda, USA).

## Immunofluorescence analysis

To detect different proteins in myotubes or cross sections (10 μm) of muscles (fast frozen with iso-methyl-butane cooled in liquid nitrogen) samples were obtained and processed as described previously [11]. Briefly, samples were fixed in paraformaldehyde 4 % by 5 min, then incubated at 4 °C for 12 h with diluted primary anti-Cx40.1 (1:100), anti-Cx43 (1:250), anti-Cx45 (1:250), anti-Panx1 (1:300), anti-TRPV2 (1:100) or anti-P2X$_7$ (1:100) antibodies followed by 4 washes with PBS 1X and then, incubated with an appropriate dilution of alexa488-conjugated goat anti-rabbit or anti-mouse IgG antibodies. Samples were rinsed with PBS 1X, mounted with fluoromount G containing DAPI on glass slides and representative images were acquired in an epifluorescence microscope Nikon *Ti* (Tokio, Japan).

## Silencing of dysferlin by small hairpin RNA

Commercial small hairpin RNA plasmid against dysferlin (Santa Cruz Biotech, Dallas, Tx), with puromycin resistance, was used to transfect RCMH myoblast through lypofectamine method [32]. The silencing of dysferlin was confirmed by western blot analysis.

## Selection of dysferlinopathy patients, mutations and biopsy studies

Following ethical guidelines, healthy volunteers and dysferlinopathy patients who participated in this study signed an informed consent approved by the local Ethics Committee in accordance with the ethical standards laid down in the 1964 Declaration of Helsinki and its later amendments. Diagnosis of dysferlinopathy was achieved based on clinical examination, muscular biopsy and mutation analysis. The mutations of the patients are published in Woudt et al. [33].

## Statistical analysis

Results are presented as mean ± standard error (SE). For multiple comparisons with a single control, a non-parametric one-way ANOVA followed by the Dunn's multiple comparison test was used. Analyses were carried out using GRAPHPAD software. $P < 0.05$ was considered statistically significant.

## Additional file

**Additional file 1: Figure S1.** TRPV2 channels are functionally expressed in absence of dysferlin. Human control myotubes, RCMH myotubes (O); human myotubes that lack dysferlin, small hairpin RNA Dysferlin (□); and myotubes 379 (▼), were loaded with Fura-2 AM dye and the Ca$^{2+}$ elevation, induced by 2-Aminoethoxydiphenyl borate (2-APB), was evaluated and compared with that of RCMH myotubes that express dysferlin. The data represents mean ± SEM. $n = 4$ cell cultures for each cell line. (JPG 371 kb)

The absence of dysferlin induces the expression of functional connexin-based hemichannels in human...

21

## Competing interests

PC, declares patent protection for the RCMH cell line. LAC, JAB, CA, AMC, AB, VM and JCS, declare no financial and non-financial competing interests.

## Authors' contribution

LAC designed research, performed experiments, analyzed data and wrote the paper. JAB, CA, AMC, AB, VM, JCS and PC designed research, analyzed the data, and wrote the paper. All authors read and approved the final manuscript.

## Acknowledgement

We thanks to Ms. Alejandra Trangulao for her technical support and to Dr. S. Spuler for providing the initial material from patients for develop of the cell lines 107, 379, AB320 and ER.

## Grants

Publication of this article was partially funded by CONICYT/PAI (Chile) Proyecto de Inserción en la Academia grant 79140023 (to LAC); CONICYT/PIA (Chile) Rings grant ACT 1121 (to PC, JAB, AMC,); Fondo Nacional de Desarrollo Científico y Tecnológico (FONDECYT) grant 1151383 (to JAB); FONDECYT grant 1150291 (to JCS); ICM-Economía P09-022-F Centro Interdisciplinario de Neurociencias de Valparaíso (to JCS); Association Française contre les Myopathies (AFM) and the jain foundation (to AB and VM).

## Author details

[1]Program of Anatomy and Developmental Biology, Faculty of Medicine, Institute of Biomedical Sciences, University of Chile, Av. Independencia #1027, Independencia, Santiago, Chile. [2]Departamento de Neurología y Neurocirugía, Hospital Clínico Universidad de Chile, Universidad de Chile, Santiago, Chile. [3]Centro Interdisciplinario de Neurociencias de Valparaíso, Universidad de Valparaíso, Valparaíso, Chile. [4]Center for Research in Myology, Sorbonne Universités, UPMC Univ Paris 06, INSERM UMRS974, CNRS FRE3617, 47 Boulevard de l'hôpital, 75013 Paris, France. [5]Departamento de Fisiología, Facultad de Ciencias Biológicas, Pontificia Universidad Católica de Chile, Santiago, Chile. [6]Programa de Farmacología Molecular y Clínica, Facultad de Medicina, Instituto de Ciencias Biomédicas, Universidad de Chile, Santiago, Chile.

## References

1. Matsuda C, Hayashi YK, Ogawa M, Aoki M, Murayama K, Nishino I, Nonaka I, Arahata K, Brown RH Jr. The sarcolemmal proteins dysferlin and caveolin-3 interact in skeletal muscle. Hum Mol Genet. 2001;10:1761–6.
2. Klinge L, Laval S, Keers S, Haldane F, Straub V, Barresi R, Bushby K. From T-tubule to sarcolemma: damage-induced dysferlin translocation in early myogenesis. FASEB J. 2007;21:1768–76.
3. Kerr JP, Ziman AP, Mueller AL, Muriel JM, Kleinhans-Welte E, Gumerson JD, Vogel SS, Ward CW, Roche JA, Bloch RJ. Dysferlin stabilizes stress-induced Ca2+ signaling in the transverse tubule membrane. Proc Natl Acad Sci U S A. 2013;110:20831–6.
4. Bansal D, Miyake K, Vogel SS, Groh S, Chen CC, Williamson R, McNeil PL, Campbell KP. Defective membrane repair in dysferlin-deficient muscular dystrophy. Nature. 2003;423:168–72.
5. Klinge L, Harris J, Sewry C, Charlton R, Anderson L, Laval S, Chiu YH, Hornsey M, Straub V, Barresi R, Lochmüller H, Bushby K. Dysferlin associates with the developing T-tubule system in rodent and human skeletal muscle. Muscle Nerve. 2010;41:166–73.
6. Cenacchi G, Fanin M, De Giorgi LB, Angelini C. Ultrastructural changes in dysferlinopathy support defective membrane repair mechanism. J Clin Pathol. 2005;58:190–5.
7. Urtizberea JA, Bassez G, Leturcq F, Nguyen K, Krahn M, Levy N. Dysferlinopathies. Neurol India. 2008;56:289–97.
8. Han R. Muscle membrane repair and inflammatory attack in dysferlinopathy. Skelet Muscle. 2011;1:10.
9. Nagaraju K, Rawat R, Veszelovszky E, Thapliyal R, Kesari A, Sparks S, Raben N, Plotz P, Hoffman EP. Dysferlin deficiency enhances monocyte phagocytosis: a model for the inflammatory onset of limb-girdle muscular dystrophy 2B. Am J Pathol. 2008;172(3):774–85.
10. Rawat R, Cohen TV, Ampong B, Francia D, Henriques-Pons A, Hoffman EP, Nagaraju K. Inflammasome up-regulation and activation in dysferlin-deficient skeletal muscle. Am J Pathol. 2010;176(6):2891–900.
11. Cea LA, Cisterna BA, Puebla C, Frank M, Figueroa XF, Cardozo C, Willecke K, Latorre R, Sáez JC. De novo expression of connexin hemichannels in denervated fast skeletal muscles leads to atrophy. Proc Natl Acad Sci U S A. 2013;110:16229–34.
12. Cea LA, Riquelme MA, Cisterna BA, Puebla C, Vega JL, Rovegno M, Sáez JC. Connexin- and pannexin-based channels in normal skeletal muscles and their possible role in muscle atrophy. J Membr Biol. 2012;245:423–36.
13. Sáez JC, Leybaert L. Hunting for connexin hemichannels. FEBS Lett. 2014;588:1205–11.
14. Cea LA, Riquelme MA, Vargas AA, Urrutia C, Sáez JC. Pannexin 1 channels in skeletal muscles. Front Physiol. 2014;5:eCollection.
15. Lostal W, Bartoli M, Roudaut C, Bourg N, Krahn M, Pryadkina M, Borel P, Suel L, Roche JA, Stockholm D, Bloch RJ, Levy N, Bashir R, Richard I. Lack of correlation between outcomes of membrane repair assay and correction of dystrophic changes in experimental therapeutic strategy in dysferlinopathy. PLoS One. 2012;7:e38036.
16. Craig SW, Pardo JV. Gamma actin, spectrin, and intermediate filament proteins colocalize with vinculin at costameres, myofibril-to-sarcolemma attachment sites. Cell Motil. 1983;3:449–62.
17. Kerr JP, Ward CW, Bloch RJ. Dysferlin at transverse tubules regulates Ca(2+) homeostasis in skeletal muscle. Front Physiol. 2014;5:89. eCollection.
18. Iwata Y, Katanosaka Y, Arai Y, Shigekawa M, Wakabayashi S. Dominant-negative inhibition of Ca2+ influx via TRPV2 ameliorates muscular dystrophy in animal models. Hum Mol Genet. 2009;18:824–34.
19. Neeper MP, Liu Y, Hutchinson TL, Wang Y, Flores CM, Qin N. Activation properties of heterologously expressed mammalian TRPV2: evidence for species dependence. J Biol Chem. 2007;282:15894–902.
20. Contreras JE, Sánchez HA, Eugenin EA, Speidel D, Theis M, Willecke K, Bukauskas FF, Bennett MV, Sáez JC. Metabolic inhibition induces opening of unapposed connexin 43 gap junction hemichannels and reduces gap junctional communication in cortical astrocytes in culture. Proc Natl Acad Sci U S A. 2002;99:495–500.
21. D'hondt C, Ponsaerts R, De Smedt H, Bultynck G, Himpens B. Pannexins, distant relatives of the connexin family with specific cellular functions? Bioessays. 2009;31:953–74.
22. Schalper KA, Sánchez HA, Lee SC, Altenberg GA, Nathanson MH, Sáez JC. Connexin 43 hemichannels mediate the Ca2+ influx induced by extracellular alkalinization. Am J Physiol Cell Physiol. 2010;299:C1504–15.
23. Proulx A, Merrifield PA, Naus CC. Blocking gap junctional intercellular communication in myoblasts inhibits myogenin and MRF4 expression. Dev Genet. 1997;20:133–44.
24. Tóth A, Fodor J, Vincze J, Oláh T, Juhász T, Zákány R, Csernoch L, Zádor E. The Effect of SERCA1b Silencing on the Differentiation and Calcium Homeostasis of C2C12 Skeletal Muscle Cells. PLoS One. 2015;10:e0123583.
25. Turner PR, Schultz R, Ganguly B, Steinhardt RA. Proteolysis results in altered leak channel kinetics and elevated free calcium in mdx muscle. J Membr Biol. 1993;133:243–51.
26. Azakir BA, Erne B, Di Fulvio S, Stirnimann G, Sinnreich M. Proteasome inhibitors increase missense mutated dysferlin in patients with muscular dystrophy. Sci Transl Med. 2014;6:250ra112.
27. Turner PR, Fong PY, Denetclaw WF, Steinhardt RA. Increased calcium influx in dystrophic muscle. J Cell Biol. 1991;115:1701–12.
28. Azakir BA, Di Fulvio S, Kinter J, Sinnreich M. Proteasomal inhibition restores biological function of mis-sense mutated dysferlin in patient-derived muscle cells. J Biol Chem. 2012;287:10344–54.
29. Ho M, Post CM, Donahue LR, Lidov HG, Bronson RT, Goolsby H, Watkins SC, Cox GA, Brown RH Jr. Disruption of muscle membrane and phenotype divergence in two novel mouse models of dysferlin deficiency. Hum Mol Genet. 2004;13:1999–2010.

30. Philippi S, Bigot A, Marg A, Mouly V, Spuler S, Zacharias U. Dysferlin-deficient immortalized human myoblasts and myotubes as a useful tool to study dysferlinopathy. Version 2. PLoS Curr. 2012;4:RRN1298.

31. Mamchaoui K, Trollet C, Bigot A, Negroni E, Chaouch S, Wolff A, Kandalla PK, Marie S, Di Santo J, St Guily JL, Muntoni F, Kim J, Philippi S, Spuler S, Levy N, Blumen SC, Voit T, Wright WE, Aamiri A, Butler-Browne G, Mouly V. Immortalized pathological human myoblasts: towards a universal tool for the study of neuromuscular disorders. Skelet Muscle. 2011;1:34.

32. Rojas G, Cárdenas AM, Fernández-Olivares P, Shimahara T, Segura-Aguilar J, Caviedes R, Caviedes P. Effect of the knockdown of amyloid precursor protein on intracellular calcium increases in a neuronal cell line derived from the cerebral cortex of a trisomy 16 mouse. Exp Neurol. 2008;209:234–42.

33. Woudt L, Di Capua GA, Krahn M, Castiglioni C, Hughes R, Campero M, Trangulao A, González-Hormazábal P, Godoy-Herrera R, Lévy N, Urtizberea J, Jara L, Bevilacqua JA. Toward an objective measure of functional disability in dysferlinopathy. Muscle Nerve. 2015. In press.

# 3

# Microvesicles derived from hypoxia/reoxygenation-treated human umbilical vein endothelial cells promote apoptosis and oxidative stress in H9c2 cardiomyocytes

Qi Zhang[1†], Man Shang[1†], Mengxiao Zhang[1], Yao Wang[1], Yan Chen[1], Yanna Wu[1], Minglin Liu[2,3], Junqiu Song[1*] and Yanxia Liu[1*]

## Abstract

**Background:** Vascular endothelial dysfunction is the closely related determinant of ischemic heart disease (IHD). Endothelial dysfunction and ischemia/reperfusion injury (IRI) have been associated with an increase in microvesicles (MVs) in vivo. However, the potential contribution of endothelial microvesicles (EMVs) to myocardial damage is unclear. Here we aimed to investigate the role of EMVs derived from hypoxia/reoxygenation (H/R) -treated human umbilical vein endothelial cells (HUVECs) on cultured H9c2 cardiomyocytes.

**Results:** H/R injury model was established to induce HUVECs to release H/R-EMVs. The H/R-EMVs from HUVECs were isolated from the conditioned culture medium and characterized. H9c2 cardiomyocytes were then incubated with 10, 30, 60 μg/mL H/R-EMVs for 6 h. We found that H9c2 cells treated by H/R-EMVs exhibited reduced cell viability, increased cell apoptosis and reactive oxygen species (ROS) production. Moreover mechanism studies demonstrated that H/R-EMVs could induce the phosphorylation of p38 and JNK1/2 in H9c2 cells in a dose-dependent manner. In addition, H/R-EMVs contained significantly higher level of ROS than EMVs generated from untreated HUVECs, which might be a direct source to trigger a cascade of myocardial damage.

**Conclusion:** We showed that EMVs released during H/R injury are pro-apoptotic, pro-oxidative and directly pathogenic to cardiomyocytes in vitro. EMVs carry ROS and they may impair myocardium by promoting apoptosis and oxidative stress. These findings provide new insights into the pathogenesis of IRI.

**Keywords:** Endothelial microvesicles, Oxidative stress, Apoptosis, H9c2 cardiomyocytes, Hypoxia/reoxygenation

## Background

Ischemic heart disease (IHD) is the major cause of death worldwide. The pathological processes leading to IHD (including myocardial infarction, angina pectoris, or both) are very complicated and closely accompanied with ischemia/reperfusion injury (IRI) [1]. It is generally accepted that oxidative stress is responsible for the damage of IRI, which is often associated with vascular dysfunction [2]. The endothelial cells that line the inner layer of blood vessels form a vital and dynamic structure that is essential for vascular hemostatic balance. These cells appear to be particularly vulnerable to the deleterious effects of both hypoxia (ischemia) and reoxygenation (reperfusion) [3].

Microvesicles (MVs) are small vesicles of 0.1 ~ 1 μm diameter released from stimulated or apoptotic cells, such as platelets, endothelial cells, lymphocytes, erythrocytes and even smooth muscle cells [4]. MVs contain a

* Correspondence: song_junqiu@126.com; liu_yanxia126@126.com
Qi Zhang and Man Shang are co-first authors.
†Equal contributors
[1]Department of Pharmacology, School of Basic Medical Sciences, Tianjin Medical University, No. 22, Qixiangtai Road, Heping District, Tianjin 300070, People's Republic of China
Full list of author information is available at the end of the article

subset of cell surface proteins derived from the plasma membrane of the original cells, which allow them to function as messengers that mediate many biological processes [5, 6]. In addition, MVs also carry various bio-active molecules, such as cytokines, RNA and DNA derived from their metrocyte, which can be transferred into target cells and mediate a series of biological effects [7–9]. Increased levels of circulating MVs have been suggested in acute coronary ischemia, myocardial infarction and other IHD, and MVs are likely contributing to endothelial dysfunction, leukocyte adhesion, platelet activation and obstruction of blood flow [10].

It has been reported that endothelial microvesicles (EMVs) may participate in inflammatory responses or angiogenesis, and propagate biological responses involved in haemostatic balance [11, 12]. Recent evidence suggests that EMVs may contribute to the oxidative injury and cell apoptosis in the course of IRI [13]. EMVs derived under pathological high glucose conditions induce adhesion protein expression in endothelial cells and subsequent monocyte adhesion in a NADPH oxidase-ROS-p38-dependent way [14]. Our group previously reported that MVs derived from hypoxia/reoxygenation-treated HUVECs impaired relaxation of rat thoracic aortic rings, and declined the production of NO and the expression of p-eNOS [15]. In this experiment, we established hypoxia/reoxygenation injury model to induce EMVs release in vitro and investigated its role on endothelial function of the aortic rings. However, the detailed mechanisms underlying EMVs-mediated cardiac damage and its relation to oxidative stress are not clear. Here we demonstrated the pathogenic roles of H/R-EMVs: (i) to cause cardiomyocytes injury directly; (ii) to promote cardiomyocytes apoptosis; (iii) to generate ROS in cardiomyocytes.

## Methods
### Cell culture
Human umbilical vein endothelial cells (HUVECs, Human EA.hy926 endothelial cells, Cell bank of Chinese Academy of Sciences, Shanghai, China) and H9c2 cells (ATCC, Manassas, VA, USA) were cultured in DMEM (Hyclone, Logan, UT, USA) with 10 % FBS (Gibco, CA, USA) under standard cell-culture conditions (37 °C, 5 % $CO_2$). All procedures were performed in accordance with the Declaration of Helsinki of the World Medical Association and the research protocol was approved by Ethics Committee of Tianjin Medical University.

### H/R-EMVs preparation
To generate endothelial microvesicles (EMVs), HUVECs were stimulated by hypoxia/reoxygenation (H/R) as previously described [15]. HUVECs of passage 5–8 were used when 70–80 % confluent. Briefly, HUVECs were subjected to hypoxic buffer (in mM: 0.9 $NaH_2PO_4$, 6.0

$NaHCO_3$, 1.0 $CaCl_2$, 1.2 $MgSO_4$, 20.0 HEPES, 98.5 NaCl, 10.0 KCl, 40.0 sodium lactate, pH 6.2) in a hypoxic chamber (95 % $N_2$ and 5 % $CO_2$, Billups-Rothenberg, Del Mar, CA, USA) for 12 h and then reoxygenated under standard cell-culture conditions for 4 h. Hypoxic buffer was collected in 15-mL centrifuge tubes and centrifuged at 2 700 g, 4 °C for 20 min to remove cell debris. Most supernatants were collected in 13.2-mL ultracentrifuge tubes and centrifuged at 33 000 rpm for 150 min to pellet H/R-EMVs. The pellet was resuspended in 100 μL PBS and kept at –20 °C.

### H/R-EMVs characterization by flow cytometry
H/R-EMVs were characterized by flow cytometry in terms of size assessment and biomarker identification. After centrifuging at 2 700 g, 4 °C for 20 min, aliquots of 90 μL supernatant without cell debris were collected and fixed with paraformaldehyde (PFA, Boster immunoleader, Wuhan, China) to a final concentration of 1 % for 1 h at room temperature (RT), then snap-frozen in liquid nitrogen and stored at –80 °C. For flow cytometry analysis, 10 μL fixed cell-free supernatants were blocked with mouse serum (Zhongkechenyu, Beijing, China) and then incubated with 5 μL anti-PE-CD144 antibody or its anti-PE Mouse IgG1 isotype (Santa Cruz, CA, USA) in dark for 30 min at RT, respectively. Latex beads of 1 μm (Molecular Probe, Invitrogen, Carlsbab, CA, USA) were used to calibrate gain setting and evaluate the size of EMVs. Events < 1 μm in diameter and CD144 positive were defined as H/R-EMVs. Each sample was analyzed with the flow cytometer (FACS Calibur, BD biosciences, Franklin Lakes, NJ, USA). Protein quantification of H/R-EMVs was performed by a BCA protein assay (Beyotime, Nanjing, China).

### Treatment with H/R-EMVs on H9c2 cells
H9c2 cells of passage 6–10 were used when 70–80 % confluent. For subsequent experiments, H9c2 cells were incubated with 10, 30, 60 μg/mL H/R-EMVs for 6 h. After H/R-EMVs treatment, culture supernatants and protein extracts of H9c2 cells were collected for further study.

### Colorimetric assay of cell viability and LDH activity
Cell viability was determined using methyl thiazolyl tetrazolium (MTT, Amresco, Solon, OH, USA) method. H9c2 cells cultured in 96-well plates at $1 \times 10^5$ cells/mL were treated with H/R-EMVs for 6 h. Then they were incubated with 10 μL 0.5 % MTT solution for 4 h at 37 °C. The supernatant was discarded after the incubation and 150 μL dimethyl sulfoxide was added to each well. The culture plate was shaken at high speed for 10 min until crystals dissolved completely. The absorbance of the blue formazan derivative was measured at a wavelength of 490 nm using a microplate reader (Bio-Rad Laboratories, CA, USA).

Lactate dehydrogenase (LDH) release detection was performed using a LDH Kit (Jiancheng Bioengineering Institute, Nanjing, China). Culture supernatants of H9c2 cells were collected after 6 h incubation with H/R-EMVs. Each supernatant (20 µL) was transferred to a fresh 96-well plate and an equal volume of freshly prepared reaction mixture was added according to the manufacturer's instruction. The absorbance was measured at a wavelength of 450 nm using the microplate reader following 15 min incubation at 37 °C. All experiments were repeated three times independently.

### H9c2 cell apoptosis assay

Apoptosis of H9c2 cells was examined by Hoechst 33258 staining, flow cytometry with Annexin V-FITC/PI staining and caspase 3 activity. After H/R-EMVs treatment, H9c2 cells in 6-well plates were washed twice with PBS and stained with 10 µg/mL Hoechst 33258 (KeyGen Biotech, Nanjing, China) at 37 °C for 20 min, and then examined under a fluorescent microscope (Nikon Melville, NY, USA) with the excitation wavelength of 350 nm for morphological changes.

To perform a quantitative analysis of cell apoptosis, flow cytometry with Annexin V-FITC/PI staining was employed. H9c2 cells were incubated with 5 µL Annexin V-FITC and PI (BD biosciences, Franklin Lakes, NJ, USA) for 15 min at RT in dark. Cells from each sample were then analyzed by FacsCalibur flow cytometer. The data was analyzed using Flowjo software.

For detection of the activity of caspase 3, H9c2 cells in 6-well plates were trypsinized and collected, then lysed at 4 °C for 15 min in a caspase 3 lysis buffer (Beyotime, Nanjing, China). Protein extracts of 10 µL were incubated with 90 µL reaction buffer containing 2 mM caspase-3 substrate (Ac-DEVD-pNA) for 2 h at 37 °C. The absorbance was measured at a wavelength of 405 nm using a multilabel reader (Bio-Tek, Winooski, VT, USA). Results were expressed as nmol/µg protein.

### Determination of lipid peroxidation level and superoxide dismutase activity

Lipid peroxidation levels in H9c2 cells were determined by estimating malondialdehyde (MDA) levels using the thiobarbituric acid reactive substance (TBARS) test (Jiancheng Bioengineering Institute, Nanjing, China). Cells were lysed by 1 % Triton-X 100 for 30 min on ice and then centrifuged at 12 000 g, 4 °C for 10 min. Protein concentration of the supernatants was determined by the BCA protein assay. Aliquots of 30 µL supernatants were incubated with reactive solutions according to the product instructions. The supernatant absorbance was measured at a wavelength of 532 nm. The results were expressed as nmol/mg protein.

The activity of superoxide dismutase (SOD) was measured in terms of inhibition of superoxide anions. Protein samples were prepared in the same way as MDA assay. Aliquots of 30 µL supernatants were incubated with reactive solutions at 37 °C water bath for 40 min. The absorbance was measured at a wavelength of 550 nm. SOD activities (U/mg protein) were calculated using the equation provided by the manufacture (Jiancheng Bioengineering Institute, Nanjing, China).

### Measurement of reactive oxygen species

Both reactive oxygen species (ROS) content in H/R-EMVs and ROS production in H9c2 cells were determined by 2′,7′-dichlorodihydrofluorescein diacetate assay (DCFH-DA, Beyotime, Nanjing, China). Pelleted H/R-EMVs and adherent MVs-treated H9c2 cells were diluted with 10 µM DCFH-DA and incubated for 20 min at 37 °C in dark, respectively. DCF intensity of MVs samples was analyzed with flow cytometry. To measure cellular ROS production, some MVs-treated H9c2 cells were washed and observed using fluorescent microscopy with the excitation wavelength of 520 nm. The other MVs-treated H9c2 cells were washed, trypsinized, pelleted and resuspended with PBS at $1 \times 10^6$ cells/mL. DCF intensity of these cell samples was also measured using flow cytometry.

### Western blot analysis of Bcl-2/Bax, p-p38 and p-JNK1/2

H9c2 cells were lysed in a lysis buffer (20 mM Tris pH 7.5, 150 mM NaCl, 1 % Triton X-100, sodium pyrophosphate, β-glycerophosphate, EDTA, $Na_3VO_4$, leupeptin, Beyotime, Nanjing, China) at 4 °C for 30 min. Protein concentration was measured using BCA assay. Equal amounts of proteins (80 µg) were loaded into 12 % SDS electrophoresis and transferred onto PVDF membranes. Blots were incubated with blocking buffer for 60 min at RT, then incubated with the relevant primary antibodies (anti-β-actin, anti-Bcl-2, anti-Bax, anti-p-JNK1/2 antibody, Santa Cruz, CA, USA; anti-p-p38 MAPK, anti-p38 MAPK, anti-JNK1/2 antibody, Cell Signaling Technology, Danvers, MA, USA) overnight at 4 °C, followed by the corresponding HRP-conjugated secondary antibodies for 120 min. Then proteins were revealed by chemiluminescence using the ECL kit (Beyotime, Nanjing, China).

### Statistical analysis

Data were expressed as mean ± standard derivation (SD). The one-way analysis of variance (ANOVA) was used for multiple comparisons. Statistical evaluation was performed using GraphPad Prism 5. The value of $P < 0.05$ was considered statistically significant. All experiments were repeated three times independently.

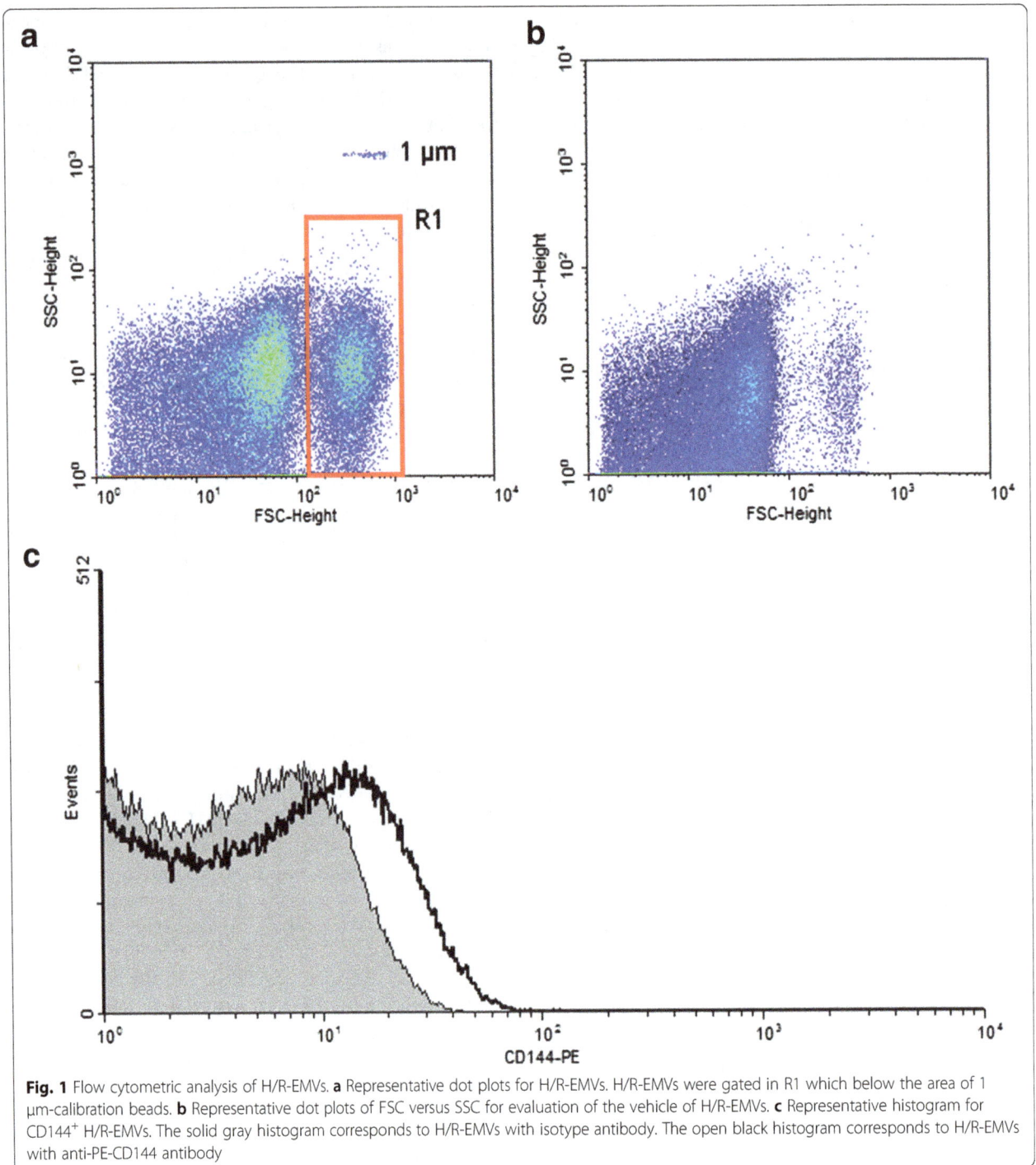

**Fig. 1** Flow cytometric analysis of H/R-EMVs. **a** Representative dot plots for H/R-EMVs. H/R-EMVs were gated in R1 which below the area of 1 μm-calibration beads. **b** Representative dot plots of FSC versus SSC for evaluation of the vehicle of H/R-EMVs. **c** Representative histogram for CD144⁺ H/R-EMVs. The solid gray histogram corresponds to H/R-EMVs with isotype antibody. The open black histogram corresponds to H/R-EMVs with anti-PE-CD144 antibody

## Results

### Characterization of H/R-EMVs

H/R stimulation of HUVECs resulted in the formation of MVs of about <1 μm in diameter as assessed by flow cytometry. HUVECs were first exposed to hypoxia (12 h) and reoxygenation (4 h). The H/R injury decreased HUVECs viability to $70.53 \pm 2.61$ % compared with control ($P < 0.001$, supplemental data). Flow cytometry analysis of H/R-

EMVs was used to determine their size and cellular origin. Using 1 μm beads as size standards, the majority of H/R-EMVs were observed around the forward scatter signal corresponding to 1 μm beads (Fig. 1a, b). Cellular origin was identified by investigating the specific surface antigens of MVs. Most of H/R-EMVs externalized their endothelial cell marker CD144 (Fig. 1c). These results indicated that H/R-EMVs had a size of <1 μm and expressed on their surface

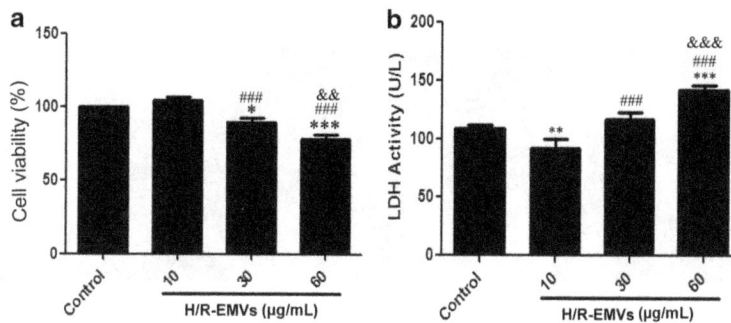

**Fig. 2** H/R-EMVs exerted cytotoxic effect on H9c2 cells. **a** MTT assays. H9c2 cells were treated with 10, 30, 60 µg/mL H/R-EMVs for 6 h. The viability of H9c2 cells was expressed as a percentage relative to non-H/R-EMVs-treated control cells. Control cells were considered to be 100 % viable. **b** LDH activity assay. H9c2 cells were treated at the concentration of 10, 30, 60 µg/mL H/R-EMVs for 6 h to detect LDH activity. *$p < 0.05$, ***$p < 0.001$ versus control; ###$p < 0.001$ versus H/R-EMVs 10 µg/mL; &&$p < 0.01$, &&&$p < 0.001$ versus H/R-EMVs 30 µg/mL

adhesion molecules of HUVECs from which they originated.

### H/R-EMVs reduced the viability of H9c2 cells

To explore whether H/R-EMVs influence the progression of myocardium damage, target H9c2 cardiomyocytes were exposed to 10, 30, 60 µg/mL H/R-EMVs for 6 h, respectively. Compared with control group, 10 µg/mL H/R-EMVs showed little effect on H9c2 cells, whereas 30 and 60 µg/mL H/R-EMVs significantly decreased H9c2 cell viability by 10 % and 20 %, respectively ($P < 0.05$, Fig. 2a). The cytotoxic effect of H/R-EMVs was further confirmed using LDH assay. H/R-EMVs of 10 µg/mL did not induce more LDH leakage than the control group. As expected, H9c2 cells exposed to 30 and 60 µg/mL H/R-EMVs exhibited more release of LDH through damaged cell membranes ($P < 0.05$, Fig. 2b). These findings indicated that H/R-EMVs displayed a dose-dependent cytotoxicity in H9c2 cardiomyocytes.

### H/R-EMVs increased H9c2 cell apoptosis

In order to investigate the mechanisms of H/R-EMVs-induced cell death, we performed Hoechst 33258 staining and Annexin V-FITC/PI staining. After being treated with H/R-EMVs for 6 h, H9c2 cells were stained with Hoechst 33258 and observed under the fluorescence microscope. The dye stains condense chromatin of apoptotic cells more brightly than that of normal cells. H/R-EMVs groups in the concentrations of 10 and 30 µg/mL showed few fragmented nuclei, while 60 µg/mL H/R-EMVs group displayed apparently increasing number of fragmented or condensed nuclei (Fig. 3a). Flow cytometry with Annexin V-FITC/PI staining showed 60 µg/mL H/R-EMVs increased the apoptotic rate of target H9c2 cells by 18 % ($P < 0.05$, Fig. 3b, c).

To confirm the co-incubation of H9c2 cells with high dose of H/R-EMVs is contributable to their apoptosis,

the activity of caspase 3 and expression of Bcl-2/Bax in target cells were determined. Consistent with the rates of Annexin V positive cells, caspase 3 activity of H9c2 cells stimulated by 60 µg/mL H/R-EMVs increased to $294.14 \pm 28.03$ nmol/µg, compared with $140.23 \pm 29.43$ nmol/µg in control group ($P < 0.05$, Fig. 3d). In addition, high dose (60 µg/mL) of H/R-EMVs also induced higher activity of caspase 3 than 10 and 30 µg/mL H/R-EMVs ($P < 0.05$, Fig. 3d). The dynamic balance of Bcl-2 and Bax determines a cell's fate. Bax levels increase but Bcl-2 levels decrease during cell apoptosis. Western blot analysis revealed that Bcl-2/Bax ratio in H9c2 cells gradually decreased with the increasing dose of H/R-EMVs (Fig. 3e). Therefore, the pro-apoptotic effect of H/R-EMVs in H9c2 cells was associated with Bcl-2 inhibition and Bax up- regulation.

### H/R-EMVs induced the oxidative damage in H9c2 cells by MDA and ROS production

It has been reported that excessive generation of ROS plays a major role in the initiation of apoptosis during acute myocardial infarction. Specific ROS such as $H_2O_2$ or superoxide have been implicated as crucial mediators of apoptotic cell death [16]. As mentioned above, H/R-EMVs were confirmed to be pro-apoptotic, next we investigated whether H/R-EMVs induce excessive production of lipid peroxidation and ROS in target cardiomyocytes. Results showed that SOD (an eliminator of free radicals) activity of H9c2 cells decreased, while MDA (an indicator of lipid peroxidation) content increased markedly in 60 µg/mL H/R-EMVs-treated group when compared with control ($P < 0.01$, Fig. 4a, b). In further experiments, ROS production was examined by fluorescent microscopy and flow cytometry in DCFH-DA-labelled H9c2 cardiomyocytes. It was observed that ROS gradually accumulated in H9c2 cells with the increasing dose of H/R-EMVs (Fig. 4c). Flow cytometry analysis showed that DCF fluorescence (ROS level) in

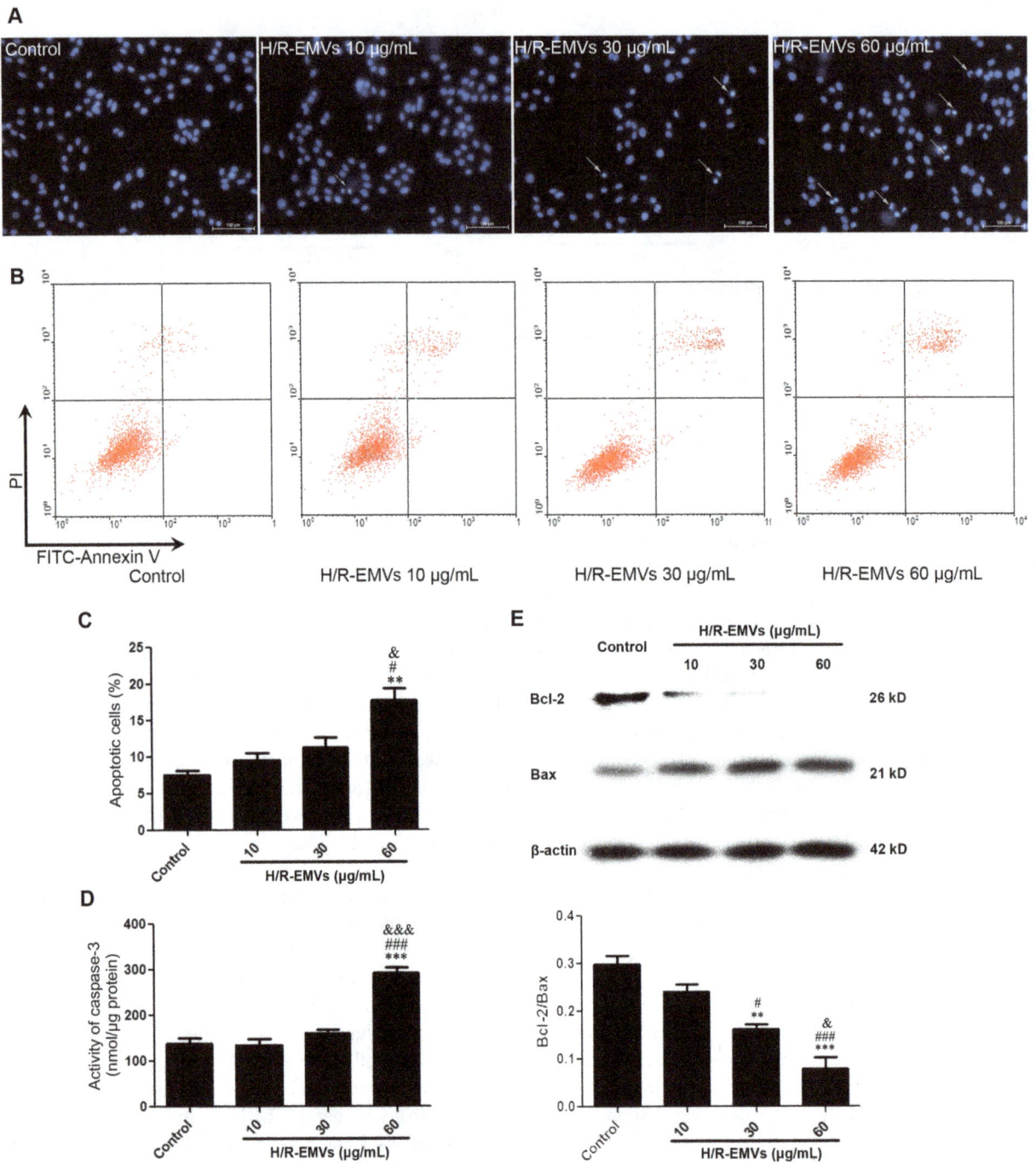

**Fig. 3** H/R-EMVs exhibited pro-apoptotic effect on H9c2 cells. **a** Hoechst 33258 staining was used to observe the apoptotic cells after treatment with 10, 30, 60 µg/mL H/R-EMVs for 6 h (100×). The arrows show apoptotic cells. The untreated cells serve as Control. **b** Annexin V-FITC/PI double staining. The results were interpreted in the following fashion: cells in the lower-left quadrant (Annexin V−/PI−) represent living cells; those in the lower-right quadrant (Annexin V+/PI−) represent early apoptotic cells; those in the upper-right quadrant (Annexin V+/PI+) represent late apoptotic cells; and those in the upper-left quadrant (Annexin V−/PI+) represent necrotic cells. **c** The apoptotic rate of H9c2 cells was quantified according to flow cytometry analysis. **d** Caspase 3 activity of H9c2 cells was measured after treatment with 10, 30, 60 µg/mL H/R-EMVs for 6 h. **e** Representative images of immunoblots with antibodies against Bcl-2, Bax, β-actin, respectively among groups. β-actin serves as a loading control. $^*p < 0.05$, $^{**}p < 0.01$, $^{***}p < 0.001$ versus Control; $^\#p < 0.05$, $^{\#\#}p < 0.01$, $^{\#\#\#}p < 0.001$ versus H/R-EMVs 10 µg/mL; $^\&p < 0.05$, $^{\&\&\&}p < 0.001$ versus H/R-EMVs 30 µg/mL

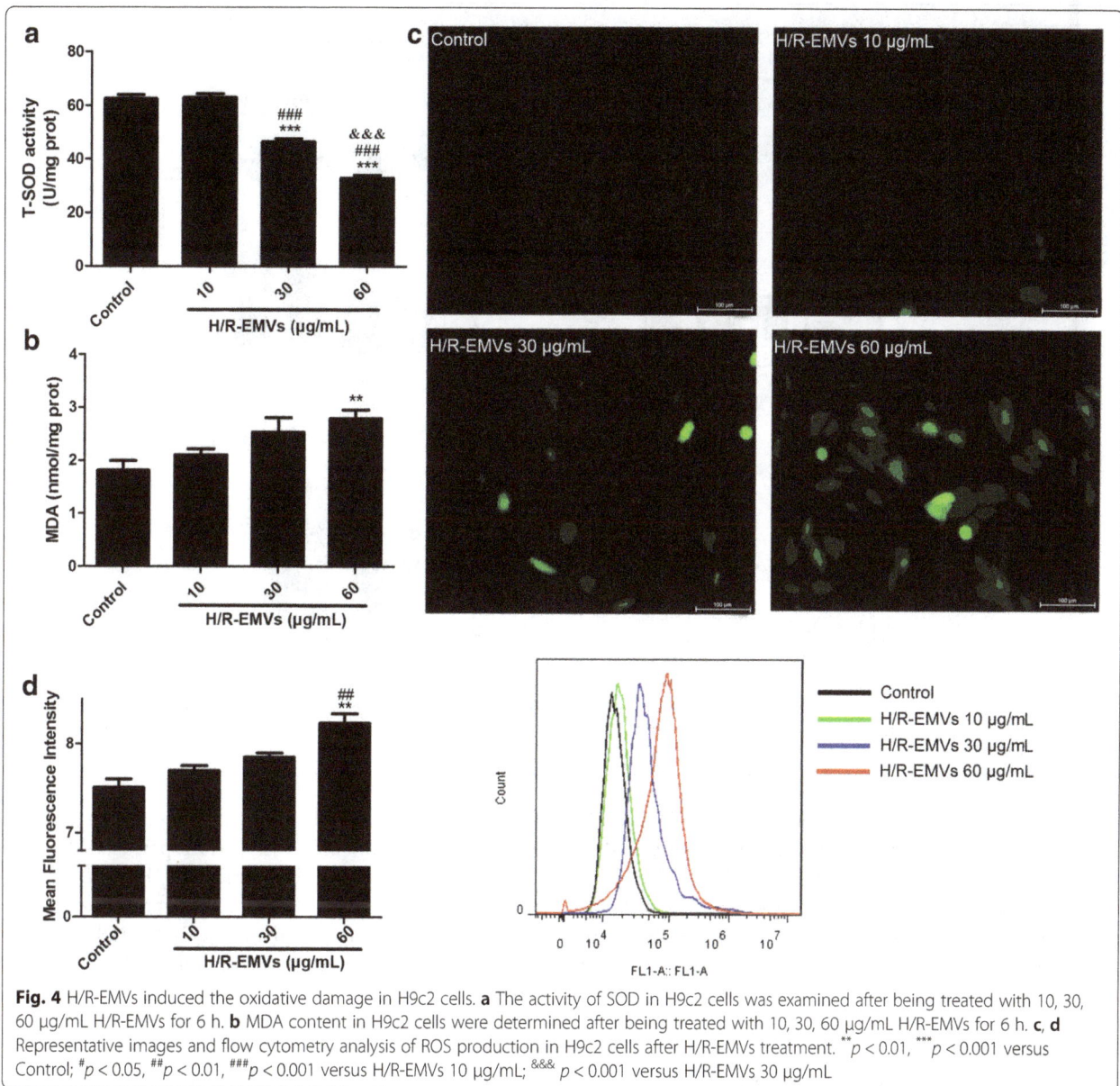

**Fig. 4** H/R-EMVs induced the oxidative damage in H9c2 cells. **a** The activity of SOD in H9c2 cells was examined after being treated with 10, 30, 60 µg/mL H/R-EMVs for 6 h. **b** MDA content in H9c2 cells were determined after being treated with 10, 30, 60 µg/mL H/R-EMVs for 6 h. **c, d** Representative images and flow cytometry analysis of ROS production in H9c2 cells after H/R-EMVs treatment. $^{**}p < 0.01$, $^{***}p < 0.001$ versus Control; $^{#}p < 0.05$, $^{##}p < 0.01$, $^{###}p < 0.001$ versus H/R-EMVs 10 µg/mL; $^{&&&}p < 0.001$ versus H/R-EMVs 30 µg/mL

60 µg/mL H/R-EMVs-treated group increased by 9.59 % compared with control ($P < 0.01$, Fig. 4d). These results suggested that H/R-EMVs trigger ROS production to induce target cell apoptosis and oxidative damage.

### H/R-EMVs up-regulated p-p38 MAPK and p-JNK1/2 expression in target H9c2 cardiomyocytes

In this study, we performed Western blot analysis on p38 and JNK1/2 activation of target H9c2 cells. The activation of p38 MAPK pathway was indicated by a significant increase of p38 phosphorylation in H9c2 cardiomyocytes treated with 30 or 60 µg/mL H/R-EMVs ($P < 0.01$, Fig. 5a). Moreover, the phosphorylation of JNK1/2 in H9c2 cells was up-regulated significantly in 30 and 60 µg/mL H/R-EMVs groups compared with

control ($P < 0.01$, Fig. 5b). And the levels of p-p38 and p-JNK1/2 were increased in target cells in a dose-dependent manner. Thus, these findings showed that exposure of H9c2 cardiomyocytes to high dose H/R-EMVs results in activation of both the p38 and JNK1/2 signaling pathways.

### H/R-EMVs carried ROS

To determine which factor in H/R-EMVs might be responsible for p38 and JNK1/2 activation in H9c2 cardiomyocytes, we assessed ROS content in H/R-EMVs. EMVs derived from normal cultured HUVECs (without hypoxia12 h/reoxygenation 4 h) were introduced as control, which was defined as C-EMVs. Flow cytometry analysis showed that the DCF intensity of H/R-EMVs was

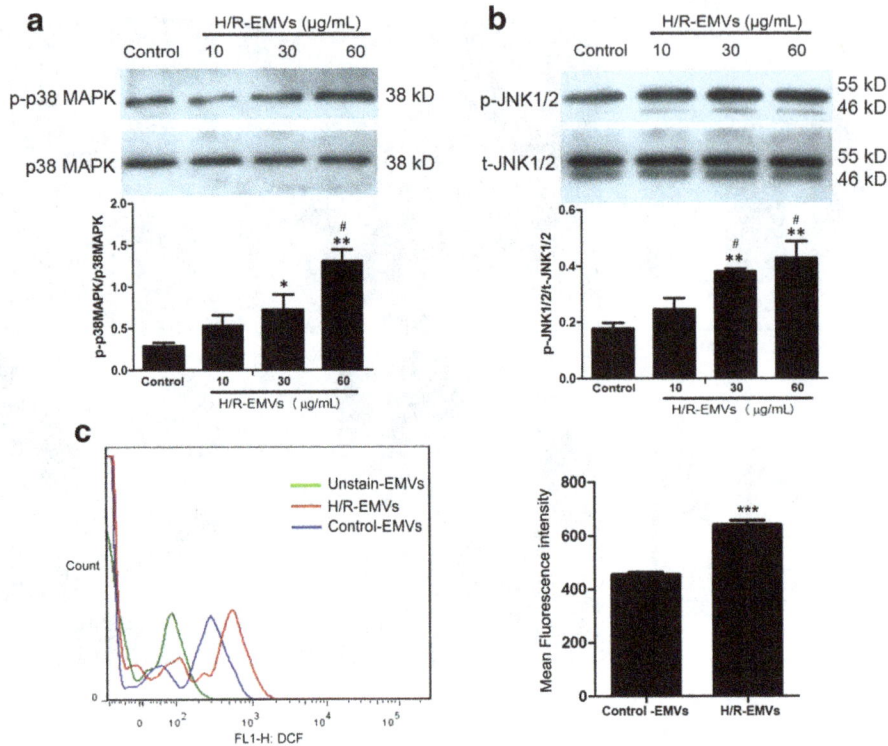

**Fig. 5** H/R-EMVs up-regulated the expression of p38 MAPK and JNK1/2 in H9c2 cells. **a** Representative images of immunoblots of p38 MAPK activation in each group. **b** JNK1/2 activation in H9c2 cells (phospho-JNK1/2). $^*p < 0.05$, $^{**}p < 0.01$ versus Control; $^{#}p < 0.05$ versus H/R-EMVs 10 μg/mL. **c** ROS content in Control-EMVs and H/R-EMVs. $^{***}p < 0.001$ versus Control-EMVs

significantly stronger than that of C-EMVs, which indicated that H/R-EMVs contained more ROS than C-EMVs ($639.83 \pm 41.03$ v.s $453.67 \pm 20.42$, $P < 0.001$, Fig. 5c). Therefore, H/R-treated HUVECs released EMVs containing ROS. The ROS-containing EMVs might contribute to ROS production and subsequent oxidative stress in H9c2 cells.

## Discussion

IHD is a fatal disease and characterized by deficiency of coronary blood supply and impaired myocardium. MVs from various cellular origins have been identified in IHD, specially, EMVs account for a large proportion [17]. Accumulating evidences demonstrate that the elevated EMVs can actively modulate disease progression mostly including atherosclerosis and myocardial infarction [11]. EMVs could be interpreted as a vascular injury marker with prognostic value [18]. However, it is technically difficult to separate MVs from different origins in circulating blood. MVs are present at relatively low concentrations in normal physiological condition, but their levels increase in pathophysiological states [19, 20]. Recent evidences showed that MVs could be generated abundantly from cells undergoing inflammation, radiation, oxidative stress and so on [6]. We have reported that EMVs could be generated from cultured endothelial

cells by calcium ionophore A23187 in the manner of magnificent calcium influx [21]. As reperfusion therapy plays a critical role in the treatment of IHD, further exploration on the mechanism of IRI is very necessary. In our research, treatments of HUVECs by H/R in vitro were used as a new approach to mimic IRI in vivo to generate EMVs. And flow cytometry analysis confirmed that the vesicles induced by H/R injury were EMVs with CD144 positive. However, MVs generated in normal physiological condition were hard to detect. Therefore, H/R-EMVs in different doses were used for functional studies.

Endothelial dysfunction is involved in the initial and core processes of the pathogenesis of IHD [22]. EMVs, released in response to endothelial cell activation or apoptosis, are significantly increased in patients with IHD, but their potential effect on myocardium is largely unknown. It has been reported that the level of EMVs among patients with myocardial infarction was positively correlated with the extent of vascular inflammation and myocardial infarct size [23]. EMVs generated from starved endothelial cells could dose-dependently suppress the endothelial cell proliferation with the dosage of $10^3$-$10^5$ EMVs/mL [24]. Interestingly our results showed that high concentrations of H/R-EMVs could significantly promote cell apoptosis.

Cardiomyocyte apoptosis is a major event in the pathogenesis of IRI. Preliminary experiments indicated that in vivo myocardial I/R treatment produced obvious myocardial infarction, and TUNEL staining showed many apoptotic myocytes in the ischemic area [25]. Apoptosis is caused by an imbalance between pro-apoptotic and anti-apoptotic signals. It has already been known that Bax, Bcl-2, and caspase-3 are downstream molecules in mitochondrial apoptotic signaling pathway [26]. In particular, activation of caspase-3 plays a central role in the initiation of apoptosis. In our study, we found that the expression ratio of Bcl-2/Bax in target cardiomyocytes was down-regulated by H/R-EMVs but the activity of caspase 3 was enhanced. These results confirmed the pro-apoptotic effects of H/R-EMVs on cardiomyocytes in vitro.

Specific ROS have been implicated as crucial mediators of apoptotic cell death. EMVs exposed to AT1-AA (Angiotensin II receptor type 1 autoantibody) or high glucose condition all greatly increased ROS production in target cells. These "injured" EMVs trigged oxidative stress and induced endothelial dysfunction [27]. In agreement with these studies, our research found that H/R-EMVs increased ROS production in terms of increasing MDA content and decreasing SOD activity in H9c2 cells.

Because of their potential relevance to cell apoptosis and oxidative damage, we aimed to determine the possible pathway in which H/R-EMVs may participate. Multiple mechanisms have been proposed to explain myocardial injury during IRI. ROS lead to cell damage either directly or through behaving as intermediates in p38 MAPK and JNK1/2 downstream signaling pathways [28]. As expected, our study found that p38 and JNK1/2 were activated after treatment of 30 or 60 µg/mL H/R-EMVs. Additionally, it has been demonstrated that specific inhibition of p38 pathway resulted in reduced monocyte adhesion, in accordance with the down-regulation of ICAM-1 and VCAM-1 in target cells [14]. The use of ROS inhibitors could abolish EMVs-induced ROS production and reduce p38 phosphorylation in target cells. Moreover, addition of "injured" MVs to primary hepatocytes induced up-regulation of pro-inflammatory COX-2 and PKC-δ protein and the activation of JNK1/2 [13]. Taken together, these findings help us speculate that the activation of p38 and JNK1/2 could be triggered by ROS accumulation, suggesting that H/R- EMVs should probably promote oxidative stress in H9c2 cardiomyocytes through p38 MAPK and JNK1/2 pathways.

IRI induces oxidative stress and intense inflammatory response resulting from the capacity of endogenous constituents. Recently, the bioactive contents carried by MVs are of great concerns. It has been discovered that MVs are not merely debris; they can carry cytokines and nuclear materials such as DNA, RNA, and microRNA from their metrocyte [7–9]. MVs from endothelial progenitor cells could transfer mRNA to endothelial cells and activate an angiogenic program [9]. Here, we found H/R-EMVs carry ROS with significantly high levels, indicating ROS content in H/R-EMVs might have a link with oxidative stress in target cells, as well as the increasing ROS production. However, the exact mechanism of the increased ROS production in target cells still needs further investigation.

## Conclusion

In this study, we first established that EMVs could be generated from H/R-treated HUVECs. Then we demonstrated that 60 µg/mL H/R-EMVs exerted pro-apoptotic and oxidative effects on H9c2 cardiomyocytes via p38 and JNK1/2 signaling pathways. ROS carried by H/R-EMVs might be the underlying pathway to explicate their roles in apoptosis and oxidative stress. These findings indicated that the connection of EMVs and cardiomyocyte death would be interpreted as a novel intervention to study IRI, suggesting that decreasing the levels of EMVs should be a new therapeutic strategy for the maintenance of endothelial homeostasis and the treatment of IHD. However, whether other bioactive molecules in EMVs are contributable to myocardial injury is not clear. Moreover, our results need to be confirmed with the study of myocardial I/R models in vivo.

## Consent

All authors have read and approved the submission of this manuscript.

### Abbreviations

DCFH-DA: 2′,7′-dichlorodihydrofluorescein diacetate; HUVECs: human umbilical vein endothelial cells; H/R: hypoxia/reoxygenation; IHD: ischemic heart disease; IRI: ischemia/reperfusion injury; LDH: lactate dehydrogenase; MDA: malondialdehyde; MTT: methyl thiazolyl tetrazolium; MVs: microvesicles; PFA: paraformaldehyde; ROS: reactive oxygen species; RT: room temperature; SOD: superoxide dismutase; TBARS: thiobarbituric acid reactive substance.

### Competing interests

The authors declare that they have no competing interests.

### Authors' contributions

Qi Zhang, Man Shang, Yanna Wu, Junqiu Song and Yanxia Liu: Study design; Qi Zhang and Man Shang: Directly participated in the performance of all the experiments; Mengxiao Zhang and Yao Wang: Establishing flow cytometry experiments; Yan Chen: Support on separation of MVs; Junqiu Song: Support on hypoxia/reoxygenation injury on HUVECs; Yanxia Liu and Minglin Liu: Supervise during the whole work. All authors approved the final manuscript.

### Acknowledgements

This work was supported by the Specialized Research Fund for the Doctoral Program of Higher Education of China (No. 20101202110005), The National Natural Science Foundation of China (No. 81102446 and No. 81370422), the Natural Science Foundation of Tianjin (No. 11JCZDJC18300), the Research Foundation of Tianjin Municipal Education Commission (No. 20110106) and the National Key Basic Research Program of China (973 Program No. 2011CB933100).

**Author details**

[1]Department of Pharmacology, School of Basic Medical Sciences, Tianjin Medical University, No. 22, Qixiangtai Road, Heping District, Tianjin 300070, People's Republic of China. [2]Section of Endocrinology, Department of Medicine, Temple University School of Medicine, 3500 North Broad Street, Room 480A, Philadelphia, PA 19140, USA. [3]Department of Dermatology, Perelman School of Medicine, University of Pennsylvania, Philadelphia PA, 19104, USA.

**References**

1. Ibanez B, Heusch G, Ovize M, Van de Werf F. Evolving therapies for myocardial ischemia/reperfusion injury. J Am Coll Cardiol. 2015;65(14):1454–71.
2. Mozaffari MS, Liu JY, Abebe W, Baban B. Mechanisms of load dependency of myocardial ischemia reperfusion injury. American journal of cardiovascular disease. 2013;3(4):180–96.
3. Feng Y, Hu L, Xu Q, Yuan H, Ba L, He Y, Che H. Cytoprotective role of alpha1-antitrypsin in vascular endothelial cell under hypoxia/reoxygenation condition. J Cardiovasc Pharmacol. 2015;66(1):96–107.
4. Curtis AM, Edelberg J, Jonas R, Rogers WT, Moore JS, Syed W, Mohler ER, 3rd. Endothelial microparticles: sophisticated vesicles modulating vascular function. Vasc Med. 2013;18(4):204–14.
5. Laresche C, Pelletier F, Garnache-Ottou F, Lihoreau T, Biichle S, Mourey G, Saas P, Humbert P, Seilles E, Aubin F. Increased levels of circulating microparticles are associated with increased procoagulant activity in patients with cutaneous malignant melanoma. J Invest Dermatol. 2014; 134(1):176–82.
6. Leroyer AS, Anfosso F, Lacroix R, Sabatier F, Simoncini S, Njock SM, Jourde N, Brunet P, Camoin-Jau L, Sampol J et al. Endothelial-derived microparticles: Biological conveyors at the crossroad of inflammation, thrombosis and angiogenesis. Thromb Haemost. 2010;104(3):456–63.
7. Jansen F, Yang X, Hoelscher M, Cattelan A, Schmitz T, Proebsting S, Wenzel D, Vosen S, Franklin BS, Fleischmann BK, et al. Endothelial microparticle-mediated transfer of MicroRNA-126 promotes vascular endothelial cell repair via SPRED1 and is abrogated in glucose-damaged endothelial microparticles. Circulation. 2013;128(18):2026–38.
8. Lannan KL, Sahler J, Kim N, Spinelli SL, Maggirwar SB, Garraud O, Cognasse F, Blumberg N, Phipps RP. Breaking the mold: transcription factors in the anucleate platelet and platelet-derived microparticles. Frontiers in immunology. 2015;6(48):1–16.
9. Deregibus MC, Cantaluppi V, Calogero R, Lo Iacono M, Tetta C, Biancone L, Bruno S, Bussolati B, Camussi G. Endothelial progenitor cell derived microvesicles activate an angiogenic program in endothelial cells by a horizontal transfer of mRNA. Blood. 2007;110(7):2440–8.
10. Montoro-Garcia S, Shantsila E, Tapp LD, Lopez-Cuenca A, Romero AI, Hernandez-Romero D, Orenes-Pinero E, Manzano-Fernandez S, Valdes M, Marin F, et al. Small-size circulating microparticles in acute coronary syndromes: relevance to fibrinolytic status, reparative markers and outcomes. Atherosclerosis. 2013;227(2):313–22.
11. Zhang J, Ren J, Chen H, Geng Q. Inflammation induced-endothelial cells release angiogenesis associated-microRNAs into circulation by microparticles. Chin Med J (Engl). 2014;127(12):2212–7.
12. Arderiu G, Peña E, Badimon L. Angiogenic Microvascular Endothelial Cells Release Microparticles Rich in Tissue Factor That Promotes Postischemic Collateral Vessel Formation. Arterioscler Thromb Vasc Biol. 2015;35(2):348–57.
13. Teoh NC, Ajamieh H, Wong HJ, Croft K, Mori T, Allison AC, Farrell GC. Microparticles mediate hepatic ischemia-reperfusion injury and are the targets of Diannexin (ASP8597). PLoS One. 2014;9(9):e104376.
14. Jansen F, Yang X, Franklin BS, Hoelscher M, Schmitz T, Bedorf J, Nickenig G, Werner N. High glucose condition increases NADPH oxidase activity in endothelial microparticles that promote vascular inflammation. Cardiovasc Res. 2013;98(1):94–106.
15. Wang SX, Zhang Q, Shang M, Wei S, Liu M, Wang YL, Zhang MX, Wu YN, Liu ML, Song JQ, Liu YX. Microvesicles derived from hypoxia/reoxygenation-treated human umbilical vein endothelial cells impair relaxation of rat thoracic aortic rings. Zhongguo Ying Yong Sheng Li Xue Za Zhi. 2014;30(6):560–6.
16. Webster KA. Mitochondrial membrane permeabilization and cell death during myocardial infarction: roles of calcium and reactive oxygen species. Future Cardiol. 2012;8(6):863–84.
17. Shah MD, Bergeron AL, Dong JF, Lopez JA. Flow cytometric measurement of microparticles: pitfalls and protocol modifications. Platelets. 2008;19(5): 365–72.
18. Nozaki T, Sugiyama S, Koga H, Sugamura K, Ohba K, Matsuzawa Y, Sumida H, Matsui K, Jinnouchi H, Ogawa H. Significance of a multiple biomarkers strategy including endothelial dysfunction to improve risk stratification for cardiovascular events in patients at high risk for coronary heart disease. J Am Coll Cardiol. 2009;54(7):601–8.
19. Ou ZJ, Chang FJ, Luo D, Liao XL, Wang ZP, Zhang X, Xu YQ, Ou SJ. Endothelium-derived microparticles inhibit angiogenesis in the heart and enhance the inhibitory effects of hypercholesterolemia on angiogenesis. Am J Physiology-endocrinology Metabolism. 2011;300(4):E661–8.
20. Combes V, Simon A-C, Grau G-E, Arnoux D, Camoin L, Sabatier F, Mutin M, Sanmarco M, Sampol J, Dignat-George F. In vitro generation of endothelial microparticles and possible prothrombotic activity in patients with lupus anticoagulant. J Clin Investigation. 1999;104(1):93–102.
21. Shang M, Zhang Q, Zhang MX, Wang Y, Chen Y, Wu YN, Song JQ, Liu ML, Liu YX. Effects of endothelial microvesicles induced by A23187 on H9c2 cardiomytes. Zhongguo Ying Yong Sheng Li Xue Za Zhi. 2013;6(29):559–64.
22. Eckers A, Haendeler J. Endothelial cells in health and disease. Antioxid Redox Signal. 2015;22(14):1209–11.
23. Radecke CE, Warrick AE, Singh GD, Rogers JH, Simon SI, Armstrong EJ. Coronary artery endothelial cells and microparticles increase expression of VCAM-1 in myocardial infarction. Thromb Haemost. 2014;113(3):605–16.
24. Mezentsev A, Merks RM, O'Riordan E, Chen J, Mendelev N, Goligorsky MS, Brodsky SV. Endothelial microparticles affect angiogenesis in vitro: role of oxidative stress. Am J Physiology-heart Circulatory Physiology. 2005;289(3): H1106–14.
25. Song JQ, Teng X, Cai Y, Tang CS, Qi YF. Activation of Akt/GSK-3beta signaling pathway is involved in intermedin(1–53) protection against myocardial apoptosis induced by ischemia/reperfusion. Apoptosis. 2009; 14(11):1299–307.
26. Jiang X, Guo CX, Zeng XJ, Li HH, Chen BX, Du FH. A soluble receptor for advanced glycation end-products inhibits myocardial apoptosis induced by ischemia/reperfusion via the JAK2/STAT3 pathway. Apoptosis. 2015;20(8): 1033–47.
27. Yang S, Zhong Q, Qiu Z, Chen X, Chen F, Mustafa K, Ding D, Zhou Y, Lin J, Yan S, et al. Angiotensin II receptor type 1 autoantibodies promote endothelial microparticles formation through activating p38 MAPK pathway. J Hypertens. 2014;32(4):762–70.
28. Xiong XX, Liu JM, Qiu XY, Pan F, Yu SB, Chen XQ. Piperlongumine induces apoptotic and autophagic death of the primary myeloid leukemia cells from patients via activation of ROS-p38/JNK pathways. Acta Pharmacol Sin. 2015; 36(3):362–74.

# 4

# Matrix metalloproteinase 9 induces endothelial-mesenchymal transition via Notch activation in human kidney glomerular endothelial cells

Ye Zhao[1,2†], Xi Qiao[1,3†], Lihua Wang[3], Tian Kui Tan[1], Hong Zhao[4], Yun Zhang[5], Jianlin Zhang[4], Padmashree Rao[1], Qi Cao[1], Yiping Wang[1], Ya Wang[1], Yuan Min Wang[6], Vincent W. S. Lee[1], Stephen I. Alexander[6], David C. H. Harris[1] and Guoping Zheng[1*]

## Abstract

**Background:** Endothelial-mesenchymal transition (EndoMT) is a major source of myofibroblast formation in kidney fibrosis. Our previous study showed a profibrotic role for matrix metalloproteinase 9 (MMP-9) in kidney fibrosis via induction of epithelial-mesenchymal transition (EMT). Inhibition of MMP-9 activity reduced kidney fibrosis in murine unilateral ureteral obstruction. This study investigated whether MMP-9 also plays a role in EndoMT in human glomerular endothelial cells.

**Results:** TGF-β1 (10 or 20 ng/ml) induced EndoMT in HKGECs as shown by morphological changes. In addition, VE-cadherin and CD31 were significantly downregulated, whereas α-SMA, vimentin, and N-cadherin were upregulated. RT-PCR revealed that Snail, a known inducer of EMT, was upregulated. The MMP inhibitor GM6001 abrogated TGF-β1-induced EndoMT. Zymography indicated that MMP-9 was also upregulated in TGF-β1-treated HKGECs. Recombinant MMP-9 (2 μg/ml) induced EndoMT in HKGECs via Notch signaling, as evidenced by increased formation of the Notch intracellular domain (NICD) and decreased Notch 1. Inhibition of MMP-9 activity by its inhibitor showed a dose-dependent response in preventing TGF-β1-induced α-SMA and NICD in HKGECs, whereas inhibition of Notch signaling by γ-secretase inhibitor (GSI) blocked rMMP-9-induced EndoMT.

**Conclusions:** Taken together, our results demonstrate that MMP-9 plays an important role in TGF-β1-induced EndoMT via upregulation of Notch signaling in HKGECs.

**Keywords:** Matrix metalloproteinase 9, Endothelial-mesenchymal transition, Human glomerular endothelial cells, TGF-β1, Notch

## Background

Kidney fibrosis is an inevitable consequence of a wide variety of progressive chronic kidney diseases (CKD) that progress to end-stage kidney failure, a devastating disorder that requires kidney replacement therapies such as dialysis or kidney transplantation. Kidney fibrosis is characterized by tubular atrophy/dilation, interstitial leukocyte infiltration, fibroblast accumulation, and increased interstitial matrix deposition [1, 2]. Although different cells are involved in kidney fibrosis, fibroblasts or myofibroblasts are considered to play a pivotal role [3]. However, the cellular origins of myofibroblasts are diverse. Resident fibroblasts [4], fibrocytes from bone marrow [4, 5], pericytes and perivascular fibroblasts [5, 6], tubular epithelial cells [4, 5, 7], podocytes [8] and endothelial cells [7, 9] have been identified as contributing to the myofibroblast population.

Endothelial-mesenchymal transition (EndoMT) is process similar to that of epithelial-mesenchymal transition (EMT).

* Correspondence: guoping.zheng@sydney.edu.au
†Equal contributors
[1]Centre for Transplant and Renal Research, Westmead Institute for Medical Research, the University of Sydney, 176 Hawkesbury Road, Sydney, NSW 2145, Australia
Full list of author information is available at the end of the article

During EndoMT, endothelial cell markers are downregulated whereas mesenchymal markers are upregulated. EndoMT is involved in organ development and various types of fibrosis. For example, EndoMT contributes to cardiac fibrosis [10–12], pulmonary fibrosis [10, 13], corneal fibrosis [14], radiation-induced pelvic disease [15] and inflammatory bowel disease-associated fibrosis [16]. EndoMT was also found in the early development of kidney interstitial fibrosis in the streptozocin (STZ)-induced diabetic nephropathy (DN) model [17, 18]. In addition, EndoMT also contributes to carcinoma-associated fibroblasts [19], atherogenesis, inflammation and hypertension [20, 21]. LeBleu et al. showed that ~10 % of the interstitial myofibroblasts co-stained with markers of endothelial cells and activated fibroblasts in unilateral ureteral obstruction (UUO) mice [7]. Zeisberg and colleagues [9] have demonstrated a role for EndoMT in several models of kidney disease. These studies demonstrated that activated fibroblasts co-express the endothelial marker CD31 as well as fibroblast markers such as fibroblast specific protein-1 (FSP-1) and α-smooth muscle actin (α-SMA). To demonstrate the presence of EndoMT-derived fibroblasts, these authors also used lineage tagged transgenic mice to trace endothelial lineage. Taken together, these findings suggest EndoMT plays a critical role in myofibroblast formation. EndoMT inhibition or reversal might be a potential target for treatment and prevention of kidney fibrosis.

Notch family is involved in podocyte and kidney tubular cell differentiation [22]. Abnormal Notch pathway activation can lead to glomerulonephritis (GN) and focal segmental glomerulosclerosis (FSGS) [23, 24]. Notch signaling is typically activated upon binding of ligands (such as Dll1, Dll3, Dll4, Jag1, and Jag2) with Notch 1–4 receptors. Then intramembrane proteolysis such as ectodomain shedding of both the ligand and the receptor, releasing the intracellular domains (ICD) of the ligand and receptor, thereby allowing Notch ICD (NICD) nuclear translocation to regulate gene expression.

We previously found that matrix metalloproteinase 9 (MMP-9) is capable of inducing tubular cell EMT and contribute to tubulointerstitial fibrosis [25, 26]. Although endothelial cells are capable of expressing MMP-9 [18], whether MMP-9 plays a role in EndoMT was unknown. In the current study, we defined a role for MMP-9 in EndoMT via Notch signaling.

## Methods
### Cell culture and treatment
Human kidney glomerular endothelial cells (HKGECs) were cultured in endothelial cell media (ScienCell; Carlsbad, CA, USA) containing vascular endothelial growth factor (VEGF; Sigma-Aldrich; St. Louis, MO, USA; 2.5–5 µg/ml) in basal medium (ScienCell), 5 % fetal bovine serum (FBS; ScienCell), 1 % endothelial cell growth

supplement (ECGS; ScienCell) and 1 % penicillin/streptomycin (P/S; ScienCell). Cells were maintained at 37 °C with 5 % $CO_2$. For treatment, HKGECs were cultured for 24 h at low density in flasks or plates precoated with fibronectin and washed in PBS. Cells were treated with TGF-β1 (10 ng/ml or 20 ng/ml; Sigma-Aldrich) alone, TGF-β1 plus 10 µM GM6001 (Calbiochem; Darmstadt, Germany), TGF-β1 plus the MMP-9 inhibitor I (Merck Chemicals; Darmstadt, Germany) at different dosages (0.05 nmol/ml, 0.25 nmol/ml, and 0.5 nmol/ml), recombinant MMP-9 (rMMP-9; 2 µg/ml; Biomol International; Plymouth Meeting, PA, USA), or rMMP-9 (2 µg/ml) plus the gamma secretase inhibitor (GSI; 5–10 µM; Merck Millipore; Billerica, MA, USA). Our study does not require any human or animal ethics approval.

### Immunofluorescence analysis
For indirect immunofluorescence, HKGECs were cultured on glass coverslips, washed twice in PBS, fixed with absolute methanol for ten minutes at –20 °C, and blocked for 1 h with 2 % BSA (Sigma) at room temperature. Cells were then incubated for 1 h at room temperature with primary antibodies against endothelial markers rabbit polyclonal anti-VE-cadherin (1:200; Alexis Biochemicals; Farmingdale, NY, USA), mouse monoclonal anti-CD31 (1:100; Cell Signaling Technology; Boston, MA, USA) and mesenchymal markers mouse monoclonal anti-α-SMA (1:200; Sigma Chemical Co), rabbit monoclonal anti-vimentin (1:200; Cell Signaling Technology) and rabbit monoclonal anti-N-cadherin (1:100, BD Bioscience; San Jose, CA, USA) in 2 % BSA. The following secondary antibodies were used: goat anti-mouse IgG2a/2b phycoerythrin (PE)-conjugated antibody (1:400; Invitrogen; Carlsbad, CA, USA) for CD31, α-SMA and goat anti-rabbit IgG2a/2b FITC-conjugated antibody (1:400; Invitrogen) for VE-cadherin, vimentin and N-cadherin. Cells were washed twice with PBS, counterstained with DAPI for 5 min, and washed twice with PBS. Samples were mounted using fluorescence mounting media. For negative isotype controls staining, rat IgG2a κ purified (eBioscience; San Diego, CA, USA) was used for VE-cadherin, vimentin and N-cadherin mouse IgG2a κ (Biolegend; San Diego, CA, USA) was used for CD31 and α-SMA, and their corresponding secondary antibodies were applied.

### RNA extraction, purification, and quantitation
Total RNA was extracted from cultured cells using 350 µl of RLT buffer and homogenized by shredding through a 0.5 ml insulin syringe 5 times. Extracted RNA was purified using RNeasy Mini Kit (Qiagen; Hilden, Germany) following the manufacturer's instruction and resuspended in 30 µl of RNAse-free water. The yield and purity of RNA was measured

spectrophotometrically by absorption at 260 nm (A260) and 280 nm (A280) using a Beckman-Coulter DU800 spectrophotometer (CA, USA).

**Real-time RT-PCR analysis**

cDNAs were synthesized using 200 ng of extracted RNA in 20 μl reaction buffer by reverse transcription using

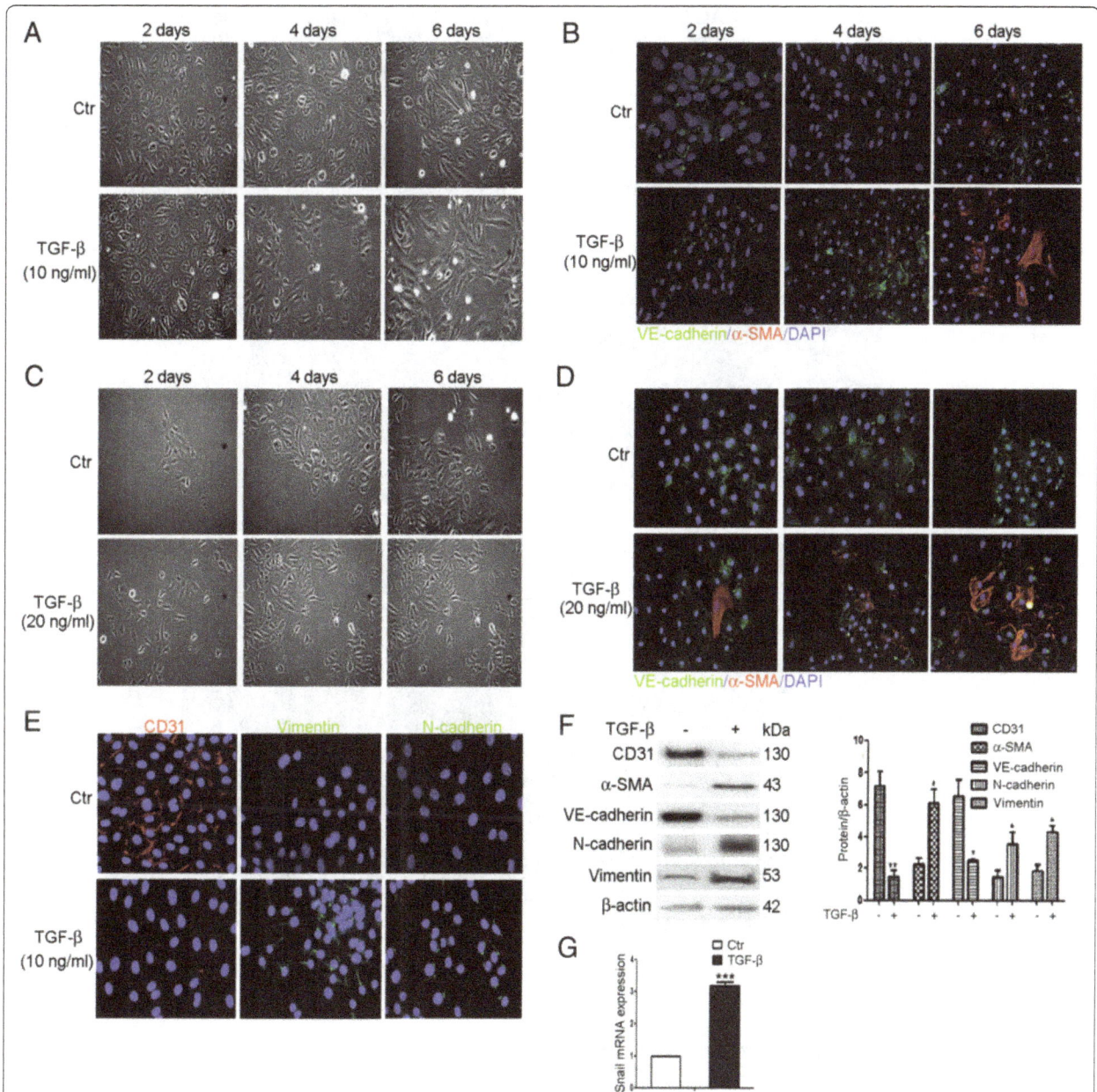

**Fig. 1** TGF-β1 induces EndoMT in HKGECs. **a** Morphological changes in HKGECs induced by TGF-β1 (10 ng/ml) were examined using phase contrast microscopy. **b** Indirect immunofluorescence staining and co-localization of VE-cadherin and α-SMA were performed in HKGECs cultured in Endothelial Cell Medium with TGF-β1 (10 ng/ml). **c** Morphological changes in HKGECs induced by TGF-β1 (20 ng/ml) were examined using phase contrast microscopy. Cells were counterstained with DAPI to visualize nuclei (*blue*). **d** Indirect immunofluorescence staining and co-localization of VE-cadherin and α-SMA were performed in HKGECs cultured in Endothelial Cell Medium with TGF-β1 (20 ng/ml). **e** Indirect immunofluorescence staining of CD31, vimentin, and N-cadherin was performed in HKGECs cultured in Endothelial Cell Medium with TGF-β1 (10 ng/ml). **f** Respective western blot analysis and quantitation of CD31, α-SMA, VE-cadherin, N-cadherin, and vimentin in HKGECs treated with TGF-β1 (10 ng/ml). β-actin was used as a loading control. **g** Snail mRNA expression from HKGECs cultured in Endothelial Cell Medium with TGF-β1 (10 ng/ml) was quantified using real-time PCR. Gene expression levels were normalized to human GADPH mRNA. Original magnification × 200. Data are expressed as mean ± SEM with $n \geq 3$ for each experimental group, $*P < 0.05$, $**P < 0.01$. Morphological and immunofluorescent are representative of experiments repeated 3 times

Superscript$^{TM}$ First Strand Synthesis System (Invitrogen) and random hexamer primers at 50 °C for 50 min. Designed primers and established primers from published papers were used for Real-time RT-PCR. The sequences of hes-1 primers used for this analysis are as follows: forward: 5′-GAC AGC ATC TGA GCA CAG AAA TG-3′ and reverse: 5′- GTC ATG GCA TTG ATC TGG GTC AT-3′ [27]. Housekeeping gene β-actin was used as the internal control. For Real-time RT-PCR, PCR mixture contained 0.5 µl of cDNA and 10 pmol/µl of each primer in a 20 µl final volume of SYBR mastermix (Invitrogen). Amplification was performed using the Rotogene-6000 Real-Time cycler thermos.

### Western blot analysis

Equal volumes or quantities of protein were loaded in 12-well NuPAGE 4–15 % Bis-Tris gels (Bio-Rad; Hercules, CA, USA) and electrophoresed under reducing conditions. After electrophoresis, proteins were transferred for 2 h to PVDF membranes using a Mini Trans-Blot Electrophoretic Transfer Cell apparatus (Bio-Rad). For immunodetection, membranes were blocked overnight at 4 °C in 5 % skim milk and incubated for 2 h at room temperature with the primary antibody against mouse monoclonal anti-α-SMA (1:300, Sigma Chemical Co.), rabbit polyclonal anti-VE-cadherin (1:1000; Alexis Biochemicals), mouse monoclonal anti-CD31 (1:1000;

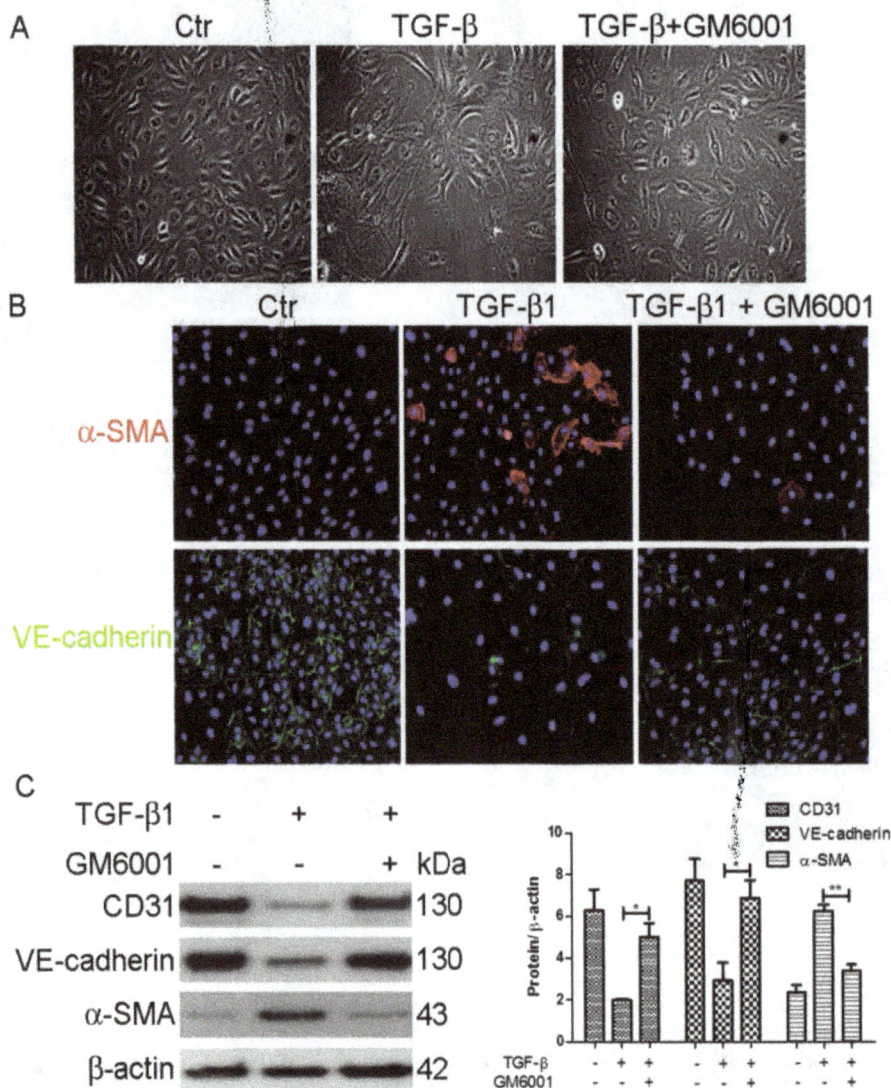

**Fig. 2** GM6001 inhibits TGF-β1-induced EndoMT in HKGECs. **a** Morphological changes in HKGECs induced by TGF-β1 (10 ng/ml) in the presence or absence of GM6001 were examined using phase contrast microscopy. **b** Indirect immunofluorescence staining of VE-cadherin and α-SMA were performed in HKGECs cultured in Endothelial Cell Medium with TGF-β1 (10 ng/ml). **c** Respective western blot analysis and quantitation of CD31, VE-cadherin and α-SMA were performed in HKGECs treated with TGF-β1 (10 ng/ml) in the presence or absence of GM6001. β-actin was used as a loading control. Original magnification × 200. Image are representative and data are expressed as mean ± SEM with $n \geq 3$ for each experimental group, *$P < 0.05$, **$P < 0.01$

Cell Signaling Technology), mouse polyclonal anti-NICD (1;1000; Merck Millipore, Cat. # 07–1232) [28], β-actin (1:3000, Sigma), rabbit monoclonal anti-Notch1 (1:1000; Cell Signaling Technology), rabbit monoclonal anti-vimentin (1:1000; Cell Signaling Technology), rabbit monoclonal anti-N-cadherin (1:500, BD BioSciences), rabbit polyclonal anti-Hes-1 (1:500, Abcam; Cambridge, UK) and rabbit polyclonal anti-Hey-1 (1:500, Abcam) prepared in blocking buffer. Membranes were washed three times (10 min per wash on a rocking platform) and incubated for 1 h with their respective horseradish peroxidase (HRP)-conjugated secondary antibodies; goat anti mouse HRP and goat anti-rabbit HRP (1:5000; Cell Signaling Technology) prepared in blocking buffer. Membranes were again washed three times (10 min per wash on a rocking platform). Bands were visualized using an enhanced chemiluminescence detection kit.

## Zymography

MMP-9 activity in medium derived from TGF-β treated HKGECs was determined by gelatin zymography. Briefly, medium was mixed with Tris-Glycine SDS Native Sample Buffer (1:1; Invitrogen) and electrophoresed through 10 % Novex Zymogram Gelatin Gels (Invitrogen) with Tris-Glycine SDS Running Buffer (Invitrogen) under constant voltage of 125 V for 120 min. After electrophoresis, gels was incubated in Zymogram Renaturing Buffer (Invitrogen) for 30 min at room temperature with gentle agitation and washed with developing buffer (Invitrogen) for 30 min. The gel was further incubated for 24 h in fresh developing buffer at 37 °C. After developing, the gel was stained with 0.5 % (w/v) Coomassie Blue R-250 (Bio-Rad) in 50 % (v/v) methanol, 10 % (v/v) acetic acid for 30 min at room temperature, and destained as described previously [25]. Gelatinolytic activity of MMP-9 was visualized as a clear band on a blue background. Band intensity was quantified by densitometry using ImageJ software. Briefly, zymogram gels were scanned using Kodak gel logic 100 imaging system and processed into gray scale. Gray scale images were quantified densitometrically by the measurement of the mean intensity of positive band multiplied by its corresponding area. The optical band intensity was then corrected by subtracting background intensity of equal area.

## Statistical analysis

Results from at least three independent experiments are expressed as mean ± SEM. Statistical significance was evaluated using two-tail $t$-test for comparison between two groups, whereas the one way analysis of variance (ANOVA) was used for comparison of multiple groups. $P < 0.05$ was considered significant.

## Results

### TGF-β1 induces EndoMT in HKGECs

To determine whether TGF-β induces EndoMT, HKGECs were treated in the presence or absence of TGF-β1. HKGECs exhibited morphological changes typical of EndoMT, as shown by phenotypic transformation from an endothelial cobblestone shape to fibroblastic spindle-shaped morphology by day 2 (Fig. 1a) and increasing numbers of fibroblasts evident on days 4 to 6 (Fig. 1a). Cells cultured in the absence of TGF-β showed fewer morphological changes. The transition from an

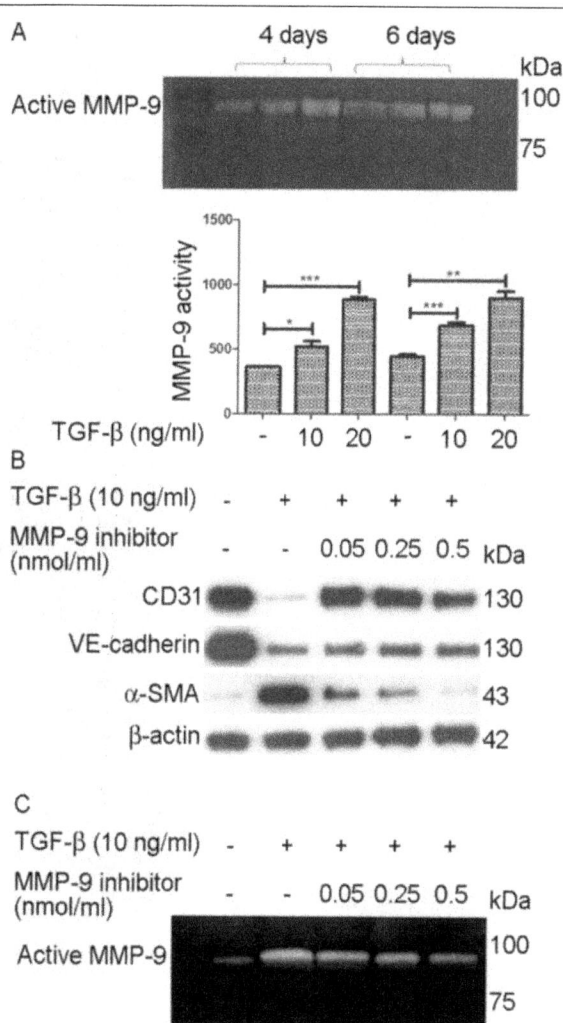

**Fig. 3** Evaluation of MMP-9 activity. **a** MMP-9 activity in response to TGF-β was examined using gelatin zymography. Relative activity was compared to control (without TGF-β treatment). **b** CD31, VE-cadherin, and α-SMA expression in HKGECs treated with TGF-β1 (10 ng/ml) and various dosages of the MMP-9 inhibitor (0.05, 0.25, and 0.5 nmol/ml) were evaluated using western blot analysis. β-actin was used as a loading control. **c** MMP-9 activity was determined using gelatin zymography. Relative activity was compared to control (without TGF-β1 treatment). Data are expressed as mean ± SEM with $n \geq 3$ for each experimental group, $*P < 0.05$, $**P < 0.01$, $***P < 0.001$

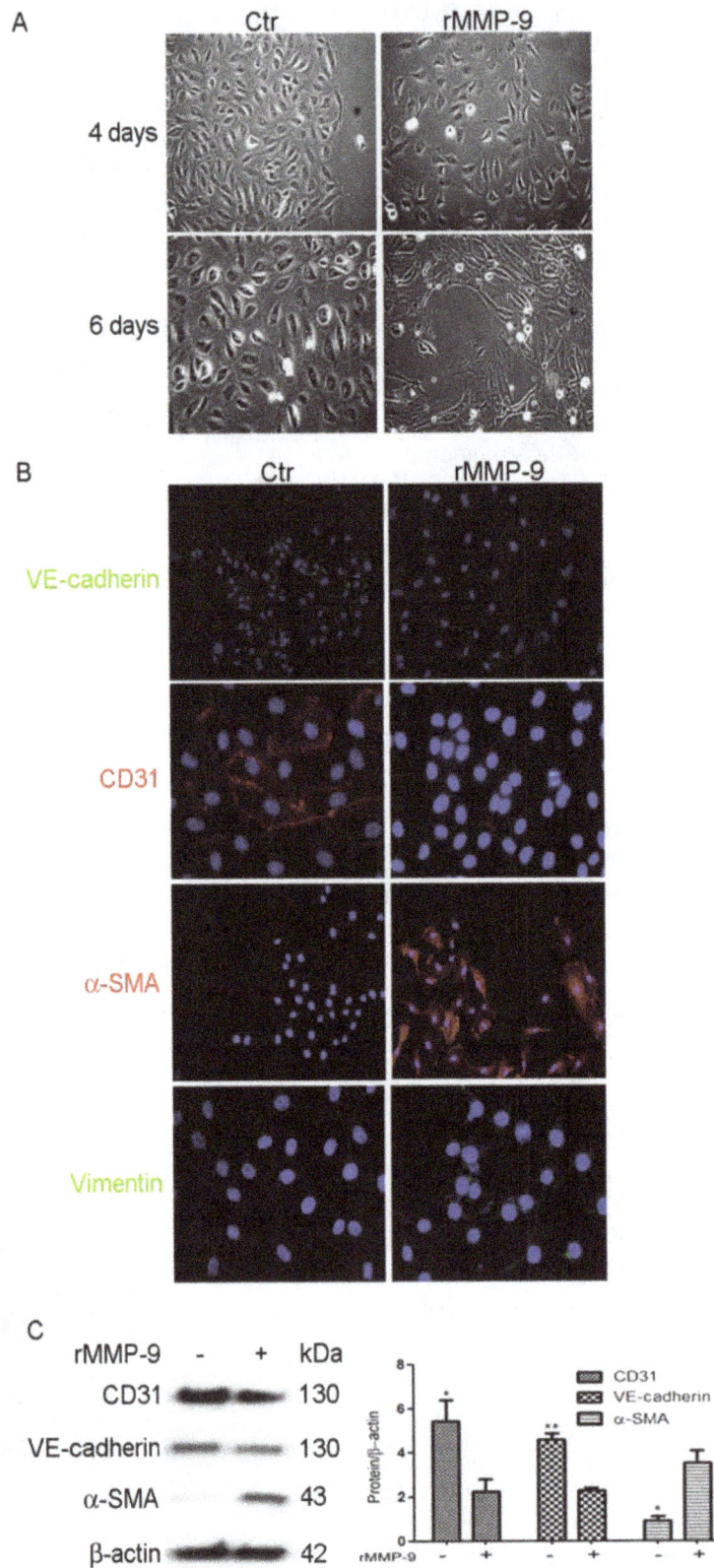

**Fig. 4** (See legend on next page.)

(See figure on previous page.)
**Fig. 4** rMMP-9 induces EndoMT in HKGECs. **a** rMMP-9-induced morphological changes in HKGECs were examined using phase contrast microscopy. **b** Indirect immunofluorescence staining of CD31, α-SMA, VE-cadherin, and vimentin was performed in HKGECs cultured in Endothelial Cell Medium in the presence or absence of rMMP-9 (2 μg/ml). **c** Western blot analysis of CD31, VE-cadherin and α-SMA in HKGECs treated or not with rMMP-9 (2 μg/ml). β-actin was used as a loading control. Original magnification × 200. Data are expressed as mean ± SEM with $n \geq 3$ for each experimental group, $^{*}P < 0.05$, $^{**}P < 0.01$

endothelial to mesenchymal phenotype was confirmed using immunofluorescent staining. Endothelial cells treated with TGF-β1 (10 and 20 ng/ml) lost VE-cadherin expression and acquired α-SMA expression (Fig. 1b and c). When treated with higher concentration of TGF-β1 (20 ng/ml), morphological changes in HKGECs typical of EndoMT was also induced (Fig. 1c). Immunofluorescence staining showed decreased expression of VE-cadherin and increased expression of α-SMA, particularly in samples that were treated with TGF-β1 for 6 days (Fig. 1d). However, 20 ng/ml TGF-β1 could not induce EndoMT in a shorter time. There were no notable differences in α-SMA-stained cells at days 2 and 4 compared to 10 ng/ml TGF-β1 treatments. These results suggest that TGF-β1 is capable of inducing EndoMT in HKGECs in 6 days. To further confirm TGF-β induction of EndoMT in HKGECs, cells were stained with the endothelial marker CD31 and the mesenchymal markers vimentin and N-cadherin. The transition from an endothelial phenotype to a mesenchymal phenotype was confirmed using immunofluorescent staining. Upon TGF-β1 treatment, endothelial cells lost CD31 expression and acquired vimentin and

N-cadherin expression (Fig. 1e). Immunofluorescent staining specificity was confirmed by negative control staining for each isotype control antibody. Consistent with the immunofluorescence staining results, western blot analysis revealed decreased levels of the endothelial markers VE-cadherin and CD31 in HKGECs treated with TGF-β1. However, there was increased expression of the mesenchymal markers α-SMA, vimentin, and N-cadherin (Fig. 1f). EndoMT was also confirmed by real-time PCR analysis. We observed upregulation of the transcription factor Snail mRNA in HKGECs treated with TGF-β1 compared to control HKGECs (Fig. 1g). Taken together, these results demonstrate that TGF-β1 induces EndoMT in HKGECs.

### The MMP inhibitor GM6001 inhibits TGF-β1-induced EndoMT in HKGECs

To determine whether MMPs contribute to TGF-β1-induced EndoMT, HKGECs were treated with the broad-spectrum MMP inhibitor GM6001. After treatment for 6 days, we found that EndoMT induced by 10 ng/ml TGF-β1 was abrogated by GM6001; the

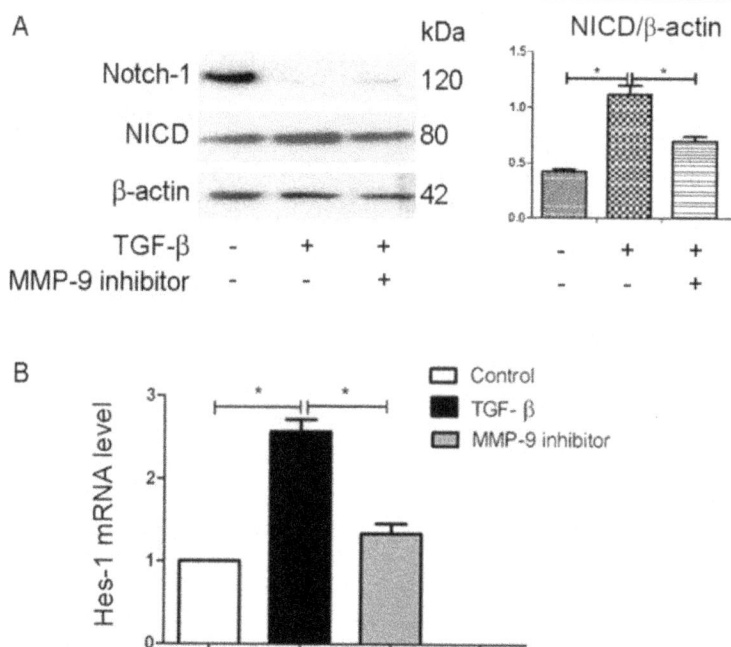

**Fig. 5** The Notch pathway is activated in TGF-β1-induced EndoMT. **a** Respective western blot analysis and quantitation of Notch-1 and NICD expression in TGF-β1 (10 ng/ml) induced EndoMT in HKGECs six days after treatment. β-actin was used as a loading control. **b** qPCR analysis of Hes-1 expression in TGF-β1-induced EndoMT in HKGECs. Data are expressed as mean ± SEM with $n \geq 3$ for each experimental group, $^{*}P < 0.05$, $^{**}P < 0.01$

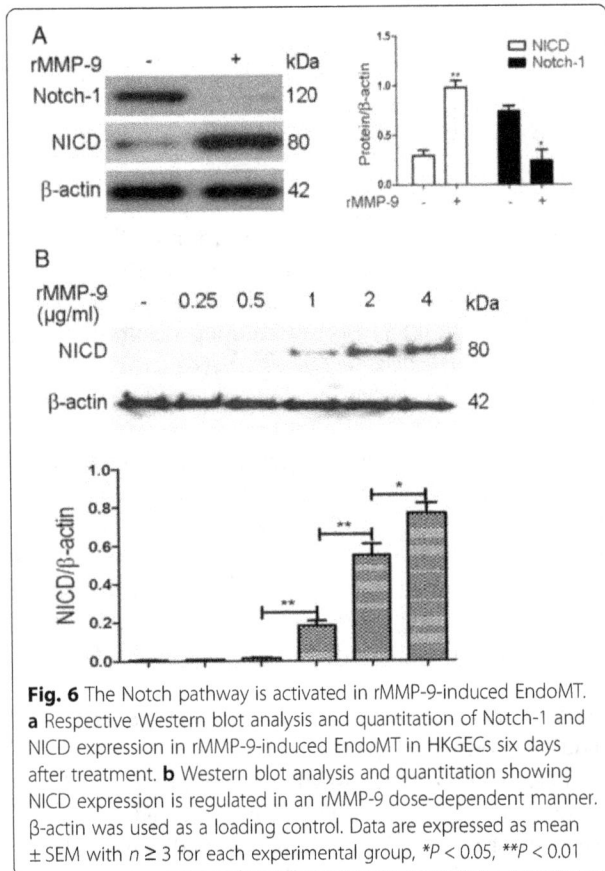

**Fig. 6** The Notch pathway is activated in rMMP-9-induced EndoMT. **a** Respective Western blot analysis and quantitation of Notch-1 and NICD expression in rMMP-9-induced EndoMT in HKGECs six days after treatment. **b** Western blot analysis and quantitation showing NICD expression is regulated in an rMMP-9 dose-dependent manner. β-actin was used as a loading control. Data are expressed as mean ± SEM with $n \geq 3$ for each experimental group, $*P < 0.05$, $**P < 0.01$

majority of HKGECs maintained typical endothelial cobblestone morphology with few cells exhibiting fibroblastic spindle-shape morphology (Fig. 2a). Consistent with the cellular morphology, immunofluorescence (Fig. 2b) and western blot analysis (Fig. 2c) indicated that GM6001 abrogated the TGF-β1-induced decrease in CD31 and VE-cadherin expression and increase in α-SMA expression. Collectively, these results suggest that MMPs are involved in TGF-β1-induced EndoMT in HKGECs.

### MMP-9 is involved in TGF-β1-induced EndoMT in HKGECs

Gelatin zymography showed that MMP-9 activity increased significantly within 4 days in the presence of conditioned media with 10 ng/ml or 20 ng/ml TGF-β1 (Fig. 3a). After the cells had been cultured for 6 days, MMP-9 activity in the media was upregulated in both the presence of 10 ng/ml or 20 ng/ml TGF-β1. No MMP-2 expression was detected in the media (Fig. 3a).

To further examine whether MMP-9 is involved in EndoMT, HKGECs were treated with TGF-β1 in presence of MMP-9 inhibitor. Our results demonstrated that the MMP-9 inhibitor reduced α-SMA expression in the TGF-β1-treated samples. In addition, when different dosages of MMP-9 inhibitor were used, the reduction in

α-SMA and recovery of VE-cadherin expression was dose dependent (Fig. 3b). To confirm that the MMP-9 inhibitor was functional, zymography was performed to evaluate MMP-9 and MMP-2 activity. The results showed that the MMP-9 inhibitor successfully inhibited MMP-9 activity (Fig. 3c).

### rMMP-9 induces EndoMT in HKGECs

To determine whether MMP-9 contributes to EndoMT, subconfluent HKGECs were treated with rMMP-9 (2 μg/ml). rMMP-9 induced HKGEC phenotypic transformation from the endothelial cobblestone shape to the fibroblastic spindle-shaped morphology within 6 days of treatment (Fig. 4a). The transition of an endothelial to mesenchymal phenotype was confirmed by immunofluorescence (Fig. 4b). Endothelial cells lost VE-cadherin and CD31 expression and acquired α-SMA and vimentin expression. In addition, VE-cadherin, CD31, and α-SMA expression was examined by western blot analysis (Fig. 4c). Consistent with the immunofluorescence results, western blot analysis revealed that in HKGECs treated with rMMP-9, levels of the endothelial markers VE-cadherin and CD31 decreased, whereas the mesenchymal marker α-SMA increased (Fig. 4c). Taken together, these results demonstrate that rMMP-9 induces EndoMT in HKGECs.

### Notch signaling is activated in TGF-β1-induced EndoMT but is inhibited by MMP-9 inhibitor

To investigate Notch signaling in TGF-β1-induced EndoMT, we examined the expression of Notch 1 protein and Notch intracellular domain (NICD) using western blot analysis. We also evaluated the Notch downstream transcriptional factors Hes-1 and Hey-1 using qPCR. We found that when HKGECs were treated with TGF-β1, Notch-1 expression was significantly down-regulated whereas NICD was significantly increased compared to control (Fig. 5a). In addition, when TGF-β1-induced cells were treated with the MMP-9 inhibitor, NICD expression decreased (Fig. 5a). The qPCR results showed that TGF-β1-induced Hes-1 expression in HKGECs was significantly reduced by MMP-9 inhibitor (Fig. 5b). These results show that MMP-9 contributes to the Notch pathway activation in TGF-β1-induced EndoMT.

### rMMP-9 activates Notch signaling

We also examined Notch pathway activation in rMMP-9-induced HKGECs using western blot analysis. After HKGECs were incubated with rMMP-9, the Notch-1 protein was cleaved and NICD was released (Fig. 6a). To examine whether rMMP-9 can activate Notch signaling, different dosages of rMMP-9 (0, 0.25, 0.5, 1.0, 2.0, and 4.0 μg/ml) were added to HKGECs. We found that

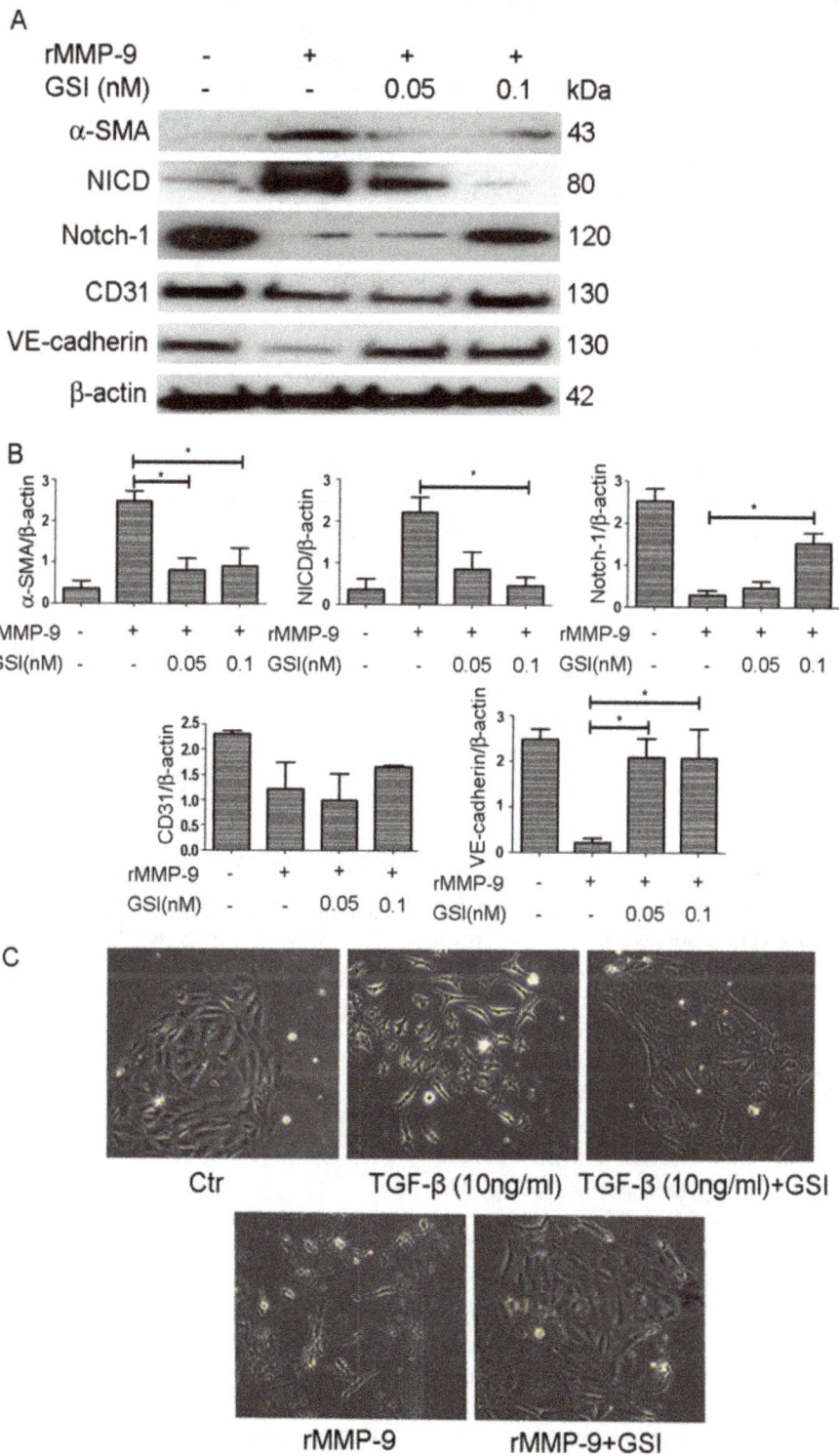

**Fig. 7** The γ-secretase inhibitor GSI-I regulates protein expression in rMMP-9-induced EndoMT. **a** Western blot analysis of α-SMA, NICD, Notch-1, CD31 and VE-cadherin expression in rMMP-9-induced EndoMT in the presence or absence of GSI-I six days after treatment. β-actin was used as a loading control. **b** Statistical analysis of protein expression in rMMP-9-induced EndoMT in the presence or absence of GSI-I. **c** Morphological changes in HKGECs induced by TGF-β1 (10 ng/ml) or rMMP-9 in the presence or absence of GSI. Data are expressed as mean ± SEM with $n \geq 3$ for each experimental group, *$P < 0.05$

NICD level was increased in a rMMP-9 dose-dependent manner (Fig. 6b).

### Inhibition of Notch signaling by γ-secretase inhibitor (GSI) blocks rMMP-9-induced EndoMT

Notch signaling was inhibited using GSI, an effective γ-secretase inhibitor (Z-Leu-Leu-Nle-CHO) [29]. Western blot results (Fig. 7a) and statistical analysis (Fig. 7b) showed that GSI significantly reduced α-SMA and NICD expression and increased CD31, VE-cadherin and Notch-1 expression of cells exposed to MMP-9 compared to cells treated with rMMP-9 alone. GSI inhibited both TGF-β1- and MMP-9-induced morphological changes (Fig. 7c). This result demonstrates that the rMMP-9-induced EndoMT is Notch signaling dependent.

### Discussion

Kidney EndoMT is a major source of fibroblast formation in kidney fibrosis, and has emerged as a potentially important mechanism in development and progression of kidney fibrosis [9]. In a recent study, LeBleu et al. found that endothelial cells play a more important role than epithelial cells and pericytes in myofibroblast formation via EndoMT [7]. Therefore, understanding and halting EndoMT is an important clinical challenge.

Kidney endothelial cells consist of glomerular, vascular and peritubular endothelial cells. Although it has been generally accepted that endothelial cells contribute to fibroblast formation in kidney, whether glomerular endothelial cells contribute to fibroblast formation remains unclear. In our recent study, we showed that mice peritubular endothelial cells have an important role in kidney EndoMT. Moreover, we found that MMP-9 was involved in peritubular EndoMT, likely via the Notch pathway. Human glomerular endothelial cells are likely to be another instructive model for studying kidney fibrosis.

Here, we demonstrate that TGF-β1 and rMMP-9 induced EndoMT in HKGECs via Notch signaling. The Notch signaling pathway is a highly conserved cascade in mammals that regulates multiple cellular processes, including proliferation, differentiation, and apoptosis [30, 31]. Notch signaling has been shown to be downstream of VEGF to regulate endothelial cell morphogenesis [32]. In addition, Notch activation induces endothelial cell morphological, phenotypic, and functional changes consistent with mesenchymal transformation [33]. Notch signaling has been found to initiate EndoMT in atrioventricular endothelial cells [34]. Inhibition of Notch signaling ameliorates EMT and tubulointerstitial fibrosis in mouse [35]. Importantly, Notch inhibition reversed podocyte injury (EMT) and kidney failure [36]. All these findings suggest a critical role for Notch signaling in both EMT and EndoMT which are

pivotal processes in kidney fibrosis. Our results demonstrate that MMP-9 activation of Notch signaling in glomerular endothelial cells is downstream of TGF-β1. Combining with observations with previous studies from our laboratory and others, a strong profibrotic role for MMP-9 is evident. It promotes EMT of tubular epithelial cells, EndoMT of peritubular endothelial cells and EndoMT of glomerular endothelial cells, thereby, leading to kidney fibrosis in both tubulointerstial and glomeruli compartments.

### Conclusion

Our data demonstrate that MMP-9 plays an important role in TGF-β1-induced EndoMT in HKGECs, via upregulation of Notch signaling. Thus, inhibition of MMP-9 or Notch signaling could be therapeutic strategies for treatment for kidney fibrosis in CKD.

#### Abbreviations
ANOVA: one way analysis of variance; CKD: chronic kidney diseases; DN: diabetic nephropathy; ECGS: endothelial cell growth supplement; EMT: epithelial-mesenchymal transition; EndoMT: endothelial-mesenchymal transition; FBS: fetal bovine serum; FSGS: focal segmental glomerulosclerosis; FSP-1: fibroblast specific protein-1; GN: glomerulonephritis; GSI: gamma secretase inhibitor; HRP: horseradish peroxidase; ICD: intracellular domains; MMP-9: matrix metalloproteinase 9; NICD: Notch intracellular domains; P/S: penicillin/streptomycin; PE: phycoerythrin; TGF-β1: transforming growth factor-beta1; UUO: unilateral ureteral obstruction; VEGF: vascular endothelial growth factor; α-SMA: α-smooth muscle actin.

#### Competing interests
The authors declare that they have no competing interests.

#### Authors' contributions
YZ, XQ, DCH and GPZ conceived and design of research; YZ, XQ, TKT, HZ, YZ, JLZ, QC, YPW, YW, YMW and VWL performed experiments; YZ and XQ analyzed data; YZ, XQ and SIA interpreted results of experiments; YZ and XQ prepared figures; YZ, XQ drafted manuscript; YZ, XQ, DCH and GPZ edited and revised manuscript; DCH and GPZ approved final version of manuscript. All authors read and approved the final manuscript.

#### Acknowledgements
This work was supported by the National Health and Medical Research Council (NHMRC) Project Grants 632688 and 1046647.

#### Author details
[1]Centre for Transplant and Renal Research, Westmead Institute for Medical Research, the University of Sydney, 176 Hawkesbury Road, Sydney, NSW 2145, Australia. [2]The School of Biomedical Sciences, Chengdu Medical College, Chengdu 610500, PR China. [3]Department of Nephrology, Second Hospital of Shanxi Medical University, Shanxi Kidney Disease Institute, WuYi Road 382, Taiyuan 030001, Shanxi, PR China. [4]Department of Biochemistry and Molecular Biology, Shanxi Medical University, Xinjian Road 56, Taiyuan 030001, Shanxi, PR China. [5]Experimental Centre of Science and Research, the First Clinical Hospital of Shanxi Medical University, Xinjian Road 382, Taiyuan 030001, Shanxi, PR China. [6]Centre for Kidney Research, Children's Hospital at Westmead, 212 Hawkesbury Road, Sydney, NSW, Australia.

#### References
1. Zeisberg M, Strutz F, Muller GA. Renal fibrosis: an update. Curr Opin Nephrol Hypertens. 2001;10(3):315–20.
2. Liu Y. Renal fibrosis: new insights into the pathogenesis and therapeutics. Kidney Int. 2006;69(2):213–7.

3. Piera-Velazquez S, Li Z, Jimenez SA. Role of endothelial-mesenchymal transition (EndoMT) in the pathogenesis of fibrotic disorders. Am J Pathol. 2011;179(3):1074–80.

4. Iwano M, Plieth D, Danoff TM, Xue C, Okada H, Neilson EG. Evidence that fibroblasts derive from epithelium during tissue fibrosis. J Clin Invest. 2002; 110(3):341–50.

5. Grgic I, Duffield JS, Humphreys BD. The origin of interstitial myofibroblasts in chronic kidney disease. Pediatr Nephrol. 2012;27(2):183–93.

6. Lin SL, Kisseleva T, Brenner DA, Duffield JS. Pericytes and perivascular fibroblasts are the primary source of collagen-producing cells in obstructive fibrosis of the kidney. Am J Pathol. 2008;173(6):1617–27.

7. LeBleu VS, Taduri G, O'Connell J, Teng Y, Cooke VG, Woda C, Sugimoto H, Kalluri R. Origin and function of myofibroblasts in kidney fibrosis. Nat Med. 2013;19(8):1047–53.

8. Li Y, Kang YS, Dai C, Kiss LP, Wen X, Liu Y. Epithelial-to-mesenchymal transition is a potential pathway leading to podocyte dysfunction and proteinuria. Am J Pathol. 2008;172(2):299–308.

9. Zeisberg EM, Potenta SE, Sugimoto H, Zeisberg M, Kalluri R. Fibroblasts in kidney fibrosis emerge via endothelial-to-mesenchymal transition. J Am Soc Nephrol. 2008;19(12):2282–7.

10. Hashimoto N, Phan SH, Imaizumi K, Matsuo M, Nakashima H, Kawabe T, Shimokata K, Hasegawa Y. Endothelial-mesenchymal transition in bleomycin-induced pulmonary fibrosis. Am J Respir Cell Mol Biol. 2010;43(2):161–72.

11. Ghosh AK, Bradham WS, Gleaves LA, De Taeye B, Murphy SB, Covington JW, Vaughan DE. Genetic deficiency of plasminogen activator inhibitor-1 promotes cardiac fibrosis in aged mice: involvement of constitutive transforming growth factor-beta signaling and endothelial-to-mesenchymal transition. Circulation. 2010;122(12):1200–9.

12. Widyantoro B, Emoto N, Nakayama K, Anggrahini DW, Adiarto S, Iwasa N, Yagi K, Miyagawa K, Rikitake Y, Suzuki T, et al. Endothelial cell-derived endothelin-1 promotes cardiac fibrosis in diabetic hearts through stimulation of endothelial-to-mesenchymal transition. Circulation. 2010;121(22):2407–18.

13. Nataraj D, Ernst A, Kalluri R. Idiopathic pulmonary fibrosis is associated with endothelial to mesenchymal transition. Am J Respir Cell Mol Biol. 2010;43(2):129–30.

14. Nakano Y, Oyamada M, Dai P, Nakagami T, Kinoshita S, Takamatsu T. Connexin43 knockdown accelerates wound healing but inhibits mesenchymal transition after corneal endothelial injury in vivo. Invest Ophthalmol Vis Sci. 2008;49(1):93–104.

15. Mintet E, Rannou E, Buard V, West G, Guipaud O, Tarlet G, Sabourin JC, Benderitter M, Fiocchi C, Milliat F, et al. Identification of endothelial-to-mesenchymal transition as a potential participant in radiation proctitis. Am J Pathol. 2015;185(9):2550–62.

16. Rieder F, Kessler SP, West GA, Bhilocha S, de la Motte C, Sadler TM, Gopalan B, Stylianou E, Fiocchi C. Inflammation-induced endothelial-to-mesenchymal transition: a novel mechanism of intestinal fibrosis. Am J Pathol. 2011;179(5):2660–73.

17. Kizu A, Medici D, Kalluri R. Endothelial-mesenchymal transition as a novel mechanism for generating myofibroblasts during diabetic nephropathy. Am J Pathol. 2009;175(4):1371–3.

18. Li J, Qu X, Bertram JF. Endothelial-myofibroblast transition contributes to the early development of diabetic renal interstitial fibrosis in streptozotocin-induced diabetic mice. Am J Pathol. 2009;175(4):1380–8.

19. Zeisberg EM, Potenta S, Xie L, Zeisberg M, Kalluri R. Discovery of endothelial to mesenchymal transition as a source for carcinoma-associated fibroblasts. Cancer Res. 2007;67(21):10123–8.

20. Tuder RM, Groves B, Badesch DB, Voelkel NF. Exuberant endothelial cell growth and elements of inflammation are present in plexiform lesions of pulmonary hypertension. Am J Pathol. 1994;144(2):275–85.

21. Kitao A, Sato Y, Sawada-Kitamura S, Harada K, Sasaki M, Morikawa H, Shiomi S, Honda M, Matsui O, Nakanuma Y. Endothelial to mesenchymal transition via transforming growth factor-beta1/Smad activation is associated with portal venous stenosis in idiopathic portal hypertension. Am J Pathol. 2009;175(2):616–26.

22. Barisoni L, Schnaper HW, Kopp JB. A proposed taxonomy for the podocytopathies: a reassessment of the primary nephrotic diseases. Clin J Am Soc Nephrol. 2007;2(3):529–42.

23. Mertens PR, Raffetseder U, Rauen T. Notch receptors: a new target in glomerular diseases. Nephrol Dial Transplant. 2008;23(9):2743–5.

24. Niranjan T, Bielesz B, Gruenwald A, Ponda MP, Kopp JB, Thomas DB, Susztak K. The Notch pathway in podocytes plays a role in the development of glomerular disease. Nat Med. 2008;14(3):290–8.

25. Tan TK, Zheng G, Hsu TT, Wang Y, Lee VW, Tian X, Cao Q, Harris DC. Macrophage matrix metalloproteinase-9 mediates epithelial-mesenchymal transition in vitro in murine renal tubular cells. Am J Pathol. 2010;176(3):1256–70.

26. Zheng G, Lyons JG, Tan TK, Wang Y, Hsu TT, Min D, Succar L, Rangan GK, Hu M, Henderson BR, et al. Disruption of E-cadherin by matrix metalloproteinase directly mediates epithelial-mesenchymal transition downstream of transforming growth factor-beta1 in renal tubular epithelial cells. Am J Pathol. 2009;175(2):580–91.

27. Ji X, Wang Z, Geamanu A, Sarkar FH, Gupta SV. Inhibition of cell growth and induction of apoptosis in non-small cell lung cancer cells by delta-tocotrienol is associated with notch-1 down-regulation. J Cell Biochem. 2011;112(10):2773–83.

28. Wakabayashi N, Shin S, Slocum SL, Agoston ES, Wakabayashi J, Kwak MK, Misra V, Biswal S, Yamamoto M, Kensler TW. Regulation of notch1 signaling by nrf2: implications for tissue regeneration. Sci Signal. 2010;3(130):ra52.

29. Curry CL, Reed LL, Golde TE, Miele L, Nickoloff BJ, Foreman KE. Gamma secretase inhibitor blocks Notch activation and induces apoptosis in Kaposi's sarcoma tumor cells. Oncogene. 2005;24(42):6333–44.

30. Baron M. An overview of the Notch signalling pathway. Semin Cell Dev Biol. 2003;14(2):113–9.

31. Kopan R, Ilagan MX. The canonical Notch signaling pathway: unfolding the activation mechanism. Cell. 2009;137(2):216–33.

32. Funahashi Y, Shawber CJ, Sharma A, Kanamaru E, Choi YK, Kitajewski J. Notch modulates VEGF action in endothelial cells by inducing Matrix Metalloprotease activity. Vasc Cell. 2011;3(1):2.

33. Noseda M, McLean G, Niessen K, Chang L, Pollet I, Montpetit R, Shahidi R, Dorovini-Zis K, Li L, Beckstead B, et al. Notch activation results in phenotypic and functional changes consistent with endothelial-to-mesenchymal transformation. Circ Res. 2004;94(7):910–7.

34. Chang AC, Fu Y, Garside VC, Niessen K, Chang L, Fuller M, Setiadi A, Smrz J, Kyle A, Minchinton A, et al. Notch initiates the endothelial-to-mesenchymal transition in the atrioventricular canal through autocrine activation of soluble guanylyl cyclase. Dev Cell. 2011;21(2):288–300.

35. Bielesz B, Sirin Y, Si H, Niranjan T, Gruenwald A, Ahn S, Kato H, Pullman J, Gessler M, Haase VH, et al. Epithelial Notch signaling regulates interstitial fibrosis development in the kidneys of mice and humans. J Clin Invest. 2010;120(11):4040–54.

36. Kretzler M, Allred L. Notch inhibition reverses kidney failure. Nat Med. 2008;14(3):246–7.

# miR-27b inhibits fibroblast activation via targeting TGFβ signaling pathway

Xiangming Zeng[1,2†], Chaoqun Huang[2,3†], Lakmini Senavirathna[2,3], Pengcheng Wang[1*] and Lin Liu[2,3*]

## Abstract

**Background:** MicroRNAs are a group of small RNAs that regulate gene expression at the posttranscriptional level. They regulate almost every aspect of cellular processes. In this study, we investigated whether miR-27b regulates pulmonary fibroblast activation.

**Results:** We found that miR-27b was down-regulated in fibrotic lungs and fibroblasts from an experimental mouse model of pulmonary fibrosis. The overexpression of miR-27b with a lentiviral vector inhibited TGFβ1-stimulated mRNA expression of collagens (COL1A1, COL3A1, and COL4A1) and alpha-smooth muscle actin, and protein expression of Col3A1 and alpha-smooth muscle actin in LL29 human pulmonary fibroblasts. miR-27b also reduced contractile activity of LL29. TGFβ receptor 1 and SMAD2 were identified as the targets of miR-27b by 3'-untranslated region luciferase reporter and western blotting assays.

**Conclusions:** Our results suggest that miR-27b is an anti-fibrotic microRNA that inhibits fibroblast activation by targeting TGFβ receptor 1 and SMAD2. This discovery may provide new targets for therapeutic interventions of idiopathic pulmonary fibrosis.

**Keywords:** miR-27b, Fibroblast activation, TGFβ, Idiopathic pulmonary fibrosis

## Background

Idiopathic pulmonary fibrosis (IPF) is a chronic, progressive, and usually fatal disease. The features of the disease are characterized by repeated injury and activation of alveolar epithelial cells, formation of myofibroblasts and excessive accumulation of extracellular matrix in the lung parenchyma [1]. The estimated prevalence of IPF in the United States is 14 to 27.9 per 100,000 people [2]. However, the etiology and molecular mechanisms of IPF initiation and progression are largely unclear.

Chronic lung inflammation had been considered preceding pulmonary fibrosis and played a major role in lung fibrogenesis. However, some evidence suggested that inflammation was not an important pathogenic event of IPF. For example, in the early or late course of IPF, including diseases similar to mild or moderate alveolitis, there was a lack of long-term effective response to anti-inflammation therapy [3]. Fibroblasts are the main source of abnormal extracellular matrix production in IPF. The TGFβ signaling is an important mediator of pulmonary fibrosis in both animals and humans. TGFβ modulates the activation of fibroblasts [1]. Alveolar epithelial cells also play an important role in the process of pathogenesis of IPF. The activated cells secrete soluble protein factors, including TGFβ, TNFα, endothelin-1 to exert their effects on fibroblasts [1].

MicroRNAs (miRNAs) are a group of endogenous non-coding small RNAs. More than 3000 mature miRNAs are found from plants to humans. miRNAs are the key regulators of many biological processes, and they function by inhibition of translation and degradation of target mRNAs to control protein expression in physiological and pathophysiological conditions [4]. miRNAs are involved in almost all aspects of cell physiology, including cell proliferation and differentiation, apoptosis, and diseases [5–8]. Based on computational analysis, the mRNA targets of miRNAs in most mammals are conserved due to selective pressure [9].

miR-27b has been reported to play a role in breast, liver, kidney or other organs. For example, antagomir of

* Correspondence: twangpc@jnu.edu.cn; lin.liu@okstate.edu
†Equal contributors
[1]Department of Immunology and Microbiology, Medical School of Jinan University, Guangdong, China
[2]Lundberg-Kienlen Lung Biology and Toxicology Laboratory, Department of Physiological Sciences, Stillwater, OK, USA
Full list of author information is available at the end of the article

miR-27b suppressed cell invasion in human breast cancer cell line, MDA-MB-231, whereas pre-miR-27b stimulated invasion in ZR75 breast cancer cells [10]. miR-27b synergized with anticancer drugs in a defined subgroup of liver and kidney cancer patients [11]. The overexpression or inhibition of miR-27b in HuH7 cells significantly decreased or increased the peroxisome proliferator-activated receptor (PPAR) alpha protein level [12]. miR-27b targets the 3'-untranslated region (3'-UTR) of PPARγ and inhibits its mRNA and protein expression in neuroblastoma cells [13].

Recently, miR-27b was identified as a major miRNA in modulating TGFβ-induced collagen I expression using a miRNA inhibitor library [14]. The inhibition of miR-27b increased COL1A1 expression. However, this study used lung epithelial cancer cell line A549 cells rather than pulmonary fibroblasts. During the course of this work, Cui et al. published a study showing that transfection of miR-27a-3p mimic into a fetal lung fibroblast cell line MRC-5 inhibited COL1A2 and alpha-smooth muscle actin (α-SMA) and this result was confirmed in IPF fibroblasts [15]. The use of a miRNA mimic could result in overwhelmed expression of a miRNA since it by-pass the cellular regulatory system for the processing of a miRNA.

In this study, we investigated a role of miR-27b in fibroblast activation using a human lung fibroblasts and a lentiviral vector expressing a primary miR-27b, which is converted into a mature miRNA via the endogenous processing system after entering cells. We found that miR-27b inhibited fibroblast activation, and TGFβ receptor 1 (TGFBR1) and SMAD2 are direct targets of miR-27b. Our results suggest that miR-27b is an anti-fibrotic miRNA in pulmonary fibroblasts.

## Methods

### Bleomycin mouse fibrosis model and isolation of primary cells

The animal procedures were approved by the Institutional Animal Care and Use Committee at the Oklahoma State University (VM-15–38). Bleomycin or saline was delivered to the lungs of C57BL/6 male mice (6–8 weeks) via nasal instillation at a dose of 3 U/kg body weight. On day 14 mice were sacrificed, and lung tissues were collected. Fibroblasts and alveolar epithelial type II cells (AEC II) were isolated from the lungs of saline or bleomycin-treated mice according to the previously described protocols [16, 17]. Alveolar epithelial type I cells (AEC I) were obtained by culturing AEC II in Dulbecco's Modified Eagle Medium (DMEM) for 5 days [18].

### RNA isolation

Total RNAs were isolated from lung tissues or cells by using Tri Reagents (Molecular Research Center,

Cincinnati, OH). The RNA concentration and quality were determined by NanoDrop ND-1000 Spectrophotometer.

### Construction of vectors

The primary hsa-miR-27b was PCR-amplified from human genomic DNA (Promega, Madison, WI) with the following primers: forward,TTTCTCGAGGGGATTACC ACGCAACCAC and reverse, TTTGAATTCGGCTAG CATTCCCAGCAGGAGA. The PCR product was inserted into a modified lentiviral vector pLVX (Clontech, Mountain View, CA) the downstream of its green fluorescent protein (GFP) at XhoI and EcoRI as described [19].

The 3'-UTRs of human TGFβ receptor 1 (TGFBR1) and SMAD2 were PCR-amplified from human genomic DNA with the following primers: TGFBR1 forward, GCTAGCTGAATATTCTCACATCAAGCTTT and reverse, GTCGACGTGAGAAATCATGTATTACAACT.

SMAD2 forward, GCTAGCTTTCTCTAGTGATAT TAAGGAACG and reverse, GTCGACACAGATGATG CACACAAATATAT. TGFBR1-UTR or SMAD2-UTR was inserted into the pmirGlo vector (Promega) the downstream of firefly luciferase gene at Nhe I and Sac I. All of the constructs were confirmed by DNA sequencing.

### Preparation of lentivirus

To produce lentivirus overexpressing miR-27b, lenti-miR-27b or its control plasmid was transfected to HEK 293 T cells along with Lenti-X HTX Packaging mix (Clontech) by using Lipofectamine 2000. After a 48-h transfection, the media containing viruses were collected. For virus titer determination, HEK 293 T cells were split into a 12-well plate at a density of $5 \times 10^5$ cells per well. Cells were infected with lentiviruses at a series of dilutions. 48 h post infection, virus titer was determined by counting GFP-positive cells (10 fields per well) under a fluorescence microscope.

### Infection of fibroblasts with a lentiviral miR-27b

LL29 lung fibroblasts were purchased from American Type Culture Collection (ATCC, Manassa, VA, CCL-134) and were maintained in F12K medium supplemented with 10% fetal bovine serum (FBS) and 1% penicillin/streptomycin (P/S). The fibroblasts were seeded on 6-well plates at a density of $2–5 \times 10^5$/well. After 24 h, cells were infected with a lentiviral miR-27b or its control at a multiplicity of infection (MOI) of 50. 48 h post infection, cells were stimulated with TGFβ (5 ng/ml). After another 48 or 72 h, cells were collected for RNA and protein analyses.

### Real-time PCR

Total RNA was reverse-transcribed into cDNA using Moloney Murine Leukemia Virus reverse transcriptase.

For miRNA quantitation, total RNA was poly (A)-tailed using an A-Plus Poly (A) Polymerase Tailing Kit (Epicentre, Madison, WI) before reverse transcription. The following primers were used: COL1A1, Forward: CGAA GACATCCCACCAATCAC, and reverse: CAGATCAC GTCATCGCACAAC; COL3A1, Forward: TGGCTA CTTCTCGCTCTGCTT, and reverse: TTCCAGACA TCTCTATCCGCATAG; COL4A1 forward: CTCTGG CTGTGGCAAATGTG, and reverse: CCTCAGGTCC TTGCATTCCA; α-SMA, forward: GTGTTGCCCC TGAAGAGCAT, and reverse: CGCCTGGATAGCCA CATACAT; microRNA-universe reverse primer: GC GAGCACAGAATTAATACGAC; RNU6 forward primer: AGAGAAGATTAGCATGGCCCCT; miR-27b forward primer: TTCACAGTGGCTAAGTTCTGC. Real-time PCR was performed using SYBR Green master mix on an ABI 7500 fast system (Applied Biosystems, Foster City, CA). The thermal temperature were: 95°C for 10 min, followed by 40 cycles of 95°C for 15 s, 60°C for 30 s, and 65°C for 30 s. The endogenous reference genes were glyceraldehyde-3-phosphate dehydrogenase (GAPDH) or RNU6. The comparative ΔCt method was used to calculate the relative mRNA and miRNA expression levels.

### Luciferase reporter assay

miR-27b or its control plasmid (150 ng) were co-transfected into HEK 293 T cells using Lipofectamine 2000 with TGFBR1-UTR or SMAD2-UTR plasmid (5 ng), which contains a *Renilla* luciferase gene for normalization. After a 48-h transfection, the cells were harvested, and luciferase activities were measured using the Dual Luciferase Reporter Assay System (Promega).

### Western blot

Proteins (10–20 μg) were separated on SDS-PAGE and transferred onto nitrocellulose membranes. After being incubated with primary antibodies, the membranes were washed with Tris-buffered saline (pH 7.5) and Tween 20 and incubated with horseradish peroxidase-conjugated anti-mouse or rabbit secondary antibodies for 1 h. The target proteins were visualized with Super Signal West Pico Chemiluminescents Substrate and analyzed with Amersham 600 Molecular Imager. The following antibodies and dilutions were used: mouse anti-α-SMA monoclonal antibody (1:1000; Sigma), rabbit anti-β-actin monoclonal antibody (1:1000; Santa Cruz), rabbit anti-TGFBR1 polyclonal antibody (1:500; Santa Cruz), rabbit anti-SMAD2 polyclonal antibody (1:500; Santa Cruz), rabbit anti-Col3A1 polyclonal antibody (1:500; Santa Cruz), goat anti-rabbit monoclonal second antibody (1:2000; Sigma), and goat anti-mouse monoclonal second antibody (1:2000; Sigma).

### Fibroblast contraction assay

LL29 fibroblasts were seeded in 6-well plates at a density of $1-2 \times 10^5$ cells per well overnight and infected with miR-27b or its control lentivirus at a MOI of 50 for 48 h. The cells were treated with TGFβ1 (5 ng/ml) for another 48 h. The cells were trypsinized and mixed with rat tail collagen I (BD Bioscience, Cat# 354236) to a final concentration of $1 \times 10^5$ cells/ml and 1 mg/ml of collagen 1, followed by the addition of 15 μl 0.5 M NaOH to 1 ml of the cells. The cells were then added to 24-well BSA-coated plates (500 μl/well). After a 30-min incubation, medium with or without TGFβ (5 ng/ml) were added (500 μl/well). Cells were incubated for 48 h and images were taken. The gel areas were quantified using Image J software.

### Statistics

All experiments were performed with at least three independent replicates. For statistical analysis, student *t*-test was used for two group comparisons and ANOVA, followed by Turkey's test for multiple group comparisons. $P < 0.05$ was considered significance.

## Results

### miR-27b is down-regulated in the lungs and fibroblasts from bleomycin-treated mice

We determined miR-27b expression levels in the fibrotic lungs induced by bleomycin in mice by real-time PCR. The expression of miR-27b in the lung tissue of bleomycin-treated mice was decreased significantly compared to that of the control mice (Fig. 1). To examine which types of the lung cells account for the decrease, we isolated fibroblasts and alveolar epithelial cells from the bleomycin-treat mice. The expression level of miR-27b in fibroblasts was much higher than alveolar epithelial type I and type II cells (AEC I and AEC II). Bleomycin treatment reduced miR-27b expression in the fibroblasts (Fig. 1), suggesting that fibroblasts are the cells responsible for the reduction of miR-27b in the fibrotic lungs.

### miR-27b inhibits pulmonary fibroblast activation

To study whether miR-27b affects fibroblasts' function, we overexpressed miR-27b in LL29 human pulmonary fibroblasts using a lentiviral vector and determined the effect of miR-27b on TGFβ1-mediated collagen and α-SMA expression by real-time PCR and western blotting. The overexpression of miR-27b in LL29 fibroblasts were confirmed (Fig. 2a). The mRNA expression of *COL1A1*, *COL3A1*, *COL4A1* and *α-SMA* was increased by TGFβ1 treatment, and this increase was suppressed by miR-27b overexpression (Fig. 2b). Moreover, the TGFβ1-induced protein expression of Col3A1 and α-SMA was also inhibited by miR-27b overexpression (Fig. 2c, d). These results indicate that miR-27b represses fibroblast activation.

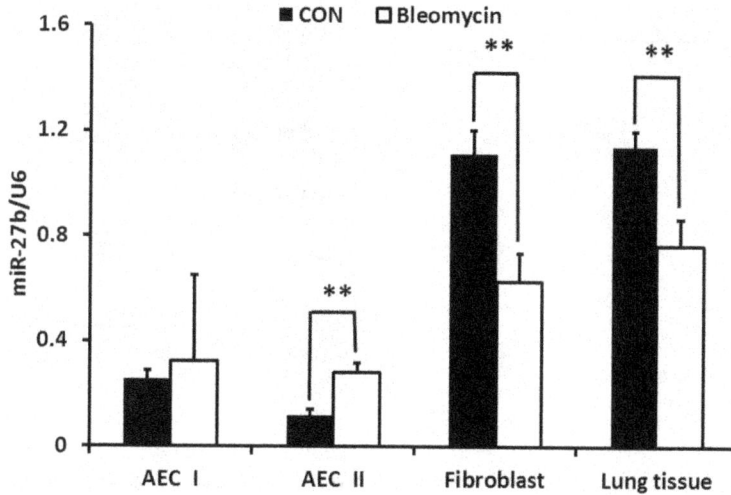

**Fig. 1** miR-27b expression in the lung tissues and fibroblasts from bleomycin-treated mice. The lung tissues were collected from saline control (*CON*) and bleomycin-treated mice. Fibroblasts and alveolar epithelial type II cells (*AEC II*) were isolated from the lungs. Alveolar epithelial type I cells (*AEC I*) were obtained via trans-differentiation by culturing AEC II for 5 days in vitro. The expression of miR-27b was determined by real-time PCR and normalized to RNU6 (U6). Data shown are means ± S.E. **$P < 0.01$. $n = 9$ for lung tissues; $n = 10$ for fibroblasts; and $n = 3$ for AEC I and AEC II. Student $t$-test

### miR-27b attenuates the contractile activity in pulmonary fibroblasts

The contraction process is known to be mediated by specialized fibroblasts in IPF and TGFβ signaling pathway regulates this process. Therefore, we investigated whether miR-27b influences the contractility of pulmonary fibroblasts. As shown in Fig. 3, miR-27b inhibited the TGFβ1-induced contractility of LL29 fibroblasts.

### miR-27b directly targets TGFBR1 and SMAD2

Because miR-27b inhibits TGFβ1-induced fibroblast activation, it likely targets the components in the TGFβ

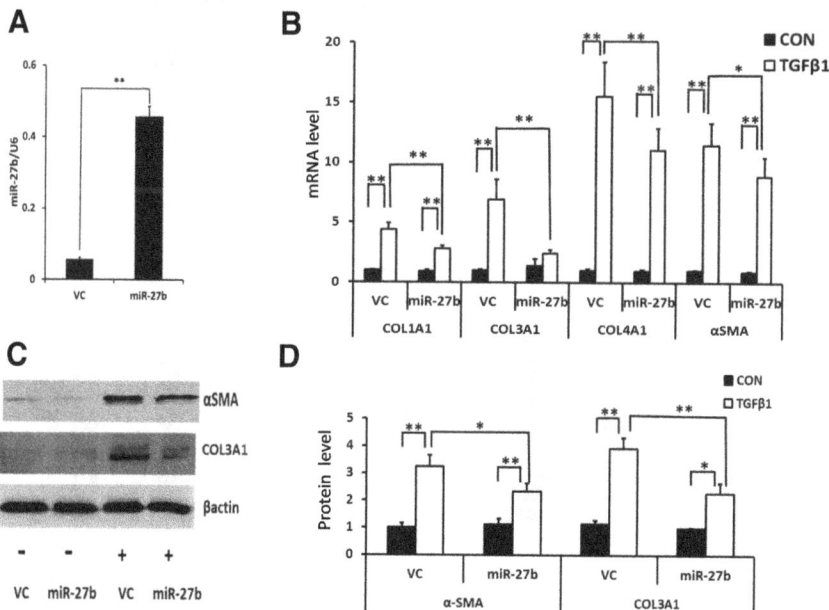

**Fig. 2** Effect of miR-27b on fibroblast activation. LL29 fibroblasts were infected with a miR-27b lentivirus or virus control (*VC*) at a MOI of 50. **a** miR-27b expression and **b** mRNA expression of COL1A1, COL3A1, COL4A1, and α-SMA were determined by real-time PCR. The expression levels of miR-27b and mRNA were normalized to U6 and GAPDH, respectively. mRNA levels were expressed as fold changes over the control group without TGFβ1 treatment (*CON*). **c, d** The protein level expression of α-SMA and COL3A1 were determined by western blotting and quantitated using Image J software. The results were normalized to β-actin and expressed as fold changes over control without TGFβ1 treatment (*CON*). Data shown are means ± S.E, *$P < 0.05$, **$P < 0.01$. $n = 3$ for **a** and **d**, $n = 6$–10 for **b**. Student $t$-test for **a** and ANOVA, followed by Turkey's test for **b** and **d**

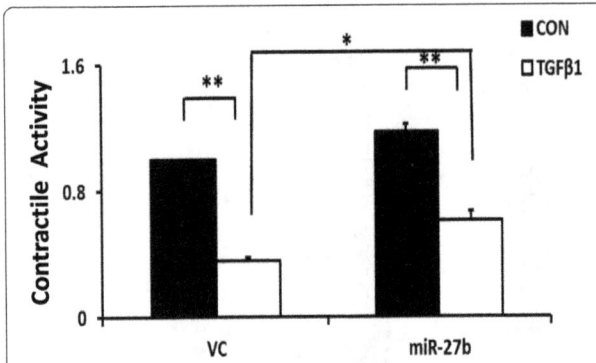

**Fig. 3** Effects of miR-27b on contractile activity of lung fibroblasts. LL29 lung fibroblasts were infected with miR-27b lentivirus or virus control (*VC*), and then treated with TGFβ1. The cells were mixed with collagen I and seeded in 24-well plates for 48 h. Images were taken and gel areas were quantified using Image J. Collagen gel contractile activity was calculated as gel surface area divided by well surface area. The results are expressed as a fold change over VC without TGFβ1 treatment (*CON*). Data shown are means ± S.E., $n = 5$, *$P < 0.05$, **$P < 0.01$. ANOVA, followed by Turkey's test

signaling pathway. Using TargetScan, TGFBR1 and SMAD2 were identified as the potential targets of miR-27b (Fig. 4a). Next we constructed 3'-UTR luciferase reporter vectors and performed the dual luciferase assay to validate whether TGFBR1 and SMAD2 are the direct targets of miR-27b. As shown in Fig. 4b, miR-27b significantly inhibited the luciferase activities of TGFBR1-UTR and SAMD2-UTR reporters. Furthermore, TGFBR1 and SMAD2 protein expression was inhibited by miR-27b in LL29 fibroblasts (Fig. 4c, d).

## Discussion

miRNAs play important roles in various biological processes, such as cell proliferation, differentiation, apoptosis, and other functions. miRNAs show different expression profiles in many diseases including IPF. Abnormal expression of miRNAs may contribute to the development of many diseases. miRNAs may be potentially utilized in the treatment of human diseases, such as infectious and metabolic diseases [20, 21]. In the present study, we demonstrated that miR-27b was downregulated in fibrotic lungs and fibroblasts from bleomycin-induced mouse pulmonary fibrosis model. In contrast, miR-27a-3p was found to be increased in the fibroblasts isolated from the lungs of the bleomycin-treated mice although the same authors found that miR-27a-3p was decreased in IPF fibroblasts compared to normal lung fibroblasts [15]. Additionally, TGFβ inhibited miR-27b expression in lung epithelial A549 cells [14], but increased miR-27a expression in MRC-5 fibroblasts [15]. The reasons for these differences remain to

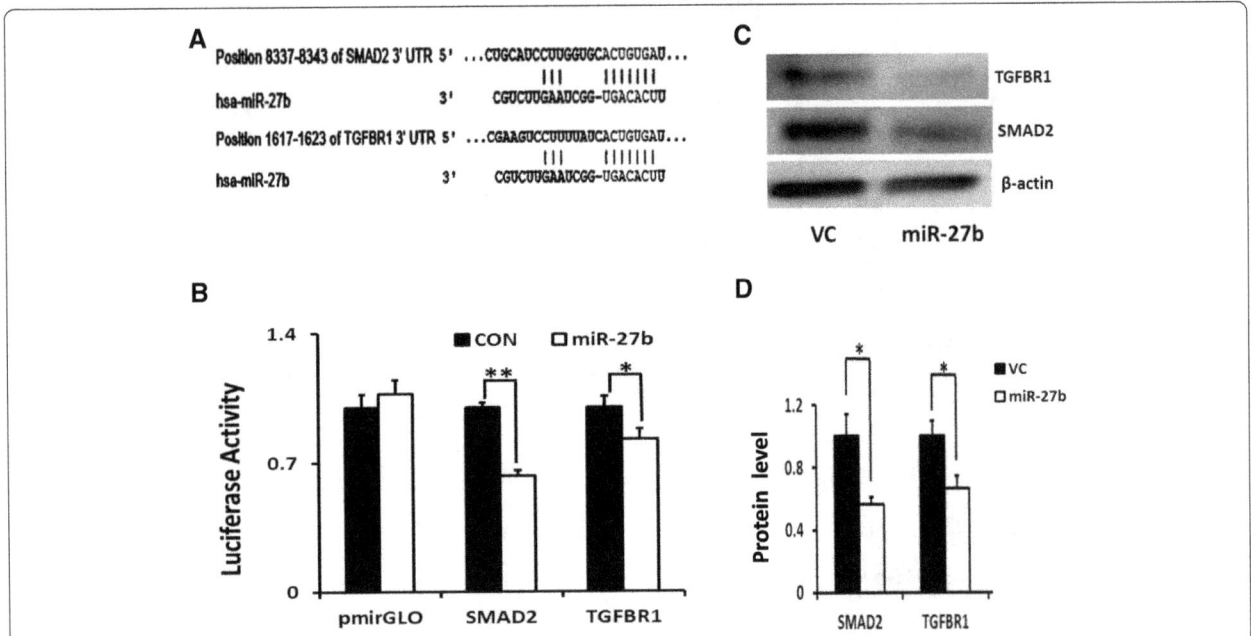

**Fig. 4** TGFBR1 and SMAD2 are the targets of miR-27b. **a** The binding sites of miR-27b on the 3'-UTRs of TGFBR1 and SMAD2 as predicted by TargetScan. **b** 3'-UTR luciferase reporter assay. HEK 293 T cells were transfected with TGFBR1-UTR or SMAD2-UTR luciferase vector together with the miR-27b expression plasmid or vector control (CON). pmiRGLO is the empty vector without any 3'-UTR sequences. Luciferase activities were determined 48 h post-transfection. The fold changes were relative to CON for each reporter construct. Data shown are means ± S.E *$P < 0.05$, **$P < 0.01$ $n = 4$. **c, d** TGFBR1 and SMAD2 protein levels in miR-27b overexpressing fibroblasts. LL29 fibroblasts were infected with a miR-27b lentivirus or virus control (VC) at a MOI of 50. After 72 h, the cells were lysed for Western blotting. Protein levels were quantitated by Image J software and normalized to β-actin. The fold changes were relative to VC. Date were represented as means ± S.E., *$P < 0.05$, $n = 3-4$. ANOVA, followed by Turkey's test

be determined, but could be due to the differences in transcription because miR-27a-3p and miR-27b are located in different chromosomes, chromosome19 and chromosome 9, respectively. However, there is only one base difference between miR-27a-3p and miR-27b.

The abnormal activation of fibroblasts is one of the major factors driving fibrotic progression in IPF [22–24]. TGFβ activates fibroblasts and enhances collagen synthesis and extracellular matrix deposition [25–27]. In the current study, we found that ectopic overexpression of miR-27b suppressed TGFβ-induced fibroblast activation, as evidenced by the decreased collagen synthesis, inhibition of α-SMA mRNA and protein expression, and enhanced contractile ability. A recent report shows that miR-27b inhibitor increased TGFβ-induced COL1A1 expression in lung epithelial A549 cells [14]. The significance of this regulation in lung epithelial cells is unclear since pulmonary fibroblasts are the major cells for extracellular matrix deposition in IPF. Most recently, miR-27a-3p has been shown to inhibit TGFβ-induced COL1A2 and α-SMA expression and gel contractility in MRC-5 fibroblasts [15]. However, a miR-27b-3p mimic was used for these studies, which may result in overwhelmed expression of miR-27a. Further, MRC-5 fibroblasts are derived from fetal lungs, which may not be the best fibroblast cell line for studying IPF, which normally occurs in patients between the ages of 50 and 70 years. Whether miR-27a-3p and miR-27b have the same functions or are redundant in pulmonary fibroblasts needs further investigations.

Several miRNAs are involved in the regulation of pulmonary fibrosis through the TGFβ/SMAD pathway, including miR-21, miR-26a and miR-29 [28–30]. We suspected that miR-27b may act in a similar mechanism in pulmonary fibrosis. Indeed, TGFBR1 and SMAD2 were identified as potential targets of miR-27b by TargetScan. The dual-luciferase reporter assay confirmed that miR-27b functioned through a direct binding to its 3-'UTR of TGFBR1 and SMAD2. The endogenous protein expressions of these targets were reduced in LL29 fibroblasts by miR-27b, which further confirmed that TGFBR1 and SMAD2 are the targets of miR-27b. Using similar approaches, Cui et al. found that miR-27a-3p targeted SMAD2/4 in fetal lung MRC-5 fibroblasts and they also identified α-SMA as an additional target of miR-27a-3p [15]. In lung epithelial A549 cells, miR-27b targets gremlin 1 [14]. In neuroblastoma cells, PPARγ is a target of miR-27b as determined by 3'-UTR luciferase reporter and endogenous protein assays [13]. In liver cells, miR-27b regulates PPARα indirectly since overexpression of miR-27 reduced the PPARα protein level, but 3'-UTR luciferase reporter assay did not confirm PPARα as a direct target protein level [12]. It appears that miR-27a-3p and miR-27b have multiple targets and which genes are the main targets may depend on cell types.

## Conclusions

To sum up, our present studies show that miR-27b plays an important role in the pulmonary fibroblast activation by regulating TGFBR1 and SMAD2.

### Abbreviations
3'-UTR: 3' Untranslated region; AEC I: Alveolar epithelial type I cells; AEC II: Alveolar epithelial type II cells; COL: Collagens; FBS: Fetal bovine serum; GAPDH: Glyceraldehyde-3-phosphate dehydrogenase; GFP: Green fluorescent protein; IPF: Idiopathic pulmonary fibrosis; miRNAs: microRNAs; MOI: Multiplicity of infection; P/S: Penicillin/streptomycin; PPAR: Peroxisome proliferator-activated receptor; TGFBR1: TGFβ receptor 1; α-SMA: Alpha-smooth muscle actin

### Acknowledgements
None.

### Funding
This work was supported by the National Heart, Lung and Blood Institute under Award Number R01 HL116876, the National Institute of General Medical Sciences under Award number P20GM103648 and the Oklahoma Center for Adult Stem Cell Research (to LL) and Guangdong Natural Science Foundation (2014A030313370), and Jinan University Innovation Foundation (21615420) (to PW).

### Authors' contributions
XZ, CH, LS carried out experiments. XZ, CH analyzed data. XZ, CH, PW, LL designed and wrote the manuscript. All authors read and approved the final manuscript.

### Competing interests
The authors declare that they have no competing interests.

### Author details
[1]Department of Immunology and Microbiology, Medical School of Jinan University, Guangdong, China. [2]Lundberg-Kienlen Lung Biology and Toxicology Laboratory, Department of Physiological Sciences, Stillwater, OK, USA. [3]Oklahoma Center for Respiratory and Infectious Diseases, Oklahoma State University, Stillwater, OK, USA.

### References
1. Wynn TA. Integrating mechanisms of pulmonary fibrosis. J exp med. 2011; 208:1339–50.
2. Nalysnyk L, Cid-Ruzafa J, Rotella P, Esser D. Incidence and prevalence of idiopathic pulmonary fibrosis: review of the literature. Eur respir rev. 2012;21: 355–61.
3. Rafii R, Juarez MM, Albertson TE, Chan AL. A review of current and novel therapies for idiopathic pulmonary fibrosis. J thorac dis. 2013;5:48–73.
4. Inui M, Martello G, Piccolo S. MicroRNA control of signal transduction. Nat rev mol cell biol. 2010;11:252–63.
5. Ambros V. The functions of animal microRNAs. Nature. 2004;431:350–5.
6. Garzon R, Calin GA, Croce CM. MicroRNAs in cancer. Annu rev med. 2009;60: 167–79.
7. Gangaraju VK, Lin H. MicroRNAs: key regulators of stem cells. Nat rev mol cell biol. 2009;10:116–25.
8. Wang Y, Stricker HM, Gou D, Liu L. MicroRNA: past and present. Front biosci. 2007;12:2316–29.
9. Friedman RC, Farh KK, Burge CB, Bartel DP. Most mammalian mRNAs are conserved targets of microRNAs. Genome res. 2009;19:92–105.
10. Wang Y, Rathinam R, Walch A, Alahari SK. ST14 (suppression of tumorigenicity 14) gene is a target for miR-27b, and the inhibitory effect of ST14 on cell growth is independent of miR-27b regulation. J biol chem. 2009;284:23094–106.

11. Mu W, Hu C, Zhang H, Qu Z, Cen J, Qiu Z, Li C, Ren H, Li Y, He X, Shi X, Hui L. miR-27b synergizes with anticancer drugs via p53 activation and CYP1B1 suppression. Cell res. 2015;25:477–95.

12. Kida K, Nakajima M, Mohri T, Oda Y, Takagi S, Fukami T, Yokoi T. PPARalpha is regulated by miR-21 and miR-27b in human liver. Pharm res. 2011;28: 2467–76.

13. Lee JJ, Drakaki A, Iliopoulos D, Struhl K. MiR-27b targets PPARgamma to inhibit growth, tumor progression and the inflammatory response in neuroblastoma cells. Oncogene. 2012;31:3818–25.

14. Graham JR, Williams CMM, Yang ZY. MicroRNA-27b targets gremlin 1 to modulate fibrotic responses in pulmonary cells. J cell biochem. 2014;115: 1539–48.

15. Cui H, Banerjee S, Xie N, Ge J, Liu RM, Matalon S, Thannickal VJ, Liu G. MicroRNA-27a-3p is a negative regulator of lung fibrosis by targeting myofibroblast differentiation. Am j respir cell mol biol. 2016;54:843–52.

16. Guo Y, Mishra A, Howland E, Zhao C, Shukla D, Weng T, Liu L. Platelet-derived Wnt antagonist dickkopf-1 is implicated in ICAM-1/VCAM-1-mediated neutrophilic acute lung inflammation. Blood. 2015;126:2220–9.

17. Bruce MC, Honaker CE, Cross RJ. Lung fibroblasts undergo apoptosis following alveolarization. Am j respir cell mol biol. 1999;20:228–36.

18. Bhaskaran M, Kolliputi N, Wang Y, Gou D, Chintagari NR, Liu L. Trans-differentiation of alveolar epithelial type II cells to type I cells involves autocrine signaling by transforming growth factor beta 1 through the Smad pathway. J biol chem. 2007;282:3968–76.

19. Zhao C, Huang C, Weng T, Xiao X, Ma H, Liu L. Computational prediction of MicroRNAs targeting GABA receptors and experimental verification of miR-181, miR-216 and miR-203 targets in GABA-A receptor. BMC res notes. 2012; 5:91.

20. Nana-Sinkam SP, Croce CM. Clinical applications for microRNAs in cancer. Clin pharmacol ther. 2013;93:98–104.

21. Hydbring P, Badalian-Very G. Clinical applications of microRNAs. F1000Res. 2013;2:136.

22. Lim MJ, Ahn J, Yi JY, Kim MH, Son AR, Lee SL, Lim DS, Kim SS, Kang MA, Han Y, Song JY. Induction of galectin-1 by TGF-beta1 accelerates fibrosis through enhancing nuclear retention of Smad2. Exp cell res. 2014;326:125–35.

23. Kottmann RM, Kulkarni AA, Smolnycki KA, Lyda E, Dahanayake T, Salibi R, Honnons S, Jones C, Isern NG, Hu JZ, Nathan SD, Grant G, Phipps RP, Sime PJ. Lactic acid is elevated in idiopathic pulmonary fibrosis and induces myofibroblast differentiation via pH-dependent activation of transforming growth factor-beta. Am j respir crit care med. 2012;186:740–51.

24. Lepparanta O, Sens C, Salmenkivi K, Kinnula VL, Keski-Oja J, Myllarniemi M, Koli K. Regulation of TGF-beta storage and activation in the human idiopathic pulmonary fibrosis lung. Cell tissue res. 2012;348:491–503.

25. Chen J, Xia Y, Lin X, Feng XH, Wang Y. Smad3 signaling activates bone marrow-derived fibroblasts in renal fibrosis. Lab invest. 2014;94:545–56.

26. Frangogiannis NG. Targeting the inflammatory response in healing myocardial infarcts. Curr med chem. 2006;13:1877–93.

27. Judge JL, Owens KM, Pollock SJ, Woeller CF, Thatcher TH, Williams JP, Phipps RP, Sime PJ, Kottmann RM. Ionizing radiation induces myofibroblast differentiation via lactate dehydrogenase. Am j physiol lung cell mol physiol. 2015;309:L879–87.

28. Gao Y, Lu J, Zhang Y, Chen Y, Gu Z, Jiang X. Baicalein attenuates bleomycin-induced pulmonary fibrosis in rats through inhibition of miR-21. Pulm pharmacol ther. 2013;26:649–54.

29. Liang H, Xu C, Pan Z, Zhang Y, Xu Z, Chen Y, Li T, Li X, Liu Y, Huangfu L, Lu Y, Zhang Z, Yang B, Gitau S, Lu Y, Shan H, Du Z. The antifibrotic effects and mechanisms of microRNA-26a action in idiopathic pulmonary fibrosis. Mol ther. 2014;22:1122–33.

30. Xiao J, Meng XM, Huang XR, Chung AC, Feng YL, Hui DS, Yu CM, Sung JJ, Lan HY. miR-29 inhibits bleomycin-induced pulmonary fibrosis in mice. Mol ther. 2012;20:1251–60.

# 6

# Release of endothelial cell associated VEGFR2 during TGF-β modulated angiogenesis *in vitro*

M. Jarad, E. A. Kuczynski, J. Morrison, A. M. Viloria-Petit and B. L. Coomber[*]

## Abstract

**Background:** Sprouting angiogenesis requires vascular endothelial proliferation, migration and morphogenesis. The process is regulated by soluble factors, principally vascular endothelial growth factor (VEGF), and via bidirectional signaling through the Jagged/Notch system, leading to assignment of tip cell and stalk cell identity. The cytokine transforming growth factor beta (TGF-β) can either stimulate or inhibit angiogenesis via its differential surface receptor signaling. Here we evaluate changes in expression of angiogenic signaling receptors when bovine aortic endothelial cells were exposed to TGF-β1 under low serum conditions.

**Results:** TGF-β1 induced a dose dependent inhibition of tip cell assignment and subsequent angiogenesis on Matrigel, maximal at 5.0 ng/ml. This occurred via ALK5-dependent pathways and was accompanied by significant upregulation of the TGF-β co-receptor endoglin, and SMAD2 phosphorylation, but no alteration in Smad1/5 activation. TGF-β1 also induced ALK5-dependent downregulation of Notch1 but not of its ligand delta-like ligand 4. Cell associated VEGFR2 (but not VEGFR1) was significantly downregulated and accompanied by reciprocal upregulation of VEGFR2 in conditioned medium. Quantitative polymerase chain reaction analysis revealed that this soluble VEGFR2 was not generated by a selective shift in mRNA isoform transcription. This VEGFR2 in conditioned medium was full-length protein and was associated with increased soluble HSP-90, consistent with a possible shedding of microvesicles/exosomes.

**Conclusions:** Taken together, our results suggest that endothelial cells exposed to TGF-β1 lose both tip and stalk cell identity, possibly mediated by loss of VEGFR2 signaling. The role of these events in physiological and pathological angiogenesis requires further investigation.

**Keywords:** Tip cell, Vascular sprouting, Notch1, Dll4, Smad signaling, Microparticle

## Background

The cytokine transforming growth factor beta (TGF-β) is a member of the TGF-β superfamily consisting of 33 members including TGF-βs (1–3) [1]. TGF-β1 is the predominant and more ubiquitous form. TGF-β ligands signal canonically through type I (ALK) and type II serine/threonine kinase receptors [2], and via the accessory (type III) receptor endoglin in vascular endothelial cells [3]. TGF-β binding leads to phosphorylation of intracellular R-Smads 1, 2, 3, 5 or 8 [2, 4], which then complex with Co-Smad4, enter the nucleus and associate with transcriptional co-activators or co-repressors to regulate the expression of target genes.

TGF-β plays a dual role in angiogenesis by orchestrating a switch from vascular inhibition to pro-angiogenic activity [5–7]. In particular, there is evidence that TGF-β1 induced angiogenesis acts with VEGF to mediate apoptosis of excessive vascular sprouts and may even be required for initial sprouting from an existing vascular network [8, 9]. The nature of the angiogenic response to TGF-β depends on the balance of ALK1 versus ALK5 signaling input, with ALK1 predominantly promoting sprouting and ALK5 favoring the resolution/stabilization phase of angiogenesis [6]. Thus, TGF-β is either pro- or anti-angiogenic, depending on which TGF-RII/Smad pathway is engaged [10]. Differences also exist in the

---

* Correspondence: bcoomber@uoguelph.ca
Department of Biomedical Sciences, Ontario Veterinary College, University of Guelph, OVC Room 3645, Guelph N1G 2W1, ON, Canada

kinetics and dose responses of these pathways: ALK1 mediated signaling is transient and maximal at low concentrations of ligand while ALK5 mediated signaling is sustained and maximal at higher doses of ligand [11]. Increased expression of endoglin led to inhibition of TGF-β/ALK5 signaling (as demonstrated by Smad and CAGA dependent reporter activity) in a dose dependent fashion [12].

Activation of vascular sprouting during angiogenesis entails the specification of endothelial cells into tip and stalk cells. Endothelial tip cells are mainly migratory and polarized with minimal proliferation while stalk cells proliferate throughout sprout establishment and form the nascent vascular lumen cells [13]. The delta-like ligand 4 (Dll4)-Notch1 signaling pathway is involved in tip-stalk cell identity [14]. Tip cells express high levels of Dll4 and vascular endothelial growth factor receptor-2 (VEGFR2), and have low levels of Notch signaling activity [13, 15, 16]. The specification of endothelial cells as tip or stalk cells is transient and its reversibility is contingent on the balance between pro-angiogenic factors. Here we quantify the effect of TGF-β1 on in vitro angiogenesis using a Matrigel cord formation assay, and determine the impact of this angiogenic cytokine on expression of molecules associated with endothelial tip and stalk cell identity.

## Methods

### Cells and culture conditions

Primary bovine aortic endothelial cells (BAEC) were isolated and characterized as described [17] and used between passage 4 and 10. Cultures were maintained in Dulbecco's Modified Eagle Medium (DMEM; Sigma-Aldrich) with 10% fetal bovine serum (FBS; Invitrogen) and 1 mM sodium pyruvate (Sigma-Aldrich) at 37 °C in 5% $CO_2$ and 95% atmospheric air. Confluent monolayers were serum starved 16 h prior to treatment.

### TGF-β1 dose-response

Serum starved BAEC were treated with 0 (Control), 0.1, 1.0, 5.0, or 10.0 ng/ml of recombinant human (rh) TGF-β1 (R&D Systems) in serum free DMEM for up to 24 h, ± 5 µM SB-431542 (an inhibitor of TGβR-I isoforms ALK-5, −4 and −7 [18]). Cell supernatant (conditioned medium) was collected and centrifuged at 350 × g for 4 min, and adherent cells lysed for protein or RNA analysis. For some assays, conditioned medium was subjected to additional ultracentrifugation (100,000 × g for 1 h at 4 °C) prior to western blotting.

### Protein isolation and western blotting

Cells were lysed on ice with cell lysis buffer (Cell Signaling) supplemented with 1% phosphatase inhibitor cocktail 2 (Sigma-Aldrich), 2 µg/ml aprotinin (Sigma-

Aldrich) and 1 mM PMSF (Sigma-Aldrich). Total protein was quantified using the Bio-Rad system and electrophoresis was performed with 20–30 µg protein plus β-mercaptoethanol; samples were heated prior to SDS-PAGE electrophoresis. Proteins were transferred to PVDF membrane, blocked in 5% skim milk or 5% BSA for 1 h at room temperature, then incubated in primary antibody overnight at 4 °C. The antibodies used were: mouse monoclonal anti-Adam 10 (sc-48400; 1:200) and goat polyclonal anti-Adam 17 (sc-6416; 1:200); both Santa Cruz Biotechnology; rabbit monoclonal anti-endoglin (#4335; 1:1000), anti-Notch1 (#4380; 1:1000), anti-VEGFR2 (#9698; 1:1000), anti-pVEGFR2 Tyr 951 (#4991; 1:1000), rabbit polyclonal anti-HSP90 (α and β isoforms; #4874; 1:1000) and anti-pSMAD2 (#3101; 1:500), mouse monoclonal anti-SMAD2 (#3103; 1:1000), all from Cell Signaling; rabbit polyclonal anti-Dll4 (ab7280; 1:1000) and anti-MMP-14 (ab73879; 1:500), rabbit monoclonal anti-VEGFR1 (ab32152; 1:10000), all from abcam; anti-α Tubulin (T5168; 1:200000, Sigma-Aldrich). Membranes were washed and incubated with a secondary antibody (goat anti-mouse HRP or goat anti-rabbit HRP; 1:10,000; both Sigma-Aldrich) for 30 min at room temperature. Bands were visualized with Luminata Classico or Luminata Forte Western HRP Substrate (Millipore) and membranes imaged using a ChemiDoc MP Imaging System (Bio- Rad). Protein loading was normalized by stripping and re-probing for α-tubulin, followed by densitometry analysis.

### RNA isolation and gene expression analysis

Cells were lysed in Ribozol (AMRESCO, VWR International, Mississauga, ON) and RNA isolated using Aurum Total RNA columns (Bio-Rad, Mississauga, ON) according to manufacturers' protocols. RNA was reverse-transcribed with 4 µL iScript (BioRad, Mississauga, ON) before amplification using primers for target and housekeeping genes (Table 1). Ssofast EvaGreen Supermix

**Table 1** Sequences of primers used for qPCR

| Gene | | Primer sequence |
|---|---|---|
| sVEGFR2[a] | F | GCTTTGCTCAGGACAGGAAGAC |
| | R | GGTCCAGAGTGACTGCCCTA |
| VEGFR2[a] | F | GCTTTGCTCAGGACAGGAAGAC |
| | R | CATGCGCTCTAGGACTGTGA |
| HPRT[b] | F | CGCGCCAGCCGGCTACGTTA |
| | R | GGCCACAATGTGATGGCCACCC |
| GAPDH[b] | F | CAGCAACAGGGTGGTGGACC |
| | R | AGTGTGGCGGAGATGGGCA |

[a]bovine ortholog of human primer sequences derived from [25]
[b]generated using
PrimerBLAST http://www.ncbi.nlm.nih.gov/tools/primer-blast/index.cgi?LINK_LOC=BlastHome

(BioRad) was used to determine primer efficiency and for quantitative PCR. No-template controls and quantitative PCR reactions were run in triplicate with an initial 2 min denaturation at 95 °C, 40 cycles of 95 °C for 5 s, 60 °C for 5 s, followed by melt curve analysis of 65 °C to 95 °C in 0.5 °C increments on the CFX96 RealTime System (Bio-Rad). Full length and soluble splice variant mRNA for VEGFR2 were first normalized to both HPRT and GAPDH housekeeping genes, and then results from TGF-β1-treated samples were expressed as relative amounts normalized to control (untreated) samples. Data were analyzed using CFX Manager (Bio-Rad).

## Matrigel cord formation assay

The impact of TGF-β1 and inhibitors on in vitro angiogenesis was determined by cord formation assay. 10 μl of growth factor containing BD Matrigel Matrix (cat # 354234; BD Biosciences) was added to each well of cold angiogenesis μ-slides (ibidi) and allowed to gel at 37 °C for 30 min. 50 μl of BAEC cell suspension containing 1 × $10^4$ cells in Media-200 made with complete supplements (GIBCO) plus or minus TGF-β1 and inhibitors were added to each well. Slides were incubated for 8 h at 37 °C, and one image/well was captured with phase contrast microscopy using a 4X objective. The generation of cord networks was quantified using the tube formation assay analysis service Wimtube (http://www.wimasis.com; ibidi); analyzed parameters were total cord length, # of branching points, and # of loops. # of tip cells was also determined by manual count of phase contrast images; tip cells were defined as free-ended cells/sprouts with visible filopodia.

## Statistical analysis

All statistical analysis and graphing were performed using GraphPad Prism 6 software (GraphPad Software). Means for at least three biological replicates were calculated and plotted with standard error bars. Data were analyzed by using the nonparametric Kruskal-Wallis test followed by Dunn's *post hoc* test for multiple comparisons, as appropriate. Differences between means were considered statistically significant when the *p* value was less than 0.05. Assays were performed in triplicate unless otherwise indicated.

## Results

Exogenous TGF-β1 induced a dose-dependent decrease in endothelial cord formation on Matrigel, with significant reductions in cord length, cord branching, loop formation and tip cell generation detectable with doses of TGF-β1 ≥ 1.0 ng/ml (Fig. 1). Western blotting of whole cell lysates from BAEC treated with exogenous TGF-β1 under serum free conditions confirmed our previous findings [19] of significant downregulation of VEGFR2,

and no alteration in VEGFR1 expression (Fig. 2 a-c). VEGFR2 was detected as a doublet, with one band approximately 230 kDa (thought to represent the full-length mature transmembrane form), and a smaller (~180–200 kDa) putative 'immature' isoform [20]. Reduction in cell associated VEGFR2 protein was seen in both mature and immature forms, and was sustained for at least 24 h after removal of exogenous TGF-β1 (Additional file 1: Figure S1). There was also significant reduction in the levels of Notch1 but no changes in its ligand Dll4 (Fig. 2 a, d, f). TGF-β1 induced a profound upregulation of the type III receptor endoglin (Fig. 2 a, e) and concomitant activation of SMAD2 signaling, as revealed by enhanced levels of phosphorylated SMAD2 (pSMAD2) (Fig. 2 a, g). SMAD2 activation was maximal at 5.0 ng/ml TGF-β1, thus subsequent assays were performed with this concentration.

The mechanisms of the observed changes in angiogenic signaling pathways were then investigated. As demonstrated by treatment with the ALK5 signaling inhibitor SB-431542 (SB, 5 μM), blockade of ALK5 signaling significantly eliminated the ability of 5.0 ng/ml TGF-β1 to inhibit Matrigel angiogenesis (Fig. 3 a). SB inhibitor alone had no significant effect compared to DMSO vehicle (Control) in these assays. Exogenous TGF-β1 led to a reduction in VEGFR2 phosphorylation (as revealed by western blotting for pVEGFR2 at residue tyrosine 951) and expression levels, but addition of the SB inhibitor rescued VEGFR2 signaling (Fig. 4 a). Similar results were obtained using SD-208, another ALK5 signaling inhibitor (Additional file 2: Figure S2). TGF-β1 activation of SMAD2 phosphorylation and endoglin upregulation was also blocked by SB (Fig. 4). Notch1 levels showed a trend towards significant differences with SB inhibitor (*p* = 0.07), but expression of Dll4 was not altered (Fig. 4). SMAD1/5 phosphorylation was also not altered in these samples. Thus, ALK5 appeared to mediate the effects of TGF-β1 on the observed altered angiogenesis signaling pathways.

The basis of VEGFR2 downregulation was further investigated. Loss of VEGFR2 from whole cell lysates of TGF-β1 treated BAEC was accompanied by concomitant increasing levels of full-length protein in serum free conditioned media (CM) collected from the same cells (Fig. 5 a). qPCR analysis showed that both full-length and the soluble VEGFR2 (sVEGFR2) splice variant mRNAs were expressed by these cells, with the full-length isoform being the predominant species, expressed approximately 16 fold higher than the sVEGFR2 message. Both isoforms of VEGFR2 message were significantly downregulated in TGF-β1 treated cells compared to control. However, the ratio of full-length to soluble VEGFR2 mRNA was not altered by 5.0 ng/ml TGF-β1 treatment (Fig. 5 b), supporting the finding from Fig. 5a that a

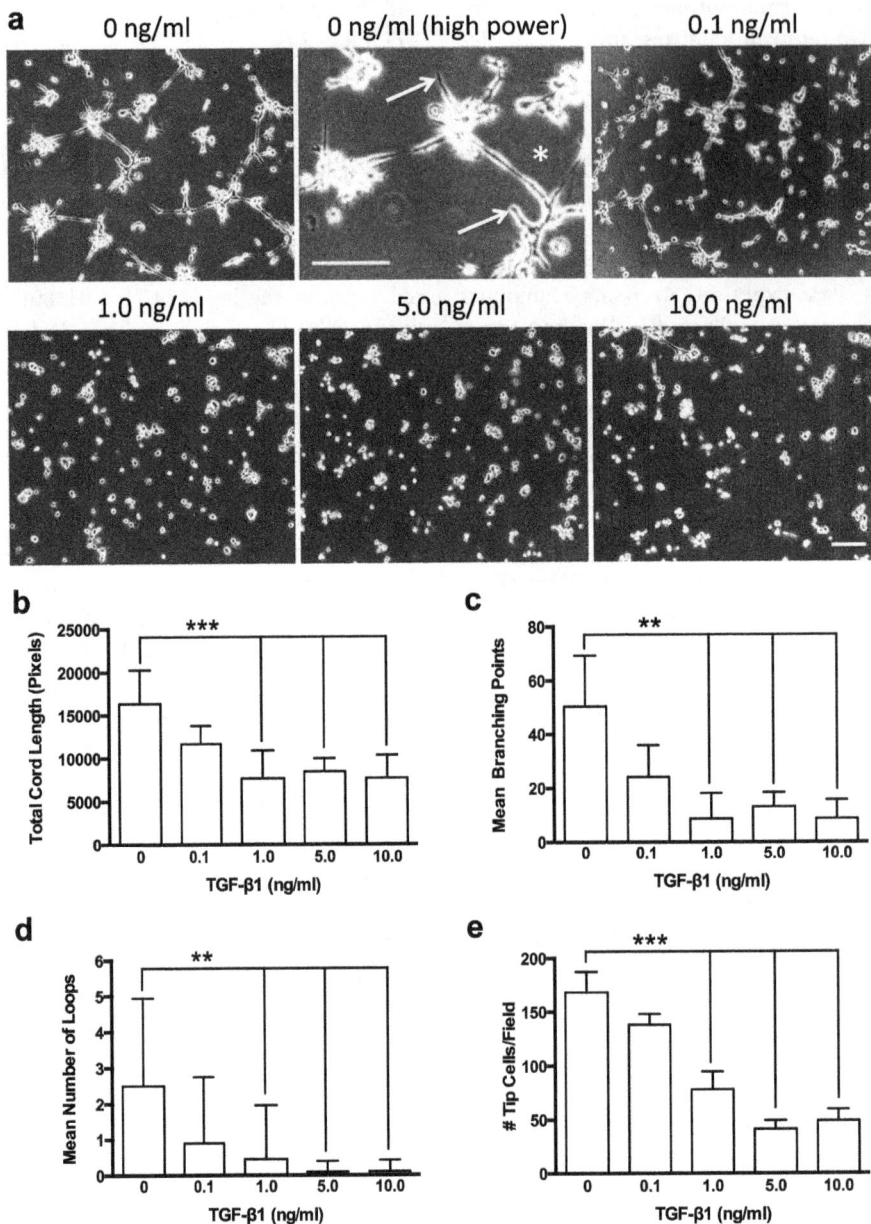

**Fig. 1** TGF-β1 induced dose-dependent reduction in endothelial cord formation. **a** Representative phase contrast images showing endothelial cell cord formation 8 h after plating on Matrigel™ in the presence of exogenous TGF-β1 (0–10 ng/ml). Note higher magnification control panel (0 ng/ml TGF-β1), showing tip cells (*arrows*) and loop of endothelial cords (*asterisk*). WimTube automated quantification of cord formation showed that TGF-β1 significantly inhibits total cord length (**b**), cord branching (**c**), formation of loops (**d**), and generation of tip cells (**e**). TGF-β1 doses of 1.0 ng/ml or higher significantly inhibited cord formation compared to control (0 ng/ml) or 0.1 ng/ml. **\*\***$p \leq 0.01$; **\*\***p $\leq 0.001$; $N = 4$; Kruskal-Wallis test. Scale bars = 300 μm

truncated soluble form of VEGFR2 protein is not released from these cells upon TGF- β1 exposure.

No change in MMP-14 (MT1-MMP) or ADAM17 protein levels were detected, and ADAM10 was weakly upregulated in TGF-β1 treated cells, suggesting enzymatic shedding was not the mechanism for increased CM-associated VEGFR2 (Fig. 5 d). In contrast, increased

levels of the extracellular vesicle/exosome-associated protein HSP90 were also detected in serum-free conditioned medium from TGF-β1 treated cells (Fig. 5 d). These data suggest that in addition to transcriptional downregulation, full-length VEGFR2 is released from TGF-β1 treated endothelial cells as a mechanism to regulate angiogenesis.

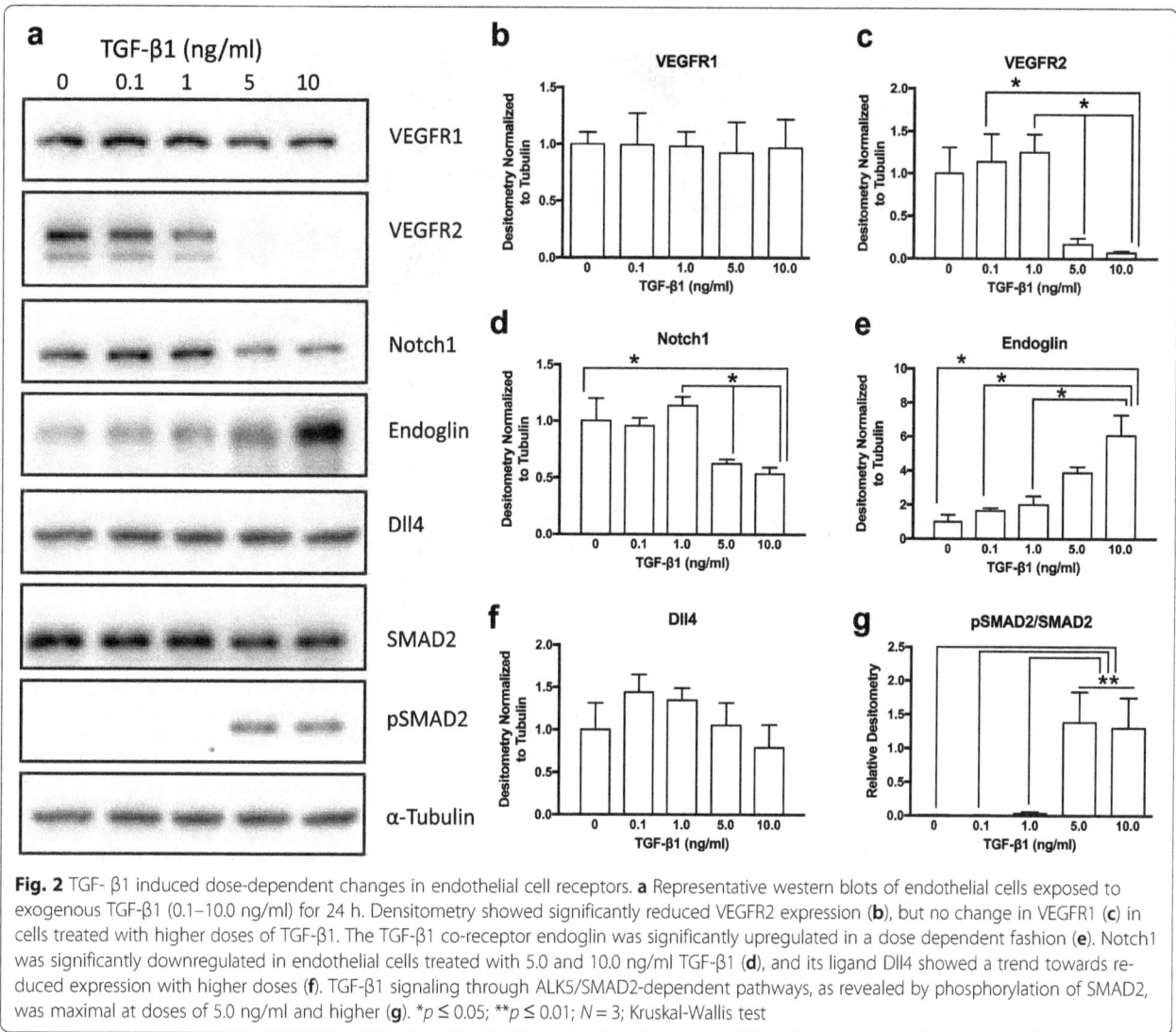

**Fig. 2** TGF- β1 induced dose-dependent changes in endothelial cell receptors. **a** Representative western blots of endothelial cells exposed to exogenous TGF-β1 (0.1–10.0 ng/ml) for 24 h. Densitometry showed significantly reduced VEGFR2 expression (**b**), but no change in VEGFR1 (**c**) in cells treated with higher doses of TGF-β1. The TGF-β1 co-receptor endoglin was significantly upregulated in a dose dependent fashion (**e**). Notch1 was significantly downregulated in endothelial cells treated with 5.0 and 10.0 ng/ml TGF-β1 (**d**), and its ligand Dll4 showed a trend towards reduced expression with higher doses (**f**). TGF-β1 signaling through ALK5/SMAD2-dependent pathways, as revealed by phosphorylation of SMAD2, was maximal at doses of 5.0 ng/ml and higher (**g**). *$p \leq 0.05$; **$p \leq 0.01$; $N = 3$; Kruskal-Wallis test

## Discussion

TGF-β1 induced a dose dependent inhibition of endothelial cord formation and modulation of endothelial angiogenic receptor expression in an ALK5 dependent fashion. We observed enhanced endoglin expression, and reduced Notch1 and VEGFR2 expression, which collectively abrogated tip cell formation and stalk elongation. During sprouting angiogenesis, combined increased VEGFR2 and reduced Notch1 signaling lead to gain tip cell phenotype, and increased Notch1 signaling promotes loss of tip cell phenotype [21, 22]. Our results suggest that TGF-β1 interferes with tip/stalk cell identity when ALK5/Smad2 signaling pathways are activated.

TGF-β-mediated ALK1 signaling serves to enhance sprouting angiogenesis in part by indirectly inhibiting TGF-β/ALK5 signaling [6]. Endoglin modulates the balance between ALK1 and ALK5 signaling, favouring a pro-angiogenic phenotype [23]. However, in our study, endoglin expression was highest in cells treated with

'anti-angiogenic' levels of TGF-β1 (5 ng/ml and higher), which were also associated with ALK5/Smad2 activation. Interestingly, activated (phosphorylated) Smad1/5 was readily detected independent of TGF-β1 treatment. Thus, in our system, tip cell identity may be the default phenotype, which is repressed upon activation of ALK5/ Smad2 signaling, possibly via loss of Notch 1 and VEGFR2 expression.

We saw concomitant loss of cell-associated full-length VEGFR2 and increased levels of CM-associated full-length VEGFR2 upon TGF-β1 exposure. Endothelial cells are known to produce an alternatively spliced VEGFR2 mRNA, leading to transcripts lacking the transmembrane domain coded for by exon 13 [24, 25]. This transcript codes for a soluble form of the receptor (sVEGFR2), which could potentially account for the presence of CM-associated VEGFR2 in our samples. However, while both full length and alternatively spliced VEGFR2 transcripts were detectable in BAEC, the full

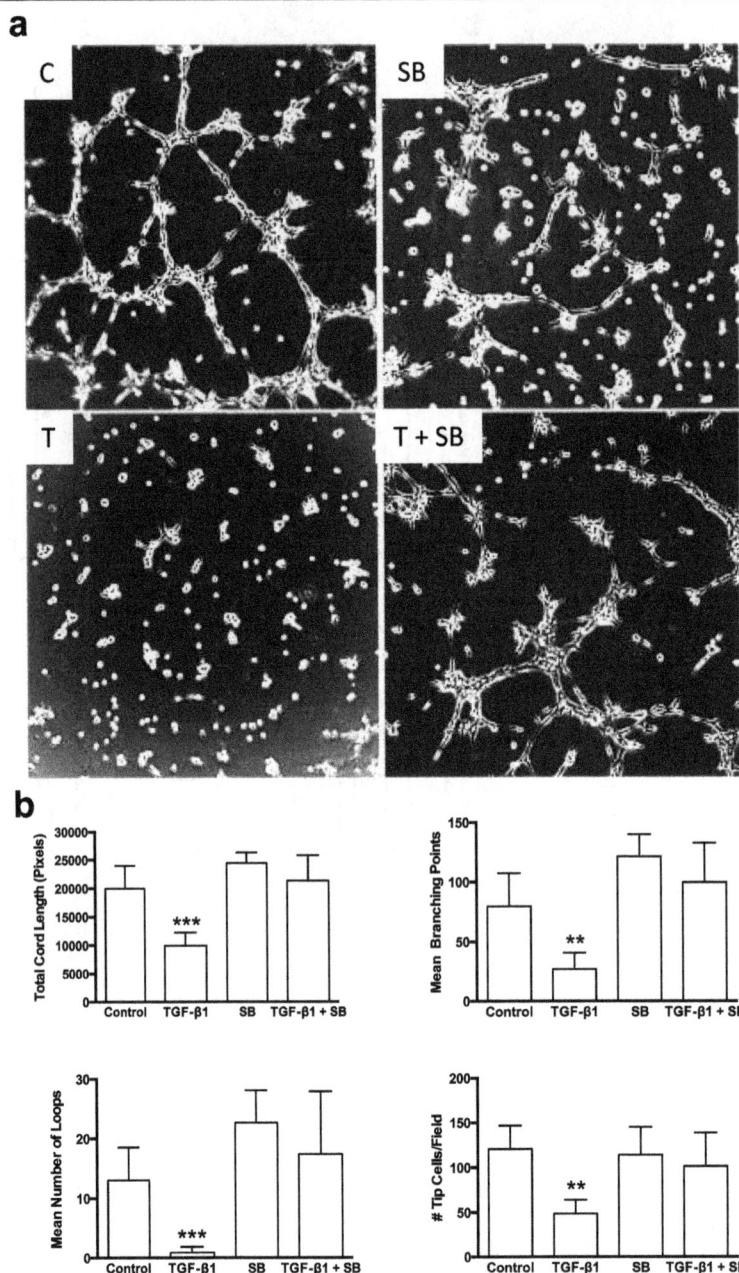

**Fig. 3** TGF-β1 induced reduction in endothelial cord formation occurs via an ALK5 related pathway. **a** Representative phase contrast images showing endothelial cell cord formation 8 h after plating on Matrigel™ under control conditions (C; DMSO vehicle), 5 μM SB-431542 (an ALK5 inhibitor) in DMSO (SB), 5 ng/ml exogenous TGF-β1 (T) or 5.0 ng/ml TGF-β1 plus 5 μM SB-431542 (T + SB). **b** WimTube automated quantification of cord formation showed SB inhibitor significantly blocked the ability of 5.0 ng/ml TGF-β1 to significantly inhibit total cord length, cord branching, formation of loops, and generation of tip cells. SB inhibitor, either alone or in combination with 5.0 ng/ml TGF-β1 was not significantly different from DMSO control. **p ≤ 0.01; ***p ≤ 0.001; $N = 4$; Kruskal-Wallis test

length VEGFR2 mRNA was by far the predominant isoform, and we found no evidence for a preferential shift in production of message for sVEGFR2 in TGF-β1 treated cells. Further, the CM-associated VEGFR2 was full-length protein as assessed by estimated molecular mass. Thus, the source of VEGFR2 found in conditioned

medium is unlikely to be due to preferential translation of a sVEGFR2 variant mRNA upon exposure to TGF-β1.

Alternatively, VEGFR2 could be cleaved from the BAEC cell surface through proteolytic activity. The matrix metalloproteinase MMP-14 (membrane type MMP; MT1-MMP) is known to cleave the TGF-β type

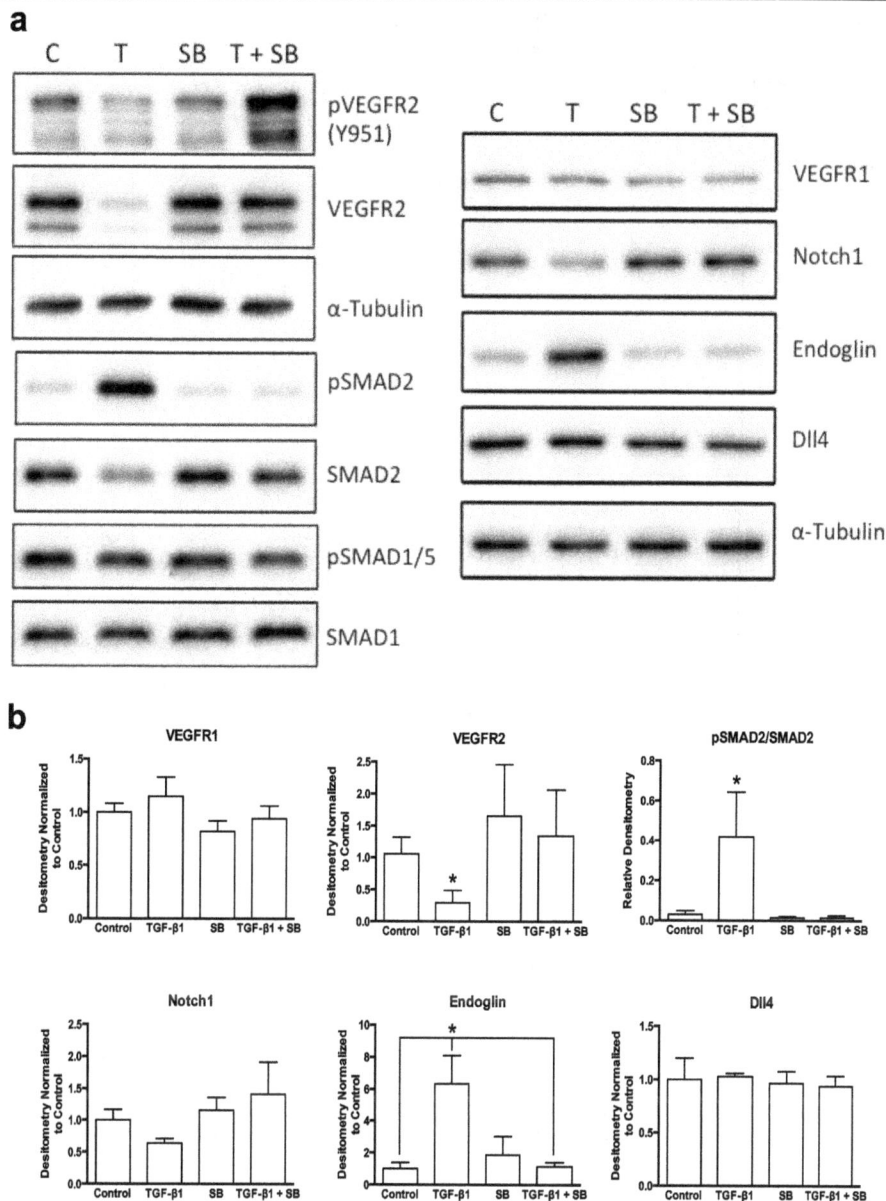

**Fig. 4** TGF-β1 induced reduction in receptor expression occurs via an ALK5 related pathway. **a)** Representative western blots and densitometry (**b**) from BAEC exposed to control conditions (C; DMSO vehicle), 5 μM SB-431542 (an ALK5 inhibitor) in DMSO (SB), 5 ng/ml exogenous TGF-β1 (T) or 5.0 ng/ml TGF-β1 plus 5 μM SB-431542 (T + SB) for 24 h. TGF-β1 leads reduced VEGFR2 expression and activation, and significantly upregulated endoglin in an ALK5/SMAD2-dependent fashion (as revealed by pSMAD2). There were no significant changes in SMAD1/5-dependent signaling, or in VEGFR1, Notch1 or Dll4 expression. *$p \leq 0.05$; $N = 3$; Kruskal-Wallis test

III receptor endoglin from endothelial cells, generating a soluble form [26]. In our study, levels of MMP-14 were weakly detectable by western blotting, and qPCR revealed $Cq$ values > 35 (data not shown). This, combined with the lack of detectable soluble endoglin in BAEC conditioned medium, suggests that MMP-14 cleavage of surface protein is not a significant determinant of CM-associated VEGFR2 in our system. A disintegrin and metalloprotease (ADAM) family of enzymes act as 'sheddases', cleaving surface bound proteins to general soluble

isoforms. ADAM-17 is known to shed VEGFR2 from endothelial [27] and non-endothelial [28] cell surfaces, and VEGFR2 is a target of ADAM-10 mediated cleavage in endothelial cells [29]. Both ADAM-10 and ADAM-17 were produced by BAECs, and there was weak induction of ADAM-10 by TGF-β1 in our cells. However, such sheddase activity would generate CM-associated VEGFR2 molecules of ~130 kDa [27], but we only detected full length (~250 kDa) VEGFR2 in conditioned medium, consistent with an alternative mechanism for

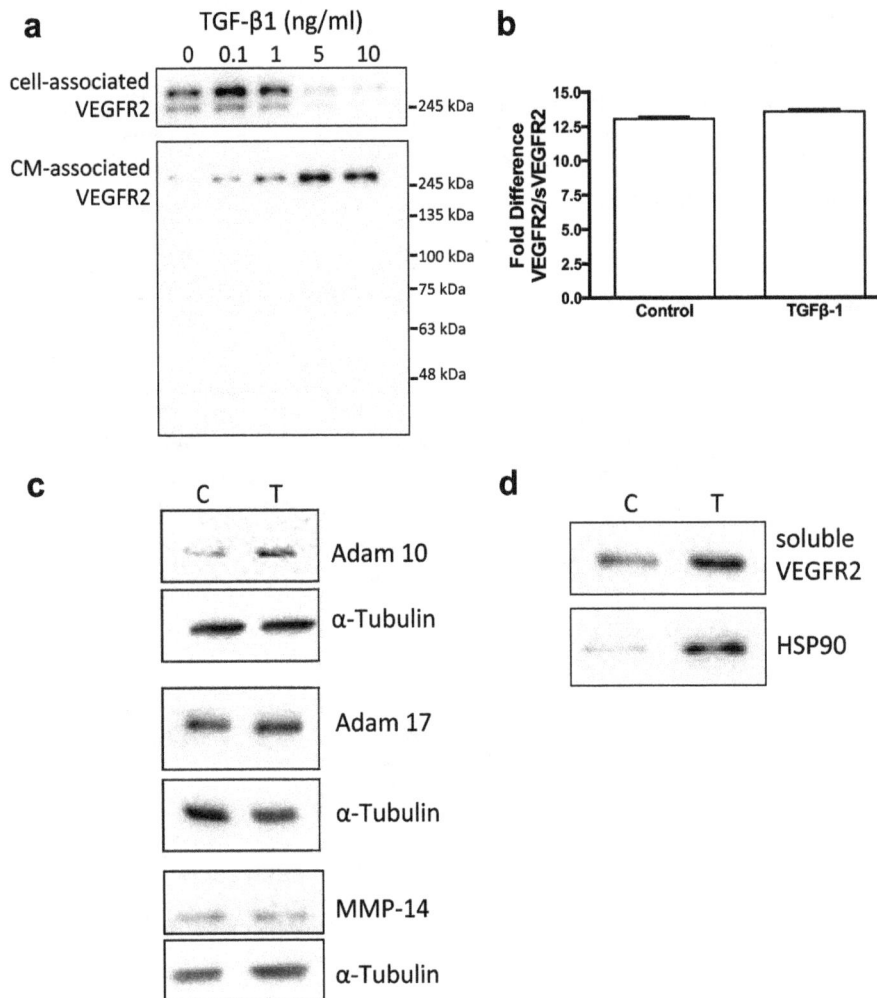

**Fig. 5** TGF-β1 reduces VEGFR2 expression via multiple mechanisms. **a** Western blot showing TGF-β1 induced loss of cell associated VEGFR2 protein and concomitant increase in full-length VEGFR2 detected in conditioned medium. **b** Quantitative PCR analysis showed that there was no change in the relative ratio of full length VEGFR2 mRNA to alternative spliced sVEGFR2 mRNA upon TGF-β1 treatment. **c** TGF-β1 treatment induced increased expression of ADAM family sheddase enzyme ADAM 10. TGF-β1 treatment had no effect on the expression of ADAM 17 or the membrane-associated metalloproteinase MMP-14. **d** Ultracentrifugation of endothelial conditioned medium demonstrated that soluble VEGFR2 is associated with the extracellular vesicle/exosome marker HSP90 in TGF-β1 treated cells

release of this receptor from endothelial cells. Recent reports suggest that prolonged surface residence of VEGFR2 in HUVECs leads to protease-mediated cleavage and the generation of a soluble fragment of ~100 kDa and a residual cell associated 130 kDa fragment [30]. As well, enhanced ubiquitination of VEGFR2 upon internalization in HUVECs leads to endosome-lysosome pathway mediated fragmentation into 160 and 120 kDa fragments [31, 32]. The absence of such VEGFR2 fragments in our samples suggests that TGF-β1 is not modulating VEGFR2 levels via these mechanisms in these BAECs.

There is growing evidence that VEGFR2 signals cells in an 'autocrine' or 'intracrine' fashion, leading to increased receptor activity within the cytosolic (endosome) and/or nuclear compartments [33–36]. Endothelial cells can also release exosomes containing sequestered molecules that then modulate neighbouring cells [37–39]. Endothelial cells expressing high levels of endoglin also upregulate autophagy [40]. Autophagy may enhance extracellular vesicle/exosome release via the mutivesicular body pathway, especially in serum starved endothelial cells [41, 42]. These pathways may also be relevant for tip/stalk cell specification during sprouting angiogenesis, as tips cells display enhanced VEGFR2 turnover via endocytosis and transfer from early endosomes to multivesicular bodies [42, 43]. Further, Dll4 containing exosomes can be shed from endothelial cells, and Dll4 containing exosomes modulate tip cell phenotype during sprouting angiogenesis [39, 44].

In agreement with the aforementioned findings, our results extend this activity to include the possibility that shedding of VEGFR2 containing extracellular vesicles (possibly exosomes) by vascular endothelium may be enhanced upon exposure to TGF-β1. We detected full length VEGFR2 in serum free conditioned medium along with the exosome marker HSP90. Retinal pigment epithelial cells also shed exosomes containing VEGFR2 which subsequently modulate endothelial cord formation in vitro [45], but to our knowledge this is the first report that VEGFR2 might be shed from endothelial cells themselves.

## Conclusions

Here we report that TGF-β1 alters levels of key angiogenic receptors, and interferes with tip/stalk cell identity when ALK5/Smad2 signaling pathways are activated. Our results suggest that downregulation of surface VEGFR2 and concomitant increase in VEGFR2 levels in endothelial cell conditioned medium may be a direct response to ALK5-mediated TGF-β signaling. Exosome shedding of VEGFR2 may rapidly limit the effects of angiogenic stimuli on the cells and stop further sprout formation, but the role of these events in physiological and pathological angiogenesis requires further investigation.

## Additional files

**Additional file 1: Figure S1.** Time course of restoration of VEGFR2 expression. A) Western blot of BAEC treated with serum free media containing 5 ng/ml TGF-β1 for 24 h, followed by recovery in serum free medium lacking TGF-β1. Samples were collected after 0, 24, 48 and 72 h recovery. Control culture (C) was treated with serum free medium lacking TGF-β1 for 24 h. B) Densitometry of western blots showing significant recovery of VEGFR2 expression by 48 h post TGF-β1 treatment. * p < 0.001; N = 2. (PDF 167 kb)

**Additional file 2: Figure S2.** ALK5 inhibitor SD-208 prevents TGF-β1 induced downregulation of VEGFR2 expression. Western blots showing impact of increases doses of SD-208 on BAEC treated with DMSO vehicle (0) or 5 ng/ml TGF-β1 for 24 h. (PDF 176 kb)

## Acknowledgements
We thank other members of the Coomber and Viloria-Petit laboratories for their helpful suggestions.

## Funding
This study was funded by the Nick Natale Innovation Grant of the Canadian Cancer Society (#2012-701069) to BLC and AVP. Infrastructure funds were provided by a Canadian Foundation for Innovation Grant #26472 to AVP. Personal support was provided to MJ by the Saudi Arabia Ministry of Higher Education. These funding bodies played no role in the design of the study, the collection, analysis, and interpretation of data or in writing the manuscript.

## Authors' contributions
MJ, EAK and JM performed cell culture experiments, western blotting and data analysis; BLC performed cord formation assays and data analysis; all authors contributed to the writing of the manuscript and approved its final content.

## Competing interests
The authors declare that they have no competing interests.

## References
1. Gordon KJ, Blobe GC. Role of transforming growth factor-beta superfamily signaling pathways in human disease. Biochim Biophys Acta. 2008;1782(4): 197–228.
2. Moustakas A, Heldin CH. The regulation of TGFbeta signal transduction. Development. 2009;136(22):3699–714.
3. Kirkbride KC, Townsend TA, Bruinsma MW, Barnett JV, Blobe GC. Bone morphogenetic proteins signal through the transforming growth factor-beta type III receptor. J Biol Chem. 2008;283(12):7628–37.
4. Ikushima H, Miyazono K. TGFbeta signalling: a complex web in cancer progression. Nat Rev Cancer. 2010;10(6):415–24.
5. Pepper MS, Vassalli JD, Orci L, Montesano R. Biphasic effect of transforming growth factor-beta 1 on in vitro angiogenesis. Exp Cell Res. 1993;204(2):356–63.
6. Holderfield MT, Hughes CC. Crosstalk between vascular endothelial growth factor, notch, and transforming growth factor-beta in vascular morphogenesis. Circ Res. 2008;102(6):637–52.
7. Mustafa DA, Dekker LJ, Stingl C, Kremer A, Stoop M, Sillevis Smitt PA, Kros JM, Luider TM. A proteome comparison between physiological angiogenesis and angiogenesis in glioblastoma. Mol Cell Proteomics. 2012; 11(6):M111.008466.
8. Ferrari G, Pintucci G, Seghezzi G, Hyman K, Galloway AC, Mignatti P. VEGF, a prosurvival factor, acts in concert with TGF-beta1 to induce endothelial cell apoptosis. Proc Natl Acad Sci U S A. 2006;103(46):17260–5.
9. Ferrari G, Terushkin V, Wolff MJ, Zhang X, Valacca C, Poggio P, Pintucci G, Mignatti P. TGF-beta1 Induces Endothelial Cell Apoptosis by Shifting VEGF Activation of p38MAPK from the Prosurvival p38beta to Proapoptotic p38alpha. Mol Cancer Res. 2012;10(5):605–14.
10. Kumar S, Pan CC, Bloodworth JC, Nixon AB, Theuer C, Hoyt DG, Lee NY. Antibody-directed coupling of endoglin and MMP-14 is a key mechanism for endoglin shedding and deregulation of TGF-beta signaling. Oncogene. 2014;33(30):3970–9.
11. Goumans MJ, Valdimarsdottir G, Itoh S, Rosendahl A, Sideras P, ten Dijke P. Balancing the activation state of the endothelium via two distinct TGF-beta type I receptors. EMBO J. 2002;21(7):1743–53.
12. Blanco FJ, Santibanez JF, Guerrero-Esteo M, Langa C, Vary CP, Bernabeu C. Interaction and functional interplay between endoglin and ALK-1, two components of the endothelial transforming growth factor-beta receptor complex. J Cell Physiol. 2005;204(2):574–84.
13. Gerhardt H, Golding M, Fruttiger M, Ruhrberg C, Lundkvist A, Abramsson A, Jeltsch M, Mitchell C, Alitalo K, Shima D, et al. VEGF guides angiogenic sprouting utilizing endothelial tip cell filopodia. J Cell Biol. 2003;161(6):1163–77.
14. Eilken HM, Adams RH. Dynamics of endothelial cell behavior in sprouting angiogenesis. Curr Opin Cell Biol. 2010;22(5):617–25.
15. Siekmann AF, Lawson ND. Notch signalling limits angiogenic cell behaviour in developing zebrafish arteries. Nature. 2007;445(7129):781–4.
16. Claxton S, Fruttiger M. Periodic Delta-like 4 expression in developing retinal arteries. Gene Expr Patterns. 2004;5(1):123–7.
17. Coomber BL. Suramin inhibits C6 glioma-induced angiogenesis in vitro. J Cell Biochem. 1995;58(2):199–207.
18. Laping NJ, Grygielko E, Mathur A, Butter S, Bomberger J, Tweed C, Martin W, Fornwald J, Lehr R, Harling J, et al. Inhibition of transforming growth factor (TGF)-beta1-induced extracellular matrix with a novel inhibitor of the TGF-beta type I receptor kinase activity: SB-431542. Mol Pharmacol. 2002;62(1):58–64.
19. Kuczynski EA, Viloria-Petit AM, Coomber BL. Colorectal carcinoma cell production of transforming growth factor beta decreases expression of endothelial cell vascular endothelial growth factor receptor 2. Cancer. 2011; 117(24):5601–11.
20. Takahashi T, Shibuya M. The 230 kDa mature form of KDR/Flk-1 (VEGF receptor-2) activates the PLC-gamma pathway and partially induces mitotic signals in NIH3T3 fibroblasts. Oncogene. 1997;14(17):2079–89.
21. Benedito R, Hellstrom M. Notch as a hub for signaling in angiogenesis. Exp Cell Res. 2013;319(9):1281–8.
22. Dejana E, Lampugnani MG. Differential adhesion drives angiogenesis. Nat Cell Biol. 2014;16(4):305–6.
23. Lebrin F, Goumans MJ, Jonker L, Carvalho RL, Valdimarsdottir G, Thorikay M, Mummery C, Arthur HM, ten Dijke P. Endoglin promotes endothelial cell proliferation and TGF-beta/ALK1 signal transduction. EMBO J. 2004;23(20):4018–28.

24. Albuquerque RJ, Hayashi T, Cho WG, Kleinman ME, Dridi S, Takeda A, Baffi JZ, Yamada K, Kaneko H, Green MG, et al. Alternatively spliced vascular endothelial growth factor receptor-2 is an essential endogenous inhibitor of lymphatic vessel growth. Nat Med. 2009;15(9):1023–30.

25. Munaut C, Lorquet S, Pequeux C, Coulon C, Le Goarant J, Chantraine F, Noel A, Goffin F, Tsatsaris V, Subtil D, et al. Differential expression of Vegfr-2 and its soluble form in preeclampsia. PLoS One. 2012;7(3):e33475.

26. Hawinkels LJ, Kuiper P, Wiercinska E, Verspaget HW, Liu Z, Pardali E, Sier CF, ten Dijke P. Matrix metalloproteinase-14 (MT1-MMP)-mediated endoglin shedding inhibits tumor angiogenesis. Cancer Res. 2010;70(10):4141–50.

27. Jin Y, Liu Y, Lin Q, Li J, Druso JE, Antonyak MA, Meininger CJ, Zhang SL, Dostal DE, Guan JL, et al. Deletion of Cdc42 enhances ADAM17-mediated vascular endothelial growth factor receptor 2 shedding and impairs vascular endothelial cell survival and vasculogenesis. Mol Cell Biol. 2013;33(21):4181–97.

28. Swendeman S, Mendelson K, Weskamp G, Horiuchi K, Deutsch U, Scherle P, Hooper A, Rafii S, Blobel CP. VEGF-A stimulates ADAM17-dependent shedding of VEGFR2 and crosstalk between VEGFR2 and ERK signaling. Circ Res. 2008;103(9):916–8.

29. Donners MM, Wolfs IM, Olieslagers S, Mohammadi-Motahhari Z, Tchaikovski V, Heeneman S, van Buul JD, Caolo V, Molin DG, Post MJ, et al. A disintegrin and metalloprotease 10 is a novel mediator of vascular endothelial growth factor-induced endothelial cell function in angiogenesis and is associated with atherosclerosis. Arterioscler Thromb Vasc Biol. 2010;30(11):2188–95.

30. Basagiannis D, Christoforidis S. Constitutive endocytosis of VEGFR2 protects the receptor against shedding. J Biol Chem. 2016;291(32):16892–903.

31. Bruns AF, Herbert SP, Odell AF, Jopling HM, Hooper NM, Zachary IC, Walker JH, Ponnambalam S. Ligand-stimulated VEGFR2 signaling is regulated by co-ordinated trafficking and proteolysis. Traffic. 2010;11(1):161–74.

32. Smith GA, Fearnley GW, Abdul-Zani I, Wheatcroft SB, Tomlinson DC, Harrison MA, Ponnambalam S. VEGFR2 trafficking, signaling and proteolysis is regulated by the ubiquitin isopeptidase USP8. Traffic. 2016;17(1):53–65.

33. Lee TH, Seng S, Sekine M, Hinton C, Fu Y, Avraham HK, Avraham S. Vascular endothelial growth factor mediates intracrine survival in human breast carcinoma cells through internally expressed VEGFR1/FLT1. PLoS Med. 2007;4(6):e186.

34. Domingues I, Rino J, Demmers JA, de Lanerolle P, Santos SC. VEGFR2 translocates to the nucleus to regulate its own transcription. PLoS One. 2011;6(9):e25668.

35. Adamcic U, Skowronski K, Peters C, Morrison J, Coomber BL. The effect of bevacizumab on human malignant melanoma cells with functional VEGF/VEGFR2 autocrine and intracrine signaling loops. Neoplasia. 2012;14(7):612–23.

36. Domigan CK, Ziyad S, Iruela-Arispe ML. Canonical and noncanonical vascular endothelial growth factor pathways: new developments in biology and signal transduction. Arterioscler Thromb Vasc Biol. 2015;35(1):30–9.

37. Haqqani AS, Delaney CE, Tremblay TL, Sodja C, Sandhu JK, Stanimirovic DB. Method for isolation and molecular characterization of extracellular microvesicles released from brain endothelial cells. Fluids Barriers CNS. 2013;10(1):4.

38. Raposo G, Stoorvogel W. Extracellular vesicles: exosomes, microvesicles, and friends. J Cell Biol. 2013;200(4):373–83.

39. Sharghi-Namini S, Tan E, Ong LL, Ge R, Asada HH. Dll4-containing exosomes induce capillary sprout retraction in a 3D microenvironment. Sci Rep. 2014;4:4031.

40. Pan CC, Kumar S, Shah N, Bloodworth JC, Hawinkels LJ, Mythreye K, Hoyt DG, Lee NY. Endoglin regulation of Smad2 function mediates Beclin1 expression and endothelial autophagy. J Biol Chem. 2015;290(24):14884–92.

41. Pallet N, Sirois I, Bell C, Hanafi LA, Hamelin K, Dieude M, Rondeau C, Thibault P, Desjardins M, Hebert MJ. A comprehensive characterization of membrane vesicles released by autophagic human endothelial cells. Proteomics. 2013;13(7):1108–20.

42. Maes H, Olmeda D, Soengas MS, Agostinis P. Vesicular trafficking mechanisms in endothelial cells as modulators of the tumor vasculature and targets of antiangiogenic therapies. FEBS J. 2016;283(1):25–38.

43. Pitulescu ME, Adams RH. Regulation of signaling interactions and receptor endocytosis in growing blood vessels. Cell Adh Migr. 2014;8(4):366–77.

44. Sheldon H, Heikamp E, Turley H, Dragovic R, Thomas P, Oon CE, Leek R, Edelmann M, Kessler B, Sainson RC, et al. New mechanism for Notch signaling to endothelium at a distance by Delta-like 4 incorporation into exosomes. Blood. 2010;116(13):2385–94.

45. Atienzar-Aroca S, Flores-Bellver M, Serrano-Heras G, Martinez-Gil N, Barcia JM, Aparicio S, Perez-Cremades D, Garcia-Verdugo JM, Diaz-Llopis M, Romero FJ, et al. Oxidative stress in retinal pigment epithelium cells increases exosome secretion and promotes angiogenesis in endothelial cells. J Cell Mol Med. 2016;20:1457–66.

# 7

# Variation in human dental pulp stem cell ageing profiles reflect contrasting proliferative and regenerative capabilities

Amr Alraies[1,2], Nadia Y. A. Alaidaroos[2,3], Rachel J. Waddington[1,2], Ryan Moseley[2,3] and Alastair J. Sloan[1,2*]

## Abstract

**Background:** Dental pulp stem cells (DPSCs) are increasingly being recognized as a viable cell source for regenerative medicine. Although significant variations in their ex vivo expansion are well-established, DPSC proliferative heterogeneity remains poorly understood, despite such characteristics influencing their regenerative and therapeutic potential. This study assessed clonal human DPSC regenerative potential and the impact of cellular senescence on these responses, to better understand DPSC functional behaviour.

**Results:** All DPSCs were negative for hTERT. Whilst one DPSC population reached >80 PDs before senescence, other populations only achieved <40 PDs, correlating with DPSCs with high proliferative capacities possessing longer telomeres (18.9 kb) than less proliferative populations (5–13 kb). High proliferative capacity DPSCs exhibited prolonged stem cell marker expression, but lacked CD271. Early-onset senescence, stem cell marker loss and positive CD271 expression in DPSCs with low proliferative capacities were associated with impaired osteogenic and chondrogenic differentiation, favouring adipogenesis. DPSCs with high proliferative capacities only demonstrated impaired differentiation following prolonged expansion (>60 PDs).

**Conclusions:** This study has identified that proliferative and regenerative heterogeneity is related to contrasting telomere lengths and CD271 expression between DPSC populations. These characteristics may ultimately be used to selectively screen and isolate high proliferative capacity/multi-potent DPSCs for regenerative medicine exploitation.

**Keywords:** Dental pulp, Stem cells, Cumulative population doublings, Telomeres, Cellular senescence, Differentiation, Multi-potency, CD271

## Background

Dental pulp stem cells (DPSCs) are increasingly becoming recognized as a viable cell source for the development of effective cell-based therapies. This is due to their accessibility, multi-lineage differentiation capabilities towards osteogenic, chondrogenic, myogenic and neurogenic lineages; and similar regenerative properties to bone marrow-derived cells [1–4]. DPSCs exhibit a fibroblast-like morphology, plastic adherence, express mesenchymal stem cell (MSC) markers (CD73, CD90 and CD105); and thus satisfy the minimal criteria for MSCs [1, 3, 5, 6]. However, similar to bone marrow stem cells, DPSCs isolated from pulpal tissues are recognised to represent a heterogeneous population, with individual isolated clones demonstrating differences in proliferative rates and their abilities to differentiate down particular lineages [1, 5, 7]. Indeed, despite heterogeneous DPSC population expansion being capable of achieving >120 cumulative population doublings (PDs) in vitro, only 20% of purified DPSCs are capable of proliferating beyond >20 PDs. Of these, only two-thirds were able to generate abundant ectopic dentine in vivo, implying that subset DPSC populations differ in their regenerative potential [5, 7]. In vitro, heterogeneous DPSCs

* Correspondence: Sloanaj@cardiff.ac.uk
[1]Mineralised Tissue Group, Oral and Biomedical Sciences, School of Dentistry, College of Biomedical and Life Sciences, Cardiff University, Heath Park, CF14 4XY, Cardiff, UK
[2]Cardiff Institute Tissue Engineering and Repair (CITER), Cardiff University, Cardiff CF14 4XY, UK
Full list of author information is available at the end of the article

can differentiate into osteoblasts, chondrocytes, adipocytes, neurocytes and myocytes, but it has been reported that there are occasions when DPSCs fail to differentiate into adipocytes, chondrocytes and myoblasts; suggested to be a consequence of the potential stem cell niches within dental pulp tissue [1].

Adult stem cells are proposed to exist in a hierarchical arrangement. Pivotal to this model is the mother stem cell, which divides slowly and asymmetrically to yield a replacement mother cell and rapidly dividing transit amplifying (TA) cells [8]. It has been proposed that as TA cells continue to divide, their proliferative capacity is reduced and they become more lineage-restricted. In contrast, newly formed TA cells possess a greater proliferative and multi-differentiation capacity. The presence of TA cells has been suggested to rise within the postnatal dental pulp, which are the first to differentiate into new odontoblast-like cells following cavity-induced injury [9]. Whilst this would indicate a strong role for TA cells in tissue repair and regeneration, the nature, origins or the relationship of DPSC populations with contrasting proliferative capacities to this hierarchical arrangement, have yet to be elucidated.

Another important requirement for the tissue engineering exploitation of stem cells is the considerable in vitro cell expansion required before sufficient cell numbers are obtained for therapeutic use. However, a significant limitation of stem cell therapy is that extensive in vitro cell expansion eventually leads to proliferative decline and cellular senescence, accompanied by altered cellular behaviour and impaired regenerative potential [10]. This feature has been particularly reported for the in vitro expansion of MSCs from human bone marrow, where no more than 4–7 PDs is recommended in preparations for therapeutic use [11]. For most cell types, in vitro expansion and subsequent cellular senescence is a consequence of replicative (telomere-dependent) senescence, characterised by progressive telomere shortening and the loss of telomeric TTAGGG repeats, due to repeated cell divisions [12]. Cellular senescence may also occur through DNA damage by p53, ionizing radiation or oxidative stress (premature or telomere-independent senescence). Either mechanism is associated with the activation of various signalling pathways, including those involving the tumour suppressor genes, p53 and retinoblastoma protein (pRb), via the cyclin-dependant kinase inhibitors, $p21^{waf1}$ and $p16^{INK4a}$, respectively [12]. However, foetal cells, germ lines, stem cells and many tumour cells are established to contain the human telomerase catalytic subunit (hTERT); a reverse transcriptase capable of the complete replication of telomere ends, which plays a major role in counteracting erosion, maintaining telomeric integrity and proliferative lifespan in these cells [13].

Although significant differences in the ex vivo expansion capabilities of individual DPSC populations has been recognized for some time, few studies have addressed the reasons behind these differences and the subsequent impact of such variations on differentiation and regenerative potential. Consequently, this study aimed to examine whether inherent differences in telomere lengths and relative susceptibilities to cellular senescence between individual DPSC populations, contributed to the significant variations in proliferative and differentiation capabilities identified for individual DPSC populations. To achieve this, proliferative lifespans, stem cell marker expression, multi-potent differentiation capabilities and senescence-related marker development, were characterized for individual DPSC populations expanded from individual human teeth derived from young adults with a similar donor age range.

## Methods
### DPSC isolation, culture and population doubling levels
Human third molar teeth were collected from three young adult patients (Patients A-C, all female, age range 18–30 years), undergoing orthodontic extractions at the School of Dentistry, Cardiff University, UK. Teeth were collected with informed patient consent and ethical approval by the South East Wales Research Ethics Committee of the National Research Ethics Service (NRES), UK. The outer surface of the teeth were sterilised with 70% ethanol. Teeth were grooved in the mesial, distal and occlusal regions with a rotary bone saw; and halved. Pulps were removed, minced on a glass slide; and the cells were released from the pulpal tissues by digestion with 4 µg/µl collagenase/dispase (1 ml, Roche, Welwyn Garden City, UK), for 45 min at 37 °C. Pulpal tissues from individual patients were prepared separately. Digests were passed through a 70 µm mesh cell strainer and the eluents collected to obtain single cell suspensions. Cells were centrifuged and further resuspended in 1 ml culture medium; α-Minimum Essential Medium (αMEM) containing ribonucleosides, deoxyribonucleosides, 4 mM L-glutamine, 100 U/ml penicillin G sodium, 0.1 µg/ml streptomycin sulphate, 0.25 µg/ml amphotericin; and 20% foetal calf serum (all Invitrogen, Paisley, UK), in addition to 100 µM L-ascorbate 2-phosphate (Sigma-Aldrich, Poole, UK).

The present study utilised a fibronectin adhesion assay to preferentially select and isolate immature DPSCs, based on their expression of high $\beta_1$-integrin levels on stem cell surfaces [14]. This fibronectin adhesion isolation method has previously been used successfully for the isolation of tissue-specific, immature stem / progenitor cells from dental pulp [15, 16], oral mucosa [17], cartilage [18]; and bone marrow [16]. Briefly, viable cell counts were calculated and cells seeded onto fibronectin-coated, 6-well plates, at 4000 cells/cm$^2$.

Plates were subsequently maintained at 37 °C / 5% $CO_2$ for 20 min, in accordance with our previously published protocol [15, 16]. Following adhesion, the culture medium and non-adherent cells were removed and replaced with 2 ml αMEM medium. Medium was replenished every 2 days.

Colonies (>32 cells) were allowed to form over 12 days in culture, with individual colonies isolated within cloning rings using 100 μl pre-warmed, accutase (PAA, Velizy-Villacoublay, France); and seeded into 1 well of a 96-well plate. Isolated colonies were subsequently expanded until the numbers were sufficient for seeding into T-75 flasks. Of the DPSC colonies established from Patients A-C, six populations (A1, A2 and A3; B1; C2 and C3) were expanded, with PDs for each DPSC population calculated from cell counts throughout their proliferative lifespans in culture until reaching senescence, as previously described [19]. DPSC senescence was identified when PDs were reduced to <0.5 PDs / week and later confirmed by the increased presence of other senescence-related markers, as described below.

### Telomere length determination

DPSCs at various PDs throughout their proliferative lifespans (A1, B1 and C2 at 8 PDs, A2 at 12 PDs, A3 at 9 PDs and 55 PDs, C3 at 16 PDs), were seeded in T-75 flasks at 5000 cells / cm$^2$ and grown to confluence. Genomic DNA was isolated from each DPSC population using QIAmp® DNA Mini Kits (Qiagen, Crawley, UK), according to manufacturer's instructions; and DNA yields quantified at 260 nm (NanoVue, GE Healthcare, Amersham, UK). Telomere length assessments were performed using the TeloTAGGG Telomere Restriction Fragment Length (TRF) Assay Kit (Roche), according to manufacturer's instructions. Briefly, each DNA sample (1 μg) was digested with HinF1 / Rsa1 (both 40 U / μl, in Kit), at 37 °C for 2 h. DPSC samples, a digoxigenin (DIG)-labelled, Molecular Weight Marker and a positive DIG-labelled, Control DNA sample (both in Kit), were separated (20 mV, overnight) on 0.8% by agarose gels (Agarose MP, Geneflow, Lichfield, UK); in 1x Tris-acetate-EDTA buffer containing 0.2 μg/ml ethidium bromide (Sigma-Aldrich). Gels were subsequently treated with 0.25 M hydrochloric acid solution (Thermo Fisher Scientific) and rinsed (x2) with water, before treatment (x2) with Denaturation Solution (0.5 M sodium hydroxide, 1.5 M sodium chloride solution), rinsing (x2); and treatment with Neutralisation Solution (0.5 M Tris-HCl buffer, pH 7.5, containing 3 M sodium chloride).

Separated products were transferred onto positively charged nylon membranes (Roche) by Southern blotting; using 20× standard sodium citrate (SSC) buffer (3 M sodium chloride / 0.3 M sodium citrate solution, pH 7.0). Nylon membranes were fixed by UV-crosslinking for

2 × 10 s (Stratalinker, Agilent Technologies, Stockport, UK) and washed with SSC (x2), prior to hybridisation with DIG Easy Hyb (in Kit), for 1 h at 42 °C. This was replaced with Hybridisation Solution mixed with Telomere Probe (2 μl, in Kit); and incubated at 42 °C for 3 h. Membranes were consecutively washed (x2) with 25 ml Stringent Wash Buffers, before incubations in Blocking Solution, Anti-DIG-AP Working Solution, Washing Buffer (x2); and Detection Buffer, respectively (all in Kit). Finally, membranes were incubated for 5 min in Chemiluminescent Substrate Solution (in Kit), before being exposed (30 s) to X-ray film (Hyperfilm°, GE Healthcare) and developed.

Average DPSC telomere lengths at each PD were calculated from the Southern blot images, using ImageJ° Software (http://rsb.info.nih.gov/ij/), to quantify the densitometic intensity profile of the entire length of the separated products within each particular lane. In line with TRF Assay Kit instructions, each lane was overlaid along its entire length with a grid consisting of 30 equal squares, where telomere-specific signal was detectable. The background intensity of each lane was calculated by selecting squares within each lane containing no telomere-specific signal and by subtracting these averaged values from the squares containing telomeric signal. For each square containing detectable telomeric signal, the total signal density within each square ($OD_i$) and the corresponding molecular weight at the midpoint of each square ($L_i$) (based on the DIG-labelled, Molecular Weight Marker bands, kb), were calculated. Average telomere lengths were subsequently calculated, using the following formula, where $OD_i$ is the chemiluminescent signal and $L_i$ is the length of the telomeres at position i:-

$$\frac{\Sigma(OD_i)}{\Sigma(OD_i / L_i)}$$

### Stem cell and senescence-related marker expression

DPSCs were examined by RT-PCR for the expression of purported stem / progenitor cell markers (CD73, CD90, CD105, CD271) and the absence of the hematopoietic stem cell marker (CD45); at various PDs throughout their proliferative lifespans (1–10 PDs, 11–23 PDs, 24–60 PDs and where applicable, >60 PDs). In addition, DPSCs at early PDs were assessed for the expression of hTERT, whilst expression of the senescence-associated marker genes, p53, p21$^{waf1}$ and p16$^{INK4a}$, were analysed for all DPSCs at PDs throughout their proliferative lifespans. DPSCs were cultured in 6-well plates until ~90% confluence. Total RNA was isolated for DPSCs using an RNeasy Mini Kit (Qiagen), according to the manufacturer's instructions. RNA was quantified by measurement of 260:280 nm ratio

(NanoVue). cDNA was synthesised from 1 µg total RNA, using 5x Moloney murine leukaemia virus (M-MLV) buffer, 0.5 µg Random Primers, 0.625 µl RNasin, 1.25 µl dNTPs (10 mM) and 1 µl M-MLV reverse transcriptase, reconstituted to 25 µl with DNase-free water (all Promega, Southampton, UK, total volume 15 µl). All reactions were performed on a G-storm™ GS1 Thermal Cycler (Genetic Research Instrumentation, Braintree, UK), at 37 °C for 1 h, followed by 95 °C for 5 min.

PCR reactions were established including cDNA (1 µl), 5x Green GoTaq™ Flexi Buffer (5 µl), 25 mM Magnesium Chloride Solution (1 µl), 10 mM PCR Nucleotide Mix (0.5 µl), 5 U / µl GoTaq™ DNA Polymerase (0.25 µl, all Promega), 0.04 µg/µl forward primer and 0.04 µg/µl reverse primer (both 1.25 µl). PCR were performed using the primer sequences and cycling conditions described in Table 1, with β-actin serving as the reference housekeeping gene in all cases. Primer replacement with DNase-free water served as the negative controls. All reaction volumes were made up to 25 µl with DNase-free water. Reactions were run on a G-storm™ GS1 Thermal Cycler, with an initial denaturing step of 95 °C (5 min), followed by 35–40 cycles at 95 °C (1 min), 1 cycle at 55–62 °C (1 min), 1 cycle at 72 °C (1 min); and 1 cycle at 72 °C (5 min). PCR products and a 100 bp DNA ladder (Promega) were separated on 2% agarose gels (125 mV, 45 min), in 1x Tris-acetate-EDTA buffer (above). Gel images were captured under UV light and analysed using a Gel Doc 3000 Scanner and Image Analysis Software (Bio-Rad, Hemel Hempstead, UK).

## DPSC differentiation capabilities

Based on the distinct proliferative differences identified above between A3 and the other DPSC populations assessed, further studies compared the respective osteogenic chondrogenic and adipogenic differentiation capabilities of A3 (at 30 PDs, 62 PDs, and 75 PDs), versus low proliferative capacity DPSC populations, A1 (17 PDs) and B1 (25 PDs). Differentiation was assessed using commercially available media (OsteoDiff, ChondroDiff and AdipoDiff, Miltenyi Biotec, Bisley, UK), according to manufacturer's protocols and following supplementation with antibiotics, as above.

For the analysis of osteogenic capabilities, DPSCs were seeded at $4.5x10^3$ cells / $cm^2$ in 6- or 12-well plates; and maintained in OsteoDiff medium at 37 °C / 5% $CO_2$ for 18 days. Medium was replenished every 3 days. At day 18, cells in 6-well plates were used for the RT-PCR

**Table 1** Details of the primer sequences and cycling conditions used for RT-PCR

| Gene marker | Primer sequence | Annealing temp. (°C) | Cycles | NCBI reference sequence |
|---|---|---|---|---|
| CD73 | F:5'GTCGCGAACTTGCGCCTGGCCGCCAAG-3' R: 5'-TGCAGCGGCTGGCGTTGACGCACTTGC-3' | 65 | 35 | NM_001204813.1 |
| CD90 | F:5'- ATGAACCTGGCCATCAGCATCG-3' R:5'- CACGAGGTGTTCTGAGCCAGCA-3' | 55 | 35 | NM_006288.3 |
| CD105 | F:5'-GAAACAGTCCATTGTGACCTTCAG-3' R: 5'-GATGGCAGCTCTGTGGTGTTGACC-3' | 65 | 35 | NM_001114753.2 |
| CD271 | F:5'- CTGCAAGCAGAACAAGCAAG-3' R:5'- GGCCTCATGGGTAAAGGAGT-3' | 55 | 35 | NM_002507.3 |
| CD45 | F:5'-GTGACCCCTTACCTACTCACACCACTG-3' R:5'-TAAGGTAGGCATCTGAGGTGTTCGCTG-3' | 65 | 35 | NM_002838.4 |
| hTERT | F:5'- CGGAAGAGTGTCTGGAGCAA-3' R:5'- GGATGAAGCGGAGTCTGGA-3' | 55 | 40 | NM_198253.2 |
| p53 | F:5'- AGACCGGCGCACAGAGGAAG-3' R:5'- CTTTTTGGACTTCAGGTGGC-3' | 55 | 35 | NM_001126118.1 |
| p21waf1 | F:5'- GGATGTCCGTCAGAACCCAT-3' R:5'- CCCTCCAGTGGTGTCTCGGTG-3' | 60 | 35 | NM_001291549.1 |
| p16INK4A | F:5'- CTTCCTGGACACGCTGGT-3' R:5'- GCATGGTTACTGCCTCTGGT-3' | 55 | 35 | NM_001195132.1 |
| OCN | F:5'-GCAGGTGCGAAGCCCAGCGGTGCAGAG-3' R: 5'-GGGCTGGGAGGTCAGGGCAAGGGCAAG-3' | 62 | 35 | NM_199173.4 |
| OPN | F:5'- ATCACCTGTGCCATACCA-3' R:5'- CATCTTCATCATCCATATCATCCA-3' | 55 | 35 | NM_001251830.1 |
| PPARγ | F:5'-GCCATCAGGTTTGGGCGGATGCCACAG-3' R: 5'-CCTGCACAGCCTCCACGGAGCGAAACT-3' | 62 | 35 | NM_138711.3 |
| LPL | F:5'-GCTGGCATTGCAGGAAGTCTGACCAATAA-3' R:5'-GGCCACGGTGCCATACAGAGAAATCTCAA-3' | 55 | 35 | NM_000237.2 |
| β-actin | F:5'-AGGGCAGTGATCTCCTTCTGCATCCT-3' R:5'- CCACACTGTGCCCATCTACGAGGGGT-3' | 65 | 35 | NM_001101.3 |

analysis of Runx2, osteocalcin (OCN) and osteopontin (OPN) expression (Table 1), whilst the 12-well plates were washed (x2) with phosphate buffered saline (PBS) and fixed with 4% paraformaldehyde (Santa Cruz, Dallas, USA) for 10 min and washed with PBS (x2). Osteogenic cultures were assessed for mineral nodule deposition detection by staining with 2% alizarin red solution, pH 4.2 (500 μl / well, Sigma-Aldrich), for 3 min. Images representing the extent of alizarin red staining were captured by light microscopy (Eclipse TS100, Nikon UK, Kingston upon Thames, UK), using a digital camera (Canon PC1234, Uxbridge, UK).

Chondrogenic capacity was analysed using single cell suspensions containing $2.5 \times 10^5$ DPSCs in 1 ml medium, which were centrifuged to produce cell pellets. Medium was replaced with 1 ml of ChondroDiff medium, prior to re-centrifugation. Pellets were maintained in ChondroDiff medium at 37 °C/5% $CO_2$ for 24 days. Medium was replenished every 3 days. At day 24, pellets were washed (x1) with PBS, prior to fixation overnight in 3.7% neutral buffered formalin. Each pellet underwent automated histological processing (Leica ASP300 S, Leica Microsystem, Milton Keynes, UK), through a graded alcohol series and paraffin wax embedding. Sections (5 μm) were cut using a sliding microtome (Leica SM2400, Leica Microsystems), collected onto poly-L-lysine coated slides (SuperFrost, Thermo Fisher Scientific); and dried. Sections were deparaffinised and rehydrate with xylene (10 min), industrial methylated spirits (5 min) and water (5 min). Pellet sections were treated with fast green FCF solution (1:5000, Sigma-Aldrich) for 5 min and washed in 1% acetic acid (Thermo Fisher Scientific). Sections were assessed for proteoglycan content by 0.1% Safranin O staining (Sigma-Aldrich) for 10 min; and washed in 95% ethanol-xylene mixture (1:1). Images representing the extent of Safranin O staining were captured, as above.

For the analysis of adipogenic capabilities, DPSCs were seeded at $7.5 \times 10^3$ cells / $cm^2$ in 6- or 12-well plates; and maintained in AdipoDiff medium at 37 °C/5% $CO_2$ for 21 days. Medium was replenished every 3 days. At day 21, cells in 6-well plates were used for the RT-PCR analysis of adipogenic marker, lipoprotein lipase (LPL) and PPARγ expression (Table 1), whilst the 12-well plates were washed (x2) with PBS and assessed for intracellular lipid-rich vacuole accumulation by Oil Red O staining (500 μl/well, Sigma-Aldrich; 3.5 g/l in isopropanol, mixed 3:2 with double-distilled water), for 10 min. Following excess stain removal with double-distilled water and 60% isopropanol, images representing the extent of Oil Red O staining were captured, as above.

### DPSC morphology and size assessment
Digital images were captured, as above, to assess changes in cellular size and morphology throughout each DPSC's proliferative lifespan. Random cells (20 cells / image, 3 images / PD) were outlined along their peripheral borders and cellular surface areas calculated, using ImageJ® Software. Surface areas were expressed as an average ± standard error of the mean (SE) for each DPSC population and PD.

### Senescence-associated β-galactosidase staining
DPSCs at defined PDs throughout their proliferative lifespans were seeded into 6-well plates at 5000 cells/$cm^2$ and maintained at 37 °C / 5% $CO_2$ for 24 h. A Senescence Cell Histochemical Kit (Sigma-Aldrich) was used to detect senescence-associated β-galactosidase activity in these cells, according to manufacturer's instructions. Culture medium was aspirated and the cells washed (x2) with PBS (in Kit). DPSCs were treated with Fixation Buffer (1.5 ml / well, in Kit) for 7 min and washed (x3) with PBS (1 ml / well). Cells were further treated with Staining Solution (1 ml / well, in Kit) and incubated at 37 °C / 5% $CO_2$, overnight. Digital images representing the extent of senescence-associated β-galactosidase staining were captured by light microscopy, as above. Random cells (100 cells in total) were counted and the percentage number of positive β-galactosidase stained cells calculated.

### Statistical analysis
Statistical analysis on the differences in cellular surface area (average ± SE) between DPSC populations throughout their proliferative lifespans, was determined by one-way ANOVA with post-test Tukey multiple comparison analysis, using GraphPad InStat 3 (GraphPad Software, La Jolla, USA). Statistical significance were considered at $p < 0.05$.

### Results
#### DPSC isolation, expansion and population doubling levels
Following the isolation of immature DPSC populations using the fibronectin adherence selection methodology [15, 16], DPSCs were allowed to form colonies of >32 cells over a 12 day period. Of the DPSC colonies established, a selection of six populations (A1, A2 and A3; B1; C2 and C3), were subsequently expanded in vitro through to senescence, confirmed when cumulative PDs fell to <0.5 PDs/week. The cumulative PDs for each DPSC population are presented in Fig. 1a. Marked variations in proliferative capacity were evident between individual DPSC populations, with each demonstrating differences in PDs, irrespective of whether DPSCs were derived from the same or different patients. This was particularly apparent with A3, which demonstrated the highest PD levels (>80 PDs) over 280 days in culture, whilst other DPSC populations from the same patient (A1 and A2) and different patients (B1, C2 and C3) only achieved between 20 and 35 PDs over 35–85 days in

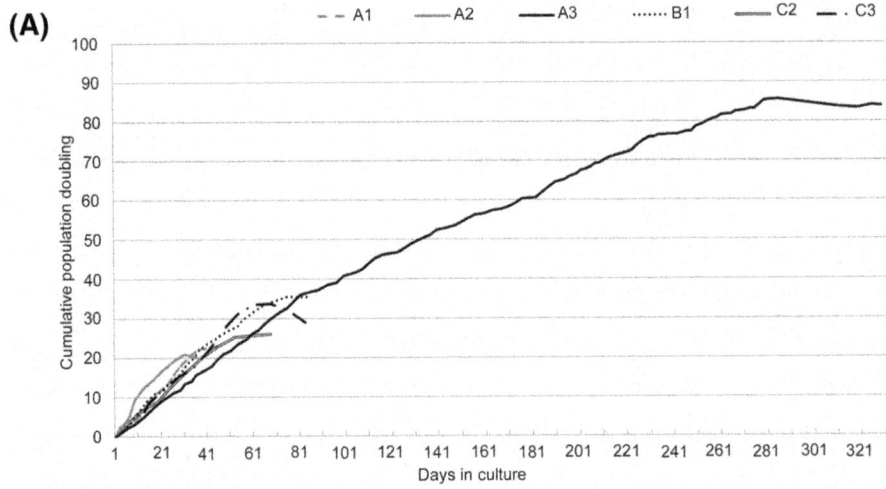

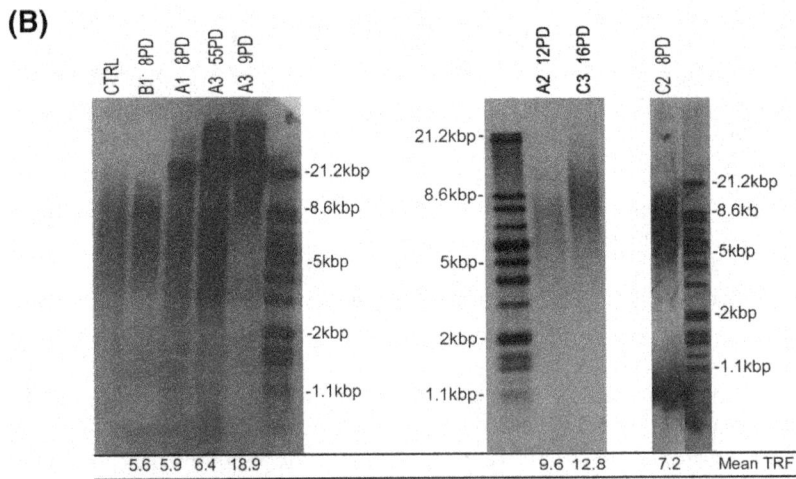

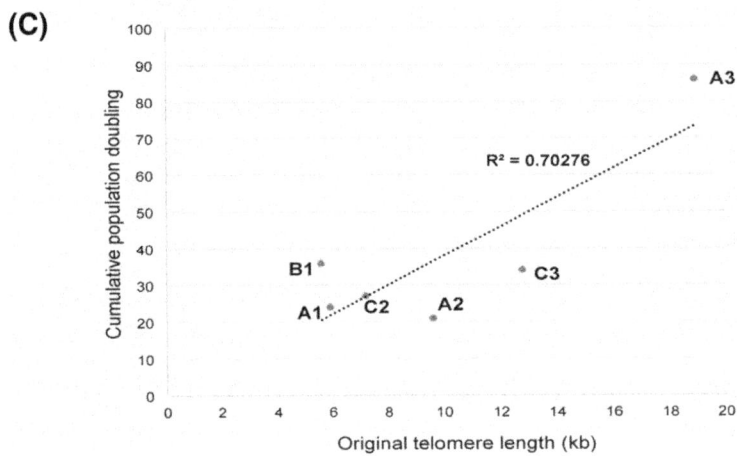

**Fig. 1** (See legend on next page.)

(See figure on previous page.)

**Fig. 1** Characterisation of DPSC populations derived from human dental pulp tissues. **a** Cumulative population doublings (PDs) were recorded for six DPSC populations during extended culture until senescence was reached, when PDs reached <0.5 PDs / week. DPSCs A1, A2 and A3 were derived from patient A, while B1, C2 and C3 were derived from patients B and C, respectively. Population A3 demonstrated high proliferative capacity, achieving >80 PDs over 280 days in culture, whilst the other DPSC populations only achieved 20–35 PDs over 35–85 days in culture. **b** Telomere length analysis for the six DPSC populations at cited PDs, by Southern blotting. Average telomere lengths were calculated using ImageJ® Software. Much longer telomeres were identified with high proliferative capacity, A3 (18.9 kb), compared with low proliferative capacity populations, A1 (5.9 kb), A2 (9.6 kb), B1 (5.6 kb), C2 (7.2 kb) and C3 (12.8 kb). A3 (18.9 kb, 9 PDs) only exhibited equivalent telomere lengths to low proliferative capacity DPSCs, at much later PDs towards the end of its proliferative lifespan (6.4 kb, 55 PDs). CTRL lanes represent the separate DIG-labelled, Control DNA sample. Average telomere lengths values obtained for each CTRL lane from left-right, were 10.6 ± 0.73 kb, 11.4 ± 0.71 kb and 11.1 ± 0.60 kb, respectively. Separated DIG-labelled telomere length standards are also included. **c** Correlation between cumulative PDs versus original telomere lengths for all DPSCs analysed

culture. In support of the contrasting PDs capabilities between these DPSC populations, telomere length analysis demonstrated much longer average telomere lengths with high proliferative capacity, A3 (18.9 kb), compared with DPSCs with low proliferative capacities at comparable PDs, A1 (5.9 kb), A2 (9.6 kb), B1 (5.6 kb), C2 (7.2 kb) and C3 (12.8 kb) (Fig. 1b and c). The consistency of analysis between separate Southern blots was confirmed by the average telomere lengths calculated for each positive Control DNA sample on each blot (i.e. 10.6 ± 0.73 kb, 11.4 ± 0.71 kb and 11.1 ± 0.60 kb, for the CTRL lanes from left-right in Fig. 1b, respectively).

### Stem cell and senescence-related marker expression
Each DPSC population was further investigated to compare stem/progenitor cell marker expression at early PDs (Fig. 2). All DPSCs demonstrated the expression of the MSC markers (CD73, CD90 and CD105) and the absence of the hematopoietic marker, CD45 [3]. Low proliferative capacity DPSCs (A1, A2, B1, C2 and C3) were also positive for the expression of nerve growth factor receptor p75 (CD271). All DPSCs analysed also demonstrated the expression of the senescence-associated marker genes, p21$^{waf1}$ and p16$^{INK4a}$, at early PDs; although p16$^{INK4a}$ expression appeared to be much less for the high proliferative capacity, A3, than with the low proliferative capacity DPSCs (A1, A2, B1, C2 and C3). Similarly, the other senescence-associated marker investigated, p53, was only expressed in the low proliferative capacity DPSCs (A1, A2, B1, C2 and C3); being undetectable in high proliferative capacity, A3. All DPSCs analysed were negative for hTERT expression.

### DPSC differentiation capabilities
Based on the contrasting PDs identified between A3 and the other DPSCs assessed, further studies evaluated whether these populations also possessed contrasting differentiation capabilities. Tri-lineage (osteogenic, chondrogenic and adipogenic) differentiation was compared between high proliferative capacity, A3 (30 PDs), versus two low proliferative capacity DPSCs (A1 at 11 PDs, B1 at 15 PDs). Only high proliferative capacity, A3,

demonstrated multi-potency. Osteogenic differentiation was demonstrated by positive staining with alizarin red and the up-regulation of OCN and OPN after 18 days in osteogenic medium (Fig. 3a). Chondrogenic differentiation was evident by positive Safranin O staining for high proteoglycan content, following 24 days in chondrogenic medium (Fig. 3b). However, mRNA extraction from cartilage pellets was difficult and thus, prevented gene expression analysis. A3 was also able to undergo adipogenic differentiation, evident by the presence of Oil Red O-positive, lipid vacuoles and the up-regulation of the late adipogenic marker, LPL, expression after 21 days

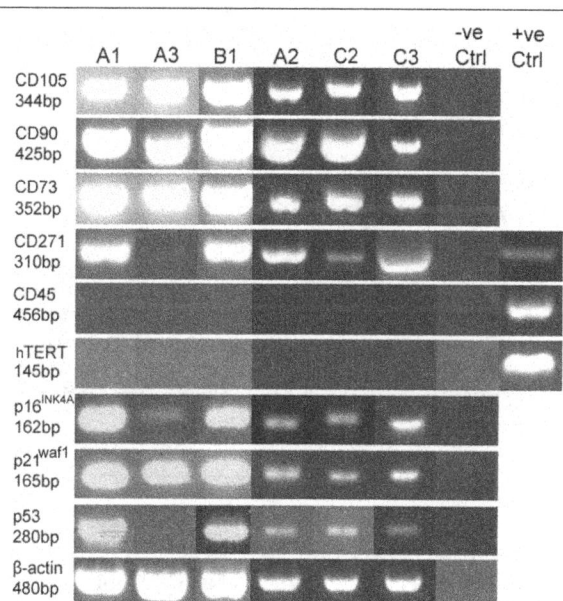

**Fig. 2** Gene expression analysis of DPSC populations by RT-PCR, at early PDs. All DPSCs showed positive expression for the MSC markers, CD73, CD90 and CD105; and the absence of the hematopoietic marker, CD45. All DPSCs analysed were also negative for hTERT. Only low proliferative capacity DPSCs (A1, A2, B1, C2 and C3) were positive for the expression of CD271 (nerve growth factor receptor p75, LNGFR) and for all 3 senescence-associated marker genes analysed, p53, p21$^{waf1}$ and p16$^{INK4a}$. Replacement of cDNA with H$_2$O served as negative controls, with β-actin serving as the reference housekeeping gene

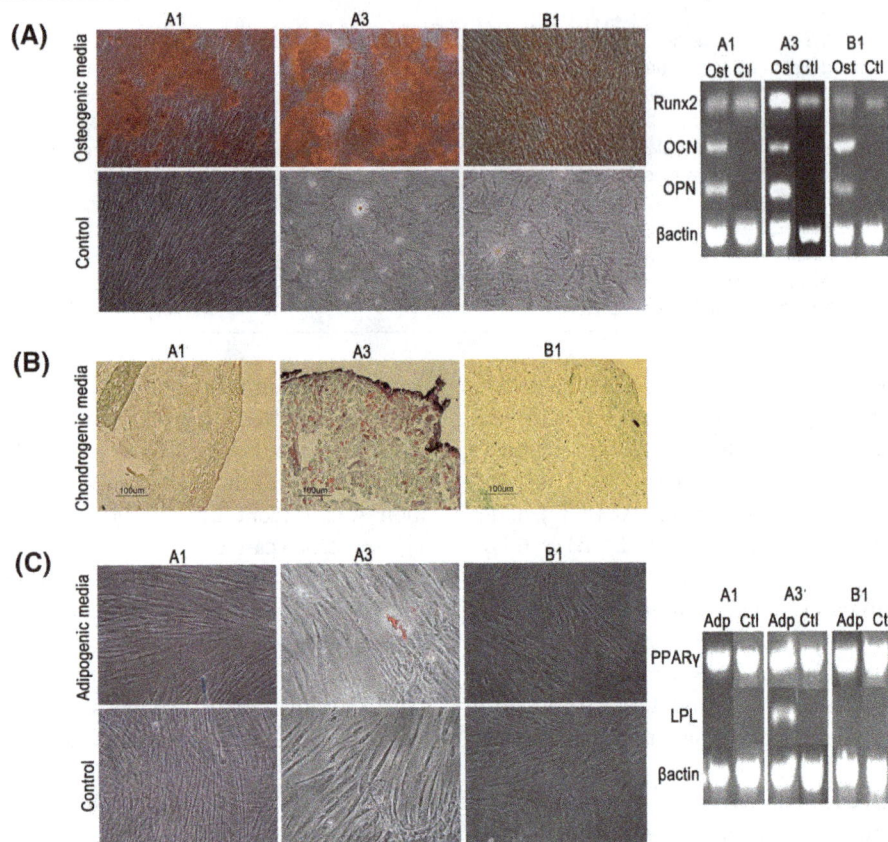

**Fig. 3** Osteogenic, chondrogenic and adipogenic differentiation of DPSCs. Tri-lineage differentiation analysis was compared between high proliferative capacity, A3 (30 PDs) and low proliferative capacity DPSCs, A1 (11 PDs) and B1 (15 PDs). **a** Osteogenesis was confirmed for each DPSC population by the detection of alizarin red staining for mineralised calcium nodules. No staining was observed in control DPSC cultures in the absence of osteogenic medium. RT-PCR analysis of osteogenic markers (*right panel*) indicated increased osteocalcin (OCN) and osteopontin (OPN) expression for DPSCs cultured in osteogenic media. Cells maintained in both osteogenic and control media expressed Runx2. **b** Chondrogenesis was only particularly evident with high proliferative capacity, A3, which exhibited positive Safranin O staining for high proteoglycan content. No Safranin O staining was observed with the low proliferative capacity DPSCs, A1 and B1. **c** Adipogenesis was only particularly evident with high proliferative capacity, A3, which exhibited positive Oil Red O staining for intracellular lipid-rich vacuole accumulation. No Oil Red O staining was observed with the low proliferative capacity DPSCs, A1 and B1; or in control cultures in the absence of adipogenic medium. RT-PCR analysis of adipogenic markers (right panel) indicated increased expression of the early adipogenic marker, PPARγ, in all DPSCs maintained in both adipogenic and control cultures; whilst the late adipogenic marker, lipoprotein lipase (LPL), was only expressed in high proliferative capacity, A3, in adipogenic media. Scale bars = 100 μm. β-actin served as the reference housekeeping gene

in adipogenic medium (Fig. 3c). In contrast, low proliferative capacity DPSCs, A1 and B1, were uni-potent for osteogenic differentiation only and to a lesser extent than high proliferative capacity, A3; as confirmed by alizarin red staining and OCN / OPN expression (Fig. 3a). Of note, all three DPSCs expressed the early osteogenic and adipogenic markers, Runx2 and PPARγ, irrespective of differentiation status (Fig. 3a and c).

**Prolonged in vitro expansion effects on DPSCs**
A decline in cellular proliferative capacity is well-established to correlate with the onset of cellular senescence; and the increased presence of characteristics associated with cellular ageing. At early (7–8 PDs), all DPSCs were morphologically similar, with the characteristic long,

spindle, bipolar shape; and were negative for senescence-associated β-galactosidase staining (Fig. 4a). Conversely, at PDs towards the later stages of their respective proliferative lifespans (examples shown for A1 at 23 PDs, A3 at 83 PDs and B1 at 34 PDs), higher percentages of cells stained positive for β-galactosidase and were larger and more stellate-like in appearance, with prominent stress fibres (Fig. 4a). This was confirmed by cell surface area measurements (Fig. 4b), which demonstrated significant increases in the surface areas of DPSCs, A1 and A3, with proliferative lifespan (for A1, approximately 6-fold between 4 PDs and 15 PDs, 12-fold between 4 PDs and 22 PDs; for A3, approximately 6-fold between 3 PDs and 21 PDs; and 12-fold between 3 PDs and 85 PDs; all $p < 0.001$). However, only when high proliferative capacity, A3, had reached

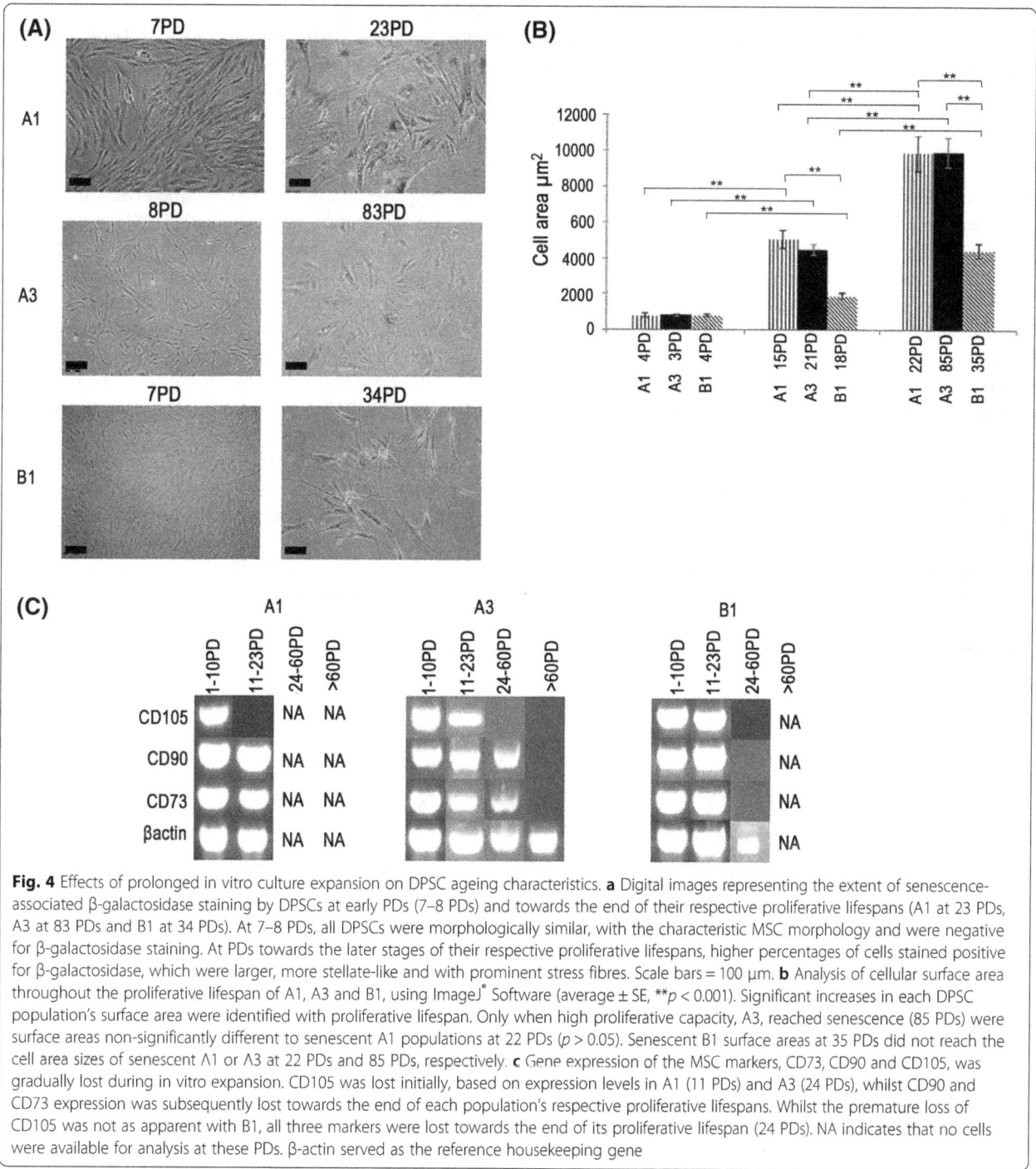

**Fig. 4** Effects of prolonged in vitro culture expansion on DPSC ageing characteristics. **a** Digital images representing the extent of senescence-associated β-galactosidase staining by DPSCs at early PDs (7–8 PDs) and towards the end of their respective proliferative lifespans (A1 at 23 PDs, A3 at 83 PDs and B1 at 34 PDs). At 7–8 PDs, all DPSCs were morphologically similar, with the characteristic MSC morphology and were negative for β-galactosidase staining. At PDs towards the later stages of their respective proliferative lifespans, higher percentages of cells stained positive for β-galactosidase, which were larger, more stellate-like and with prominent stress fibres. Scale bars = 100 μm. **b** Analysis of cellular surface area throughout the proliferative lifespan of A1, A3 and B1, using ImageJ® Software (average ± SE, **$p < 0.001$). Significant increases in each DPSC population's surface area were identified with proliferative lifespan. Only when high proliferative capacity, A3, reached senescence (85 PDs) were surface areas non-significantly different to senescent A1 populations at 22 PDs ($p > 0.05$). Senescent B1 surface areas at 35 PDs did not reach the cell area sizes of senescent A1 or A3 at 22 PDs and 85 PDs, respectively. **c** Gene expression of the MSC markers, CD73, CD90 and CD105, was gradually lost during in vitro expansion. CD105 was lost initially, based on expression levels in A1 (11 PDs) and A3 (24 PDs), whilst CD90 and CD73 expression was subsequently lost towards the end of each population's respective proliferative lifespans. Whilst the premature loss of CD105 was not as apparent with B1, all three markers were lost towards the end of its proliferative lifespan (24 PDs). NA indicates that no cells were available for analysis at these PDs. β-actin served as the reference housekeeping gene

senescence at 85 PDs did it exhibit similar non-significant, cell surface area values to senescent low proliferative capacity, A1 populations, at 22 PDs ($p > 0.05$). B1 also demonstrated similar trends of increasing cell surface area with proliferative lifespan, with significant increases in the surface areas with proliferative lifespan (approximately 2.5-fold between 4 PDs and 18 PDs; and 5-fold between 4 PDs and 35 PDs (both $p < 0.001$). However, senescent B1 surface areas at 35 PDs did not reach the cell area sizes of senescent A1 or A3 at 22 PDs and 85 PDs, respectively.

Although the stem/progenitor cell markers, CD73, CD90 and CD105, were previously shown to be expressed by DPSCs, A1, A3 and B1, at early PDs, the expression of these markers was gradually lost during in vitro expansion (Fig. 4c). CD105 appeared to be lost initially, based on expression levels in A1 (11 PDs) and A3

(24 PDs). CD90 and CD73 expression was subsequently lost from these DPSC populations towards the end of their respective proliferative lifespans. Whilst the premature loss of CD105 was not apparent with B1, all three markers were lost towards the end of its proliferative lifespan.

## Prolonged in vitro expansion effects on DPSC differentiation

In light of the superior proliferative capacity and multi-potent differentiation properties of the A3 population at early PDs, we also determined whether these preferential differentiation capabilities were diminished with prolonged in vitro expansion, in line with those of low proliferative capacity DPSCs, A1 and B1. A3 telomere lengths decreased from 18.9 kb (9 PD) to 6.4 kb (55 PD) (Fig. 1b). In line with such decreases in telomere length, the osteogenic potential of A3 was greatly reduced following culture expansion from 30 PDs to 62 PDs and 75 PDs, apparent by a reduction in alizarin red staining and the loss of osteogenic marker, Runx2, OCN and OPN, expression (Fig. 5a). However, subsequent analyses indicated that the clear loss of A3 osteogenic differentiation potential at 75 PDs, was concomitant with the increased presence of enlarged intracellular lipid vesicles (Oil Red O staining) and the increased expression of early and late adipogenic markers, PPARγ and LPL (Fig. 5b).

## Discussion

Similar to other tissue sources such as bone marrow, MSC populations in dental pulp are recognised to be heterogeneous populations, with isolated DPSCs demonstrating differences in proliferative rates and ability to differentiate along various lineages. The present study was successful in isolating DPSCs from adult human dental pulp tissues, in order to characterize the relative expansion potentials of individual DPSC populations through to cellular senescence; and the impact of these proliferative lifespan differences on stem/progenitor cell characteristics and multi-potent differentiation potential. Although many animal and human studies have established the effects of increasing donor chronological age on the impairment of stem/progenitor cell regenerative capabilities [20–24], to our knowledge, this is the first study to report inherent differences in the telomere lengths of individual DPSC populations from within a young donor age group (18–30 years); and their correlations to the distinct differentiation capabilities of each population. Indeed, despite DPSCs from young adults with similar donor age ranges having previously been reported to exhibit no significant differences in proliferative or differentiation potential [23, 25], this study identified differences between high (A3) and low (A1 and A2) proliferative capacity DPSC populations, even from the same pulpal tissue sample.

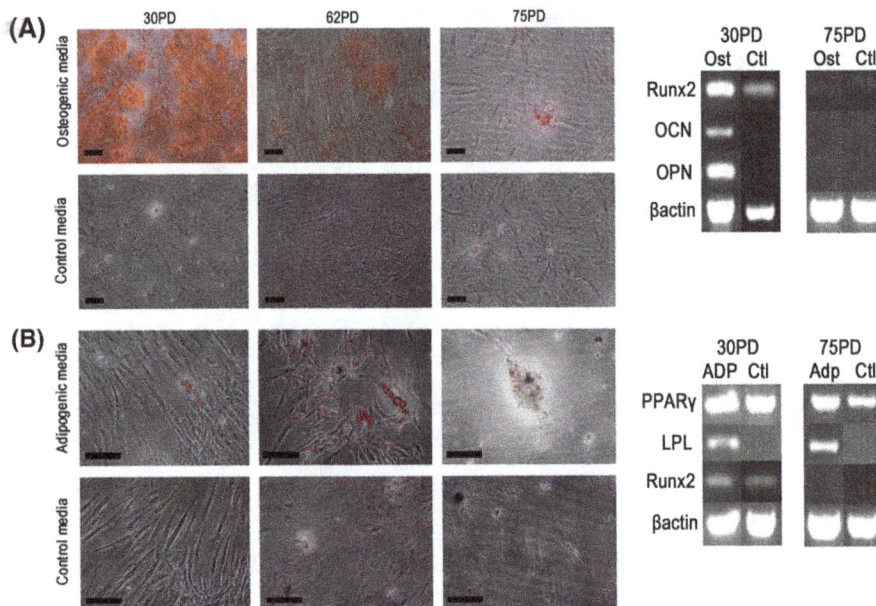

**Fig. 5** Effects of prolonged in vitro culture expansion on DPSC differentiation capabilities. **a** Osteogenic differentiation potential of high proliferative capacity, A3, was greatly reduced following culture expansion from 30 PDs, to 62 PDs and 75 PDs; apparent by a reduction in alizarin red staining and the loss of osteogenic marker Runx2, osteocalcin (OCN) and osteopontin (OPN) expression (*right panel*). **b** Loss of A3 osteogenic differentiation potential was concomitant with the increased presence of enlarged intracellular lipid vesicles (Oil Red O staining) and the increased expression of early and late adipogenic markers, PPARγ and lipoprotein lipase (LPL, right panel). No staining was observed in control cultures in the absence of osteogenic or adipogenic media. Scale bars = 100 μm

DPSC population, A3, possessed an extensive proliferative capacity (>80 PDs), which was 2-4× the proliferative lifespans of other DPSC populations expanded (20–35 PDs). In support of these findings, A3 was further shown to be morphologically smaller, possessed fewer senescence-associated β-galactosidase positive cells and lacked the expression of p53 and p16$^{INK4a}$, at PDs where low proliferative capacity DPSCs (A1, A2, B1, C2 and C3) demonstrated increased detection of these senescence-associated markers. Such conclusions are consistent with only 20% of purified DPSCs being able to proliferate beyond 20 PDs, whilst it can be surmised that high proliferative capacity DPSCs, such as A3, are responsible for the extensive expansion potential of heterogeneous DPSC populations (>120 PDs) in vitro [5, 7], as DPSCs with less proliferative potential are selectively lost from the mixed population during extended subculture [15, 16]. A consequence of A3 possessing a superior proliferative lifespan was the retention of stem cell characteristics, evident by the maintenance of marker expression (CD73, CD90 and CD105) for longer periods in culture (≥23 PDs), compared to the other DPSCs analysed. A3 was also the only population capable of multi-potent osteogenic, chondrogenic and adipogenic differentiation, at least up to 30 PDs in culture. In contrast, other DPSCs appeared lineage-restricted to osteogenic differentiation only. Therefore, there appears to be a clear association between DPSC PDs and the relative differentiation abilities of these populations. However, as DPSCs derived from adult teeth are generally regarded to possess lower proliferation rates and longer population doubling times than those isolated from human exfoliated deciduous teeth (SHEDs) [26–29], it is conceivable that SHEDs are still likely to have equivalent or even greater proliferative potential than adult DPSCs.

The principle reason identified to be behind these contrasting proliferative responses were the average telomere lengths of the DPSC populations. Although it is possible that these differences reflect the relative number of cell divisions undertaken by each stem cell before isolation, such contrasting telomere lengths may also suggest the existence of sub-populations of DPSCs within the stem cell population, with superior telomere dynamic characteristics. Indeed, at early PDs, A3 (18.9 kb) demonstrated telomere lengths 1.5–3.5× longer than low proliferative DPSCs (~5–13 kb). A3 only exhibited equivalent telomere lengths to those of the low proliferative DPSCs at much later PDs towards the end of its proliferative lifespan (6.4 kb, 55 PDs), corresponding to the increased presence of senescence-associated markers. Actively proliferating stem cells are well-known to possess longer telomeres (10–20 kb), compared to somatic cells (5–15 kb) [30], whilst previous studies have reported DPSC telomere lengths between 9.4 and 12.1 kb [31]. Therefore, the contrasting telomere lengths determined for high and low proliferative capacity DPSCs are generally in line with those previously identified in MSCs from other sources.

Based on the correlation comparison between cumulative PDs versus original telomere lengths for all DPSCs analysed, it appears that this relationship is a little more complex than expected, particularly between the low proliferative capacity DPSCs (A1, A2, B1, C2 and C3). For instance, C3 (12.8 kb) had an initial telomere length 2–3× longer than A1 (5.9 kb) and B1 (5.6 kb), yet exhibited similar proliferative capacities (20–35 PDs). The extrinsic and intrinsic factors which influence the respective telomere lengths and PD capabilities of individual DPSC populations remain to be elucidated, although inherent differences due to inter-patient donor variation are likely to play a significant role. As this initial proof-of-concept study has only compared 6 separate DPSC populations (5 low proliferative and 1 high proliferative) from 3 separate donors, we acknowledge that the present study is somewhat limited, particularly in terms of number of highly proliferative DPSCs with longer telomeres assessed. Consequently, further research is warranted with much larger numbers of high and low proliferative DPSCs from different donors, in order to fully establish the relationship between telomere lengths and how these impact on the overall PD capabilities and multi-potent differentiation capabilities of individual DPSC populations.

Although certain studies have reported hTERT expression in human DPSCs [31, 32], we and others have shown that human DPSCs possess no or negligible hTERT expression [33–35]. Therefore, it appears that hTERT may not be wholly responsible for maintaining telomere lengths in high proliferative capacity DPSCs or for the significant variations in DPSC proliferative capabilities, implying that other intrinsic telomere protective mechanisms exist [36, 37]. Nonetheless, the later onset of senescence in A3 was supported by the delayed expression of senescence-associated genes, most notably p53 and p16$^{INK4a}$, at PDs where these and the other senescence marker gene (p21$^{waf1}$) assessed, were highly expressed in low proliferative capacity DPSCs. MSC replicative senescence is acknowledged to be a multi-step process driven by p53, which promotes growth arrest by inducing p21$^{waf1}$ expression, thereby inhibiting $G_1$-S phase progression [12]. MSC telomere erosion can also initiate pRb / p16$^{INK4a}$ checkpoints, triggering permanent senescent states by preventing pRB phosphorylation and suppressing cell proliferation. Alternatively, senescence mediators, such as p53, p16$^{INK4a}$ and pRb, are also associated with premature/telomere-independent senescence [12]. The proliferative, telomere length reductions and senescence-related marker detection with increasing proliferative lifespan presented herein, firmly suggest that the senescence of high proliferative capacity DPSCs, such as A3, is a consequence of replicative senescence; although it is plausible that both $G_1$-S phase inhibition and pRb/$p16^{INK4a}$ checkpoint initiation contribute to A3 senescence [12]. In contrast, as p53 and $p16^{INK4a}$ are also implicated in the onset of premature senescence, we can only speculate at present which is the principle mechanism of early senescence and the reasons underlying the pre-existing shortened telomeres in low proliferative capacity DPSCs, following their isolation and short-term in vitro expansion. Nonetheless, as p53 and $p16^{INK4a}$ have both been demonstrated to be the principal mediators of senescence in MSCs [23, 34, 38, 39], as $p21^{waf1}$ also has a role in maintaining stem cell renewal due to its positive effects on cell cycle progression [40]; the contrasting p53 and $p16^{INK4a}$ expression between high (A3) and low (A1, A2, B1, C2 and C3) proliferative capacity DPSCs further confirms the earlier onset of senescence with less proliferative populations at early PDs, compared to A3.

Although telomere length maintenance is recognized to facilitate cell division in stem cell populations [41], a notable observation within this study was the impact that contrasting telomere dynamics between high and low proliferative capacity DPSCs has on lineage differentiation capabilities. Whilst A3 exhibited multi-potency towards osteogenic, chondrogenic and adipogenic lineages at early PDs, low proliferative capacity DPSCs, such as A1 and B1, were only uni-potent for osteogenesis. Only following extensive culture expansion beyond 30 PDs did the osteogenic potential of A3 decline, with a corresponding increase in adipogenic differentiation. Such findings concur with previous reports of reduced multi-potency in DPSCs and MSCs from other sources, with extensive proliferative expansion and senescence [23, 34, 42, 43]. Evidence indicates that osteogenesis and adipogenesis are inversely regulated by transcription factors, Runx2 and PPARγ, with changes to cellular osteogenic/adipogenic potential with age being a consequence of transcription factor dysregulation [41, 43].

Another key consideration relating to the findings, is whether the contrasting regenerative properties between high and low proliferative capacity DPSC populations reflect their isolation from different mesodermal or neuro-ectodermal origins [44, 45], stem cell niches within the dental pulp [2], or are from the same origin, but are at different stages within the proposed hierarchical model for adult stem cells [8]. Although the nature and origins of these DPSCs within dental pulpal tissues have yet to be fully elucidated, the absence of cell surface marker, CD271 (nerve growth factor receptor p75, LNGFR), expression in high proliferative capacity and multi-potent DPSCs, such as A3, may suggest that such populations are not neural crest- or sub-odontoblast layer-derived, as

$CD271^+$ cells are regarded as being of neural crest origin and have been located within the cell-rich, sub-odontoblast region of dental pulp [44, 46]. In line with the uni-potentiality of low proliferative capacity DPSCs, CD271 has been proposed to inhibit DPSC adipogenic, chondrogenic, myogenic and osteogenic potential [47, 48], although not all studies have demonstrated complete inhibition of multi-potent differentiation in CD271-expressing DPSCs [44, 45]. Nonetheless, positive CD271 expression in low proliferative capacity DPSCs (A1, A2, B1, C2 and C3), may provide further credence to these cells being more lineage restricted than high proliferative capacity DPSCs, such as A3.

The prominence of low proliferative capacity and less potent DPSC populations in the present study may imply that high proliferative capacity/multi-potent DPSCs with longer telomere and lacking CD271 expression represent a minor population within the heterogeneous DPSCs in dental pulp tissues; meaning that highly proliferative/long telomere DPSC populations are more difficult to identify and isolate from pulpal tissues, although donor variability is also likely to contribute to this. However, the identification of telomere length and CD271 expression differences between high and low proliferative capacity and multi-/uni-potent DPSCs does advocate their use as potential phenotypic biomarkers for the identification and selective isolation of superior proliferative capacity DPSC populations from dental pulp tissues for regenerative medicine purposes. Variations in telomere lengths have previously been used as selective markers for the isolation of particular stem/progenitor cell populations from tissues, such as bone marrow and cartilage [49, 50]. Furthermore, CD271 has recently been identified to be highly expressed in $STRO-1^+/c-Kit^+/CD34^+$ DPSC populations, possessing slow proliferation rates, reduced stemness and early-onset senescence; compared to their $STRO-1^+/c-Kit^+/CD34^-$ counterparts [45]. However, despite differences in CD271 expression, both DPSC sub-populations exhibited similar osteogenic, myogenic and adipogenic differentiation, although $STRO-1^+/c-Kit^+/CD34^+$ DPSCs expressing CD271 demonstrated greater neurogenic lineage commitment. High $CD90^+$ expressing DPSCs with low CD271 expression have also been demonstrated to possess superior colony-forming efficiencies, prolonged in vitro proliferative/multi-lineage potential; and enhanced in vivo bone formation capabilities, versus $CD90^+$ / $CD271^+$ DPSCs [48].

## Conclusions

This study has reiterated the significant variability in the proliferative and differentiation capabilities for individual DPSC populations expanded from individual human teeth from donors within a similar age range; and even identified inherent differences between DPSC populations derived from the same patient. The findings presented also strongly support the use of telomere length and CD271 expression as viable markers of high proliferative capacity and multi-potent DPSC populations. Consequently, if such superior DPSC populations are to be fully exploited for regenerative medicine, we must utilise these and other potential markers of DPSC proliferation and senescence, such as Bmi-1 and SSEA-4 [34, 51]. Such criteria will allow the optimization of population selection by selectively screening and isolating better quality DPSCs from whole dental pulp tissues for in vitro expansion and assessment, aiding the translational development of more effective DPSC-based therapies for clinical evaluation and application.

## Abbreviations

DPSCs: Dental pulp stem cells; hTERT: Human telomerase catalytic subunit; LBL: Lipoprotein lipase; M-MLV: Moloney murine leukaemia virus; MSC: Mesenchymal stem cell; NRES: National Research Ethics Service; OCN: Osteocalcin; OPN: Osteopontin; PBS: Phosphate buffered saline; PDs: Population doublings; PPARγ: Peroxisome proliferator-activated receptor γ; pRb: Retinoblastoma protein; RT-PCR: Reverse transcription polymerase chain reaction; SSC: Sodium citrate; TA: Transit amplifying; TRF: Telomere restriction fragment length; αMEM: α-minimum essential medium

## Acknowledgements

The authors acknowledge for the receipt of PhD studentship funding for Dr Amr Alraies to undertake this study, from Albawani Company, Saudi Arabia. This study was funded by a PhD studentship, provided by Albawani Company, Saudi Arabia.

## Authors' contributions

AA: Performed most of the experimental work, participated in study design and preparation of the final manuscript. NA: Performed the remainder of the experimental work not undertaken by AA. RW: Advisor to the study and participated in the preparation of the final manuscript. RM: Project Supervisor, participated in study design and preparation of the final manuscript. AS: Project Supervisor, participated in study design and preparation of the final manuscript. All authors read and approved the final manuscript.

## Competing interest

The authors declare that they have no competing interests.

## Author details

[1]Mineralised Tissue Group, Oral and Biomedical Sciences, School of Dentistry, College of Biomedical and Life Sciences, Cardiff University, Heath Park, CF14 4XY, Cardiff, UK. [2]Cardiff Institute Tissue Engineering and Repair (CITER), Cardiff University, Cardiff CF14 4XY, UK. [3]Stem Cells, Wound Repair and Regeneration, Oral and Biomedical Sciences, School of Dentistry, College of Biomedical and Life Sciences, Cardiff University, Heath Park, Cardiff CF14 4XY, UK.

## References

1. Huang GT, Gronthos S, Shi S. Mesenchymal stem cells derived from dental tissues vs. those from other sources: Their biology and role in regenerative medicine. J Dent Res. 2009;88(9):792–806.
2. Sloan AJ, Waddington RJ. Dental pulp stem cells: What, where, how? Int J Paediatr Dent. 2009;19(1):61–70.
3. Kawashima N. Characterisation of dental pulp stem cells: A new horizon for tissue regeneration? Arch Oral Biol. 2012;57(11):1439–58.
4. Tatullo M, Marrelli M, Shakesheff KM, White LJ. Dental pulp stem cells: function, isolation and applications in regenerative medicine. J Tissue Eng Regen Med. 2015;9(11):1205 16.
5. Gronthos S, Brahim J, Li W, Fisher LW, Cherman N, Boyde A, et al. Stem cell properties of human dental pulp stem cells. J Dent Res. 2002;81(8):531–5.
6. Shi S, Gronthos S. Perivascular niche of postnatal mesenchymal stem cells in human bone marrow and dental pulp. J Bone Miner Res. 2003;18(4):696–704.
7. Gronthos S, Mankani M, Brahim J, Robey PG, Shi S. Postnatal human dental pulp stem cells (DPSCs) in vitro and in vivo. Proc Natl Acad Sci U S A. 2000; 97(25):13625–30.
8. Stocum DL. Developmental mechanisms of regeneration. In: Atala A, Lanza R, Thomson JA, Nerem RA, editors. Foundations of regenerative medicine. San Diego: Academic; 2010. p. 100–25.
9. Téclès O, Laurent P, Zygouritsas S, Burger AS, Camps J, Dejou J, About I. Activation of human dental pulp progenitor/stem cells in response to odontoblast injury. Arch Oral Biol. 2005;50(2):103–8.
10. Wagner W, Ho AD, Zenke M. Different facets of aging in human mesenchymal stem cells. Tissue Eng Part B Rev. 2010;16(4):445–53.
11. Banfi A, Muraglia A, Dozin B, Mastrogiacomo M, Cancedda R, Quarto R. Proliferation kinetics and differentiation potential of ex vivo expanded human bone marrow stromal cells: Implications for their use in cell therapy. Exp Haematol. 2000;28(6):707–15.
12. Campisi J, d'Adda di Fagagna F. Cellular senescence: When bad things happen to good cells. Nat Rev Mol Cell Biol. 2007;8(9):729–40.
13. Serakinci N, Graakjaer J, Kolvraa S. Telomere stability and telomerase in mesenchymal stem cells. Biochimie. 2008;90(1):33–40.
14. Jones PH, Watt FM. Separation of human epidermal stem cells from transit amplifying cells on the basis of differences in integrin function and expression. Cell. 1993;73(4):713–24.
15. Waddington RJ, Youde SJ, Lee CP, Sloan AJ. Isolation of distinct progenitor stem cell populations from dental pulp. Cells Tissues Organs. 2009;189(1–4):268–74.
16. Harrington J, Sloan AJ, Waddington RJ. Quantification of clonal heterogeneity of mesenchymal progenitor cells in dental pulp and bone marrow. Connect Tissue Res. 2014;55 Suppl 1:62–7.
17. Davies LC, Locke M, Webb RD, Roberts JT, Langley M, Thomas DW, et al. A multipotent neural crest-derived progenitor cell population is resident within the oral mucosa lamina propria. Stem Cells Dev. 2010;19(6):819–30.
18. Dowthwaite GP, Bishop JC, Redman SN, Khan IM, Rooney P, Evans DJ, et al. The surface of articular cartilage contains a progenitor cell population. J Cell Sci. 2004;117(6):889–97.
19. Cristofalo VJ, Allen RG, Pignolo RJ, Martin BG, Beck JC. Relationship between donor age and the replicative lifespan of human cells in culture: A re-evaluation. Proc Natl Acad Sci U S A. 1998;95(18):10614–9.
20. Ma D, Ma Z, Zhang X, Wang W, Yang Z, Zhang M, et al. Effect of age and extrinsic microenvironment on the proliferation and osteogenic differentiation of rat dental pulp stem cells in vitro. J Endo. 2009;35(11):1546–53.
21. Bressan E, Ferroni L, Gardin C, Pinton P, Stellini E, Botticelli D, et al. Donor age-related biological properties of human dental pulp stem cells change in nanostructured scaffolds. PLoS One. 2012;7(11):e49146.
22. Feng X, Xing J, Feng G, Sang A, Shen B, Xu Y, et al. Age-dependent impaired neurogenic differentiation capacity of dental stem cell is associated with Wnt/β-catenin signalling. Cell Mol Neurobiol. 2013;33(8):1023–31.
23. Feng X, Xing J, Feng G, Huang D, Lu X, Liu S, et al. p16[INK4A] mediates age-related changes in mesenchymal stem cells derived from human dental pulp through the DNA damage and stress response. Mech Ageing Dev. 2014;141-142C:46–55.
24. Horibe H, Murakami M, Iohara K, Hayashi Y, Takeuchi N, Takei Y, et al. Isolation of a stable subpopulation of mobilized dental pulp stem cells (MDPSCs) with high proliferation, migration and regeneration potential is independent of age. PLoS One. 2014;9(5):e98553.
25. Kellner M, Steindorff MM, Strempel JF, Winkel A, Kühnel MP, Stiesch M. Differences of isolated dental stem cells dependent on donor age and consequences for autologous tooth replacement. Arch Oral Biol. 2014; 59(6):559–67.
26. Nakamura S, Yamada Y, Katagiri W, Sugito T, Ito K, Ueda M. Stem cell proliferation pathways comparison between human exfoliated deciduous teeth and dental pulp stem cells by gene expression profile from promising dental pulp. J Endod. 2009;35(11):1536–42.
27. Lee S, An S, Kang TH, Kim KH, Chang NH, Kang S, Kwak CK, Park HS. Comparison of mesenchymal-like stem/progenitor cells derived from supernumerary teeth with stem cells from human exfoliated deciduous teeth. Regen Med. 2011;6(6):689–99.

28. Wang X, Sha XJ, Li GH, Yang FS, Ji K, Wen LY, Liu SY, Chen L, Ding Y, Xuan K. Comparative characterization of stem cells from human exfoliated deciduous teeth and dental pulp stem cells. Arch Oral Biol. 2012;57(9):1231–40.

29. Abdullah MF, Abdullah SF, Omar NS, Mahmood Z, Fazliah Mohd Noor SN, Kannan TP, Mokhtar KI. Proliferation rate of stem cells derived from human dental pulp and identification of differentially expressed genes. Cell Biol Int. 2014;38(5):582–90.

30. Cheng PH, Snyder B, Fillos D, Ibegbu CC, Huang AH, Chan AW. Postnatal stem/progenitor cells derived from the dental pulp of adult chimpanzee. BMC Cell Biol. 2008;9:20–30.

31. Jeon BG, Kang EJ, Kumar BM, Maeng GH, Ock SA, Kwack DO, et al. Comparative analysis of telomere length, telomerase and reverse transcriptase activity in human dental stem cells. Cell Transplant. 2011; 20(11–12):1693–705.

32. Hakki SS, Kayis SA, Hakki EE, Bozkurt SB, Duruksu G, Unal ZS, et al. Comparison of mesenchymal stem cells isolated from pulp and periodontal ligament. J Periodontol. 2015;86(2):283–91.

33. Egbuniwe O, Idowu BD, Funes JM, Grant AD, Renton T, Di Silvio L. P16/p53 expression and telomerase activity in immortalized human dental pulp cells. Cell Cycle. 2011;10(22):3912–9.

34. Mehrazarin S, Oh JE, Chung CL, Chen W, Kim RH, Shi S, et al. Impaired odontogenic differentiation of senescent dental mesenchymal stem cells is associated with loss of Bmi-1 expression. J Endod. 2011;37(5):662–6.

35. Murakami M, Horibe H, Iohara K, Hayashi Y, Osako Y, Takei Y, et al. The use of granulocyte-colony stimulating factor induced mobilization for isolation of dental pulp stem cells with high regenerative potential. Biomaterials. 2013;34(36):9036–47.

36. Liu J, Cao L, Finkel T. Oxidants, metabolism and stem cell biology. Free Radic Biol Med. 2011;51(12):2158–62.

37. Shyh-Chang N, Daley GQ, Cantley LC. Stem cell metabolism in tissue development and aging. Development. 2013;140(12):2535–47.

38. Shibata KR, Aoyama T, Shima Y, Fukiage K, Otsuka S, Furu M, et al. Expression of the p16INK4A gene is associated closely with senescence of human mesenchymal stem cells and is potentially silenced by DNA methylation during in vitro expansion. Stem Cells. 2007;25(9):2371–82.

39. Muthna D, Soukup T, Vavrova J, Mokry J, Cmielova J, Visek B, et al. Irradiation of adult human dental pulp stem cells provokes activation of p53, cell cycle arrest, and senescence but not apoptosis. Stem Cells Dev. 2010;19(12):1855–62.

40. Ju Z, Choudhury AR, Rudolph KL. A dual role of p21 in stem cell aging. Ann N Y Acad Sci. 2007;1100:333–44.

41. Allen ND, Baird DM. Telomere length maintenance in stem cell populations. Biochim Biophys Acta. 2009;1792(4):324–8.

42. Cheng H, Qiu L, Ma J, Zhang H, Cheng M, Li W, et al. Replicative senescence of human bone marrow and umbilical cord derived mesenchymal stem cells and their differentiation to adipocytes and osteoblasts. Mol Biol Rep. 2011;38(8):5161–8.

43. Kim M, Kim C, Choi YS, Kim M, Park C, Suh Y. Age-related alterations in mesenchymal stem cells related to shift in differentiation from osteogenic to adipogenic potential: Implication to age-associated bone diseases and defects. Mech Ageing Dev. 2012;133(5):215–25.

44. Alvarez R, Lee HL, Hong C, Wang CY. Single CD271 marker isolates mesenchymal stem cells from human dental pulp. Int J Oral Sci. 2015;18(7):205–12.

45. Pisciotta A, Carnevale G, Meloni S, Riccio M, De Biasi S, Gibellini L, et al. Human dental pulp stem cells (hDPSCs): isolation, enrichment and comparative differentiation of two sub-populations. BMC Dev Biol. 2015;15:14.

46. Martens W, Wolfs E, Struys T, Politis C, Bronckaers A, Lambrichts I. Expression pattern of basal markers in human dental pulp stem cells and tissue. Cells Tissues Organs. 2012;96(6):490–500.

47. Mikami Y, Ishii Y, Watanabe N, Shirakawa T, Suzuki S, Irie S, et al. CD271/p75NTR inhibits the differentiation of mesenchymal stem cells into osteogenic, adipogenic, chondrogenic and myogenic lineages. Stem Cells Dev. 2011;20(5):901–13.

48. Yasui T, Mabuchi Y, Toriumi H, Ebine T, Niibe K, Houlihan DD, et al. Purified human dental pulp stem cells promote osteogenic regeneration. J Dent Res. 2016;95(2):206–14.

49. Williams R, Khan IM, Richardson K, Nelson L, McCarthy HE, Analbelsi T, et al. Identification and clonal characterisation of a progenitor cell sub-population in normal human articular cartilage. PLoS One. 2010;5:e13246.

50. Samsonraj RM, Raghunath M, Hui JH, Ling L, Nurcombe V, Cool SM. Telomere length analysis of human mesenchymal stem cells by quantitative PCR. Gene. 2013;519(2):348–55.

51. Kawanabe N, Murata S, Fukushima H, Ishihara Y, Yanagita T, Yanagita E, et al. Stage-specific embryonic antigen-4 identifies human dental pulp stem cells. Exp Cell Res. 2012;318(5):453–63.

# 8

# Bmp-12 activates tenogenic pathway in human adipose stem cells and affects their immunomodulatory and secretory properties

Weronika Zarychta-Wiśniewska[1], Anna Burdzinska[1*], Agnieszka Kulesza[1], Kamila Gala[1], Beata Kaleta[2], Katarzyna Zielniok[3], Katarzyna Siennicka[4], Marek Sabat[1] and Leszek Paczek[1,5]

## Abstract

**Background:** Cell-based therapy is a treatment method in tendon injuries. Bone morphogenic protein 12 (BMP-12) possesses tenogenic activity and was proposed as a differentiating factor for stem cells directed to transplantation. However, BMPs belong to pleiotropic TGF-β superfamily and have diverse effect on cells. Therefore, the aim of this study was to determine if BMP-12 induces tenogenic differentiation of human adipose stem cells (hASCs) and how it affects other features of this population.

**Results:** Human ASCs from 6 healthy donors were treated or not with BMP-12 (50 or 100 ng/ml, 7 days) and tested for gene expression (COLL1, SCX, MKH, DCN, TNC, RUNX2), protein expression (COLL1, COLL3, MKH), proliferation, migration, secretory activity, immunomodulatory properties and susceptibility to oxidative stress. RT-PCR revealed up-regulation of SCX, MKH and RUNX2 genes in BMP-12 treated cells (2.05, 2.65 and 1.87 fold in comparison to control, respectively, $p < 0.05$) and Western Blot revealed significant increase of COLL1 and MHK expression after BMP-12 treatment. Addition of BMP-12 significantly enhanced secretion of VEGF, IL-6, MMP-1 and MPP-8 by hASCs while had no effect on TGF-β, IL-10, EGF and MMP-13. Moreover, BMP-12 presence in medium attenuated inhibitory effect of hASCs on allo-activated lymphocytes proliferation. At the same time BMP-12 displayed no influence on hASCs proliferation, migration and susceptibility to oxidative stress.

**Conclusion:** BMP-12 activates tenogenic pathway in hASCs but also affects secretory activity and impairs immunomodulatory potential of this population that can influence the clinical outcome after cell transplantation.

**Keywords:** Adipose stem cells, Mesenchymal stem cells, BMP-12, Tenogenic differentiation, Secretory activity

## Background

Tendon injuries are common musculoskeletal disorders that clinicians address daily, however, the question of optimal treatment is still unanswered. Naturally occurring tendon healing is far from satisfactory, because tendons after injury rarely return to their full mechanical strength [1]. Therefore, trials to improve tendon recovery by local supplementation of different bioactive components are undertaken [2]. One of the proposed approaches is to implement cellular therapy into treatment protocols both for acute tendon injuries and for chronic tendinopathies [3]. Mesenchymal stem cells (MSCs) are considered as the most promising population in those applications because of several reasons: 1) they can undergo tenogenic differentiation [4], 2) they are suitable for both autologous and allogeneic transplantation, 3) they possess immunomodulatory properties [5], 4) they stimulate tissue regeneration via paracrine effect [6]. Another concept to accelerate tendon healing is to use local delivery of growth factors which have tenogenic activity. At present, the most often proposed candidate for such a treatment is a

* Correspondence: anna.burdzinska@wum.edu.pl; aniaburdzia@interia.pl
[1]Department of Immunology, Transplantology and Internal Medicine, Transplantation Institute, Medical University of Warsaw, Nowogrodzka str. 59, 02-006 Warsaw, Poland
Full list of author information is available at the end of the article

group of highly conserved bone morphogenic proteins (BMPs): BMP-12, BMP-13 and BMP-14 (also called growth and differentiation factor 7 (GDF-7), GDF-6 and GDF-5, respectively) [7]. It was originally reported that subcutaneous or intramuscular administration of mentioned BMPs resulted in ectopic formation of tendon-like structure in contrast to BMP-2 which is well known osteogenic inductor [8]. Later, it was demonstrated that delivery of BMP-12,-13,-14 (either in form of a gene or a protein) into experimentally injured tendon improved healing parameters i.e., tensile strength [9–11]. Finally, it was shown that these BMPs, especially BMP-12, are able to activate tenogenic pathway in rat and human bone marrow (BM) derived MSCs [4, 12, 13]. Thus, based on previously published studies, the idea of using BMP-12 (-13, -14) treated MSCs or co-administration of MSCs and discussed BMPs seems to be a promising approach for tendon injuries treatment. Recently, much attention is focused on adipose derived MSCs, which are also called adipose-derived stem cells or adipose stem (stromal) cells (ASCs). They constitute attractive alternative to BM-MSCs as were demonstrated to share most important features with bone marrow counterparts. Moreover, fat harvesting is a less invasive procedure than bone marrow aspiration. However, recently it was shown in rat cells, that adipose derived MSCs possess weaker tenogenic activity than bone marrow derived counterparts [4]. Therefore, the primary aim of the present study was to evaluate if BMP-12 activate tenogenic pathway in human ASCs. However, the MSCs (ASCs) differentiation potential is only one of several clinically beneficial features of these cell populations. BMPs belong to the TGF-β superfamily and are pleiotropic molecules involved in regulation of multiply fundamental cellular functions [14]. Moreover, BMP signals influence various kinds of stem cells with very diverse outcomes [15]. Although the tenogenic activity of BMP-12 was previously studied and demonstrated [4, 16], it is not clear how this cytokine influence on other MSCs (ASCs) features. Therefore, we aimed to extend current knowledge and the secondary goal of this study was to determine whether treatment of hASCs with BMP-12 affects other potentially beneficial cells traits like proliferative and migrative capacity, secretory activity, immunomodulatory properties and susceptibility to oxidative stress.

## Methods

### Human adipose stem cells (hASCs) - isolation and identification

Adipose tissue was collected from healthy donors by liposuction. Donors were informed and agreed to participate in this study. The procedure was approved by the Local Bioethics Committee. In order to remove red blood cells, 400 ml fat tissue was mixed 2:1 vol/vol with buffered physiological salt solution (Phosphate-buffered saline- PBS) and shaken every 15 min. Following phases separation, PBS with red blood cells were discarded. Purification process was repeated three times. Afterwards, 0.075% collagenase solution from Clostridium histolyticum (Sigma – Aldrich) in PBS was added to adipose tissue (1:2 vol/vol), shaken and incubated in temp. 37 °C for 1.5 h in order to digest the tissue. The fat and collagenase mixture were shaken every 15 min. After obtaining a homogenous suspension, human albumin (20% concentration) was added (final concentration 2%) to stop the digestion reaction. Mixture was centrifuged ($400\,g$) for 10 min at room temperature (RT). The liquid fat and salt interphases were discarded and the cell pellet was suspended in PBS. Cell suspension was filtered through 100 μm nylon filter, washed in PBS and centrifuged ($300\,g$) for 10 min, RT. The amount and viability of the cells were determined and cells were seeded into plastic flasks at a density of $8 \times 10^4$ cells/cm$^2$ for further cultured in growth medium (GM) composed of DMEM-LG (Dulbecco's modified Eagle's Medium with low glucose; Sigma-Aldrich) supplemented with fetal calf serum (FCS; 15%; Invitrogen) and antibiotic–antimycotic solution (Penicillin-streptomycin-amfoterycin; 1.5%; Invitrogen) and incubated under standard cell culture conditions (37 °C, 5% CO$_2$, 95% humidity). When primary cultures reached subconfluency, cells were detached by exposure to trypsin (0.25% trypsin with 1 mM EDTA; Invitrogen) and replated at a density of $5.0 \times 10^3$ cells/cm$^2$ for subsequent passage. After passage 3 cells were identified by flow cytometry and multilineage differentiation capacity and were frozen in liquid nitrogen. For experiments refrozen hASCs from 6 independent donors at passage 4–7 were used.

### In vitro osteogenic differentiation

Osteogenic differentiation was performed at the third passage. Cells were cultured in hMSC Osteogenic Differentiation BulletKit™ Medium (Lonza) for 3 weeks. The medium was changed every 3 days. Osteogenic differentiation was characterized by identification of mineral depositions in extracellular matrix. At 3 weeks, the plated cells were fixed for 15 min with 4% formaldehyde and stained with Alizarin Red (Sigma-Aldrich). After staining, the wells were rinsed with distilled water and visualized by standard light microscopy.

### In vitro adipogenic differentiation

Adipogenic differentiation was performed at the third passage. Cells were cultured in hMSC Adipogenic Differentiation BulletKit™ Medium (Lonza) for 3 weeks. Adipogenic differentiation was assessed using Oil Red O (Sigma-Aldrich) stain as an indicator of intracellular lipid accumulation. Prior to staining, plastic-adherent cells were fixed for 45 min with 10% formaldehyde and then for 5 min with 60% isopropanol. After fixation and staining, the wells were rinsed with distilled water and visualized by standard light microscopy.

### In vitro chondrogenic differentiation

To induce chondrogenic differentiation, three-dimensional pellet culture was performed. In a 15 ml tube, $3 \times 10^5$ cells were pelleted by centrifugation. Unsuspended cell pellets were cultured for 19 days in chondrogenic medium (Lonza) composed of basic medium supplemented with dexamethasone, ascorbate, ITS + supplement, pyruvate, proline, GA-1000, L-glutamine and recombinant human transforming growth factor-β3. For histological analysis, pellets were immersed in paraffin, sectioned and stained with Masson trichrome method.

### Flow cytometry analysis

The surface antigen profiles of adipose derived MSCs at the third passage were characterized by flow cytometry. A total of $2,5 \times 10^6$ cells were incubated with the following phycoerythrin (PE)-conjugated anti-mouse antibodies: CD29, CD34, CD45, CD73, CD90 and CD105 (Becton Dickinson) for 30 min, RT in the dark. Nonspecific PE-conjugated IgG was substituted as an isotype control. The fluorescence intensity of cells was evaluated using BD FACScalibur flow cytometer equipped with CellQuest Pro software (Becton Dickinson).

### Study design

Cells were grown in Petri dishes (∅ 3.5, 6 or 10 cm, depending on the experiment). At 80% confluence cells were exposed to growth medium supplemented with human recombinant BMP-12 (Sigma-Aldrich, SRP4572) in the concentrations of 50 ng/ml and/or 100 ng/ml (depending on the test). Cells from the same donors cultured at the same time in standard GM without BMP-12 served as a control. Media were changed every 2 or 3 days. After 7 days cells were harvested by trypsinisation, counted and directed either to RNA/protein isolation, or to functional tests on microplates (proliferation, migration, oxidative stress susceptibility, mixed lymphocyte reaction). If certain test required further culturing, the medium containing or not BMP-12 was used respectively. Experiments were always conducted on cells from each donor separately. The cells from different donors were not pooled in this study. This approach allowed for detection inter-individual variations. Unless it stated differently, all experiments were performed on cells from 6 different donors $n = 6$. The scheme of study design is presented in Additional file 1.

### RNA isolation

For gene expression analysis cells were treated for 7 days with/without 100 ng/ml of BMP-12. At least $3 \times 10^5$ cells were used for this procedure. Isolation of total RNA was performed using RNeasy Mini Kit (Qiagen) according to the manufacturer's instructions. RNA concentration and purity was assessed by spectrophotometer at 260 nm using NanoDrop (ND-1000 Spectrophotometer, Nano-Drop Technologies, Inc).

### Real-time PCR analysis

Real-Time PCR was performed on ABI Prism 7500 Sequence Detection System using TaqMan® RNA-to-$C_T$™ 1-Step Kit (Applied Biosystems, Foster City, USA). Specific primer and probe set was purchased from Applied Biosystems: Collagen, type I, alpha 1 (Col1α1) Hs00164004_m1, Scleraxis (SCX) Hs03054634_g1, Mohawk homeobox (MKX) Hs00543190_m1, Tenascin (TNC) Hs01115665_m1, Decorin (DCN) Hs00370385_m1, Runt-related transcription factor 2 (RunX) Hs01047973_m1,. GAPDH (4333764 T) gene was used for normalization. Duplicates of each sample were performed. The relative expression of mRNA expression was calculated by $2^{-\Delta\Delta Ct}$ method. The result was presented as a fold change of gene expression in relation to the calibrator. Statistical analysis was performed by comparison of dCt values using non-parametric test for related data (control versus treated cells from the same population).

### Immunocytochemistry (ICC)

To assess the effect of BMP-12 treatment on expression of collagen type I and type III ICC staining was performed. For this analysis cells were seeded on Nunc™ Lab-Tek™ II CC2™ 8-Chamber Slide System. First, cells were cultured for 7 day with or without 50 or 100 ng/ml BMP-12. For ICC quantification, the incubation time of was shortened to 5 days in order to avoid full confluence which would hinder subsequent analysis). At the end of experiment, hASCs were fixed with 4% paraformaldehyde (10 min, RT), permeabilized with 70% methanol (15 min, -20 °C), treated with blocking solution composed of 5% normal donkey serum, 1% of bovine serum albumin in PBS and probed overnight in 4 °C with Rabbit polyclonal Anti-Collagen I antibody (Abcam, ab34710, 1:300) or Rabbit polyclonal Anti-Collagen III antibody (Abcam, ab7778, 1:150) followed by secondary Alexa Fluor 594- conjugated Donkey Anti- Rabbit antibody (1:150, Jackson ImmunoResearch, 1 h, rt). The nuclei were visualized with DAPI staining (20 ng/mL of DAPI solution for 4 min, RT). The result was evaluated with fluorescence microscopy (Olympus IX51 and CellSens™ Microscope Imaging Software). The expression of collagens (type I and III) was measured as the area of specific fluorescence per cell [μm$^2$]. At least 900 cells per well were analyzed from 10 randomly selected fields of view. Immunocytochemistry was performed on hASCs from two different donors.

## Western blot (WB)

Human ASCs were cultivated with or without BMP-12 (100 ng/ml) for 7 days on Ø 100 mm. culture dishes. Collected cell pellets were lysed with RIPA buffer (50 mM Tris, pH 7.5, 150 mM NaCl, 1 mM EDTA, 1% NP-40, 0.25% Na-deoxycholate, and 1 mM PMSF) supplemented with protease inhibitor cocktail and phosphatase inhibitor cocktail (Sigma-Aldrich) for 30 min at 4°C in order to isolate protein extracts. Lysates were cleared for 20 min at 14000 rpm, and supernatants were collected. The total protein concentration was determined using Bio-Rad protein assay dye reagent according to the producer's instructions (Bio-Rad Laboratories Inc., Hercules, CA, USA). Proteins (35 μg of total protein per well) were resolved by SDS-PAGE and transferred onto PVDF membrane (Sigma-Aldrich). For immunostaining membranes were blocked with 5% nonfat dry milk in TBS (20 mM Tris-HCl, 500 mM NaCl) containing 0.5% Tween20. The membranes were incubated with Rabbit polyclonal Anti-Collagen I antibody (Abcam, ab34710, 1:500) or Rabbit polyclonal Anti-Collagen III antibody (Abcam, ab7778, 1:1000) or Rabbit polyclonal Anti-Mohawk antibody (LSBio, aa46-75, 1:1000) or Goat polyclonal Anti-Actin (Santa Cruz Biotechnology, C-11, sc1615, 1:1000) primary antibodies. Next the blots were washed three times for 15 min and incubated with appropriate secondary antibodies conjugated with IR fluorophores: IRDye 680 or IRDye 800 CW (purchased from LICOR Biosciences; Lincoln, NE, USA) at 1:5000 dilution. Odyssey Infrared Imaging System (LI-COR Biosciences) was used to analyze the protein expression. Scan resolution of the instrument was set at 169 μm and the intensity at 5. Quantification of the integrated optical density (IOD) was performed with the analysis software provided with the Odyssey scanner (LI-COR Biosciences). Immunoblot analysis for cells from each donor was performed on samples from three independent electrophoreses. For the purpose of publication the color immunoblot images were converted into black and white images in the Odyssey software.

## Mixed lymphocyte reaction (MLR)

Ten milliliters of venous blood was collected in heparinized tubes from healthy blood donors after obtaining informed consent. Separation of peripheral blood mononuclear cells (PBMCs) was performed within 2 h of withdrawal of blood. Blood samples were taken into preservative-free heparin (20 units/ml) tubes, and PBMCs were isolated by centrifugation on Histopaque-

1077 (Sigma-Aldrich) of the blood diluted 1:1 with Sodium Chloride 0.9% (0.9% NaCl, Fresenius Kabi). PBMCs were taken up in Parker medium (Biomed) supplemented with 2 mM L-glutamine (Sigma-Aldrich), 0.1 mg/ml gentamycin (KRKA), β-mercaptoethanol (Sigma), 0.23% Hepes (Sigma) and 10% fetal bovine serum (FBS, Gibco). Half of the isolated PBMCs were inactivated by gamma-irradiation for 90 min.

hASCs after 7 days culture with BMP-12 were collected and seeded onto 96-well flat-bottom plate (Greiner) in a concentration of $0.8 \times 10^4$/well. Each time, cells from the same donor cultured in parallel without BMP-12 were seeded on the same plate in the identical scheme. Cells were left overnight to attach. For the MLR, $2 \times 10^5$ PBMCs ($1 \times 10^5$ cells/well from a first donor and $1 \times 10^5$ cells/well from the second donor) were co-seeded with hASCs in the following combinations: $XX_{ir}$, $YY_{ir}$, $XY_{ir}$, $YX_{ir}$ (X – first donor's PBMCs, $X_{ir}$ – irradiated first donor's PBMCs, Y – second donor's PBMCs, $Y_{ir}$ – irradiated second donor's PBMCs). PBMC cultures without hASCs were used as controls. Cells were cultured for 5 days at 37 °C in a humidified atmosphere with 5% $CO_2$. After 5 days cells were pulsed with 1 μCi/well of 3H-thymidine (113 Ci/nmol, NEN) for the last 18 h of the incubation and harvested with an automated cell harvester (Skatron). The amount of 3H-thymidine incorporated into the cells was measured using a Wallac Microbeta scintillation counter (Wallac), giving the level of radioactivity as 'Corrected Counts per Minute' (CCPM). For this experiments hASCs from 5 different donors were used ($n = 5$), but the influence of hASCs from each donor was tested on PBMCs from two different blood donors (X and Y). Therefore, 10 separate experiments were performed, each in triplicate.

## Cell proliferation assay

Cell proliferation assay was conducted using a colorimetric BrdU proliferation ELISA immunoassay (Roche). hASCs after 7 days culture with or without BMP-12 (50 or 100 ng/ml) were collected and seeded onto 96-well flat-bottom plate in a concentration of $0.8 \times 10^4$/well in experimental or control medium respectively. The cells were allowed to grow at 37 °C and 5% CO2 for the next 36 h. Afterwards, BrdU labeling reagent was added to each well. Then, the cells were further incubated for 12 h, and the pyrimidine analogue BrdU was incorporated in place of thymidine into the DNA in proliferating cells. Next, the immunostaining was performed according to manufacturer instructions. The absorbance was measured at wavelength 450 nm (BioTek PowerWave XS). Experiments were performed in triplicates for each sample.

## Cell migration assay

The migration assay was performed using trans-well inserts with 8 μm pore membrane (BD Biosciences, San Jose, CA, USA). The wells of the 24-well glass- bottom plates (SensoPlate, Grainer) was filled with 1 ml of different culture medium: standard GM, GM containing 50 ng/mL BMP-12, or GM containing 100 ng/mL BMP-12. Human ASCs after 7 days culture with or without BMP-12 were seeded to the upper compartment of the cell culture inserts at a density $1.5 \times 10^4$ cells per insert and then inserts were placed into the proper wells. To allow cell migration from the inserts to the wells, plates were incubated at 37 °C and 5% $CO_2$ for the next 72 h. Afterwards, medium was removed and the inserts were peeled off the cells that migrated to the bottom side of the membrane using 0.25% tripsin- EDTA solution (Sigma). After 24 h, when the cells adhere to the bottom of the wells, hASCs were fixed with 70% methanol and the nuclei were visualized with DAPI staining (20 ng/mL of DAPI solution for 4 min). Results were assessed with a cell imaging multi-mode microplate reader Cytation™ 3 (BioTek) which allowed to specify the number of objects giving a signal of blue fluorescence (DAPI, wavelength: 377–477 nm). Experiments were performed in duplicates for each sample.

## hASCs viability under oxidative stress in vitro

hASCs after 7 days culture with or without BMP-12 (100 ng/ml) were collected and seeded onto 96-well flat-bottom plate in a concentration of $0.8 \times 10^4$/well in respective experimental or control medium. Cells were left to attach and after 24 h were exposed to oxidative stress by adding a medium with hydrogen peroxide in increasing concentrations (0, 750, 1000, 1500 μM). After 22 h of incubation, 20 μl of 3-(4,5-dimethylthiazol-2-yl)-2,5-diphenyl tetrazolium bromide (MTT, Sigma-Aldrich) at the concentration of 5 mg/ml in PBS were added and incubated for another 2 h. The medium was discarded and 100 μl of DMSO was added. The plate was shaken for 10 min using microplate shaker. Absorbance, corresponding to the mitochondrial dehydrogenases activity of viable cells was measured colorimetrically at wavelength 570 nm (BioTek PowerWave XS).

## Secretory activity

Like in other experiments, hASC were cultured with or without BMP-12 (100 ng/ml) for 7 days. For the last 48 h the serum was withdrawn from the medium and cells were incubated with DMEM-LG supplemented with 4% bovine serum albumin and antibiotics (1%) with or without BMP-12 respectively. At the end of incubation period, the supernatants were collected and frozen in -80 °C. Luminex multiplex assays (Procarta) were used for measuring concentration of the following molecules: IL-6, TNF-α, IL-10, EGF, VEGF, MMP-1, MMP-8 and MMP-13. TGF-β1 concentration was evaluated with ELISA kit (R&D) according to the manufacturer's instruction. All experiments were done in duplicates.

## Statistical analysis

For data analysis STATISTICA software (StatSoft®-Polska) was used. Data are presented as medians, quartiles and min-max or means ± SEM. Differences between groups were analyzed by non-parametric Wilcoxon test for related data or U Mann-Whitney if non-related data were compared. Student $T$-test was used if in compared groups consisted of at least 10 values with normal distribution confirmed by Shapiro-Wilk test. The $p$ value less than 0.05 was considered as statistically significant.

## Results

Cells were successfully isolated from 6 donors. All populations were able to form colonies and adhere to plastic surface. To characterize hASCs populations the surface antigen profiles were examined. Flow cytometry analysis demonstrated the expression of CD29 (median 93% of positive cells), CD73 (median 96%), CD90 (median 96%) and CD105 (median 87%) and revealed no (median less than 1%) expression of hematopoietic cell lines marker-CD45. Human ASCs showed the ability to differentiate into adipocytes, osteocytes and chondrocytes which was confirmed by specific staining. The representative effect of differentiation procedure are presented in Fig. 1.

## The effect of BMP-12 on tenogenic differentiation of hASCs

Treatment of hASCs with BMP-12 for 7 days resulted in an increase of *SCLERAXIS* and *MOHAWK* genes expression ($p < 0.05$ for both transcription factors). The mean change in relative expression amounted respectively 2.05 and 2.65 folds in comparison to control, untreated samples. BMP-12 treatment induced also up-regulation of osteogenic factor- *RUNX2* (mean change in relative expression was 1.87 fold, $p < 0.05$). Although the changes of dCt values for all analyzed transcription factors were statistically significant, it is worth to stress that the response of cells from independent donors was diverse what illustrates Fig. 2. The gene expression of *COLL1α1*, *TNC* and *DCN* did not differ significantly. Immunocytochemical method revealed that untreated hASCs expressed both collagen type I and type III in both tested populations (Fig. 3a, b). Quantification of ICC results demonstrated no significant differences between groups (CTRL vs BMP-12 treated cells, Fig. 3c). To verify those results, Western blot analysis was additionally performed and this technique indicated that BMP-12 treatment (100 ng/ml, 7 days) increased expression of collagen type

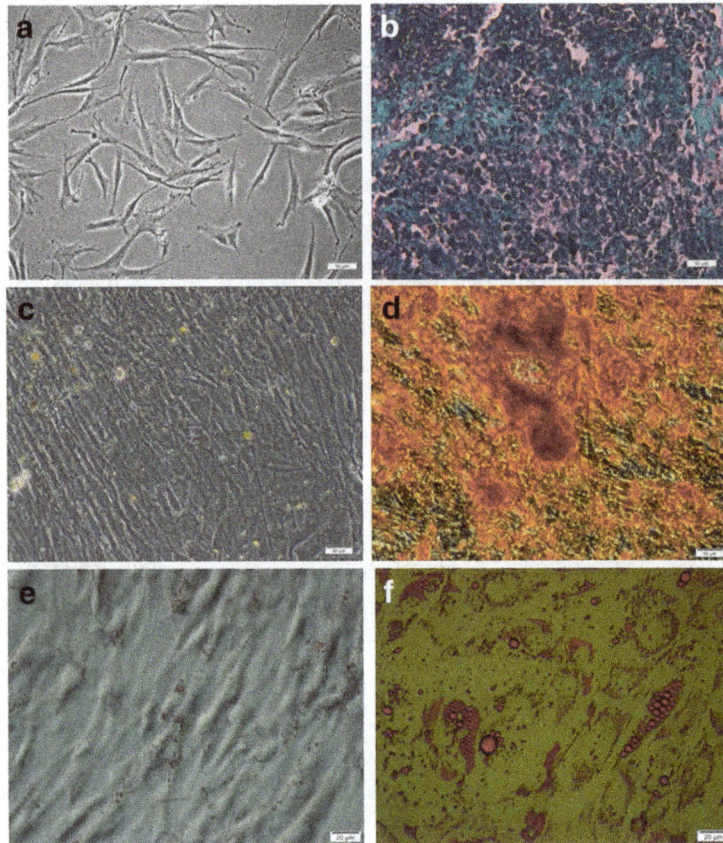

**Fig. 1** Human ASCs differentiation potential. light microscopy: **a** morphology of undifferentiated hASCs, **b** hASCs after chondrogenic differentiation, chondropellet stained with Masson's Trichrome method, collagen deposits are *blue*; **c, d** Alizarin Red staining (calcium deposits are *red*) of hASCs cultured in standard (**c**) or osteogenic medium (**d**); Oil Red O staining (lipid droplets are *red*) of hASCs cultured in standard (**e**) and adipogenic (**f**) medium. Representative data from one donor are presented. Scale bars: 50 μm (**a**, **b**, **c**, **d**); 20 μm (**e**, **f**)

I and mohawk in hASCs (mean 1.8 and 1.5 folds, respectively, both $p < 0.05$). The expression of collagen type III was also increased by mean 1.4 fold, but this change was not statistically significant (Fig. 3d).

### BMP-12 do not affect proliferation rate and migration capacity of hASCs

There were no significant differences between proliferative rate of hASCs treated with BMP-12 in comparison to the non-treated cells regardless of the applied BMP-12 dose (Fig. 4a). Likewise, migration capacity of BMP-12-treated cells was not significantly different from those of internal control (Fig. 4b). Those results were consistent among all tested populations ($n = 6$). Combined results from all donors are presented in Fig. 4.

### BMP-12 significantly increases secretion of VEGF, MMP1, MMP8 and IL6 in hASCs

BMP-12 (100 ng/ml) treatment (7 days) affected significantly hASCs secretory activity in regard to IL-6, VEGF, MMP-1 and MMP-8. All concentrations were analyzed using non-parametric Wilcoxon test for related values - treated sample in comparison to the control sample (untreated cells from the same donor). The mean increase of IL-6 concentration in supernatants after BMP-12 treatment amounted 37% ($p = 0.027$). The mean increase of pro-angiogenic VEGF after BMP-12 treatment was 17% ($p = 0.046$), whereas the mean increase of MMP-1 and MMP-8 concentrations were 41% ($p = 0.027$) and 21% ($p = 0.027$) respectively (Fig. 5). In case of IL-6 and MMP-8 an increase in secretion after BMP-12 treatment was detected in all analyzed populations ($n = 6$), whereas in case of VEGF and MMP-1 an increase in secretion after BMP-12 treatment was noted in 5/6 tested populations. As both proliferation rate (BrdU assay) and metabolic activity (MTT assay) was no affected by BMP-12 treatment in the same cells, we conclude that noted differences were the effect of BMP-12 action on analyzed hASCs. No significant changes in secretion of EGF, IL-10, TGF-b I MMP-13 were observed. Combined results from all donors are presented in Fig. 5.

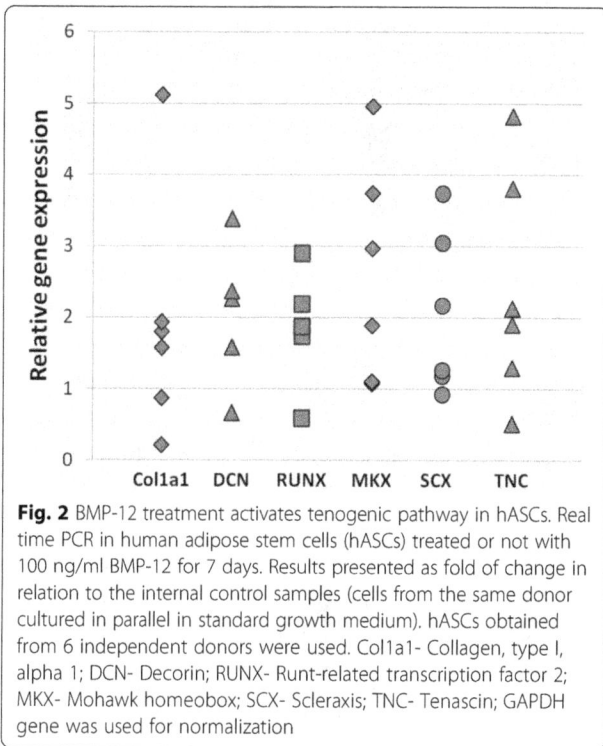

**Fig. 2** BMP-12 treatment activates tenogenic pathway in hASCs. Real time PCR in human adipose stem cells (hASCs) treated or not with 100 ng/ml BMP-12 for 7 days. Results presented as fold of change in relation to the internal control samples (cells from the same donor cultured in parallel in standard growth medium). hASCs obtained from 6 independent donors were used. Col1a1- Collagen, type I, alpha 1; DCN- Decorin; RUNX- Runt-related transcription factor 2; MKX- Mohawk homeobox; SCX- Scleraxis; TNC- Tenascin; GAPDH gene was used for normalization

## BMP-12 significantly reduces the ability of hASCs to suppress lymphocyte proliferation

ASCs are known for their immunomodulatory properties. Mixed lymphocyte reactions (MLR) were performed to determine whether treatment of hASCs with BMP-12 changes theirs in vitro immunosuppressive activity (Fig. 6). As expected, the addition of allogeneic irradiated PBMCs from donor Y to PBMCs from donor X resulted in a massive increase in lymphocytes proliferation ($p < 0.01$ in comparison to auto-stimulated cells). The addition of hASCs to the mixture of PBMCs ($X + Y_{ir}$ or opposite) significantly inhibited allo-activated lymphocytes proliferation ($p = 0.016$) to the level which was not significantly different than in control lymphocytes (auto-stimulated). This inhibition was significantly decreased when hASCs were previously exposed to 50 ng/ml BMP-12 for 7 days ($p = 0.037$ between $X + Y_{ir} + hASCs$ and $X + Y_{ir} + hASCs + BMP-12$ (50), a decrease observed in 8/10 observations). Similar effect (a decrease in inhibition in 8/10 observations) was noted in reaction with cells treated with 100 ng/ml BMP-12, but the difference between $X + Y_{ir} + hASCs$ and $X + Y_{ir} + hASCs + BMP-12(100)$ did not reach statistical significance. In reactions, where hASCs were treated with BMP-12 (both 50 and 100 ng/ml) lymphocyte proliferation became significantly higher than in control auto-stimulated ones. Figure 6 presents combined data from 5 hASCs donors (each paired with 2 different PBMCs donors, giving together 10 observations).

## BMP-12 do not change the susceptibility of hASCs to oxidative stress

The effect of oxidative stress on cells viability was determined in MTT test. In both control and BMP-12 treated cells there were significant differences in viability after 24 h exposure to growing concentrations of hydrogen peroxide compared to $H_2O_2$ untreated sample. Differences were analyzed sing non-parametric Wilcoxon test. BMP-12 treatment had no effect on hASCs resistance to oxidative stress - in both, control and BMP-12 exposed cells, 750 μM of $H_2O_2$ was required to significantly ($p < 0.05$) impair cell viability (Fig. 7). Additionally, the comparison of BMP-12 untreated and treated hASCs viability was performed within certain level of oxidative stress (750, 1000, 1500 μM). The analysis was performed using U Mann Whitney test. No significant differences were detected, the results were consistent among tested populations ($n = 6$) and in combined version are presented in Fig. 7.

## Discussion
Mesenchymal stem cells are considered as a candidate population for cell therapy in many different applications including orthopedic disorders like non-union bone fractures, osteoarthritis or more recently, tendon injuries [17]. In orthopedics, cell administration is usually local (not systemic) and it is excepted that cellular transplantation will improve regeneration of target tissue. Nevertheless, there are at least three different mechanisms by which MSCs potentially ameliorate tissue healing: 1) via differentiating, 2) via paracrine activation of endogenous progenitor cells and promoting angiogenesis, 3) via controlling inflammatory response and directing macrophages into M2 phenotype. The first, most obvious and primarily described mechanism is acting through differentiating into the target tissue cells. The capacity of MSCs (regardless of their source) to enter osteogenic and chondrogenic pathway is undoubtful and became one of identification requirement [18]. The ability to differentiate into tenocytes is not so well documented and not so easy to proof. Presented herein results demonstrate that BMP-12 induced up-regulation of genes for main transcription factors associated with tenogenesis. Previously, BMP-12 was shown to activate tenogenic pathway in rat and human bone marrow derived MSCs [12, 13]. In regard to adipose derived stem cells, the effect of BMP-12 treatment was studied on canine [16] and human cells [19]. It is important to stress that it is difficult to clearly state if a cell (especially in 2D culture) is already a tenocyte or not. There is no unequivocal method to recognize completed tenogenic differentiation as it is a case in osteogenesis (extracellular calcium deposits formation) or adipogenesis (intracellular lipid droplets). Tenocytes produce large amount of ECM

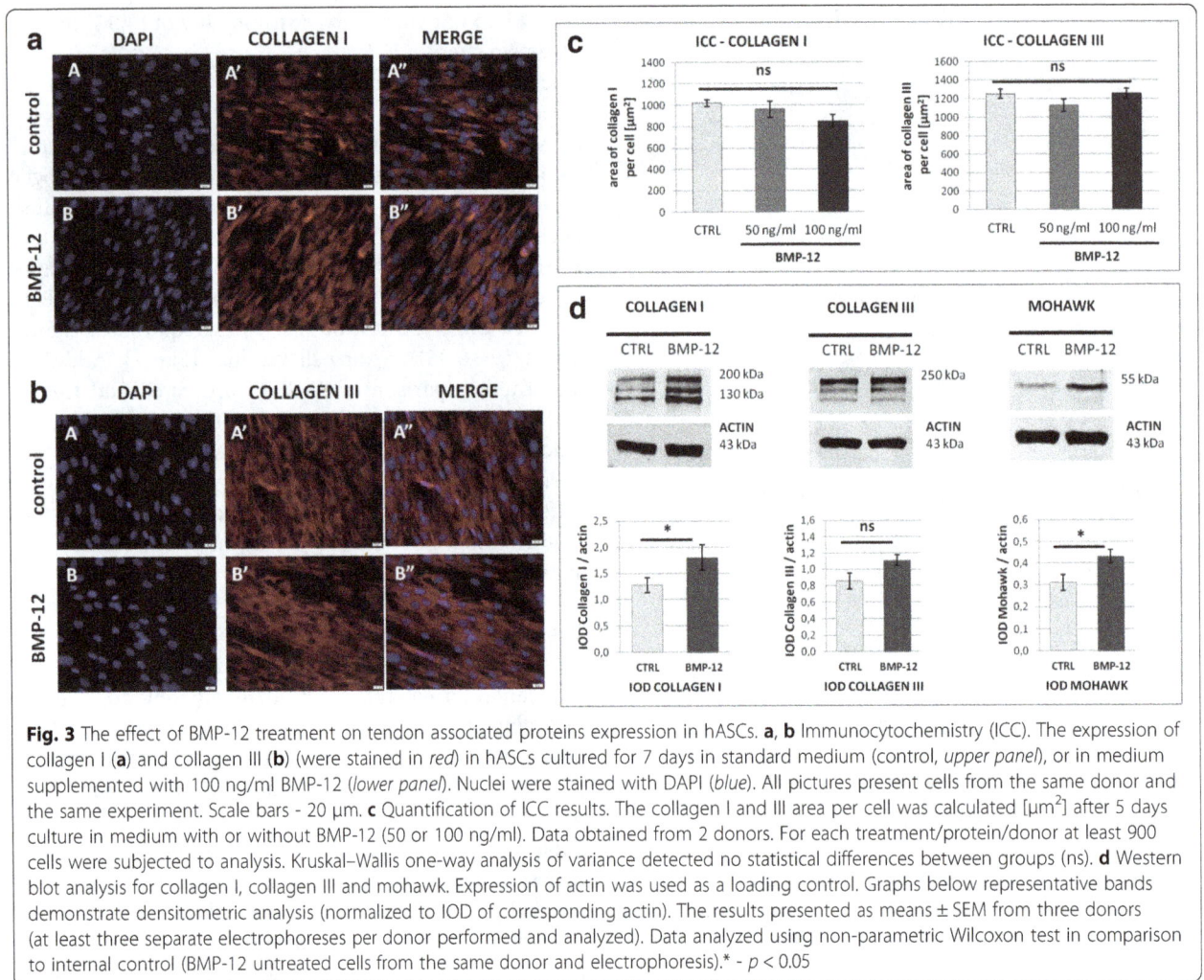

**Fig. 3** The effect of BMP-12 treatment on tendon associated proteins expression in hASCs. **a**, **b** Immunocytochemistry (ICC). The expression of collagen I (**a**) and collagen III (**b**) (were stained in *red*) in hASCs cultured for 7 days in standard medium (control, *upper panel*), or in medium supplemented with 100 ng/ml BMP-12 (*lower panel*). Nuclei were stained with DAPI (*blue*). All pictures present cells from the same donor and the same experiment. Scale bars - 20 µm. **c** Quantification of ICC results. The collagen I and III area per cell was calculated [µm$^2$] after 5 days culture in medium with or without BMP-12 (50 or 100 ng/ml). Data obtained from 2 donors. For each treatment/protein/donor at least 900 cells were subjected to analysis. Kruskal–Wallis one-way analysis of variance detected no statistical differences between groups (ns). **d** Western blot analysis for collagen I, collagen III and mohawk. Expression of actin was used as a loading control. Graphs below representative bands demonstrate densitometric analysis (normalized to IOD of corresponding actin). The results presented as means ± SEM from three donors (at least three separate electrophoreses per donor performed and analyzed). Data analyzed using non-parametric Wilcoxon test in comparison to internal control (BMP-12 untreated cells from the same donor and electrophoresis).* - $p < 0.05$

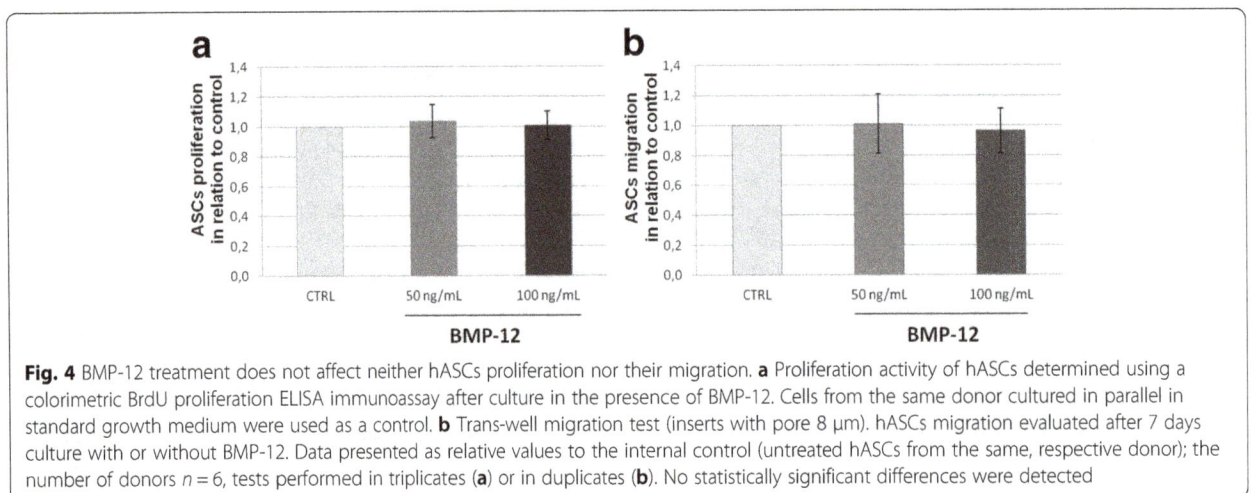

**Fig. 4** BMP-12 treatment does not affect neither hASCs proliferation nor their migration. **a** Proliferation activity of hASCs determined using a colorimetric BrdU proliferation ELISA immunoassay after culture in the presence of BMP-12. Cells from the same donor cultured in parallel in standard growth medium were used as a control. **b** Trans-well migration test (inserts with pore 8 µm). hASCs migration evaluated after 7 days culture with or without BMP-12. Data presented as relative values to the internal control (untreated hASCs from the same, respective donor); the number of donors $n = 6$, tests performed in triplicates (**a**) or in duplicates (**b**). No statistically significant differences were detected

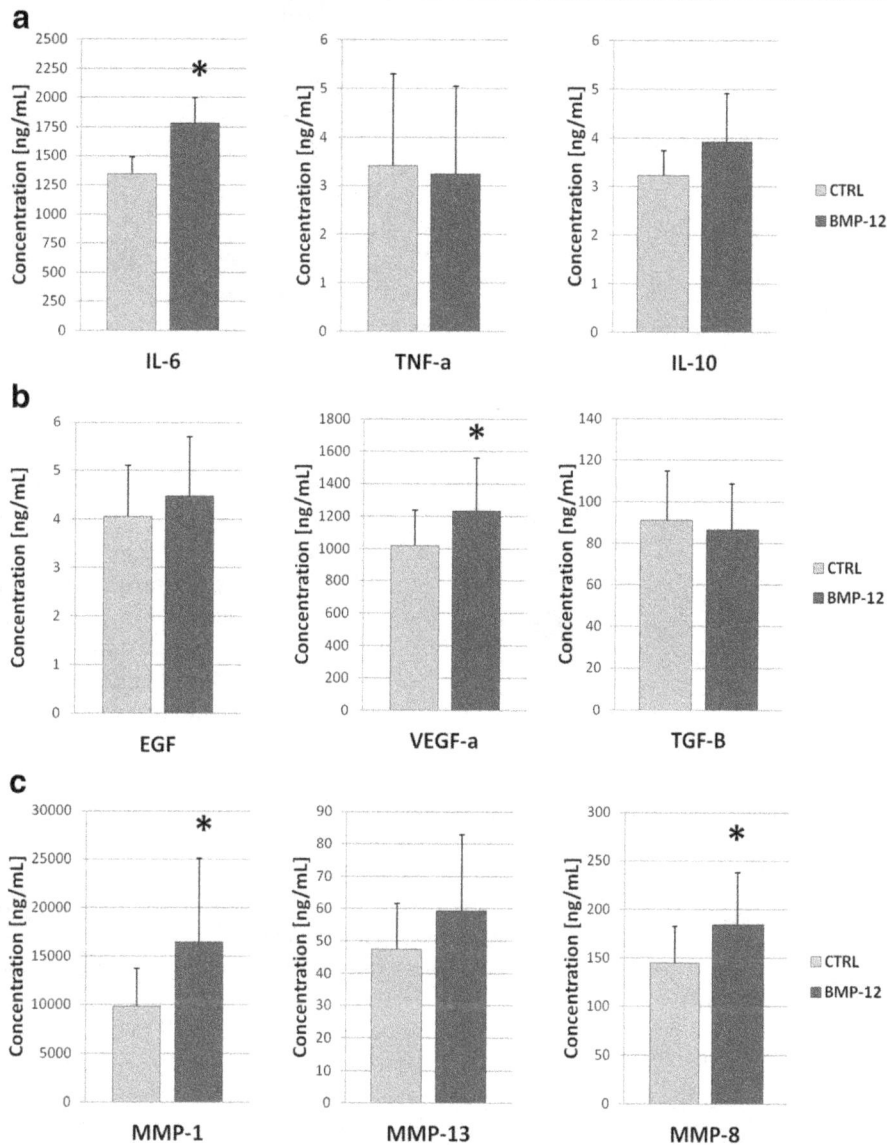

**Fig. 5** BMP-12 treatment affects secretory activity of hASCs. Cytokines (**a**), growth factors (**b**) and metalloproteinases (**c**) determined in serum free supernatants collected from above hASCs treated or not with 100 ng/ml BMP-12 (7 days, supernatants collected from last 48 h). Data presented as mean +/- SEM. Results analyzed using non-parametric Wilcoxon test in comparison to internal control (BMP-12 untreated cells from the same donor).* - $p < 0.05$. All tests were performed in duplicates; $n = 6$ donors

composed predominantly of collagen type I which is the most abundant ECM component in the whole organism and many cell types including undifferentiated MSCs express this protein. Other ECM molecules associated with tenocytes are collagen type III, decorin, tenascin C, tenomodulin, however, none of them is exclusively expressed by tenocytes. Herein, it was confirmed that undifferentiated hASCs produce both key tendon associated collagens: type I and III (Fig. 3a,b). The effect of BMP-12 on protein expression of collagen type I and III was analyzed in this study with two methods - quantified ICC and WB. In regard to collagen type I expression the results were not consistent - ICC revealed lack of

significant difference between control and treated cells whereas WB demonstrated statistically significant increase of collagen type I after BMP-12 treatment (Fig. 3d). As all cells expressed collagen type I, the ICC analysis could reflect rather cell size than the actual protein expression. Therefore, WB in this case seems to be a superior method. Other factors identified as key tenogenic markers are transcription factors: scleraxis and mohawk [12, 20], nonetheless those proteins are also expressed by undifferentiated MSCs [16]. Therefore, an in vitro evaluation of tenogenesis in two dimensional culture is always based only on relative changes in set of molecules expression. In our study we noticed an

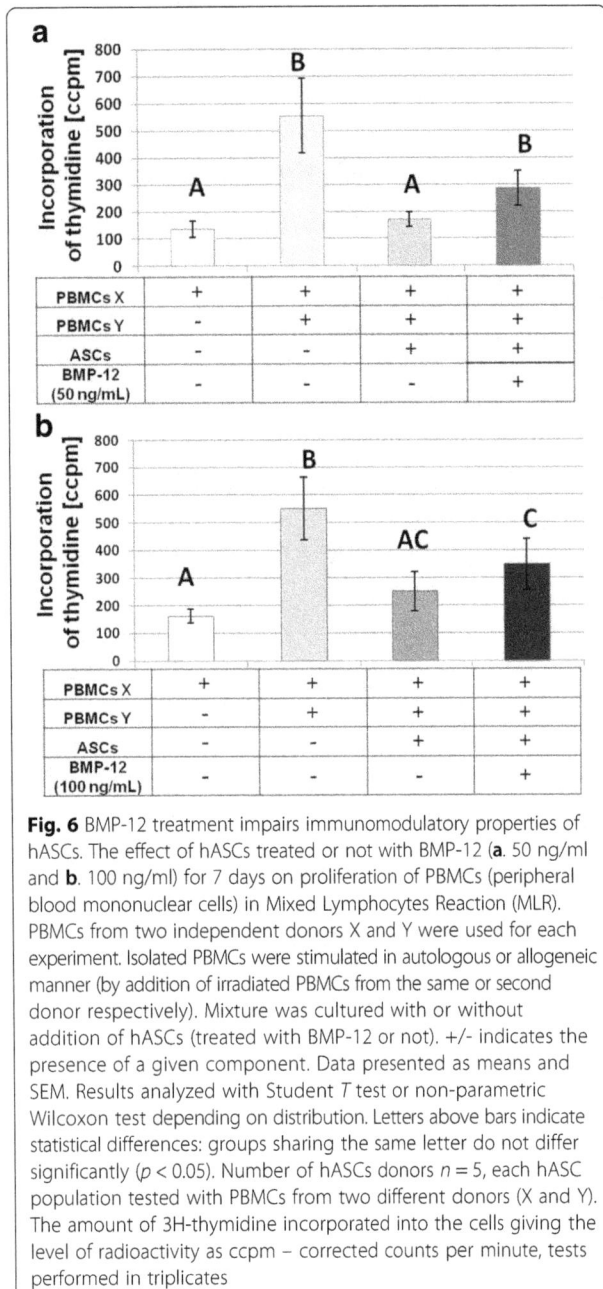

**Fig. 6** BMP-12 treatment impairs immunomodulatory properties of hASCs. The effect of hASCs treated or not with BMP-12 (**a**. 50 ng/ml and **b**. 100 ng/ml) for 7 days on proliferation of PBMCs (peripheral blood mononuclear cells) in Mixed Lymphocytes Reaction (MLR). PBMCs from two independent donors X and Y were used for each experiment. Isolated PBMCs were stimulated in autologous or allogeneic manner (by addition of irradiated PBMCs from the same or second donor respectively). Mixture was cultured with or without addition of hASCs (treated with BMP-12 or not). +/- indicates the presence of a given component. Data presented as means and SEM. Results analyzed with Student $T$ test or non-parametric Wilcoxon test depending on distribution. Letters above bars indicate statistical differences: groups sharing the same letter do not differ significantly ($p < 0.05$). Number of hASCs donors $n = 5$, each hASC population tested with PBMCs from two different donors (X and Y). The amount of 3H-thymidine incorporated into the cells giving the level of radioactivity as ccpm – corrected counts per minute, tests performed in triplicates

addition of BMP-12. In the present study, the differences in tenogenic transcription factors expression achieved statistical significance, however, the mean change (2-fold increase in case of *SCLERAXIS*) was very similar to results presented by Stanco et al. [19]. Presented herein results demonstrate also up-regulation of *MOHAWK* after BMP-12 treatment what is in agreement with data reported by Otabe et al [12]. In our study we further confirmed increased MOHAWK expression in response to BMP-12 by Western blot analysis. Nevertheless, in cells treated with BMP-12 we also noticed up-regulation of osteogenic transcription factor *RUNX2* ($p < 0.05$). Although it was demonstrated that BMP-12,-13 possess significantly lower osteogenic activity than classical bone associated BMP-2 [21], certain level of osteogenic BMP-12 potential was presented in other studies [16]. Ectopic tissue calcification is one of the concerns associated with MSC-based therapy. Moreover, tenogenic and osteogenic intracellular signaling pathways are very similar. Therefore, a risk of unfavorable differentiation in case of BMP-12 treatment should be further investigated.

Another described mechanism by which MSC can improve tissue regeneration is to promote angiogenesis through secreted molecules. The key mediators of new blood vessel formation are vascular endothelial growth factors. Mesenchymal stem cells are known to be a stable source of VEGF [22]. Presented herein results show that BMP-12 treatment significantly enhances secretion of VEGF by hASCs. It was previously demonstrated that VEGF secretion can be a crucial mechanism of MSCs action in nerve repair [23], wound healing [24], myocardium regeneration [25] or recovery from acute kidney injury repair [26]. In regard to tendon associated disorders, the role of VEGF is not so clear. In chronic tendinopathies, neovascularisation is perceived as unfavorable process which leads to painful neoinnervation [27]. On the other hand, in acute tendon injuries VEGF is rather a beneficial player. Its secretion was shown to be important in improvement of tendon graft maturation and biomechanical strength during anterior cruciate ligament healing after stem cell transplantation [28]. Similarly, administration of exogenous VEGF to the experimentally injured tendon enhanced healing parameters in comparison to the control, untreated animals [29, 30]. Finally, the relatively low level of endogenous VEGF mRNA following injury, supports its potentially beneficial role as exogenous modulator to optimize tendon healing and strength [31]. The relationship between BMP-12 (-13, -14) treatment and VEGF secretion has not been previously demonstrated, however, it was reported that BMP-4 significantly stimulated VEGF synthesis in osteoblast-like MC3T3-E1 cells [32]. The authors later proofed that this interaction was MAP kinase- dependent [33]. The release of VEGF by MSCs was

increase of tenogenic transcription factors in hASCs after BMP-12 treatment what is in agreement with previous studies performed on MSCs from other sources or species [12, 16, 19]. One previous research on human ASCs [19] showed increased expression of genes after BMP-12 treatment, but only an increase in *DECORIN* expression was statistically significant. In the mentioned study tenogenesis inducing medium contained reduced serum concentration (1% FBS) compared to control growth medium (10% FBS). For this reason cannot be unambiguously assess whether the decrease in the serum concentration can induce tenogenic path without the

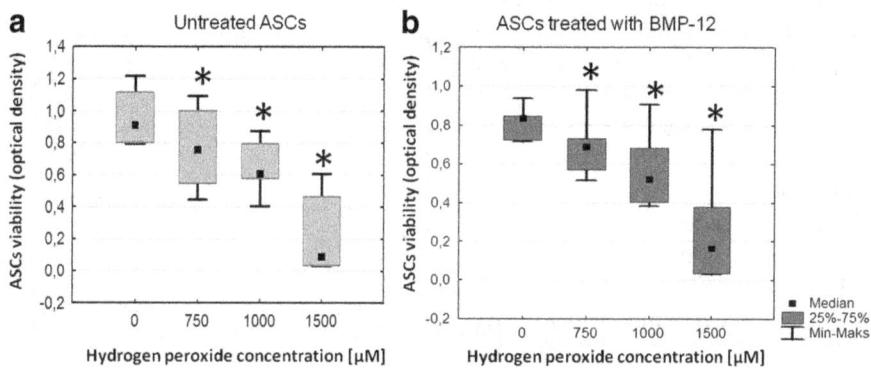

**Fig. 7** BMP-12 treatment does not affect susceptibility of hASCs to oxidative stress in vitro. MTT test. Metabolic activity (related to viability) of hASCs after exposure to growing concentrations of hydrogen peroxide (24 h exposure). **a** control hASCs (cells from the same donor cultured without BMP-12); **b** hASCs after BMP-12 treatment (100 ng/ml, 7 days before exposure to $H_2O_2$). Data presented as medians, quartiles and min-max. Results analyzed using non-parametric Wilcoxon test in comparison to $H_2O_2$ untreated sample.* - $p < 0.05$. All tests were performed in triplicates; number of hASCs donors $n = 6$

demonstrated to be mediated by both STAT3 and p38 MAPK [34]. Generally, the basic signaling process for bone morphogenic proteins goes through Smads activation (Smad 1/5/8) [15]. Activation of this particular pathway was also confirmed specifically for BMP-12 in canine ASCs [16]. Nevertheless, it was shown by Nakashima et al. that interaction between STAT3 and Smads can occur resulting in a synergistic signaling effect [35]. This phenomenon could be one of possible explanations BMP- VEGF secretion interrelation.

Interestingly, another molecule which was herein shown to be affected by BMP-12 treatment is IL-6. This cytokine was initially described as a classical pro-inflammatory molecule because its serum concentration was found to be significantly elevated in patients with various inflammatory diseases [36]. Later, it was shown, that IL-6 is a pleiotropic cytokine which can exert both pro- and anti-inflammatory effect [37]. It is believed that the way of action depends on the signaling pathway: either it is classic signaling thought membrane bound IL-6R (anti-inflammatory effect) or trans-signaling thought soluble IL-6R which interact with gp130 protein (pro-inflammatory effect) [38]. Mesenchymal stem cells are known to be an abundant source of IL-6 [39] which was confirmed in the presented study. We additionally shown that BMP-12 treatment elevates the secretion rate of this cytokine by hASCs. Because of mentioned dual role of IL-6 in inflammatory reaction it is difficult to predict what consequences would have an increased level of IL-6 in in vivo situation after cell transfer - probably the effect will depend on local inflammatory status of a host tissue. As interleukin-6 acts intracellularly through STAT-3, it might be that elevated extracellular level of IL-6 induces STAT-3 dependent pathway in ASCs and by previously mentioned interaction with Smads can enhance VEGF secretion. It is also possible

that increased level of IL-6 is related to the decreased inhibition of PBMCs proliferation by ACSs after BMP-12 treatment. It is known that IL-6 in combination with TGF-β triggers differentiation of naive Th cells towards Th17 pathway and inhibits the generation of Foxp3+ T regulatory cells induced by TGF-beta [40]. On the other hand, one of MSCs immunomodulatory mechanisms is to cause an increase in the proportion of regulatory T cells in PBMCs [41]. The TGF-β level secreted by ASCs persists unchanged and the IL-6 release is elevated under BMP-12 exposure. Therefore it is possible that more naive T cells in co-culture with BMP-12 treated ASCs differentiate toward Th17 than into T reg cells in comparison to the co-culture with untreated ASCs. This hypothesis remains to be verified. However, regardless of the mechanism, our results suggest that BMP-12 pretreatment or co-administration can be contraindicated in allogeneic ASCs (MSCs) therapies.

The key molecular feature of tendon tissue is the composition of extracellular matrix and its specific spatial arrangement. Therefore, effective remodeling of tendon ECM after injuries is particularly important. The crucial players in this process are matrix metalloproteinases, including those responsible for degradation of type I collagen which are MMP-1, MMP-8 and MMP-13 investigated in this study. All of them are released by mesenchymal stem cells irrespective of their source [42] and we demonstrated herein that BMP-12 treatment significantly elevated the secretion of MMP-1 and MMP-8 by human adipose derived stem cells. MMPs were shown to have role in physiological tendon healing [43], but on the other hand it is suggested they are involved in pathogenesis of chronic tendinopathies i.e., rotator cuff disease [44]. Tendon tissue is constantly turned over with higher rates at sites exposed to high level strain. It is believed the lost of balance between MMPs and their inhibitors can predispose to a chronic tendinopathies. Among collagenases, MMP-1 and MMP-13 were shown to be up-regulated in rotator cuff tear [45, 46]. Therefore, our results showing increased release of MMP-1 by hASCs after BMP-12 treatment suggest that using this protein in combination with cell therapy can be not beneficial in chronic tendinopathies.

Potential use of BMP-12 in combination with MSCs (ASCs) therapy required examination of BMP-12 effect on basic cellular activities like proliferation and migration. BMPs belongs to TGF-β superfamily which generally are known to inhibit cell proliferation [47]. However, it can be concluded from different studies that effect of BMPs on cell proliferation vary between both cell types and different BMPs. For example, BMP-3 were shown to promotes MSCs proliferation [48] whereas BMP-9 inhibited bFGF induced proliferation of endothelial cells [49]. Our results demonstrated that BMP-12 (50 or 100 ng/ml) had no influence on hASCs division rate what is in agreement with data presented by Haddad-Weber et al. [50]. Also, BMP-14 (GDF-5) was demonstrated to not affect human MSCs proliferation [51]. This finding is in accordance with the weak effect of BMP-12 on hASCs tenogenic differentiation which was observed in this study. It is generally accepted that the process of differentiation into relatively stable tissues like muscle, bone or tendon is associated with the significant decrease of proliferation in progenitor cells. In regard to cell migration, the effect of BMPs is not well recognized. In the discussed study, we demonstrated that BMP-12 had no effect on hASCs migration capacity what seems to be beneficial in term of potential local cell administration.

## Conclusions
In summary, MSC-based therapy is a promising treatment method in many orthopedic conditions including different tendon disorders. One of proposed approach in stem cell transplantation is to pre-differentiate cells in vitro before injection to activate certain pathways and facilitate direct regeneration of disabled tissue. In the present study we demonstrated that BMP-12 induces tenogenic pathway in human hASCs as it up-regulates key tenogenic transcription factors (SCLERAXIS and MOHAWK). However, BMP-12 treatment affected not only differentiation process in hASCs. It caused also significant changes in secretory activity of treated cells and impaired their immunomodulatory properties. The enhanced secretion of VEGF and collagenases could possibly improve regeneration process in acute tendon injuries, but in chronic tendinopathies, neoangiogenesis and MMP over-activity are rather markers of pathologic processes. Therefore, our results suggest that BMP-12 could be a candidate for cell pretreatment in cases of acute tendon injuries, but not in chronic tendinopathies.

## Additional file

Additional file 1: Study design scheme. Diagram shows the steps in the course of the study. After 7 days of culture with or without BMP-12 supernatants were collected and cells harvested. Depending on the requirements of each experiment cells were plated in 96- or 24-well plate. For analysis of gene expression RNA was isolated immediately after the end of treatment. (PDF 686 kb)

### Abbreviations
ASCs: Adipose stem cells; BMP: Bone morphogenic protein; GDF: Growth and differentiation factor; ICC: Immunocytochemistry; MLR: Mixed lymphocyte reaction; MSCs: Mesenchymal stem cells; PBMCs: Peripheral blood mononuclear cells; WB: Western blot

### Funding
This study was supported by the National Centre for Research and Developments (Grant No. STRATEGMED1/233224/10/NCBR/2014; Project START).

### Authors' contribution
WZW designed study, performed experiments, analyzed data, wrote manuscript, AB designed study, performed experiments, analyzed data, wrote manuscript, AK performed experiments, analyzed data, wrote manuscript, KG performed experiments, analyzed data, wrote manuscript, BK performed experiments, analyzed data, KZ performed WB experiments, analyzed data, wrote manuscript, KS isolated and identified ASCs, wrote manuscript, MS performed experiments, analyzed data, LP designed study, supervised experiments, interpreted data, reviewed manuscript. All authors read and approved the final manuscript.

### Competing interests
The authors declare that they have no competing interests.

### Author details
[1]Department of Immunology, Transplantology and Internal Medicine, Transplantation Institute, Medical University of Warsaw, Nowogrodzka str. 59, 02-006 Warsaw, Poland. [2]Department of Clinical Immunology, Transplantation Institute, Medical University of Warsaw, Warsaw, Poland. [3]Department of Physiological Sciences, Faculty of Veterinary Medicine, Warsaw University of Life Sciences, Warsaw, Poland. [4]Department of Regenerative Medicine, Maria Sklodowska-Curie Memorial Cancer Center, Warsaw, Poland. [5]Department of Bioinformatics, Institute of Biochemistry and Biophysics, Polish Academy of Sciences, Warsaw, Poland.

### References
1. Sharma P, Maffulli N. Tendinopathy and tendon injury: the future. Disabil Rehabil. 2008;30(20-22):1733–45.
2. Valencia Mora M, Ruiz Iban MA, Diaz Heredia J, Barco Laakso R, Cuellar R, Garcia Arranz M. Stem cell therapy in the management of shoulder rotator cuff disorders. World J Stem Cells. 2015;7(4):691–9.
3. Gaspar D, Spanoudes K, Holladay C, Pandit A, Zeugolis D. Progress in cell-based therapies for tendon repair. Adv Drug Deliv Rev. 2015;84:240–56.
4. Dai L, Hu X, Zhang X, Zhu J, Zhang J, Fu X, Duan X, Ao Y, Zhou C. Different tenogenic differentiation capacities of different mesenchymal stem cells in the presence of BMP-12. J Transl Med. 2015;13:200.
5. Aktas E, Chamberlain CS, Saether EE, Duenwald-Kuehl SE, Kondratko-Mittnacht J, Stitgen M, Lee JS, Clements AE, Murphy WL, Vanderby R. Immune modulation with primed mesenchymal stem cells delivered via biodegradable scaffold to repair an Achilles tendon segmental defect. J Orthop Res. 2016.
6. Linero I, Chaparro O. Paracrine effect of mesenchymal stem cells derived from human adipose tissue in bone regeneration. PLoS One. 2014;9(9), e107001.
7. Zhang X, Lin YC, Rui YF, Xu HL, Chen H, Wang C, Teng GJ. Therapeutic roles of tendon stem/progenitor cells in tendinopathy. Stem Cells Int. 2016;2016: 4076578.

8. Wolfman NM, Hattersley G, Cox K, Celeste AJ, Nelson R, Yamaji N, Dube JL, DiBlasio-Smith E, Nove J, Song JJ, et al. Ectopic induction of tendon and ligament in rats by growth and differentiation factors 5, 6, and 7, members of the TGF-beta gene family. J Clin Invest. 1997;100(2):321–30.

9. Bolt P, Clerk AN, Luu HH, Kang Q, Kummer JL, Deng ZL, Olson K, Primus F, Montag AG, He TC, et al. BMP-14 gene therapy increases tendon tensile strength in a rat model of Achilles tendon injury. J Bone Joint Surg Am. 2007;89(6):1315–20.

10. Lou J, Tu Y, Burns M, Silva MJ, Manske P. BMP-12 gene transfer augmentation of lacerated tendon repair. J Orthop Res. 2001;19(6):1199–202.

11. Aspenberg P, Forslund C. Enhanced tendon healing with GDF 5 and 6. Acta Orthop Scand. 1999;70(1):51–4.

12. Otabe K, Nakahara H, Hasegawa A, Matsukawa T, Ayabe F, Onizuka N, Inui M, Takada S, Ito Y, Sekiya I, et al. Transcription factor Mohawk controls tenogenic differentiation of bone marrow mesenchymal stem cells in vitro and in vivo. J Orthop Res. 2015;33(1):1–8.

13. Lee JY, Zhou Z, Taub PJ, Ramcharan M, Li Y, Akinbiyi T, Maharam ER, Leong DJ, Laudier DM, Ruike T, et al. BMP-12 treatment of adult mesenchymal stem cells in vitro augments tendon-like tissue formation and defect repair in vivo. PLoS One. 2011;6(3):e17531.

14. Zhang J, Li L. BMP signaling and stem cell regulation. Dev Biol. 2005;284(1):1–11.

15. Wagner TU. Bone morphogenetic protein signaling in stem cells–one signal, many consequences. FEBS J. 2007;274(12):2968–76.

16. Shen H, Gelberman RH, Silva MJ, Sakiyama-Elbert SE, Thomopoulos S. BMP12 induces tenogenic differentiation of adipose-derived stromal cells. PLoS One. 2013;8(10), e77613.

17. Steinert AF, Rackwitz L, Gilbert F, Noth U, Tuan RS. Concise review: the clinical application of mesenchymal stem cells for musculoskeletal regeneration: current status and perspectives. Stem Cells Transl Med. 2012;1(3):237–47.

18. Dominici M, Le Blanc K, Mueller I, Slaper-Cortenbach I, Marini F, Krause D, Deans R, Keating A, Prockop D, Horwitz E. Minimal criteria for defining multipotent mesenchymal stromal cells. The International Society for Cellular Therapy position statement. Cytotherapy. 2006;8(4):315–7.

19. Stanco D, Vigano M, Perucca Orfei C, Di Giancamillo A, Peretti GM, Lanfranchi L, de Girolamo L. Multidifferentiation potential of human mesenchymal stem cells from adipose tissue and hamstring tendons for musculoskeletal cell-based therapy. Regen Med. 2015;10(6):729–43.

20. Li Y, Ramcharan M, Zhou Z, Leong DJ, Akinbiyi T, Majeska RJ, Sun HB. The role of scleraxis in fate determination of mesenchymal stem cells for tenocyte differentiation. Sci Rep. 2015;5:13149.

21. Berasi SP, Varadarajan U, Archambault J, Cain M, Souza TA, Abouzeid A, Li J, Brown CT, Dorner AJ, Seeherman HJ, et al. Divergent activities of osteogenic BMP2, and tenogenic BMP12 and BMP13 independent of receptor binding affinities. Growth Factors. 2011;29(4):128–39.

22. Kagiwada H, Yashiki T, Ohshima A, Tadokoro M, Nagaya N, Ohgushi H. Human mesenchymal stem cells as a stable source of VEGF-producing cells. J Tissue Eng Regen Med. 2008;2(4):184–9.

23. Man AJ, Kujawski G, Burns TS, Miller EN, Fierro FA, Leach JK, Bannerman P. Neurogenic potential of engineered mesenchymal stem cells overexpressing VEGF. Cell Mol Bioeng. 2016;9(1):96–106.

24. Wu Y, Chen L, Scott PG, Tredget EE. Mesenchymal stem cells enhance wound healing through differentiation and angiogenesis. Stem Cells. 2007;25(10):2648–59.

25. Zhang J, Wu Y, Chen A, Zhao Q. Mesenchymal stem cells promote cardiac muscle repair via enhanced neovascularization. Cell Physiol Biochem. 2015;35(3):1219–29.

26. Togel F, Weiss K, Yang Y, Hu Z, Zhang P, Westenfelder C. Vasculotropic, paracrine actions of infused mesenchymal stem cells are important to the recovery from acute kidney injury. Am J Physiol Renal Physiol. 2007;292(5):F1626–35.

27. Rees JD, Stride M, Scott A. Tendons–time to revisit inflammation. Br J Sports Med. 2014;48(21):1553–7.

28. Takayama K, Kawakami Y, Mifune Y, Matsumoto T, Tang Y, Cummins JH, Greco N, Kuroda R, Kurosaka M, Wang B, et al. The effect of blocking angiogenesis on anterior cruciate ligament healing following stem cell transplantation. Biomaterials. 2015;60:9–19.

29. Tang JB, Wu YF, Cao Y, Chen CH, Zhou YL, Avanessian B, Shimada M, Wang XT, Liu PY. Basic FGF or VEGF gene therapy corrects insufficiency in the intrinsic healing capacity of tendons. Sci Rep. 2016;6:20643.

30. Zhang F, Lei MP, Oswald TM, Pang Y, Blain B, Cai ZW, Lineaweaver WC. The effect of vascular endothelial growth factor on the healing of ischaemic skin wounds. Br J Plast Surg. 2003;56(4):334–41.

31. Berglund ME, Hart DA, Reno C, Wiig M. Growth factor and protease expression during different phases of healing after rabbit deep flexor tendon repair. J Orthop Res. 2011;29(6):886–92.

32. Kozawa O, Matsuno H, Uematsu T. Involvement of p70 S6 kinase in bone morphogenetic protein signaling: vascular endothelial growth factor synthesis by bone morphogenetic protein-4 in osteoblasts. J Cell Biochem. 2001;81(3):430–6.

33. Tokuda H, Hatakeyama D, Shibata T, Akamatsu S, Oiso Y, Kozawa O. p38 MAP kinase regulates BMP-4-stimulated VEGF synthesis via p70 S6 kinase in osteoblasts. Am J Physiol Endocrinol Metab. 2003;284(6):E1202–9.

34. Wang M, Zhang W, Crisostomo P, Markel T, Meldrum KK, Fu XY, Meldrum DR. STAT3 mediates bone marrow mesenchymal stem cell VEGF production. J Mol Cell Cardiol. 2007;42(6):1009–15.

35. Nakashima K, Yanagisawa M, Arakawa H, Kimura N, Hisatsune T, Kawabata M, Miyazono K, Taga T. Synergistic signaling in fetal brain by STAT3-Smad1 complex bridged by p300. Science. 1999;284(5413):479–82.

36. Kishimoto T. IL-6: from its discovery to clinical applications. Int Immunol. 2010;22(5):347–52.

37. Kyurkchiev D, Bochev I, Ivanova-Todorova E, Mourdjeva M, Oreshkova T, Belemezova K, Kyurkchiev S. Secretion of immunoregulatory cytokines by mesenchymal stem cells. World J Stem Cells. 2014;6(5):552–70.

38. Scheller J, Chalaris A, Schmidt-Arras D, Rose-John S. The pro- and anti-inflammatory properties of the cytokine interleukin-6. Biochim Biophys Acta. 2011;1813(5):878–88.

39. Blaber SP, Webster RA, Hill CJ, Breen EJ, Kuah D, Vesey G, Herbert BR. Analysis of in vitro secretion profiles from adipose-derived cell populations. J Transl Med. 2012;10:172.

40. Bettelli E, Carrier Y, Gao W, Korn T, Strom TB, Oukka M, Weiner HL, Kuchroo VK. Reciprocal developmental pathways for the generation of pathogenic effector TH17 and regulatory T cells. Nature. 2006;441(7090):235–8.

41. Aggarwal S, Pittenger MF. Human mesenchymal stem cells modulate allogeneic immune cell responses. Blood. 2005;105(4):1815–22.

42. Amable PR, Teixeira MV, Carias RB, Granjeiro JM, Borojevic R. Gene expression and protein secretion during human mesenchymal cell differentiation into adipogenic cells. BMC Cell Biol. 2014;15:46.

43. Oshiro W, Lou J, Xing X, Tu Y, Manske PR. Flexor tendon healing in the rat: a histologic and gene expression study. J Hand Surg [Am]. 2003;28(5):814–23.

44. Del Buono A, Oliva F, Longo UG, Rodeo SA, Orchard J, Denaro V, Maffulli N. Metalloproteases and rotator cuff disease. J Shoulder Elbow Surg. 2012;21(2):200–8.

45. Yoshihara Y, Hamada K, Nakajima T, Fujikawa K, Fukuda H. Biochemical markers in the synovial fluid of glenohumeral joints from patients with rotator cuff tear. J Orthop Res. 2001;19(4):573–9.

46. Lo IK, Marchuk LL, Hollinshead R, Hart DA, Frank CB. Matrix metalloproteinase and tissue inhibitor of matrix metalloproteinase mRNA levels are specifically altered in torn rotator cuff tendons. Am J Sports Med. 2004;32(5):1223–9.

47. Huang SS, Huang JS. TGF-beta control of cell proliferation. J Cell Biochem. 2005;96(3):447–62.

48. Stewart A, Guan H, Yang K. BMP-3 promotes mesenchymal stem cell proliferation through the TGF-beta/activin signaling pathway. J Cell Physiol. 2010;223(3):658–66.

49. Scharpfenecker M, van Dinther M, Liu Z, van Bezooijen RL, Zhao Q, Pukac L, Lowik CW, ten Dijke P. BMP-9 signals via ALK1 and inhibits bFGF-induced endothelial cell proliferation and VEGF-stimulated angiogenesis. J Cell Sci. 2007;120(Pt 6):964–72.

50. Haddad-Weber M, Prager P, Kunz M, Seefried L, Jakob F, Murray MM, Evans CH, Noth U, Steinert AF. BMP12 and BMP13 gene transfer induce ligamentogenic differentiation in mesenchymal progenitor and anterior cruciate ligament cells. Cytotherapy. 2010;12(4):505–13.

51. Tan SL, Ahmad RE, Ahmad TS, Merican AM, Abbas AA, Ng WM, Kamarul T. Effect of growth differentiation factor 5 on the proliferation and tenogenic differentiation potential of human mesenchymal stem cells in vitro. Cells Tissues Organs. 2012;196(4):325–38.

# 9

# Hinokitiol induces DNA demethylation via DNMT1 and UHRF1 inhibition in colon cancer cells

Jung Seon Seo[1], Young Ha Choi[1], Ji Wook Moon[2], Hyeon Soo Kim[1] and Sun-Hwa Park[1*]

## Abstract

**Background:** DNA hypermethylation is a key epigenetic mechanism for the silencing of many genes in cancer. Hinokitiol, a tropolone-related natural compound, is known to induce apoptosis and cell cycle arrest and has anti-inflammatory and anti-tumor activities. However, the relationship between hinokitiol and DNA methylation is not clear. The aim of our study was to explore whether hinokitiol has an inhibitory ability on the DNA methylation in colon cancer cells.

**Results:** MTT data showed that hinokitiol had higher sensitivity in colon cancer cells, HCT-116 and SW480, than in normal colon cells, CCD18Co. Hinokitiol reduced DNA methyltransferase 1 (DNMT1) and ubiquitin-like plant homeodomain and RING finger domain 1 (UHRF1) expression in HCT-116 cells. In addition, the expression of ten-eleven translocation protein 1 (TET1), a known DNA demethylation initiator, was increased by hinokitiol treatment. ELISA and FACS data showed that hinokitiol increased the 5-hydroxymethylcytosine (5hmC) level in the both colon cancer cells, but 5-methylcytosine (5mC) level was not changed. Furthermore, hinokitiol significantly restored mRNA expression of $O^6$-methylguanine DNA methyltransferase (*MGMT*), carbohydrate sulfotransferase 10 (*CHST10*), and B-cell translocation gene 4 (*BTG4*) concomitant with reduction of methylation status in HCT-116 cells.

**Conclusions:** These results indicate that hinokitiol may exert DNA demethylation by inhibiting the expression of DNMT1 and UHRF1 in colon cancer cells.

**Keywords:** Hinokitiol, DNA methylation, Anti-tumor activities, DNA methylation inhibitor, Colonic neoplasm

## Background

Epigenetic modifications are responsible for the initiation and maintenance of gene silencing. Hypermethylation of a CpG island in a promoter region is the most well-established epigenetic alteration. Aberrant promoter methylation causes transcriptional inactivation of many genes involved in tumor suppression, cell cycle regulation, apoptosis, and DNA repair. Modified DNA methylation is found in various types of cancer during carcinogenesis [1, 2]. DNA methylation is mediated by DNA methyltransferases (DNMTs), which catalyzes the transfer of methyl groups to $C^5$ of cytosine from *S*-adenosyl methionine [3]. DNMT1 maintains DNA methylation and possesses *de novo* methyltransferase activity during DNA replication, and DNMT3A and DNMT3B play an important role as *de novo* methyltransferases. DNMTs interact with transcriptional repression factors and histone deacetylases (HDACs) and thus directly causes transcription inactivation [4]. DNMT1 is recruited by replication foci via its interaction with the ubiquitin-like plant homeodomain and RING finger domain 1 (UHRF1). It was well known that UHRF1 is involved in *de novo* methylation of DNMT3A and DNMT3B and plays a pivotal role in carcinogenesis through gene silencing mechanisms and co-operating with HDAC1, which activates the DNMTs and recruited by methyl CpG binding proteins [5]. On the other hand, recent evidence demonstrates that human ten-eleven translocation (TET) enzymes have catalytic activity capable to convert 5-methylcytosine (5mC) to 5-hydroxymethylcytosine (5hmC), resulting in an initiation of DNA demethylation [6].

---
* Correspondence: parksh@korea.ac.kr
[1]Department of Anatomy, Institute of Human Genetics, Korea University College of Medicine, 73, Inchon-ro, Seongbuk-gu, Seoul 02841, Republic of Korea
Full list of author information is available at the end of the article

Currently, targeting enzymes that modify DNA methylation is considered an attractive therapeutic strategy for cancer treatment. Indeed, DNMT inhibition blocks the methylation of newly synthesized DNA strands, resulting in the reversion of the methylation status and the reactivation of silenced genes, such as tumor suppressors [7]. Several DNMT inhibitors, including 5-aza-2′-deoxycytidine (5-aza-dC), zebularine, and (–)-epigallocatechin-3-gallate (EGCG), reduce DNA methylation and re-express silenced genes. Thus, they have been suggested as potential anticancer drugs in various cancer cells *in vitro* and *in vivo*, but side effects such as DNA mutagenesis and cytotoxicity are still a cause for concern [7–9].

Hinokitiol (4-isopropyltropolone) is a component of essential oils extracted from *Chymacyparis obtusa* and has anti-infective, anti-oxidative effects, and anti-tumor activities. The anti-tumor activity of hinokitiol has been demonstrated in several types of cancer cells by inhibiting cell growth and inducing apoptosis [10–12]. However, the relevant molecular mechanisms of hinokitiol regarding anti-cancer effects are still unclear.

The goal of this study was to investigate a possible mechanism of hinokitiol on DNA methylation in human colon cancer cell lines. Our data demonstrated that hinokitiol decreased DNMT1 and UHRF1 expression and increased the level of TET1 in colon cancer cell line HCT-116. Furthermore, hinokitiol altered the methylation status of 10 hypermethylated genes in colon cancer cells and significantly reactivated the mRNA expression of O$^6$-methylguanine DNA methyltransferase (*MGMT*), carbohydrate sulfotransferase 10 (*CHST10*), and B-cell translocation gene 4 (*BTG4*), which are involved in cell proliferation or biological oxidation [13–15].

## Results

### Hinokitiol inhibits colon cancer cell growth in a dose- and time-dependent manner

To gain insight into the anti-proliferation effects of hinokitiol in colon cancer cells, we treated HCT-116 and SW480 cells with hinokitiol of different concentrations and times. Using cell morphological observation, we found that the number of cells decreased with increasing concentrations of hinokitiol (Additional file 1A). To compare the effects of hinokitiol on the viability of colon cancer and normal colon cells, a MTT assay was performed. As shown in Fig. 1, hinokitiol affected the viability of HCT-116 and SW480 cells in dose- and time-dependent manners. In contrast, the viability of normal colon cells was maintained at concentrations over 5 μM of hinokitiol. Our data showed that hinokitiol had higher sensitivity in colon cancer cells than in normal colon cells (P < 0.001).

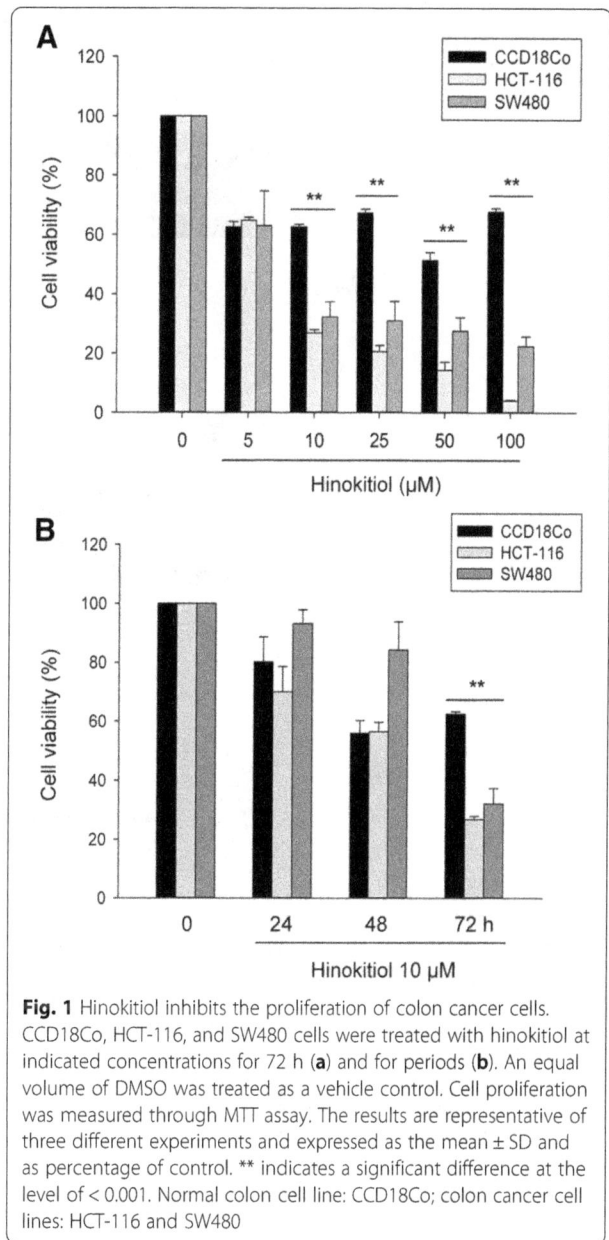

**Fig. 1** Hinokitiol inhibits the proliferation of colon cancer cells. CCD18Co, HCT-116, and SW480 cells were treated with hinokitiol at indicated concentrations for 72 h (**a**) and for periods (**b**). An equal volume of DMSO was treated as a vehicle control. Cell proliferation was measured through MTT assay. The results are representative of three different experiments and expressed as the mean ± SD and as percentage of control. ** indicates a significant difference at the level of < 0.001. Normal colon cell line: CCD18Co; colon cancer cell lines: HCT-116 and SW480

### Hinokitiol inhibits the expression of DNMT1 in HCT-116 cells

The expression patterns of DNMT1 in normal colon cells and colon cancer cells were analyzed by RT-PCR and qRT-PCR. In Fig. 2a and b, DNMT1 mRNA is highly expressed in colon cancer cells, especially in HCT-116 cells, while only slightly expressed in CCD18Co cells. To gain insight into the role of hinokitiol on DNA methylation, we measured DNMT1 expression after the hinokitiol treatment. HCT-116 cells were exposed to the indicated concentrations of hinokitiol for up to 72 hrs, and their DNMT1 expression was analyzed every 24 h using qRT-PCR. We found that hinokitiol reduced

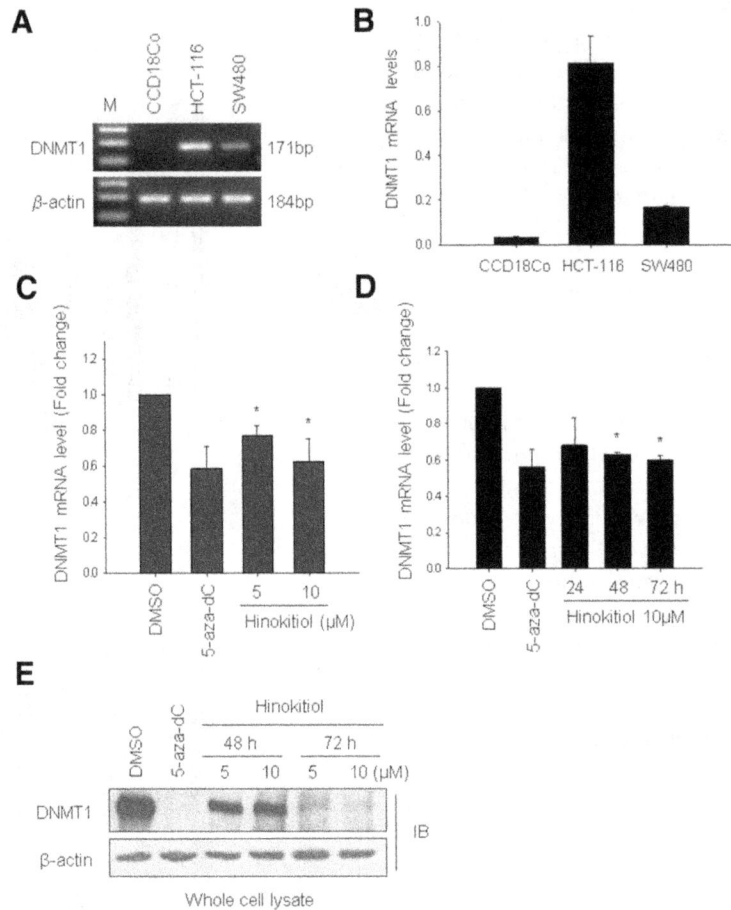

**Fig. 2** Hinokitiol decreases DNMT1 mRNA and protein expression in HCT-116 cells. The level of DNMT1 mRNA was determined in HCT-116 and SW480 cells using RT-PCR (**a**) and quantitative real-time PCR (qRT-PCR) (**b**). PCR products were gel-run in 2% agarose and visualized in ultraviolet (UV). β-actin was used as a quantitative control. The quantitative DNMT1 mRNA level was measured using qRT-PCR after hinokitiol treatment at indicated concentrations and times in HCT-116 cells (**c** and **d**). Total protein was isolated from cells treated with indicated concentrations of hinokitiol for 72 h and western blotting was performed to detect DNMT1 expression (**e**). The results were representative of three independent experiments. Cells treated with DMSO and 5-aza-dC were used as negative and positive controls, respectively. The β-actin was used as a loading control. * indicates a significant difference at the level of < 0.05. M, 100 bp DNA ladder

DNMT1 mRNA expression in time- and dose-dependent manners (P < 0.05) (Fig. 2c and d). In addition, the effect of hinokitiol on the expression of DNMT1 protein was confirmed using Western blotting. As shown in Fig. 2e, hinokitiol (5 and 10 μM) treatment decreased the protein level of DNMT1 in a time-dependent manner. Furthermore, the expression levels of DNMT1 mRNA and protein with 10 μM of hinokitiol for 72 h were similar to that of the 5-aza-dC treatment. We also showed the decreased expression of DNMT1 protein in SW480 cells treated with hinokitiol for 72 h (Additional file 1B). These results suggest that hinokitiol may be a promising agent for DNMT1 inhibition in colon cancer cells.

**Hinokitiol inhibits the expression of UHRF1 in HCT-116 cells**

UHRF1 forms a complex with DNMT1 to maintain DNA methylation. In order to investigate the involvement of hinokitiol on UHRF1, the level of UHRF1 protein was measured in nuclear extracts of HCT-116 cells. Like in Fig. 2e, hinokitiol decreased level of DNMT1 protein (Fig. 3a). In addition, the level of UHRF1 protein was reduced in HCT-116 cells exposed to 5 and 10 μM of hinokitiol, whereas the control cells were not (Fig. 3b). Furthermore, knockdown DNMT1 expression by siRNA reduced UHRF1 protein level (Fig. 3c). These results suggest that hinokitiol may be associated with DNA demethylation pathways in HCT-116 cells.

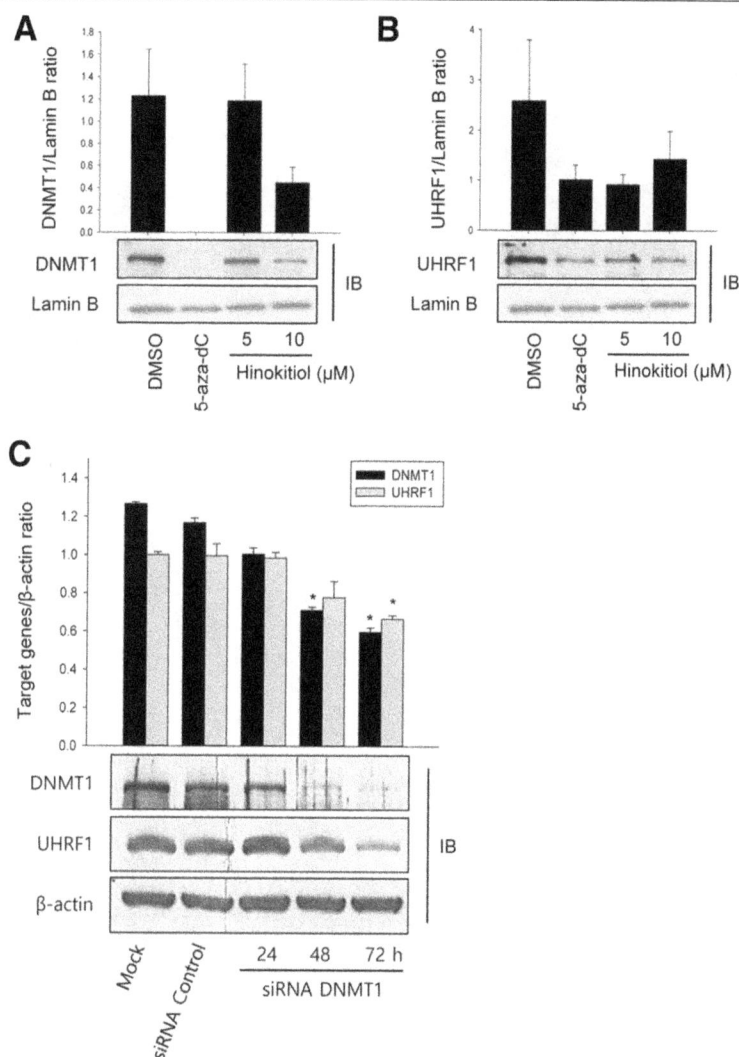

**Fig. 3** Hinokitiol decreases UHRF1 protein expression in HCT-116 cells. HCT-116 cells were treated with indicated concentrations of hinokitiol for 72 h. Nuclear protein was extracted from the cells for western blot analysis with indicated antibodies (**a-b**). Lamin B was used for loading control. HCT-116 cells were seeded into culture plates. After 24 h cultivation, cells were transiently transfected with siRNA DNMT1 and siRNA Control for 24, 48, 72 h, collected, and total protein was extracted for western blot analysis using indicated antibodies (**c**). Experiments were done twice and representative blots were shown. The expression level of each protein was quantified with the Image Studio Lite program, using Lamin B as a loading control. The histogram shows the quantification expressed as ratio of the intensity of target gene/Lamin B. * indicates a significant difference at the level of < 0.05

### Hinokitiol enhances TET1 activity in HCT-116 cells

To confirm the effect of hinoikitiol on DNA demethylation, TET1 protein expression in nuclear fraction of cells was evaluated by Western blotting. Hinokitiol increased TET1 protein expression in a dose-dependent manner and its level was higher than that of 5-aza-dC treatment (Fig. 4a). In addition, the alteration of 5hmC level by hinokitiol treatment was assessed by using ELISA in HCT-116 cells. Pretreatment with 10 μM hinokitiol increased the 5hmC level of total DNA about 1.87-fold (from 0.040 to 0.075) compared to the control (Fig. 4b). TET1 activity was also analyzed by measuring the levels of 5mC and 5hmC using

FACS analysis. Hinokitiol treatment induced a significant enhancement of 5hmC level (fluorescence intensity, FI:50) (P < 0.05). The 5mC level was not affected in either treatment (Fig. 4c). These results indicate that hinokitiol may cause DNA demethylation through the downregulation of DNMT1 as well as the upregulation of TET1 without reducing the level of 5mC in colon cancer cells.

### Hinokitiol restores the mRNA expression of *MGMT*, *BTG4*, and *CHST10* via demethylation

To verify the effect of demethylation and restoration of hinokitiol on silenced genes resulting from DNA

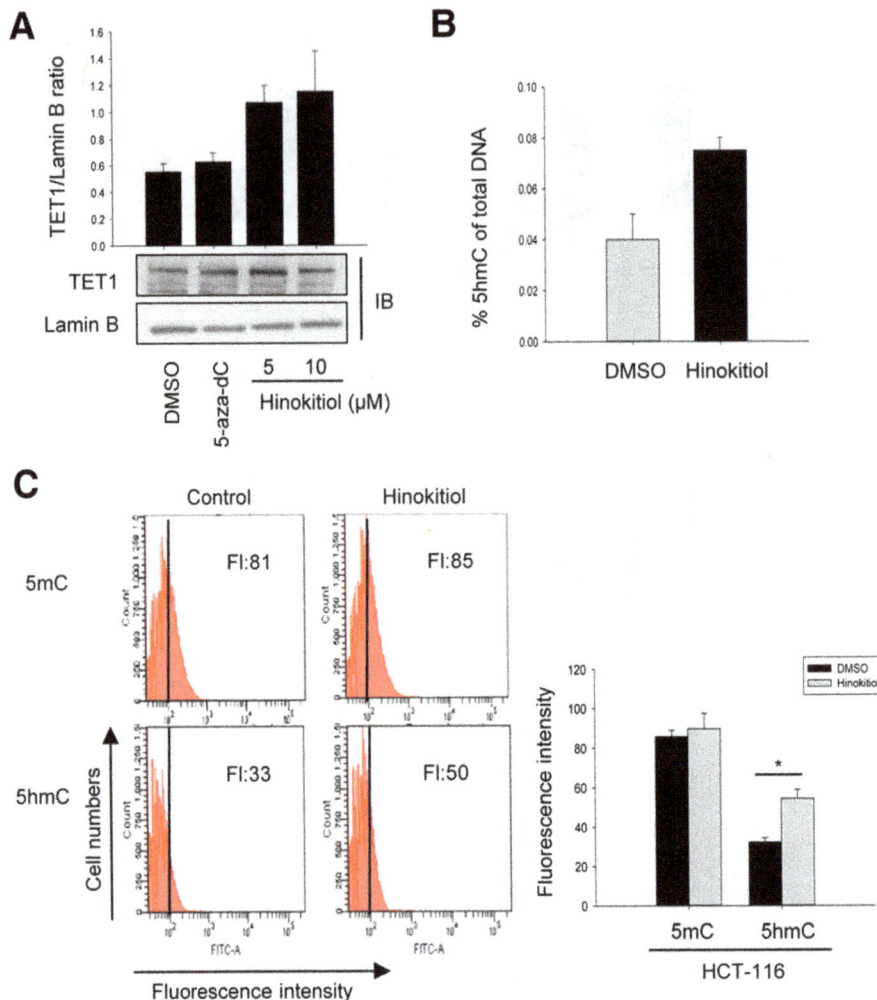

**Fig. 4** Hinokitiol increases TET1 expression via enhancement of 5hmC level in HCT-116 cells. HCT-116 cells were treated with 5 and 10 μM of hinokitiol for 72 h and nuclear protein isolated from the cells was used to detect TET1 expression using western blot analysis (**a**). The expression level of each protein was quantified with the Image Studio Lite program, using Lamin B as a loading control. The histogram shows the quantification expressed as ratio of the intensity of target gene/Lamin B. Experiments were done twice and representative blots were shown. Contents of 5hmC measured from cells treated with 10 μM of hinokitiol for 72 h using ELISA-based methylflash hydroxymethylated DNA quantification kit (**b**). The levels of 5mC and 5hmC were confirmed using flow cytometry analysis (**c**). All data are representative of three independent experiments performed in duplicate. The results were representative of three independent experiments. Data are the means ± SE of results from at least three independent experiments. * indicates a significant difference at the level of < 0.05

methylation, the levels of methylation and mRNA of three CIMP markers and seven candidate genes in colon cancer cells were analyzed by using QMSP and qRT-PCR, respectively. In our previous study, we observed that three CIMP markers (*NEUROG1*, *TAC1*, and *MGMT*) and seven novel methylation candidate genes (*AKR1B1*, *CHST10*, *BTG4*, *ELOVL4*, *EYA4*, *SPG20*, and *UNC5C*) were hypermethylated with the loss of the respective mRNA expression in HCT-116 cells but not in CCD18Co cells [16]. Here, the methylation status of these genes was reduced after treating the cells with 10 μM of hinokitiol for 72 h, though not completely

reduced, and hinokitiol-reversed methylation levels of those genes were similar to that produced by a 5-aza-dC (Fig. 5a). Moreover, hinokitiol significantly reactivated the mRNA expression of *MGMT*, *CHST10*, and *BTG4* (P < 0.05) (Fig. 5b).

## Discussion

The key findings of this study are that there were significant differences in sensitivity to hinokitiol between both colon cancer cells and normal colon cells. This result was similar to those of other reports showing that hinokitiol inhibits cell viability in colon cancer cells

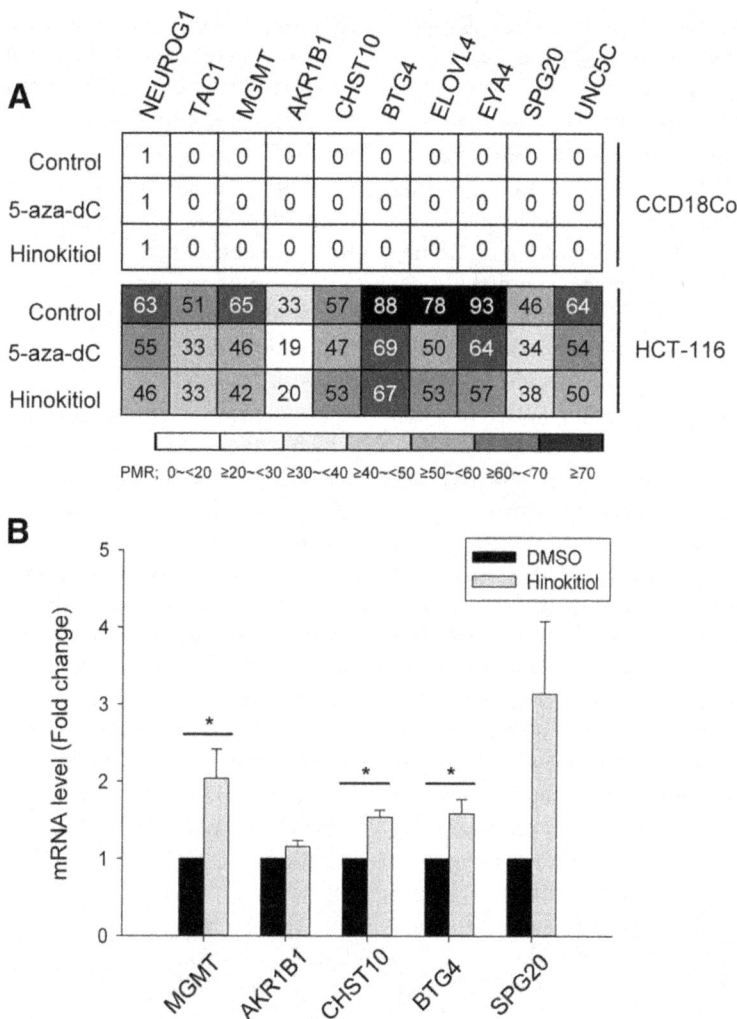

**Fig. 5** Hinokitiol reduces methylation status and restores mRNA expression of *MGMT, CHST10,* and *BTG4* genes. The effects of hinokitiol on the methylation status of hypermethylated genes in HCT-116 cells compared with CCD18Co, were assessed using QMSP. The ratio of methylation intensity was determined by the percentage of methylated reference (PMR). Cells treated with DMSO or 5-aza-dC at the same condition were used as negative and positive controls, respectively (**a**) The mRNA expressions of *MGMT, AKR1B1, CHST10, BTG4,* and *SPG20* genes were measured using qRT-PCR (**b**). Genomic DNA and total RNA were extracted from cells treated with 10 μM of hinokitiol for 72 h. Data are the means ± SE of results from at least three independent experiments. * indicates a significant difference at the level of < 0.05

(HCT116 and SW620) but not in normal colon cells (CCD112CoN) [12]. Importantly, we demonstrated that hinokitiol decreased the protein expression of DNMT1 as well as UHRF1 in HCT-116 cells, and its effect was similar with that of 5-aza-dC. Hinokitiol is a natural tropolone-based monoterpenoid which is found in *cupressacceous* plants. It has been shown that hinokitiol possesses potent anti-tumor effects in various cancer cell lines, including colon cancer, lung adenocarcinoma, breast cancer, and melanoma cells, by inducing cell cycle arrest or apoptosis [12, 17, 18]. However, previous studies have focused on elucidating the molecular mechanism of hinokitiol-induced

anticancer effects through apoptosis pathways, and the mechanisms underlying its effects are not yet fully understood. In the present study, our results demonstrated that hinokitiol has potential as a novel DNMT inhibitor and could be associated with DNA methylation and thus provide hinokitiol as new therapeutic candidate of colon cancer.

Epigenetic alterations, including DNA methylation and histone modifications, play crucial roles in carcinogenesis and show possible targets for cancer treatment and prevention. DNMT1 is most abundantly found in mammalian cells. Increased DNMT expression has been proposed as a mechanism for the increased methylation that occurs in the promoter region of tumors. In DNA methylation, the role of DNMTs is more complex and may involve changes in the expression of mRNA or protein. Previous studies have reported that DNMT1 inhibition correlates with reduction in tumorigenicity and increased expression of tumor suppressor genes, such as p16 $^{INK4a}$ or p14 $^{ARF}$ [19, 20]. Thus, DNMT1 represents one of the most attractive targets for development of anticancer drugs. For example, azacytidine and decitabine are well known as DNMT inhibitors and the most successful epigenetic modulators. Zebularine is another DNMT inhibitor with concomitant inhibitory activity towards cytidine deaminase. EGCG has dual actions involving both DNA demethylation and posttranslational histone modifications in cancer cells [7, 9, 21, 22]. However, there are still restrictions because of their toxicity and poor stability or lack of information linked to the possible clinical applications.

In the present study, we determined that hinokitiol decreased UHRF1 proteins, which was similar or lower with those of 5-aza-dC in HCT-116 cells and demonstrated that such UHRF1 inhibition was caused by DNMT1 reduced via transient transfection of siRNA DNMT1. It has been reported that increased UHRF1 is associated with cellular proliferation and has observed in various types of cancers such as colorectal cancer. In addition, UHRF1 binds to methylated promoters of many tumor suppressor genes by forming complexes with DNMTs and HDAC1, resulting ultimately in the formation of cancer [23, 24]. Therefore, our results indicate that hinokitiol may directly regulate the expression of DNA methylation-related genes and may inhibit cellular proliferation in colon cancer. We also measured the expression of HDAC1 protein to investigate whether hinokitiol could affect histone deacetylation. Incubation of cells to hinokitiol for 72 h resulted in a dose-dependent decrease of HDAC1 protein levels in the same cells, and its expression was lower than that of 5-aza-dC (Additional file 2).

Ten-eleven translocation (TET) enzymes are have been identified as key players in demethylation of cytosine. TET family genes are frequently observed in human cancers [25]. Our data demonstrated that hinokitiol increased the expression of TET1 protein well known as DNA demethylation initiator in HCT-116 and SW480 cells (Additional file 3) but did not affect to the 5mC level. TET1 is an iron-dependent α-ketoglutarate dioxygenase enzyme, and both fully methylated and hemimethylated DNA in a CG or non-CG context can serve as substrates for it. TET1 is responsible for converting 5mC to 5hmC methylcytosine because of causing demethylation. Inactivation of TET1 is associated with aberrant DNA methylation in cancers. TET1 methylation is connected with CpG island methylator phenotype (CIMP) in colorectal cancer [26, 27]. On the other hand, several studies reported that various chemical agents such as GSH, vitamin B1, and vitamin E, appeared not to affect 5mC oxidation patterns despite DNA demethylation effects via an increase of 5hmC level in cultured cells [28, 29]. We also observed whether hinokitiol affects both TET2 and TET3 protein levels. Our data showed that hinokitiol had little effect on the expression of TET3 protein in HCT-116 and SW480 cells (Additional file 4). On the other hand, TET2 protein expression did not detected in both cells (data not shown). It might be because TET2 has a higher expression level in hematological cells than TET1 or TET3 and is recruited to genomic DNA by distinct CXXC domain-independent mechanism unlike TET1 and TET3 [30]. Therefore, our results suggest that hinokitiol may cause TET1-mediated DNA modifications in colon cancer cells without 5mC reduction.

Furthermore, our data demonstrates that hinokitiol alters the methylation status of hypermethylated genes in HCT-116 cells. Three CIMP markers and seven new methylation candidates, which were reported in our previous study [16], were selected randomly and assessed by using QMSP for this study. In this experiment, hinokitiol reduced methylation status of ten genes and induced the significant restoration of MGMT, CHST10, and BTG4 mRNA expression. MGMT is a DNA repair enzyme involved in direct repair of alkylation damage products. CHST10 is also known to inhibit the invasiveness of melanoma cells, and BTG4 has anti-proliferative properties [13, 15, 31]. The evidence for promoter methylation patterns of those genes has been shown in various cancers, including colon cancer [31–33]. Further studies are required to clarify these relationships, particularly that hinokitiol induces effectively the reactivation of epigenetic-silencing genes in colon cancer cells through combination treatment with HDAC inhibitors or chemotherapeutic drugs, including 5-FU and oxaliplatin.

## Conclusions

This study has demonstrated the relationship between hinokitiol and DNA methylation in colon cancer cells and provided evidence for the potential of hinokitiol as a novel DNMT1 inhibitor for colon cancer treatment.

## Methods

### Cell culture and chemical treatment

The human colon cancer cell lines (HCT-116 and SW480) and a normal colon cell line (CCD18Co) were purchased from Korean Cell Line Bank (KCLB,

Seoul, Korea) and the American Type Culture Collection (ATCC, Manassas, VA, USA). CRC cell lines were maintained in RPMI-1640 medium (WELGENE, Daejeon, Korea), and CCD18Co cell line was maintained in Eagle's minimum essential medium (MEM, WELGENE) containing 10% FBS and 1% penicillin-streptomycin in a 5% $CO2$ atmosphere at 37 °C. Hinokitiol ($\beta$-thujaplicin) and 5-aza-dC were dissolved in dimethyl sulfoxide (DMSO). All chemicals were purchased from Sigma-Aldrich (St. Louis, MO, USA). As a control, 10 μM of 5-aza-dC was added freshly every 24 h for 3 days.

### Cell viability assay and morphologic

Cell viability was evaluated using MTT (3-(4, 5-Dimethylthiazol-2-yl)-2, 5-diphenyltetrazolium bromide) assay. Briefly, cells were cultured in 96-well plates for 24 h and then incubated with different concentrations of hinokitiol (5–100 μM) for 72 h in a 5% $CO_2$ incubator at 37 °C. At the indicated times, 0.5 mg/ml MTT solution was added to the wells. After a further 3 h of incubation, the supernatant was discarded, and DMSO was added to the plates. The color intensity was measured at 570 nm using microplate reader (SpectraMax Puls 384, Molecular Devices, USA). The cell viability of each sample was presented as a percentage of the viability of culture treated with control. Cells treated with DMSO were used as a control, which was considered to be 100% viable. These tests were performed on all samples at least three times in the same assay. Changes in the morphology of cells cultured with hinokitiol were observed by using the EVOS® Image Systems (Thermo Fisher Scientific, Hudson, NH, USA).

### RNA extraction and reverse transcription (RT)-PCR

RNA was isolated from cells using the QIAzol lysis reagent (Qiagen, Hilden, Germany), according to the manufacturer's instructions. cDNA synthesis was performed in mixtures containing 2 μg of RNA, oligo-dT, and AMV-reverse transcriptase (Promega, Madison, WI), following the manufacturer's instructions. PCR was performed using a SureCycler 8800 Thermal Cycler (Agilent Technologies, Santa Clara, CA, USA). The primers used in the PCR reactions are summarized in Table 1. The PCR program was initiated at 95 °C for 10 min, followed by 35 cycles of 95 °C for 15 s, 60 °C for 20 s, and 72 °C for 45 s. The amplified products were separated on a 2% agarose gel, stained with Safe Shine Green (Biosesang, Korea), and visualized using BioDoc-It Imaging Systems (An Analytik Jena Company, CA, USA). $\beta$-Actin was used to normalize the amount of cDNA.

**Table 1** Primers for mRNA expression

| Genes | Primer sequences (5'→ 3') | Annealing temp. (°C) | Product size (bp) |
|---|---|---|---|
| DNMT1 | F:AGACTACGCGAGATTCGAGTC | 60 | 171 |
| | R:TTGGTGGCTGAGTAGTAGAGG | | |
| MGMT | F:ACCGTTTGCGACTTGGTACT' | 54 | 268 |
| | R:CGGGGAACTCTTCGATAGCC | | |
| AKR1B1 | F:CCCATGTGTACCAGAATGAGAA' | 58 | 362 |
| | R:CTGGAGATGGTTGAAGTTGGAG' | | |
| CHST10 | F:GTGTGATTGGACACCACGAG' | 58 | 176 |
| | R:ATACAGGCGTCGGATGTCTC' | | |
| BTG4 | F:GCAAGGAACCTCGTGTCATT | 58 | 159 |
| | R:GCGAGCCATGGTAAGTGTTT' | | |
| SPG20 | F:CCAACTGGAACAGAGCAGAAG' | 58 | 160 |
| | R:TTAGCCCTTTTTCCACGTTTT' | | |
| $\beta$-actin | F:AGAGCTACGAGCTGCCTGAC | 60 | 184 |
| | R:AGCACTGTGTTGGCGTACAG | | |

### Quantitative real-time PCR (qRT-PCR)

To detect the mRNA level of DNMT1 and 10 hypermethylated genes in HCT-116 cells, qRT-PCR was performed using ABI 7500 Real-Time PCR system (Applied Biosystems, Foster City, CA, USA). The sets of primers indicated (Table 1) and primer sequences were designed using Primer3 ver. 0.4.0 (http://primer3.ut.ee/). The experiments were repeated at least three times, and each sample was analyzed in duplicate. The relative amount of target genes was normalized by the amount of $\beta$-actin used as the internal control in the same sample and described as the ratio of each target gene/$\beta$-actin. The cDNA of a known concentration was prepared by serial dilutions, which were used as the standard curve for quantification.

### Genomic DNA extraction

Genomic DNA was extracted using the Wizard genomic DNA purification kit (Promega), according to the manufacturer's recommendations. The cells were lysed in 600 μl of nuclei lysis solution, RNase solution was added, and the mixture was incubated at 37 °C for 30 min. The samples were mixed with 200 μl of protein precipitation solution and centrifuged at 13,200 g for 5 min at room temperature. The supernatants were transferred to fresh tubes, and the genomic DNA was precipitated with isopropanol and washed with 70% ethanol. The DNA eluted in 800 μl of DNA rehydration solution was quantified using a NanoDrop ND-1000 spectrophotometer (Thermo Fisher Scientific).

## Western blot analysis

To extract total protein, cells were lysed in RIPA buffer containing protease inhibitors (15 mM PMSF, 1 mM NaF, and 1 mM Na3VO4) and stored at −80 °C until use. In addition, a nuclear protein was isolated from cells using the EpiQuik nuclear extraction kit (Epigentek Group Inc), according to the manufacturer's instructions. The supernatants were collected as total protein and nuclear protein extracts, respectively. Aliquots of the supernatants (50 μg of total protein or nuclear extracts) were subjected to sodium dodecyl sulfate polyacrylamide gel electrophoresis (SDS-PAGE). The proteins were separated and then transferred onto a nitrocellulose membrane (GE Healthcare Life Science, Pittsburgh, PA, USA). The membranes were incubated with anti-DNMT1 (sc-20701), UHRF1 (Abcam, Cambridge, MA, USA), TET1, TET2, TET3, Lamin B (Santa Cruz Biotechnology, CA, USA), and $\beta$-actin (Sigma-Aldrich) antibodies. After washing, the membranes were incubated with horseradish peroxidase-conjugated secondary antibodies (Santa Cruz Biotechnology). The bound antibodies were visualized using the SuperSignal West Dura Extended Duration Substrate (Thermo-Fisher Scientific). $\beta$-actin for total protein and Lamin B for a nuclear protein were used to confirm comparable loading. The expression level of each protein was quantified with the Image J program using Lamin B or $\beta$-actin as a loading control.

## Transient transfection of DNMT1 small RNA interference (siRNA)

Cells were seeded at $0.3 \times 10^5$ cells/well in 6-well plates and allowed to reach approximately 60% confluence on the day of transfection. The siRNA constructs synthesized were: a mismatched siRNA control and an siRNA against DNMT1 (Cosmogenetech, Seoul, Korea). Cells were transfected with 50–100 nM siRNA using lipofectamine 3000 based on the manufacturer's protocol. Cells were harvested and examined by qRT-PCR and western blot analysis at 24, 48, and 72 h after transfection.

## Quantification of 5mC and 5hmC

For quantification of 5hmC, the ELISA-based methylflash hydroxymethylated DNA quantification kit (Epigentek Group Inc., NY, USA) was used. The initial incubation time was 90 min, and the final developing time was 10 min. The absolute quantification of standard curves was generated by plotting the concentration of the positive control supplied with the assay against the optical density at 450 nm after performing the assay. The amount of 5hmC was assessed by absolute quantification. For detection of the levels of 5mC and

5hmC by flow cytometry, cells were washed with PBS supplemented with 0.1% Tween-20 and 1% bovine serum albumin (PBST-BSA) and fixed with 0.25% paraformaldehyde at 37 °C for 10 min and 88% methanol at −20 °C for 1 h. Cells were permeated with 10% Triton-X 100 at room temperature for 10 min and then treated with 2 N HCL at 37 °C for 30 min and neutralized with 0.1 M sodium borate (pH 8.5). The cells were blocked with 10% FBS in PBST-BSA at 37 °C for 20 min, incubated with anti-5mC and anti-5hmC (Active Motif, Carlsbad, CA, USA) antibodies at 37 °C for 50 min, followed by staining with Cy3- or Alexa 488-conjugated secondary antibodies (Jackson ImmunoResearch Laboratories, PA, USA). The cells were washed with PBS twice and analyzed by using a flow cytometer (FACS Canto II, BD-Science, San Jose, CA, USA).

## Bisulfite treatment and quantitative real-time methylation-specific PCR (QMSP)

Bisulfite modification of DNA was performed using the EpiTect Fast Bisulfite Conversion kits (Qiagen), following the manufacturer's directions. The reaction was prepared by mixing 85 μl of bisulfite solution and 35 μl of DNA protect buffer in PCR tubes at room temperature. The modified DNA was purified using Qiaquick gel extraction kit (Qiagen) and eluted with water. The methylation status of *NEUROG1*, *TAC1*, *MGMT*, *AKR1B1*, *CHST10*, *BTG4*, *ELOVL4*, *EYA4*, *SPG20*, and *UNC5C* genes was determined by QMSP using ABI 7500 Real-time PCR Systems (Applied Biosystems). Primers for QMSP were designed for potential CpG islands near the translation start site using the NCBI database and summarized in Table 2. Methylation primers were designed using MethPrimer software (http://www.urogene.org/methprimer/). The PCR program was initiated at 95 °C for 5 min, followed by 40 cycles of 95 °C for 15 s and 54–60 °C for 1 min. The experiments were repeated at least three times, and each sample was analyzed in duplicates. Relative quantification of the amplified gene levels in bisulfite-converted genomic DNA sample was performed by measuring the threshold cycle (CT) values of target genes and $\beta$-actin. The relative amount of target genes was normalized by the amount of $\beta$-actin used as the internal control in the same sample and described as the ratio of each target gene/$\beta$-actin. The bisulfite-converted genomic DNA of a known concentration was prepared by serial dilutions, which were used as the standard curve for quantification. The modified genomic DNA by CpG methyltransferase M. *Sss*I (NEB, Ipswich, MA, USA) was used as a positive control. DNA methylation according to M.*Sss*I was verified using the restriction enzyme *BstU*I (NEB).

**Table 2** Primers for quantitative methylation-specific PCR (QMSP)

| Gene | Primer sequences (5′ → 3′) | Annealing temp. (℃) | Product size (bp) | Location | Gene bank |
|------|---------------------------|---------------------|-------------------|----------|-----------|
| NEUROG1 | F:TATGTAAATATTCGGGCGTTGTAC | 58 | 199 | (−155 ~ +43) | NC_000005.9 |
| | R:GATCTCCTAAATAATATCGCCGAC | | | | |
| TAC1 | F:TTAGATTTGTAGACGGAAGTAGGTC | 58 | 139 | (−135 ~ +3) | NC_000007.13 |
| | R:GTAATTAAAAATTTCCGAAACGAT | | | | |
| MGMT | F:GTTTGTATTGGTTGAAGGGTTATTT | 58 | 109 | (−292 ~ −184) | NC_000010.10 |
| | R:CTAAAACAATCTACACATCCTCACT | | | | |
| AKR1B1 | F:CGGAAGAAGTATTTTCGTCGA | 58 | 166 | (−120 ~ +45) | NC_000007.13 |
| | R:CAATACGATACGACCTTAACCG | | | | |
| CHST10 | F:TTTTGTAGCGGTAGAAAGGGAGATTCG | 58 | 198 | (−125 ~ +72) | NC_000002.11 |
| | R:GACTTTAAAAACCAAAACGCCGAC | | | | |
| BTG4 | F:GTATAATACGCGTAGTTGGGTTAGC | 58 | 171 | (−422 ~ −252) | NC_0000011.9 |
| | R:AAAAAAACGAAAAAAACCTAAACG | | | | |
| ELOVL4 | F:GAGTTTAGGTGTTTCGTTTTCGTTC | 58 | 114 | (−158 ~ −42) | NC_000006.11 |
| | R:CCTCCCTCCCTAATATTAAAACTCG | | | | |
| EYA4 | F:GTTATTCGAGGTTAAATAAAAACGG | 58 | 149 | (−198 ~ −50) | NC_000006.11 |
| | R: ACTTACGCAAAAAAATAAAACGAA | | | | |
| SPG20 | F:GTCGAGTAGTCGACGTGGTC | 58 | 168 | (−246 ~ −78) | NC_000013.10 |
| | R:AATAATACGTAAAAAAACGTCCGTC | | | | |
| UNC5C | F:GTTTAGGTTTGGCGTATCGC | 58 | 219 | (−463 ~ −245) | NC_000004.11 |
| | R:GCCAAAAAAACGTAAAAAACG | | | | |
| β-actin | F:TGGTGATGGAGGAGGTTTAGTAAGT | 58 | 132 | (−1645 ~ −1513) | NC_000007.13 |
| | R: AACCAATAAAACCTACTCCTCCCTTAA | | | | |

## Statistical analysis

Results are presented as means ± SE. For statistical analyses, Student's t-test for independent samples was used. A value of P < 0.01 or < 0.05 was considered to indicate statistical significance. All analyses were performed using SigmaPlot12.0.

## Additional files

**Additional file 1:** Hinokitiol affects cell morphology and DNMT1 protein expression in colon cancer cells. HCT-116 cells were treated with 10 μM of hinokitiol and cultured for 72 h. Changes in cell morphology was observed under the phase contrast microscope by using EVOS Image System (a). SW480 cells were treated with 10 μM of hinokitiol for 72 h. Nuclear protein was isolated from the cells and 50 μg of protein was separated on SDS-PAGE. After then, western blot analysis was performed with anti-DNMT1 and anti-Lamin B antibodies. Lamin B was used for loading control (b). Cells treated with DMSO or 5-aza-dC were used as negative and positive controls, respectively. (JPG 289 kb)

**Additional file 2:** Hinokitiol decreases HDAC1 protein expression in HCT-116 cells. HCT-116 cells were treated with indicated concentrations of hinokitiol for 72 h. Nuclear protein was extracted from the cells for western blot analysis with anti-HDAC1 and anti-Lamin B antibodies (Upper panel). Lamin B was used for loading control. Experiments were done twice and representative blots were shown. The expression level of each protein was quantified with the Image Studio Lite program, using Lamin B as a loading control. The histogram shows the quantification expressed as ratio of the intensity of target gene/Lamin B (lower panel). (JPG 170 kb)

**Additional file 3:** Hinokitiol enhances of 5hmC level in SW480 cells. SW480 cells were treated with 5 and 10 μM of hinokitiol for 72 h and nuclear protein was isolated. Contents of 5hmC was measured using ELISA-based methylflash hydroxymethylated DNA quantification kit (a). The levels of 5mC and 5hmC were confirmed using flow cytometry analysis (b). All data are representative of three independent experiments performed in duplicate. The results were representative of three independent experiments. Data are the means ± SE of results from at least three independent experiments. (JPG 246 kb)

**Additional file 4:** Hinokitiol has no affect TET3 protein expression in colon cancer cells. Colon cancer cells (HCT-116 or SW480) were treated with 10 μM of hinokitiol for 72 h. Nuclear protein was extracted from the cells and then western blot analysis was performed with anti-TET3 and anti-Lamin B antibodies. Lamin B was used for loading control. Experiments were done twice and representative blots were shown. Cells treated with DMSO or 5-aza-dC were used as negative and positive controls, respectively. (JPG 102 kb)

## Abbreviations

5-aza-dC: 5-aza-2′-deoxycytidine; 5-hmC: 5-hydroxymethylcytosine; 5-mC: 5-methylacytosine; AKR1B1: Aldo-keto reductase 1; BTG4: B-cell translocation gene 4; CHST10: carbohydrate sulfotransferase 10; CpG: Cytosine-phosphate-guanine; DMSO: Dimethyl sulfoxide; DNMT: DNA methyltransferease; ELOVL4: Elongation of very long chain fatty acids protein 4; EYA4: Eyes absent homolog 4; GSH: Glutathione; HDAC: Histone deacethylase; MGMT: O⁶-methylguanine DNA methyltransferase; NEUROG1: Neurogenin-1; SPG20: Spastic paraplegia 20; TAC1: Protachykinin-1; TETs: Ten-eleven translocation proteins; UHRF1: Ubiquitin-like plant homeodomain and RING finger domain 1; UNC5C: Unc-5 homolog C

## Acknowledgments

We thank Prof. Jin Won Hyun (School of Medicine, Jeju National University) for TET1, TET2, and TET3 antibodies.

## Funding

This study was supported by Korea University Grant (K1507751).

## Author's contributions

JSS and Prof. SHP developed the concept of this study. JSS designed, carried out experiments, analyzed data, created figures, and wrote the manuscript. YHC prepared RNA and genomic DNA samples. JWM participated in the analysis of methylation results. Prof. HSK and Prof. SHP revised the manuscript. All authors read and approved the final manuscript.

## Competing interests

The authors declare that there are no competing interests.

## Author details

[1]Department of Anatomy, Institute of Human Genetics, Korea University College of Medicine, 73, Inchon-ro, Seongbuk-gu, Seoul 02841, Republic of Korea. [2]Department of Pathology, Korea University College of Medicine, 73, Inchon-ro, Seongbuk-gu, Seoul 02841, Republic of Korea.

## References

1. Shen H, Laird PW. Interplay between the cancer genome and epigenome. Cell. 2013;153:38–55.
2. Baylin SB. The cancer epigenome: its origins, contributions to tumorigenesis, and translational implications. Proc Am Thorac Soc. 2012;9:64–5.
3. Robertson KD. DNA methylation, methyltransferases, and cancer. Oncogene. 2001;20:3139–55.
4. Fuks F, Burgers WA, Brehm A, Hughes-Davies L, Kouzarides T. DNA methyltransferase Dnmt1 associates with histone deacetylase activity. Nat Genet. 2000;24:88–91.
5. Liu X, Gao Q, Li P, Zhao Q, Zhang J, Li J, et al. UHRF1 targets DNMT1 for DNA methylation through cooperative binding of hemi-methylated DNA and methylated H3K9. Nat Commun. 2013;4:1563.
6. Ciesielski P, Jóźwiak P, Krześlak A. TET proteins and epigenetic modifications in cancers. Postepy Hig Med Dosw (Online). 2015;69:1371–83.
7. Jung Y, Park J, Kim TY, Park JH, Jong HS, Im SA, et al. Potential advantages of DNA methyltransferase 1 (DNMT1)-targeted inhibition for cancer therapy. J Mol Med (Berl). 2007;85:1137–48.
8. Foulks JM, Parnell KM, Nix RN, Chau S, Swierczek K, Saunders M, et al. Epigenetic drug discovery: targeting DNA methyltransferases. J Biomol Screen. 2012;17:2–17.
9. Palii SS, Van Emburgh BO, Sankpal UT, Brown KD, Robertson KD. DNA methylation inhibitor 5-Aza-2'-deoxycytidine induces reversible genome-wide DNA damage that is distinctly influenced by DNA methyltransferases 1 and 3B. Mol Cell Biol. 2008;28:752–71.
10. Li LH, Wu P, Lee JY, Li PR, Hsieh WY, Ho CC, et al. Hinokitiol induces DNA damage and autophagy followed by cell cycle arrest and senescence in gefitinib-resistant lung adenocarcinoma cells. PLoS One. 2014;9:e104203.
11. Shih YH, Chang KW, Hsia SM, Yu CC, Fuh LJ, Chi TY, et al. In vitro antimicrobial and anticancer potential of hinokitiol against oral pathogens and oral cancer cell lines. Microbiol Res. 2013;168:254–62.
12. Lee YS, Choi KM, Kim W, Jeon YS, Lee YM, Hong JT, et al. Hinokitiol inhibits cell growth through induction of S-phase arrest and apoptosis in human colon cancer cells and suppresses tumor growth in a mouse xenograft experiment. J Nat Prod. 2013;76:2195–202.
13. Zhao X, Graves C, Ames SJ, Fisher DE, Spanjaard RA. Mechanism of regulation and suppression of melanoma invasiveness by novel retinoic acid receptor-gamma target gene carbohydrate sulfotransferase 10. Cancer Res. 2009;69:5218–25.
14. Winkler GS. The mammalian anti-proliferative BTG/Tob protein family. J Cell Physiol. 2010;222:66–72.
15. Asiaf A, Ahmad ST, Malik AA, Aziz SA, Rasool Z, Masood A, et al. Protein expression and methylation of MGMT, a DNA repair gene and their

    correlation with clinicopathological parameters in invasive ductal carcinoma of the breast. Tumour Biol. 2015;36:6485–96.
16. Huang CH, Jayakumar T, Chang CC, Fong TH, Lu SH, Thomas PA, et al. Hinokitiol Exerts Anticancer Activity through Downregulation of MMPs 9/2 and Enhancement of Catalase and SOD Enzymes: In Vivo Augmentation of Lung Histoarchitecture. Molecules. 2015;20:17720–34.
17. Liu S, Yamauchi H. p27-Associated G1 arrest induced by hinokitiol in human malignant melanoma cells is mediated via down-regulation of pRb, Skp2 ubiquitin ligase, and impairment of Cdk2 function. Cancer Lett. 2009;286:240–9.
18. Subramaniam D, Thombre R, Dhar A, Anant S. DNA methyltransferases: a novel target for prevention and therapy. Front Oncol. 2014;4:80.
19. Esteller M. Epigenetics in cancer. N Engl J Med. 2008;358:1148–59.
20. Bradbury J. Zebularine: a candidate for epigenetic cancer therapy. Drug Discov Today. 2004;9:906–7.
21. Nandakumar V, Vaid M, Katiyar SK. (−)-Epigallocatechin-3-gallate reactivates silenced tumor suppressor genes, Cip1/p21 and p16INK4a, by reducing DNA methylation and increasing histones acetylation in human skin cancer cells. Carcinogenesis. 2011;32:537–44.
22. Zhou L, Shang Y, Jin Z, Zhang W, Lv C, Zhao X, Liu Y, Li Y, Li N, Liang J. UHRF1 promotes proliferation of gastric cancer via mediating tumor suppressor gene hypermethylation. Cancer Biol Ther. 2015;16:1241–51.
23. Pacaud R, Brocard E, Lalier L, Hervouet E, Vallette FM, Cartron PF. The DNMT1/PCNA/UHRF1 disruption induces tumorigenesis characterized by similar genetic and epigenetic signatures. Sci Rep. 2014;4:4230.
24. Huang Y, Rao A. Connections between TET proteins and aberrant DNA modification in cancer. Trends Genet. 2014;30:464–74.
25. Scourzic L, Mouly E, Bernard OA. TET proteins and the contol of cytosine demethylation in cancer. Genome Med. 2015;7:9.
26. Ichimura N, Shinjo K, An B, Shimizu Y, Yamao K, Ohka F, et al. Aberrant TET1 Methylation Closely Associated with CpG Island Methylator Phenotype in Colorectal Cancer. Cancer Prev Res (Phila). 2015;8:702–11.
27. Blaschke K, Ebata KT, Karimi MM, Zepeda-Martinez JA, Goyal P, Mahapatra S, et al. Vitamin C induces Tet-dependent DNA demethylation and a blastocyst-like state in ES cells. Nature. 2013;500:222–6.
28. Yin R, Mao SQ, Zhao B, Chong Z, Yang Y, Zhao C, et al. Ascorbic acid enhances Tet-mediated 5-methylcytosine oxidation and promotes DNA demethylation in mammals. J Am Chem Soc. 2013;135:10396–403.
29. Moon JW, Lee SK, Lee JO, Kim N, Lee YW, Kim SJ, et al. Identification of novel hypermethylated genes and demethylating effect of vincristine in colorectal cancer. J Exp Clin Cancer Res. 2014;33:4.
30. Rasmussen KD, Helin K. Role of TET enzymes in DNA methylation, development, and cancer. Genes Dev. 2016;1(30):733–50.
31. Mori Y, Olaru AV, Cheng Y, Agarwal R, Yang J, Luvsanjav D, et al. Novel candidate colorectal cancer biomarkers identified by methylation microarray-based scanning. Endocr Relat Cancer. 2011;18:465–78.
32. Qu Y, Dang S, Hou P. Gene methylation in gastric cancer. Clin Chim Acta. 2013;424:53–65.
33. Jin J, Xie L, Xie CH, Zhou YF. Aberrant DNA methylation of MGMT and hMLH1 genes in prediction of gastric cancer. Genet Mol Res. 2014;13:4140–5.

# 10

# Mesenchymal stem cells and myoblast differentiation under HGF and IGF-1 stimulation for 3D skeletal muscle tissue engineering

R. Witt[1†], A. Weigand[1†], A. M. Boos[1], A. Cai[1], D. Dippold[2,3], A. R. Boccaccini[2], D. W. Schubert[3], M. Hardt[1], C. Lange[4], A. Arkudas[1], R. E. Horch[1] and J. P. Beier[1*]

## Abstract

**Background:** Volumetric muscle loss caused by trauma or after tumour surgery exceeds the natural regeneration capacity of skeletal muscle. Hence, the future goal of tissue engineering (TE) is the replacement and repair of lost muscle tissue by newly generating skeletal muscle combining different cell sources, such as myoblasts and mesenchymal stem cells (MSCs), within a three-dimensional matrix. Latest research showed that seeding skeletal muscle cells on aligned constructs enhance the formation of myotubes as well as cell alignment and may provide a further step towards the clinical application of engineered skeletal muscle.

In this study the myogenic differentiation potential of MSCs upon co-cultivation with myoblasts and under stimulation with hepatocyte growth factor (HGF) and insulin-like growth factor-1 (IGF-1) was evaluated. We further analysed the behaviour of MSC-myoblast co-cultures in different 3D matrices.

**Results:** Primary rat myoblasts and rat MSCs were mono- and co-cultivated for 2, 7 or 14 days. The effect of different concentrations of HGF and IGF-1 alone, as well as in combination, on myogenic differentiation was analysed using microscopy, multicolour flow cytometry and real-time PCR. Furthermore, the influence of different three-dimensional culture models, such as fibrin, fibrin-collagen-I gels and parallel aligned electrospun poly-ε-caprolacton collagen-I nanofibers, on myogenic differentiation was analysed. MSCs could be successfully differentiated into the myogenic lineage both in mono- and in co-cultures independent of HGF and IGF-1 stimulation by expressing desmin, myocyte enhancer factor 2, myosin heavy chain 2 and alpha-sarcomeric actinin. An increased expression of different myogenic key markers could be observed under HGF and IGF-1 stimulation. Even though, stimulation with HGF/IGF-1 does not seem essential for sufficient myogenic differentiation. Three-dimensional cultivation in fibrin-collagen-I gels induced higher levels of myogenic differentiation compared with two-dimensional experiments. Cultivation on poly-ε-caprolacton-collagen-I nanofibers induced parallel alignment of cells and positive expression of desmin.

(Continued on next page)

* Correspondence: Justus.beier@uk-erlangen.de
†Equal contributors
[1]Department of Plastic and Hand Surgery and Laboratory for Tissue Engineering and Regenerative Medicine, University Hospital of Erlangen, Friedrich-Alexander University of Erlangen-Nürnberg (FAU), Krankenhausstraße 12, 91054 Erlangen, Germany
Full list of author information is available at the end of the article

(Continued from previous page)

**Conclusions:** In this study, we were able to myogenically differentiate MSC upon mono- and co-cultivation with myoblasts. The addition of HGF/IGF-1 might not be essential for achieving successful myogenic differentiation. Furthermore, with the development of a biocompatible nanofiber scaffold we established the basis for further experiments aiming at the generation of functional muscle tissue.

**Keywords:** IGF-1, HGF, Mesenchymal stem cells, Myogenic differentiation, PCL-collagen nanofibers, Skeletal muscle tissue engineering

## Background

Approximately one-half of our body consists of skeletal muscle, which is responsible for executing every single action we undertake [1]. Skeletal muscle has the ability to regenerate in response to damage by activating satellite cells resting beneath the basal lamina of adult skeletal muscle [2, 3]. However, this specific regeneration capacity is limited to only small wounds, whereas volumetric muscle loss caused by trauma or surgery requires remarkable efforts, such as free autologous muscle flap transplantation, which always come along with inevitable morbidity at the donor site [4–6]. This is where skeletal muscle tissue engineering (TE) might be a future goal, trying to mimic the structure and function of skeletal muscle [4–6].

For successfully generating muscle tissue in vivo, not only easily expandable cells but also a suitable biocompatible matrix needs to be generated. Muscle satellite cells offer the best characteristics for muscle TE, being capable of self-renewal and regeneration upon a variety of stimuli [7, 8]. However, multiple passaging decreases their differentiation capacity making their clinical applicability as single cell source difficult [9, 10]. Mesenchymal stem cells (MSCs) from the bone marrow may represent a promising alternative cell source for muscle TE since they can easily be harvested, expanded widely without losing their differentiation ability and autologous transplantation for future clinical applications does not come along with any risk of rejection [11, 12]. It has been described that MSCs can be differentiated towards the myogenic lineage by expressing muscle specific markers, even though their myogenic potential is limited [13–15]. Myogenic differentiation of MSCs alone might not be sufficiently satisfying, but they still represent an attractive cell source for co-cultivation with myoblasts. The application of MSCs co-cultivated with myoblasts has previously been investigated and it was shown that MSCs are able to fuse with myoblasts and contribute to the muscle regeneration process [13]. Moreover, it has been demonstrated that the addition of human MSCs to skeletal myoblasts cell-sheet in the ischemic cardiomyopathy model intensifies the release of different cytokines such as HGF and VEGF [16]. MSCs are not only known to secrete several growth factors involved in the muscle regeneration process such as basic fibroblast growth factor (bFGF), hepatocyte growth factor (HGF) or insulin-like growth factor 1 (IGF-1), but they also stimulate myoblast migration, proliferation, differentiation and cell survival upon co-cultivation [13, 16, 17]. Previous studies showed that stimulation with different supplements such as bFGF and dexamethasone potentiates MSC and myoblast differentiation capacity [18]. However, the effects of HGF and IGF-1 regarding the myogenic differentiation of MSC and myoblast co-cultures still require further investigation. It is well known that HGF activates satellite cells binding to the c-met tyrosine kinase receptor and stimulating different downstream targets [19]. While HGF primarily induces the proliferation of satellite cells, IGF-1 both activates proliferation and differentiation through binding to the IGF-1 receptor (IGF-1.R) [20]. The majority of circulating IGF-1 is bound to specific IGF-binding proteins (IGFBPs), a family of secreted proteins binding IGF-1 with greater affinity than IGF-1.R [21, 22]. There are different isoforms of IGFBPs and their exact roles are not clarified yet: While IGFBP4 mostly inhibits IGF stimulation, IGFBP5 acts through and independently of IGF and can therefore even potentiate or inhibit myogenic differentiation, and IGFBP6 is mostly expressed in proliferating cells [20, 21, 23, 24].

As mentioned above, successful generation of skeletal muscle needs both a suitable cell source as well as a biocompatible matrix. For optimally mimicking the in vivo structure of skeletal muscle and creating an applicable system for TE, a three-dimensional (3D) construct is needed. Different matrices have been studied for muscle TE applications, e.g. Heher et al. developed aligned fibrin fibrils in a 3D scaffold by applying static mechanical strain, demonstrating aligned myotube formation of myogenic precursor cells [25]. Further, Choi et al. demonstrated that cultivation of human skeletal muscle cells on unidirectional electrospun poly-ε-caprolacton (PCL)-collagen nanofiber meshes enhances myotube formation as well as skeletal muscle cell organization [26]. In previous studies, comparing fibrin-collagen-I gels with electrospun collagen nanofibers, good proliferation as well as differentiation of myoblasts could be shown, with parallel oriented nanofibers representing the most promising matrix [27].

One aim of this study is to investigate the influence of different concentrations as well as the combination of HGF and IGF-1 on myogenesis using co-cultures of MSCs and myoblasts as well as MSC monocultures. The three major myogenic key differentiation markers analysed in this study are, amongst others, myocyte enhancer factor 2 (MEF2), myosin heavy chain 2 (MyHC2) and alpha-sarcomeric actinin (ACTN2). MEF2 is a transcription factor, which interacts with members of the MyoD family of basic helix–loop–helix (bHLH) proteins to activate the skeletal muscle differentiation program. It plays a central role in activating pathways responsible for cell division, differentiation and death [28]. MEF2 is upregulated especially when cells enter the differentiation pathway and required in response to injury for adult myogenesis [18, 29–31]. However MEF2 seems to play a crucial role in myogenesis, its effects on MSC myoblast co-cultures have not been investigated profoundly so far. ACTN2, a cytoskeletal protein, stabilises the muscle contractile apparatus and is essential for developing the sarcomere. Such is, MyHC2, which constitutes sarcomere thick filaments and functions as a molecular motor protein in skeletal muscle. Both factors are indispensable for the formation of differentiated skeletal muscle [18, 32–34]. Their expression is proof for generating skeletal muscle. Even though, their behaviour concerning myogenic differentiation in MSC and myoblast co-cultures is not sufficiently studied and therefore of high interest.

We further analysed the behaviour of MSC-myoblast co-cultures in 3D fibrin and fibrin-collagen-I gels, especially in light of myogenic differentiation. As a final step, parallel-aligned electrospun PCL-collagen-I nanofibers were developed and cultivated with MSC-myoblast co-cultures stimulated with HGF and IGF-1 for testing the applicability for future in vivo studies.

## Methods
### Myoblast cell culture
Satellite cells were isolated from hind limb muscles of male Lewis rats (Charles River, Wilmington, Massachusetts, USA) as described previously [18]. For cell culture, Ham's F-10 medium (Gibco, Carlsbad, California, USA) containing 25% FCS (Biochrom GmbH, Berlin, Germany), 1.25% Penicillin/Streptomycin (Biochrom GmbH) and 2.5 ng/ml bFGF (Peprotech, Hamburg, Germany) was used. The medium was changed every second day. Myoblasts of passage 3 were used for all experiments. To verify the myogenic phenotype of isolated cells, staining with the highly muscle-specific MyoD nuclear protein (5.8.A, Abcam, Cambridge, UK) was performed (see Additional file 1) [35].

### MSC cell culture
Rat MSCs were isolated from the bone marrow of male Lewis 1WR2 rats as described previously [36]. MSCs were stably transduced with green fluorescent protein (GFP) for cell labelling, and GFP-positive clones were expanded as described before by Lange et al. [36, 37]. Phenotype was assessed by their ability to differentiate into chondrocytes, adipocytes and osteocytes [36, 37]. MSCs were cultured in growth medium (DMEM Ham's F-12, 10% FCS, 1% L-Glutamin, 1% P/S; all from Biochrom GmbH) and were used at passage 11 and 12 for all experiments. Medium was changed every second day.

### Differentiation conditions
Basic differentiation medium (DMEM/Ham's F-12 + 2% donor horse serum (DHS) + 1% L-Glutamin + 1% P/S (Biochrom GmbH) + 0.4 µg/ml dexamethasone (Sigma Aldrich, St. Louis, Missouri, USA) + 1 ng/ml bFGF (Peprotech)) was supplemented with different concentrations of HGF (R&D Systems, Minneapolis, Minnesota, USA; 10, 30, 60, 100 ng/ml) and IGF-1 (Peprotech; 5, 10, 30, 60 ng/ml) and the combination of 10 ng/ml HGF + 10 ng/ml IGF-1. Cells were differentiated in mono- and co-cultures of myoblasts and MSCs for 2, 7 and 14 d (d = day). For co-culture experiments, cells were seeded in a ratio of 1:1 in 12-well culture plates at a density of $6 \times 10^4$ cells in expansion medium (DMEM Ham's F-12, 10% FCS, 1% L-Glutamin, 1% P/S). After 24 h, medium was replaced by differentiation medium. Medium was changed every second day.

For each experiment, myoblasts from three different isolations were used.

### Multicolour flow cytometry
Multicolour flow cytometry was carried out on a FACSCalibur cytometer with cell Quest software and analysed with Flowjo software (Tree Star, Ashland, Oregon, USA).

Mono- and co-cultures of myoblasts and MSCs were seeded at a density of $1.5 \times 10^5$ in a 25-cm$^2$ flask (Greiner, Frickenhausen, Germany) and cultured with differentiation media containing HGF 10 ng/ml and IGF-1 10 ng/ml and stimulated for 2 d and 14 d. Cells were detached and blocked in 5% FCS for 15 min. The pellet was picked up in 100-µl Cytofix/Cytoperm solution (Cytofix/Cytoperm Fixation/Permeabilization Kit; BD Biosciences, San Jose, California, USA) and incubated for 20 min at 4 °C. Cells were washed with BD Perm/Wash buffer. The cell pellet was incubated for 30 min at 4 °C with primary antibodies solved in 100-µl BD Perm/Wash Buffer in a concentration of 1:50 (anti-alpha-sarcomeric actinin (EA-53, Abcam), anti-MEF2 (MEF2A, B-4, Santa Cruz Biotechnology, Dallas, Texas, USA), all mouse-anti-rat IgG1). As a secondary antibody, PE anti-mouse IgG1 (BD Biosciences) was used (1:50, for 30 min at 4 °C). For further flow cytometry analysis, cells were picked up in PBS (Biochrom GmbH) with 2%

FCS and 0.1% NaN$_3$. Controls included unstained cells for negative and L6-myoblasts (L6-Mb) cell line (American Type Culture Collection, ATCC, Manassas, Virginia, USA) for positive control. As the isotype control, PE-labelled anti-mouse IgG1 (BD Biosciences) was used. For MSC and myoblast co-cultures as well as myoblast monocultures, myoblasts of three different isolations were used. Experiments with MSC monocultures were performed once.

### Immunocytochemistry
Cells of each group were seeded at a density of $1.2 \times 10^4$ cells in expansion medium. After 24 h, the medium was replaced by differentiation medium. After fixation with ice-cold methanol, slides were washed and incubated in blocking buffer consisting of PBS with 1.5% FCS and 0.25% TritonX (Carl Roth GmbH, Karlsruhe, Germany) for 1 h at room temperature. After washing with TBS-T buffer (100 mM Tris and 60 mM NaCl in distilled water, 1 ml Tween20 per 1 L; pH 7.6), slides were covered with primary antibodies (anti-desmin (AB-1 (D33), Thermo Fisher Scientific, Runcorn, Cheshire, UK), anti-alpha-sarcomeric actinin (EA-53, Abcam), anti-MEF2 (MEF2A, B-4, Santa Cruz Biotechnology), anti-myosin heavy chain 2 (MYSN02, MyHC2, Thermo Fisher Scientific)) and diluted 1:50 in blocking buffer for 1 h at room temperature. As secondary antibody, Alexa Fluor 594 goat-anti-mouse IgG1 (Invitrogen, Karlsruhe, Germany) was used at 1:200 for 30 min at room temperature. Probes were counterstained with DAPI 1:1000 (Diamidine-phenylindole-dihydrochloride, Applied Science/Roche, Indianapolis; Indiana, USA) for 5 min. Slides were subsequently analysed and digitally photographed with a fluorescence microscope (IX83, cellSens software, Olympus, Hamburg, Germany). L6-Mb served as the positive control. An isotype control was performed using mouse IgG1 (BD Biosciences).

### RNA isolation and quantitative PCR analysis
In each group the expression rate of *DES (desmin)*, *MYOG (myogenin)*, *MEF2D (myocyte enhancer factor 2D)*, *MyHC2 (myosin heavy chain 2)*, *ACTN2 (alpha actinin skeletal muscle 2)*, *IGFBP4, 5, 6* was analysed. As housekeeping gene *RPL13a (ribosomal protein L13a)* was used. RNA of all probes was extracted using the RNeasy Mini Kit (Qiagen GmbH, Hilden, Germany) according to the manufacturer's protocols. RNA was reverse-transcribed into cDNA using a QuantiTect Reverse Transcription Kit and a Sensiscript Reverse Transcription Kit (both from Qiagen GmbH). cDNA was amplified through quantitative real-time PCR using SsoAdvanced Universal SYBR Green PCR Supermix (Bio-Rad, Hercules, California, USA) and Light Cycler (Bio-Rad iCycler iQ5). Probes were analysed in triplicates

and variations of more than 1.5 threshold cycles were dismissed. Data evaluation was performed using the $2^{-\Delta\Delta Ct}$ method. The primer sequences used are given in Table 1.

### Cell culture in 3D fibrin and fibrin-collagen-I gels
Fibrinogen and thrombin (Tisseel VH, S/D kit, Baxter AG, Vienna, Austria) were dissolved according to the manufacturer's instructions. Collagen (rat tail collagen type I, BD Biosciences) for the fibrin-collagen gels was equilibrated to pH 7 prior to use. A co-culture of 100.000 CM-DiI (Invitrogen) labelled rat myoblasts and GFP-transduced rat MSCs at a ratio of 1:1 was mixed with either a fibrinogen-medium solution or fibrinogen-collagen-medium solution. Cell suspensions were mixed 1:1 with thrombin (final concentration of 6 IU) in a 24-well plate. Each gel had a total volume of 700 μl with a fibrin concentration of either 2.5 or 5 mg/ml. In the fibrin-collagen gels the collagen concentration was 0.25 mg/ml. The gels were finally covered with 400 μl of differentiation medium containing 0.1 TIU/ml aprotinin. After 2 and 7 d, gels were frozen in liquid nitrogen and minced with mortar and electrical mixer (IKA Werke, Staufen, Germany). Gels were further homogenised with Trizol (Life Technologies, Carlsbad, California, USA) and chloroform, and RNA was purified as described previously. A differentiation medium with and without HGF/IGF-1 was used.

### Electrospinning of PCL-collagen-I nanofibers and cell seeding
PCL (Sigma Aldrich) was dissolved at a ratio of 2:1 with bovine collagen type 1 (Symatese, Chaponost, France) in ethanol (VWR, Darmstadt, Germany) 90% at a concentration of 10% w/v (distance needle tip counter electrode: 20 cm). Parallel nanofibers were electrospun on a counter electrode consisting of two parallel arranged beams (distance between the beams: 3 cm) on a standard electrospinning machine (Linari, Pisa, Italy). Afterwards,

**Table 1** Primer sequences

|  | Forward primer | Reverse primer |
|---|---|---|
| *DES* | ATACCGACACCAGATCCAGTCC | TCCCTCATCTGCCTCATCAAGG |
| *MYOG* | TGAGAGAGAAGGGAGGGAAC | ACAATACACAAAGCACTGGAA |
| *MEF2D* | TGCTGCTCTCACTGTCACTAC | TTCACGACTTGGGGACACTG |
| *MyHC2* | TGACTTCTGGCAAAATGCAG | CCAAAGCGAGAGGAGTTGTC |
| *ACTN2* | TCACTGAGGCCCCTTTGAAC | AGACAGCACCGCCTGAATAG |
| *IGFBP4* | CAGCGTGCTTGCTAACTTCC | GCTTAGAGAACCAGACCCGG |
| *IGFBP5* | CCCTGCACCTGAGATGAGAC | TCACAGTTGGGCAGGTACAC |
| *IGFBP6* | AAGGCCCAGTCCTGTTCAAG | TGAGGTCACAGTTTGGCACA |
| *RPL13a* | CTCATGAGGTCGGGTGGAAG | AGAGCTGCTTCTTCTTCCGG |

fibres were collected from the beams using glass plates (1 cm diameter). Nanofibers were electrospun with a voltage of 20 kV and a flow rate of 1 ml/h. Twelve hours before cell seeding, probes sterilised in 70% ethanol, washed with PBS afterwards and soaked in DMEM Ham's F-12 for approximately 1 h at 37 °C. Scaffolds were seeded with 100 μl expansion medium containing 50,000 MSCs and myoblasts at a ratio of 1:1. After an incubation time of 3 h at 37 °C, wells were filled with 1 ml of expansion medium. After 24 h, scaffolds were transferred into new well plates and stimulated with basic differentiation medium containing HGF and IGF-1 for 7 d. To analyse cell morphology and orientation, scanning electron microscopy and phase contrast microscopy (Olympus, Hamburg, Germany) was used.

### Scanning electron microscopy
Microstructural analysis of the scaffolds was performed using an Auriga Fib-SEM (Zeiss, Oberkochen, Germany). For this, the fibres were placed on aluminium stubs of 8 mm diameter. The probes were then sputter-coated with gold for 1 min using an EMITECH-K550 sputter coater at an operating pressure of $7 \times 10^2$ bar and a deposition current of 20 mA. The SEM images were taken at an acceleration voltage of 2 kV and a working distance of approximately 8 mm.

### Time-lapse microscopy
GFP-MSC and CM-DiI-myoblasts were seeded in a ratio of 1:1 in 12-well culture plates at a density of $6 \times 10^4$ cells in expansion medium. After 6 h, the medium was replaced by basic differentiation medium containing dexamethasone and bFGF. Culture plate was placed in an Olympus cell vivo microscopy system (IX83/cellVivo, cellSens software, Olympus, Hamburg, Germany). Co-cultures were cultivated under 37 °C and 5% $CO_2$ for approximately 5 d. Four different positions were determined using the Olympus cellSens software. A picture of each position was taken every 10 min.

### Statistical analysis
Data are expressed as a mean–standard deviation. Statistical analysis was performed using SPSS 21.0 for Windows (SPSS, Chicago, Illinois, USA).

Results were statistically interpreted by one-way analysis of variance (ANOVA) and Tukey HSD test as a post hoc test. Normal distribution was confirmed using the Shapiro Wilk test. In the case of no normal distribution, the nonparametric Kruskal-Wallis test and the Mann-Whitney $U$-test were used. For comparing samples over different time points, ANOVA for repeated measurements was used.

The level of statistical significance was set to $p \leq 0.05$. A $p$-value $\leq 0.01$ was considered to be highly significant.

## Results
### Effects of HGF on mRNA level of different myogenic markers in MSCs co-cultivated with primary myoblasts and in monocultures
MSC and primary rat myoblasts were co- and monocultured in basic differentiation medium containing HGF and in control medium without HGF for 2 d and 7 d. Expression of different myogenic markers could be observed under all conditions. In co-cultures, the early stimulation (2 d) with HGF demonstrated significant and highly significant upregulations of *MEF2* using 10, 30 and 60 ng/ml compared with late stimulation (7 d). A dose-dependent decrease of *MEF2* could be demonstrated after 2 d (Fig. 1a). Both *MEF2* and *ACTN2* expressions were equal or upregulated during early stimulation compared with unstimulated control groups (Fig. 1a–b). In MSC monocultures, the strongest *MEF2* expression ($1.6 \pm 0.6$-fold) could be achieved with 10 ng/ml HGF after stimulation for over 7 d. Except in groups with 30 ng/ml HGF, long-term stimulation achieved almost equal or higher levels of *MEF2* and *ACTN2* in MSCs compared with controls (Fig. 1c–d). Varying results were observed in myoblast monocultures: Early stimulation with 10–60 ng/ml HGF induced a concentration-dependent upregulation of *MEF2* and *ACTN2* (Fig. 1e–f). Comparing the three different cell groups, it could be demonstrated that early stimulation with HGF increased the levels of myogenic markers especially in co-cultures and myoblast monocultures, whereas in MSCs this occurred during long-term stimulation.

### Effects of IGF-1 on mRNA level of different myogenic markers in MSC co-cultivated with primary myoblasts and in monocultures
MSC and primary rat myoblasts were co- and monocultured in basic differentiation medium containing IGF-1 and in control medium without IGF-1 for 2 d and 7 d. Expression of *MEF2* and *ACTN2* could be observed under all conditions. A highly significant and significant higher expression of *MEF2* could be detected in MSC and myoblast co-cultures after 2 d compared with 7 d in the 10 and 30 ng/ml IGF-1 group (Fig. 2a). *ACTN2* expression was upregulated during early stimulation compared with unstimulated control groups (Fig. 2b). In MSC monocultures, stimulation with 60 ng/ml IGF-1 over 2 d induced the strongest upregulation of *MEF2* ($1.2 \pm 0.4$-fold) (Fig. 2c). *ACTN2* expression was overall increased during early stimulation (Fig. 2d). Mb monocultures were influenced positively by early stimulation with IGF-1: An overall increase of *MEF2* and *ACTN2* was observed after 2 d. Long-term stimulation showed no increase of myogenic markers (Fig. 2e–f). During early stimulation, *MEF2* expression increased in co-

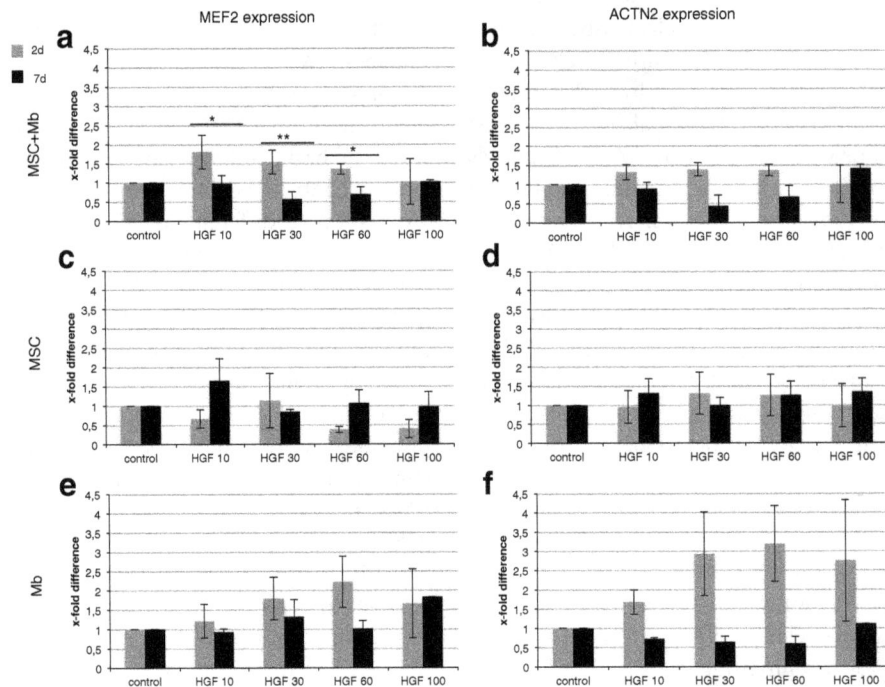

**Fig. 1** Expression of *MEF2* and *ACTN2* under different concentrations of HGF. Real-time PCR of MSC and myoblast (Mb) mono- and co-cultures under HGF stimulation as well as in unstimulated controls. Expressions are demonstrated in x-fold difference compared with unstimulated cells cultivated in basic differentiation medium (control = 1) using the $2^{-\Delta\Delta Ct}$ method. Markers are presented with mean +/- SD. **a** Significant and highly significant higher expression of *MEF2* in co-cultures after 2 d compared with 7 d using 10, 30 and 60 ng/ml HGF. **b** In co-cultures, *ACTN2* expression was upregulated during early stimulation compared with unstimulated control groups. 100 ng/ml HGF over 7 d induced the strongest *ACTN2* expression. **c** Strongest *MEF2* expression in MSC monocultures could be achieved with 10 ng/ml HGF after stimulation for over 7 d. **d** Seven-day stimulation with 10, 60 and 100 ng/ml HGF induced almost equal or higher levels of *ACTN2* in MSC monocultures compared with 2 d stimulation. **e-f** In Mb, a dose-dependent increase in *MEF2* (**e**) and *ACTN2* (**d**) expression was demonstrated from 10 to 60 ng/ml HGF during early stimulation. Increased levels of *MEF2* and *ACTN2* under HGF during early stimulation in co-cultures and Mb monocultures compared with unstimulated controls. Mb of three different isolations were used in three independent experiments. Three replicates of each were used. (** = $p \leq 0.01$). (* = $p \leq 0.05$)

cultures and myoblast monocultures the most. Highest *ACTN2* expressions were seen in myoblast monocultures. The expression of *ACTN2* in MSC monocultures and co-cultures was similar.

### Influence of the combined stimulation of HGF and IGF-1 on mRNA level

In MSC monocultures, as well as in co-cultures with myoblasts, higher expressions of *DES* compared with myoblasts (=1) could be observed. Cultivation in HGF/IGF-1 free medium achieved highest levels of *DES* (MSC: 218.5 ± 219-fold; MSC + Mb: 64.5 ± 62-fold) after 2 d. During early stimulation, MSC monocultures showed overall higher levels of *DES* than co-cultures (Fig. 3a). *MYOG* expression was upregulated under HGF stimulation compared with myoblasts under HGF after 14 d. *MYOG* could only be detected in one out of three experiments and merely in co-cultures (Fig. 3b). After 14 d, in co-cultures, highest levels of *ACTN2* (5.9 ± 14-fold) and *MyHC2* (4.8 ± 2.3-fold) could be observed under IGF-1 stimulation. *IGFBP4* expression increased

in co-and monocultures with growth factors compared with cultivation in control medium without HGF/IGF-1 (Fig. 3e). *IGFBP5* and *-6* expression in co-cultures was elevated in the IGF-1 group compared with groups without growth factors. In MSC monocultures, hardly any expression of *IGFBP5* and *-6* could be detected (Figs. 3f–g).

### Myogenic differentiation and fusion of MSCs co-cultured with primary myoblasts under the influence of HGF and IGF-1

With fluorescence microscopy, the myogenic differentiation potential of MSCs mono- and co-cultured with myoblasts was analysed under stimulatory and non-stimulatory effects after 7 and 14 d. A positive staining for muscle specific marker MyHC2, could be detected under HGF effects as well as the other tested conditions (Fig. 4a, b). MSC involvement in the formation of possibly multinucleated cells was verified by their green fluorescence protein expression, as these cells had been stable transduced prior to co-cultivation (Fig. 4b; arrows)

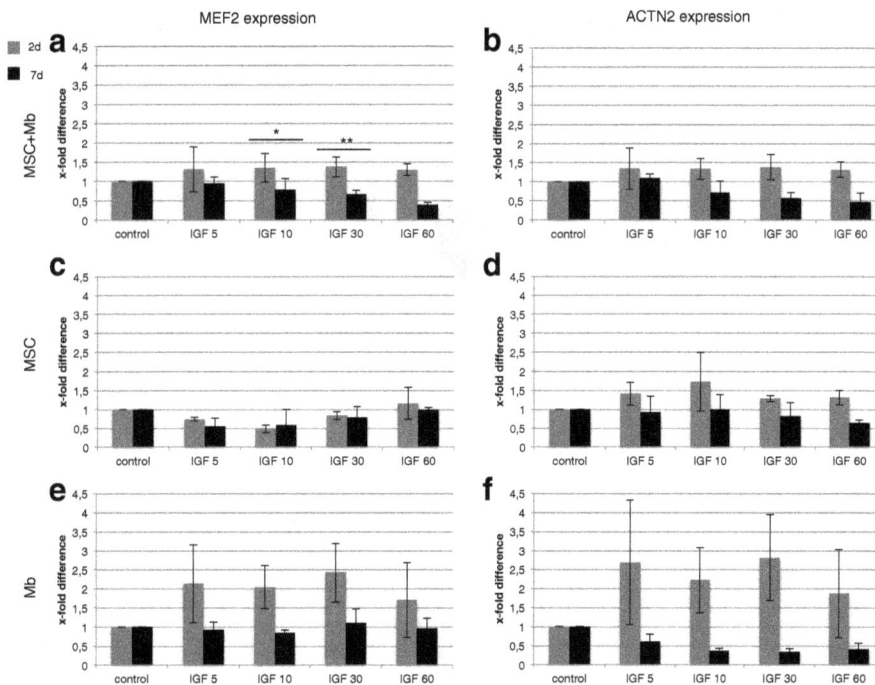

**Fig. 2** Expression of *MEF2* and *ACTN2* under different concentrations of IGF-1. Real-time PCR of MSC and myoblast (Mb) mono- and co-cultures under IGF-1 stimulation as well as in unstimulated controls. Expressions are demonstrated in x-fold difference compared with unstimulated cells cultivated in basic differentiation medium (control = 1) using the $2^{-\Delta\Delta Ct}$ method. Markers are presented with mean +/- SD. **a** Overall higher expressions of *MEF2* in co-cultures under the different IGF-1 concentrations compared with unstimulated conditions after 2 d. Significant and highly significant higher levels of *MEF2* after 2 d compared with 7 d using 10 and 30 ng/ml IGF-1. **b** Overall higher expressions of *ACTN2* in co-cultures under the different IGF-1 concentrations compared with unstimulated conditions after 2 d. **c** Stimulation with 60 ng/ml IGF-1 over 2 d induced the strongest upregulation of *MEF2* in MSC monocultures. **d** Overall increased *ACTN2* expression in MSC monocultures was observed during early stimulation, with highest levels under 10 ng/ml IGF-1. **e-f** Early stimulation with IGF-1 induced higher *MEF2* (**e**) and *ACTN2* (**f**) expressions in Mb monocultures compared with controls, with strongest expression under 30 ng/ml IGF-1. Increased MEF2 expression in co-cultures and Mb monocultures after 2 d compared with unstimulated controls. The highest *MEF2* and *ACTN2* levels were detected in Mb monocultures. Mb of three different isolations were used in three independent experiments. Three replicates of each were used. (** = $p \leq 0.01$). (* = $p \leq 0.05$)

(For single stainings see Additional files 2 and 3). Furthermore, myogenic differentiation could be demonstrated via positive staining for MEF2, expressed especially during muscle differentiation. Here, data of unstimulated controls are shown (Fig. 5) (For single stainings see Additional file 4).

Further myogenic differentiation was evaluated with flow cytometry analysis of MEF2 and ACTN2 (Fig. 6). MSC and myoblast co-cultures, MSC monocultures, myoblast monocultures and L6-Mb as positive controls were stimulated with HGF + IGF-1 or cultivated in control medium. After 14 d of cultivation, MEF2 expression in HGF + IGF-1 co-cultures showed a highly significant increase from 72.3% after 2 d up to 93.6% after 14 d. Expression of MEF2 in control groups did not increase significantly from 79% (2 d) to 91.3% (14 d). In MSC monocultures and L6-Mb, MEF2 expression was also upregulated after 14 d. Co-cultures and MSC monocultures achieved equal levels of MEF2 compared with L6-Mb after 14 d, both in HGF + IGF-1 and in control medium. Myoblast monocultures showed a decrease of MEF2 over time (Fig. 6a).

The expression of ACTN2 in co-cultures with HGF + IGF-1 was highly significantly upregulated (1.5-fold) from 2 to 14 d. Controls also showed a highly significant upregulation of ACTN2 from 50.6% after 2 d up to 67% after 14 d, but lower than the HGF + IGF-1 groups after 14 d (73.9%). In MSC monocultures, an increased ACTN2 expression was observed after 14 d compared with 2 d. Comparable to MEF2 expression, myoblasts showed a decrease in the myogenic marker over time. Co-cultures and MSC monocultures achieved equal levels of ACTN2 compared with L6-Mb after 14 d, both in HGF + IGF-1 and in the control medium (Fig. 6b).

Via microscope, we recorded the cell behaviour of MSC and myoblast co-cultures over a time period of 5 d. Signs of cell fusion between both cell sources could be seen (see Additional file 5). Considering that muscle repair and newly formation of skeletal muscle

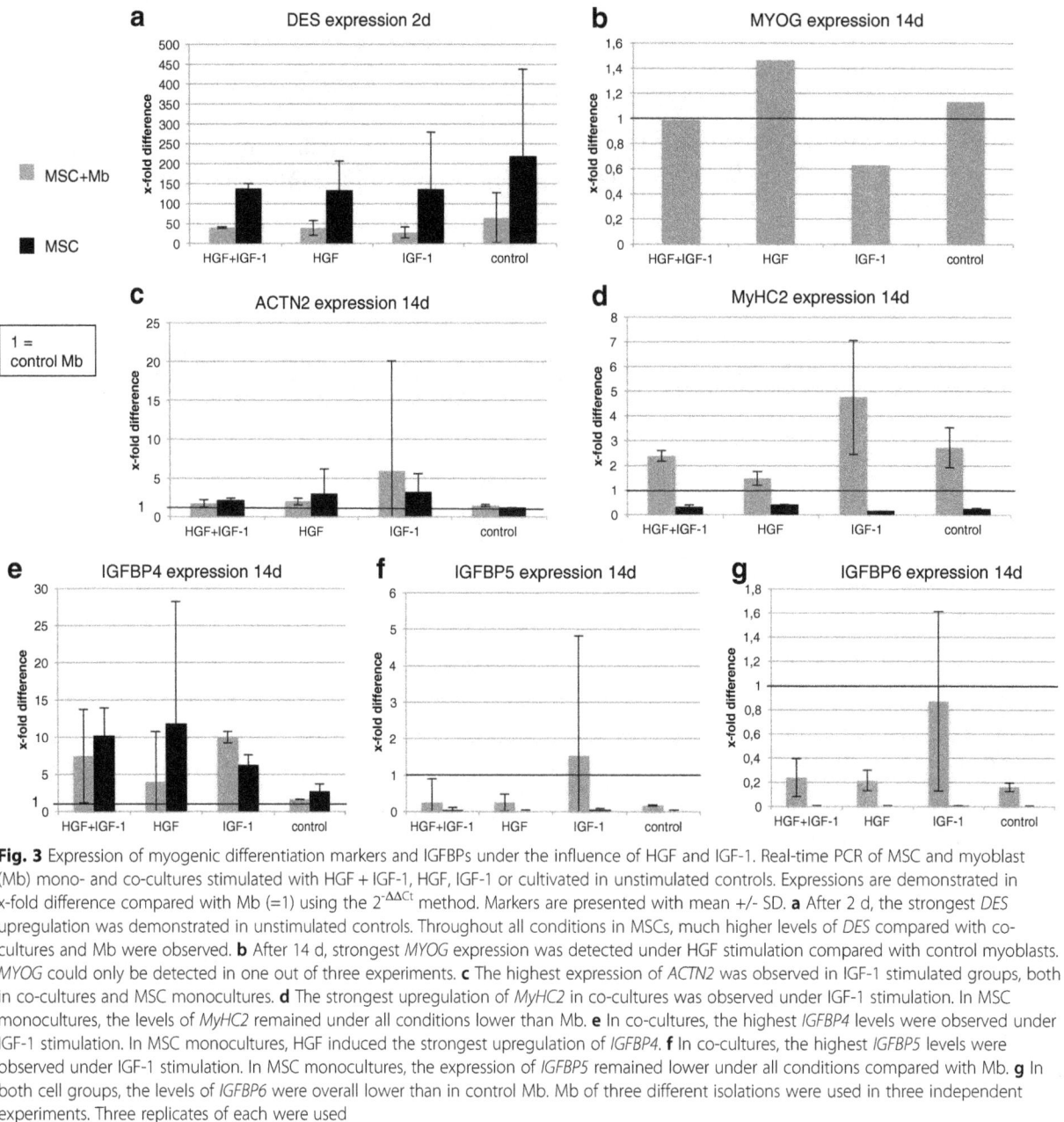

**Fig. 3** Expression of myogenic differentiation markers and IGFBPs under the influence of HGF and IGF-1. Real-time PCR of MSC and myoblast (Mb) mono- and co-cultures stimulated with HGF + IGF-1, HGF, IGF-1 or cultivated in unstimulated controls. Expressions are demonstrated in x-fold difference compared with Mb (=1) using the $2^{-\Delta\Delta Ct}$ method. Markers are presented with mean +/- SD. **a** After 2 d, the strongest *DES* upregulation was demonstrated in unstimulated controls. Throughout all conditions in MSCs, much higher levels of *DES* compared with co-cultures and Mb were observed. **b** After 14 d, strongest *MYOG* expression was detected under HGF stimulation compared with control myoblasts. *MYOG* could only be detected in one out of three experiments. **c** The highest expression of *ACTN2* was observed in IGF-1 stimulated groups, both in co-cultures and MSC monocultures. **d** The strongest upregulation of *MyHC2* in co-cultures was observed under IGF-1 stimulation. In MSC monocultures, the levels of *MyHC2* remained under all conditions lower than Mb. **e** In co-cultures, the highest *IGFBP4* levels were observed under IGF-1 stimulation. In MSC monocultures, HGF induced the strongest upregulation of *IGFBP4*. **f** In co-cultures, the highest *IGFBP5* levels were observed under IGF-1 stimulation. In MSC monocultures, the expression of *IGFBP5* remained lower under all conditions compared with Mb. **g** In both cell groups, the levels of *IGFBP6* were overall lower than in control Mb. Mb of three different isolations were used in three independent experiments. Three replicates of each were used

usually happens upon fusion of myoblasts, this might be a further step towards the generation of muscle tissue [17].

### Effect of 3D scaffolds on myogenic differentiation of MSC and myoblast co-cultures

MSCs and myoblasts were cultivated in different 3D gels, consisting of either 5 or 2.5 mg/ml fibrin alone or in combination with collagen I. After 2 and 7 d, gene expression analysis of myogenic differentiation markers was performed. *MEF2* expression decreased significantly in 5 mg/ml fibrin gels and 2.5 mg/ml fibrin-collagen-I gels and highly significantly in 2.5 mg/ml fibrin gels over time, but not in 5 mg/ml fibrin-collagen-I gels, which experienced a highly significant increase (Fig. 7a).

Comparable to *MEF2*, *ACTN2* expression decreased over time in fibrin gels and increased highly significantly in 5 mg/ml fibrin-collagen-I gels (Fig. 7b).

A range of myogenic markers (*MEF2*, *MyHC2*, *ACTN2*, *MYOG*, *IGFBP4*, *-5*, *-6*) was analysed in co-cultures cultivated in HGF + IGF-1 for 2 d and compared with a differentiation medium without HGF/IGF-1. A

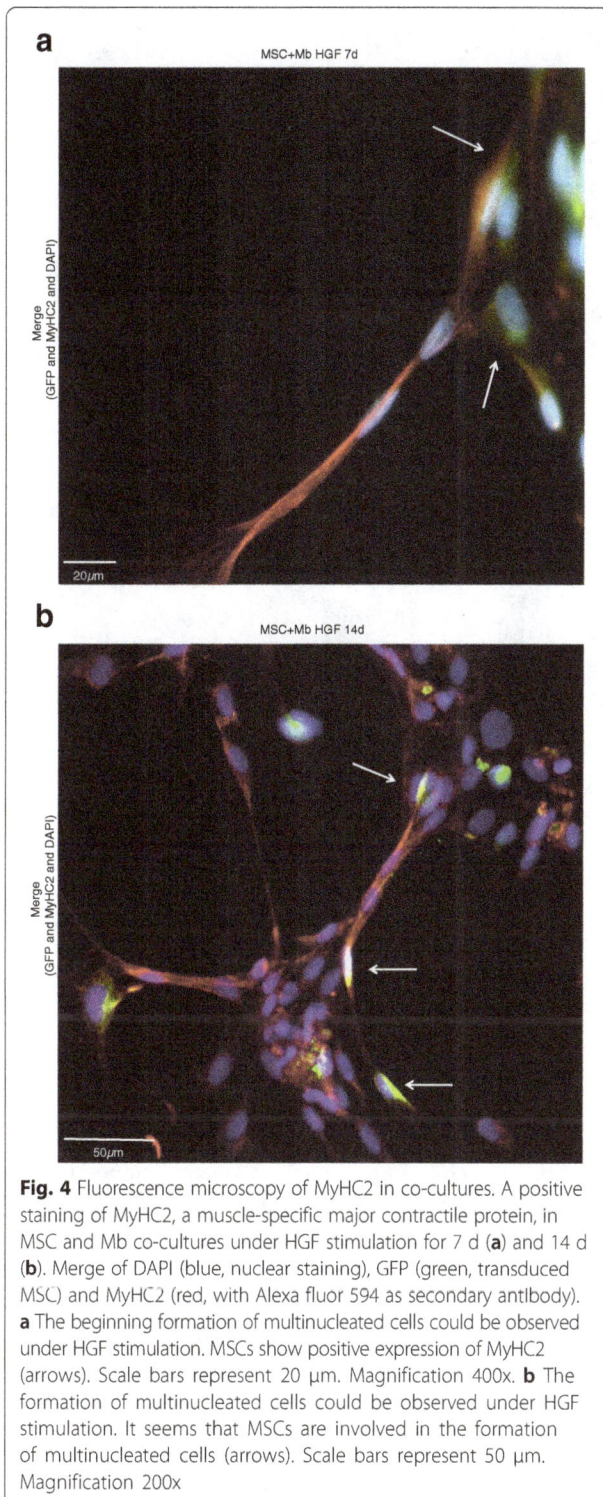

**Fig. 4** Fluorescence microscopy of MyHC2 in co-cultures. A positive staining of MyHC2, a muscle-specific major contractile protein, in MSC and Mb co-cultures under HGF stimulation for 7 d (**a**) and 14 d (**b**). Merge of DAPI (blue, nuclear staining), GFP (green, transduced MSC) and MyHC2 (red, with Alexa fluor 594 as secondary antibody). **a** The beginning formation of multinucleated cells could be observed under HGF stimulation. MSCs show positive expression of MyHC2 (arrows). Scale bars represent 20 μm. Magnification 400x. **b** The formation of multinucleated cells could be observed under HGF stimulation. It seems that MSCs are involved in the formation of multinucleated cells (arrows). Scale bars represent 50 μm. Magnification 200x

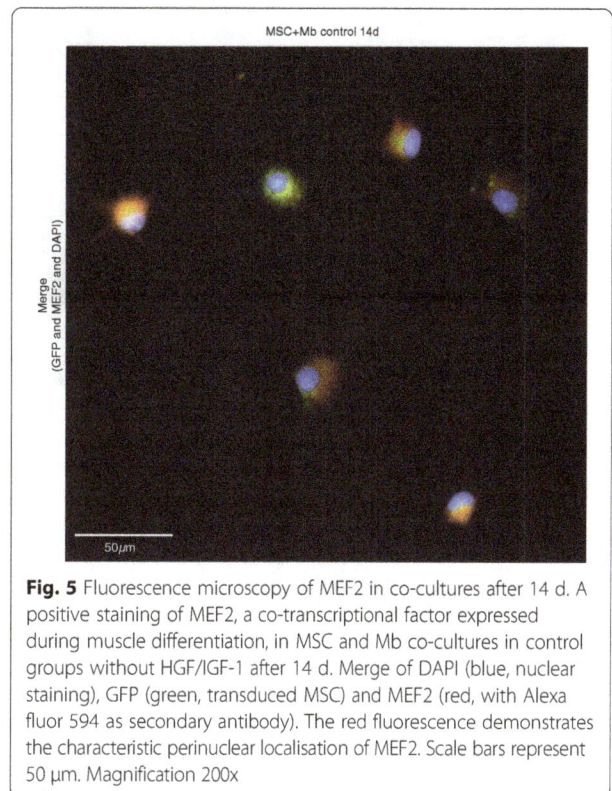

**Fig. 5** Fluorescence microscopy of MEF2 in co-cultures after 14 d. A positive staining of MEF2, a co-transcriptional factor expressed during muscle differentiation, in MSC and Mb co-cultures in control groups without HGF/IGF-1 after 14 d. Merge of DAPI (blue, nuclear staining), GFP (green, transduced MSC) and MEF2 (red, with Alexa fluor 594 as secondary antibody). The red fluorescence demonstrates the characteristic perinuclear localisation of MEF2. Scale bars represent 50 μm. Magnification 200x

PCL-collagen-I nanofibers showed a parallel-orientated scaffold (Fig. 8a, b). Using SEM and fluorescence microscopy, the attachment, proliferation and parallel alignment of the cells could be observed (Fig. 8c–d). Positive myogenic differentiation of cells growing on the scaffold could be demonstrated with desmin immunocytochemistry (Fig. 8e).

## Discussion

The aim of this study was to examine the effects of HGF and IGF-1 on MSC and myoblast co-cultures, as well as monocultures, and to investigate the cell–cell interactions in a 3D-matrix.

### Time-dependent effect of HGF and IGF-1 on myogenic differentiation of mono- and co-cultures

We investigated the influence of different concentrations of HGF and IGF-1 on MSC and myoblast co-cultures, as well as monocultures, compared with cells cultivated in HGF/IGF-1 free medium, analysing *MEF2* and *ACTN2* expression. Due to sometimes high standard deviations, some of the comparisons showed no statistically significant differences. An explanation for higher standard deviations might be that sometimes we could only isolate small amounts of RNA that perhaps do not fully reflect the total RNA of the analysed group and therefore caused variation among the experiments.

slight, significant for *MyHC2*, upregulation of gene expression compared with control was detected (Fig. 7c).

MSC-myoblast co-cultures were further cultivated on parallel-aligned PCL-collagen-I-nanofiber scaffolds for 7 d and stimulated with HGF + IGF-1. SEM images of the

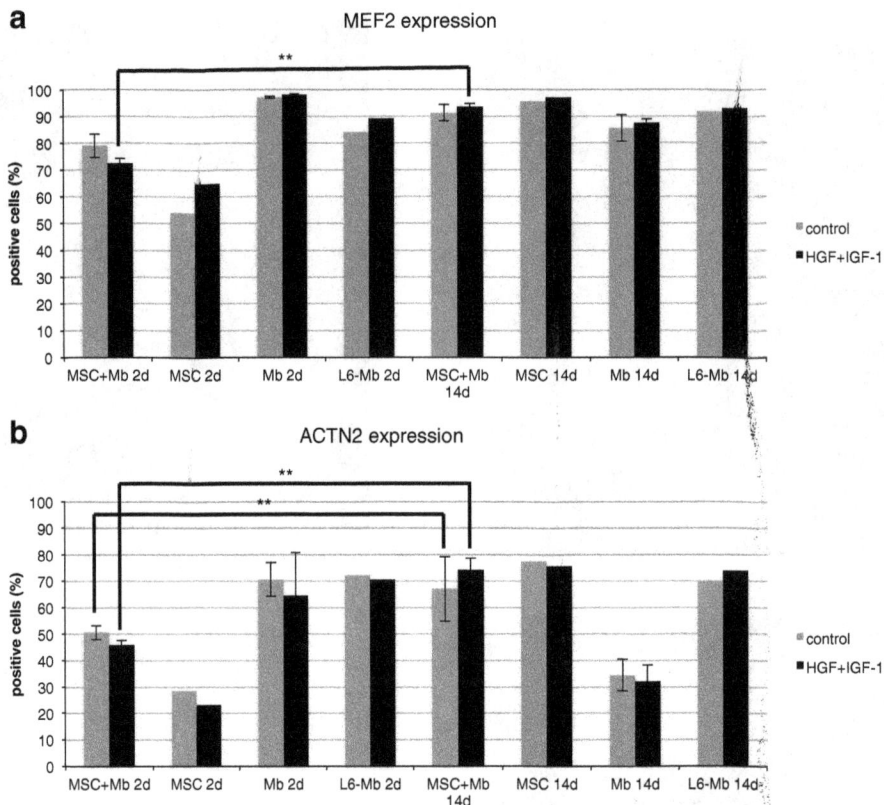

**Fig. 6** Flow cytometry analysis of MEF2 and ACTN2 in MSC and myoblast co-cultures, MSC and myoblast monocultures and L6-myoblasts. Markers are presented with mean +/- SD. **a** Highly significant upregulation of MEF2 in co-cultures from 2 to 14 d of stimulation with HGF + IGF-1. Higher levels of MEF2 in MSC monocultures could be observed in stimulated and control groups compared with L6-myoblasts. The expression of MEF2 was slightly downregulated after 14 d in myoblast (Mb) monocultures. **b** Highly significant upregulation of ACTN2 in co-cultures both under HGF + IGF-1 and in control groups after 14 d compared with 2 d. After 2 d of cultivation, the lowest levels of ACTN2 were demonstrated in MSC monocultures. A 2.7-fold upregulation in unstimulated controls and 3.2-fold under HGF + IGF-1 was observed in MSC monocultures after 14 d. The expression of ACTN2 was downregulated in stimulated and control Mb monocultures. Higher expression of ACTN2 was observed when Mb were cultivated in control groups. (** = $p \leq 0.01$). Mb of three different isolations as well as three replicates of each were used. One replicate of MSC and L6 was used

In our study, stimulation with different concentrations of the growth factors, revealed, especially under HGF, no clear trend regarding the expression of myogenic markers. It is well known that HGF both plays a role in proliferation as well as differentiation of skeletal muscle cells [38]. Yamada et al. described low expression of myogenin mRNA under 2.5 ng/ml as well as under extremely high (500 ng/ml) HGF concentrations, suggesting decreased myogenic differentiation [38, 39]. Walker et al. detected decreased Myosin Heavy Chain expression under 2 ng/ml HGF, but increased levels at 10 ng/ml, while Gal-Levi et al. showed lower MyHC levels with increased HGF concentrations (20–50 ng/ml) [38, 40]. A variety of studies have been made, trying to define the influence of different concentrations of HGF on skeletal muscle development, which has not been clarified yet.

In our study, early stimulation with HGF or IGF-1 achieved almost equal or increased levels of *MEF2* and *ACTN2* in co-cultures and myoblast monocultures, under all concentrations. Focussing on the temporal course, *MEF2* expression decreased significantly and highly significantly under HGF and IGF-1 influence in co-cultures. According to previously published findings, it may be suggested that too high concentrations of either of the added growth factors could negatively influence myogenic differentiation [41–44]. Differences were detected regarding MSC monocultures: especially long-term stimulation with HGF resulted in higher myogenic marker expression compared with unstimulated controls. During early stimulation in MSCs, only 30 ng/ml HGF achieved elevated levels of *MEF2* compared to unstimulated controls, being opposite to the other groups. Regarding that the result was not statistically significant, together with a high range of standard deviation and the fact that sometimes only small amounts of RNA could be isolated, these results might be of limited reliability.

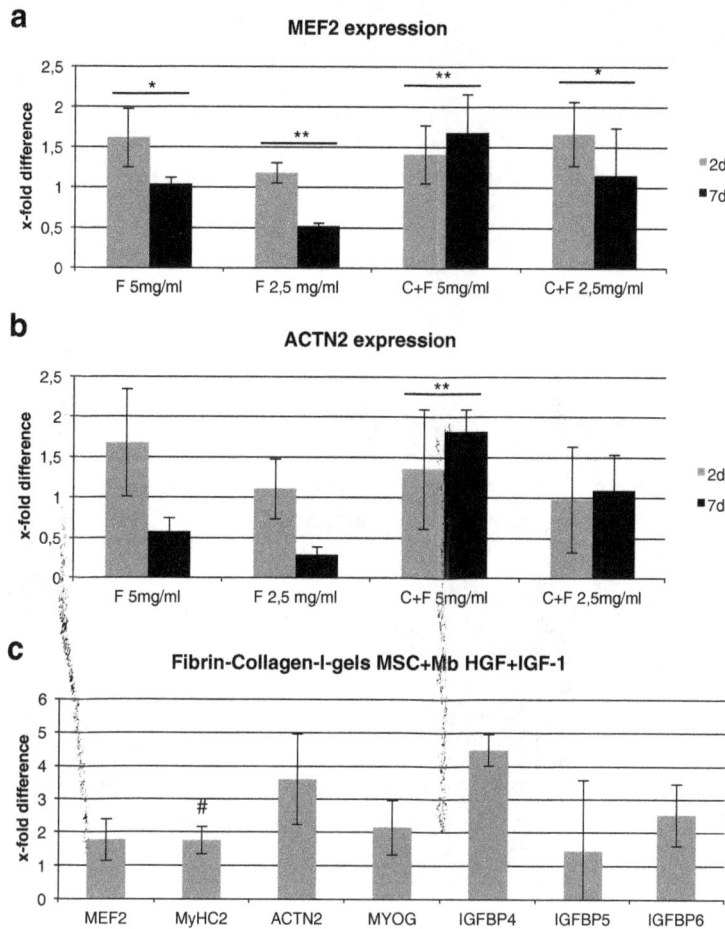

**Fig. 7** Myogenic differentiation in fibrin/fibrin-collagen-I gels. Markers are presented with mean +/- SD. **a** Real-time PCR of *MEF2* in MSC and myoblast (Mb) co-cultures cultivated in fibrin and fibrin-collagen-I gels. The expression of *MEF2* was highly significantly upregulated over time in 5-mg/ml fibrin-collagen-I gels. *MEF2* expression was significantly and highly significantly downregulated in other gel conditions. Expressions are shown in x-fold difference compared with co-cultures cultivated in 2D in control medium. **b** Real-time PCR of *ACTN2* in MSC and Mb co-cultures cultivated in fibrin and fibrin-collagen-I gels. The expression of *ACTN2* was highly significantly upregulated over time in 5-mg/ml fibrin-collagen I-gels. *ACTN2* expression was downregulated in other conditions, except 5-mg/ml fibrin-collagen-I gels with similar expression compared with the control. Expressions are shown in x-fold difference compared with co-cultures cultivated in 2D in control medium. **c** Real-time PCR of different myogenic markers (*DES, MEF2, MyHC2, ACTN2*) and *IGFBPs* (*IGFBP-4, -5, -6*) in co-cultures cultivated in fibrin-collagen-I gels and stimulated with HGF and IGF-1 for 2 d. Expressions are demonstrated in x-fold difference compared with unstimulated cells cultivated in control (=1). Upregulation of all myogenic markers under HGF + IGF-1 stimulation compared with unstimulated controls, MyHC2 significantly. (** = $p \leq 0.01$). (* = $p \leq 0.05$). (# = $p \leq 0.05$ compared with unstimulated controls). Mb of three different isolations as well as three replicates of each were used

Furthermore, IGF-1 stimulation did not increase *MEF2* at all and *ACTN2* only during early time points. Comparing the level of myogenic differentiation amongst the three different cell groups, the co-cultures and MSC monocultures showed lower levels of myogenic marker expressions compared with myoblast monocultures, especially during early growth factor stimulation. The requirement of longer differentiation time periods for MSCs might be a possible explanation for lower myogenic marker expression during early time points. This is in accordance with previously published work, in which time periods up to 6 weeks were used [45, 46].

Expression of MEF2 and ACTN2 leads towards the formation of skeletal muscle [18, 29–33].

### Beginning myogenic differentiation of MSCs upon co-cultivation with myoblasts and under IGF-1 stimulation

We further analysed the effect of combined stimulation with HGF + IGF-1 compared with HGF or IGF-1 only, as well as under unstimulated conditions in a basal differentiation medium containing DHS, L-Glutamin, dexamethasone and bFGF (Fig. 3). Co-cultures and MSC monocultures were directly related to myoblast monocultures. During early stimulation, MSCs showed overall

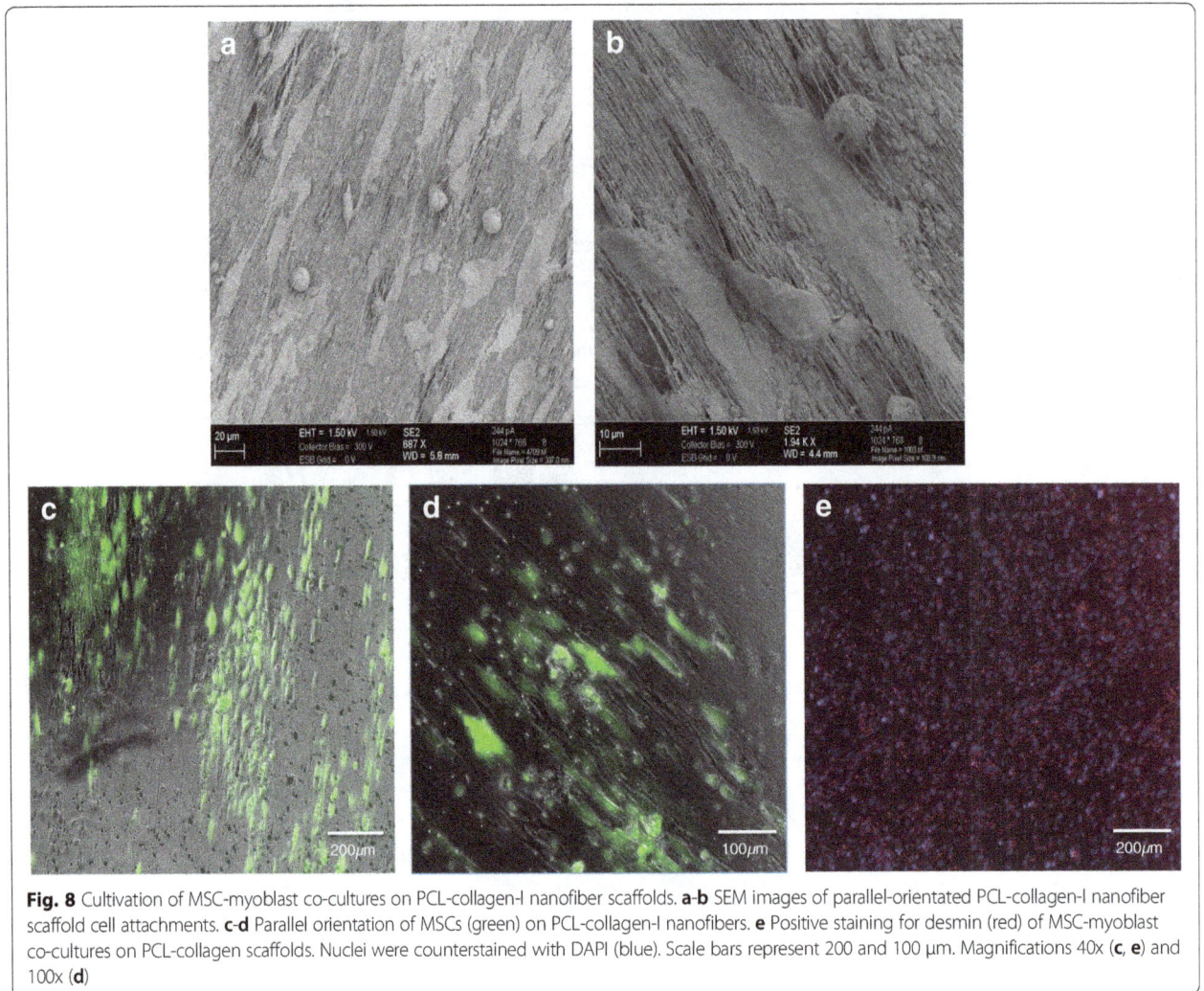

**Fig. 8** Cultivation of MSC-myoblast co-cultures on PCL-collagen-I nanofiber scaffolds. **a-b** SEM images of parallel-orientated PCL-collagen-I nanofiber scaffold cell attachments. **c-d** Parallel orientation of MSCs (green) on PCL-collagen-I nanofibers. **e** Positive staining for desmin (red) of MSC-myoblast co-cultures on PCL-collagen scaffolds. Nuclei were counterstained with DAPI (blue). Scale bars represent 200 and 100 μm. Magnifications 40x (**c, e**) and 100x (**d**)

higher levels of *DES* compared with co-cultures, probably because of desmin being a MSC marker [47]. *MYOG* expression could only be detected in one out of three experiments and was limited to co-cultures, with highest levels under HGF stimulation. Long-term stimulation with IGF-1 showed increased levels of *ACTN2* and *MyHC2* in co-cultures, higher than in MSC and myoblast monocultures, indicating that MSCs might need longer differentiation periods [13, 48, 49]. Cell–cell contact between myoblasts and MSCs could be a possible explanation for increased myogenic marker expression in co-cultures, comparable to previous findings by Beier et al., in which elevated levels of MEF2 and ACTN2 were detected upon direct co-cultivation of MSCs and myoblasts [18]. In previous studies it has been described that MSC influence myoblast differentiation in a paracrine way [13, 16, 17].

With multicolour flow cytometry, the myogenic differentiation on a protein level was further analysed. Elevated levels of MEF2 and ACTN2 were detected in co-cultures and MSC monocultures after 14 d of stimulation with HGF + IGF-1 as well as in unstimulated controls indicating an increasing myogenic differentiation [30–33]. Under fluorescence microscopy, positive staining for MyHC2, part of the myosin motor protein and therefore responsible for skeletal muscle contraction, revealed further myogenic differentiation [50]. Through stable transduced GFP expression, it was possible to detect the involvement of MSCs in the formation of multinucleated cells (Fig. 4). Cultivation in HGF/IGF-1 free medium almost always achieved similar levels of myogenic differentiation than under HGF + IGF-1 stimulation. We were not able to detect significant differences between our stimulation and controls groups. Based on this observation, these growth factors may not necessarily be needed for sufficient myogenic differentiation [48, 49]. But then – what might be the explanation for adequate myogenic differentiation in HGF/IGF-1 free environment?

First of all, our HGF/IGF-1 free medium (or control medium) contains already dexamethasone and bFGF, two factors known to influence myogenic differentiation [18, 51, 52]. Furthermore, as mentioned earlier, both MSC and myoblasts are known to secrete several growth factors involved in the muscle regeneration process [13, 16, 17, 43]. Herein, autocrine and paracrine stimulation might lead to myogenic differentiation. Nonetheless, successful muscle generation depends on more than secreted factors: cell-cell contact is crucial for a satisfactory differentiation. Previous works by Singaravelu and Padanilam compared the differentiation of MSC in conditioned medium with co-cultivation of MSC and injured renal cells. Cultivation in conditioned medium did not induce differentiation, but co-cultivation led to differentiation [53].

In summary, we successfully differentiated MSCs into the myogenic lineage both under HGF/IGF-1 stimulation and in a control medium, compared with myoblasts on the mRNA level (*MEF2, ACTN2, DES*) as well as after 14 d on the protein level (MEF2, ACTN2, MyHC2). Upon co-cultivation with myoblasts and under IGF-1 stimulation, additional expression of key myogenic marker *MyHC2* could be detected.

Comparing the myogenic potential of MSCs upon co-cultivation with myoblasts with MSC monocultures could be a promising future prospect. We already tried to separate MSCs by their GFP signal with fluorescence-activated cell sorting, but the yield was not enough for further analysis. In future experiments, it may be possible to sort MSCs for evaluation of myogenic potential cultivated in co- compared with monocultures, using higher cell numbers. Furthermore, only very low levels of *MYOG* could be detected in co-cultures in our study. MSC monocultures did not express *MYOG* at any time point. Because myogenin is mostly expressed during terminal stages of myogenic differentiation [54, 55], longer cultivation periods of at least up to 28 d would be one future goal. Although cell detachment after 28 d of cultivation made it impossible to analyse the gene expression during longer observation periods so far, coating with collagen type I or Maxgel™ (consisting of an undefined composition of human extracellular matrix components) may be a possibility to overcome cell detachment in future experiments [56]. Furthermore, myogenin is known to peak at some point of myoblast differentiation and then decline to lower expression afterwards. By the time we analysed myogenin expression, it might be possible that its expression was already starting to decline [57, 58].

### Possible involvement of IGFBPs in myogenic differentiation

IGFBPs are a family of secreted proteins binding IGF-1 and either potentiating or inhibiting IGF-1 actions on myogenic differentiation [20, 22]. In our study, in co-cultures, increased expression of *IGFBP4, - 5* and *-6* goes along with higher *ACTN2* and *MyHC2* expression under IGF-1 stimulation compared with myoblasts and MSC monocultures, accompanied by lower expressions of *DES* and *MYOG* (*MYOG* was only detected in one out of three experiments in co-cultures). *IGFBP5* and *-6* showed a similar expression pattern amongst all conditions in co-cultures, suggesting that these genes might have equal effects on myogenic differentiation and are regulated alike. Furthermore, the expression of *MyHC2* and *ACTN2* appears to correlate with *IGFBP5* and *-6* in co-cultures, indicating that they might have a positive influence on the expression of those myogenic markers.

In MSC monocultures both under HGF + IGF-1 and HGF stimulation, elevated levels of *IGFBP4* as well as *ACTN2* were observed compared with co-cultures and myoblast controls, whereas *IGFBP5, -6* and *MyHC2* expressions were almost undetectable.

Depending on the culture conditions (co-/monocultures), growth factor stimulation and the analysed myogenic markers, different effects could be detected. In co-cultures under IGF-1, increased expression of *IGFBPs* was observed together with elevated levels of *ACTN2* and *MyHC2*, and MSC monocultures showed different results under the same conditions. Hence, the function of the different IGFBPs might vary among different surrounding conditions. So far, we presume that IGFBPs play a role during myogenic differentiation.

Even though there is still no uniform opinion concerning the exact function of the IGFBPs, IGFBP4 was both identified as a positive influencer during muscle regeneration and as a potent inhibitor of muscle growth and IGF-1 actions [20, 21, 59, 60]. IGFBP5 could inhibit IGF-1 actions, potentiate IGF-1 effects or act in an IGF-independent way [24, 61]. IGFBP6 may not act primarily during the myogenic differentiation process [20].

Using ELISA or Western Blot, the concentrations of IGFBPs in the cell lysate or the supernatant could be estimated more precisely. Inhibiting IGFBPs through IGFBP antibodies could be another approach to gain more information about these binding proteins and their effects on IGF-1 and myogenic differentiation.

### Three-dimensional environment enhances myogenic differentiation of MSCs and myoblasts

Regarding the matrix evolution in TE, we previously demonstrated that 3D collagen-I gels had a stimulatory effect on myoblasts [27]. According to these findings, we investigated the effect of 3D systems on co-cultures of MSCs and myoblasts with HGF + IGF-1, and observed a strong upregulation of myogenic key markers compared with unstimulated groups. However, fibrin-collagen-I gels cannot provide the needed spatial orientation for muscle tissue. Therefore, we developed an electrospun, parallel-aligned, PCL-collagen-I nanofiber scaffold as the basis for further generation of muscle tissue. Parallel alignment of fibres stimulates myotube formation, and the combination of PCL and collagen provides strength, elasticity and compliance, which is essential for the formation of functional tissue [26, 62]. Cultivating MSCs and myoblasts on parallel-oriented PCL-collagen-I nanofibers and stimulating with HGF + IGF-1 for 7 d leads to parallel alignment of the cells in this study, indicating that this scaffold is a promising matrix for generation of muscle tissue in vitro. Jana et al. cultivated C2C12 myoblasts on aligned chitosan-PCL hybrid nanofiber scaffolds, showing formation of a compact assembly of myotube cells [63]. Zhao et al. used aligned electrospun PCL/collagen hybrid scaffolds for diaphragmatic repair in rats, demonstrating muscle cell migration and tissue formation [64]. For further investigation of our results in vivo, the newly developed arteriovenous loop model combined with nervous innervation through the saphenous nerve might offer a promising possibility for the functionalization of skeletal muscle [65].

Although promising results for engineering of vascularised tissue have already been achieved in the case of bone reconstruction [66], free autologous muscle flap transplantation still remains the gold standard for muscle reconstruction, in particular for complex soft tissue defects [67]. However, in the future, TE of skeletal muscle may help to overcome the donor site problem of harvesting large muscles of the human body.

## Conclusions

The generation of functional skeletal muscle tissue for future in vivo applications is still challenging. In this study we demonstrated that MSCs in monocultures and in co-cultivation with myoblasts are able to differentiate into the myogenic lineage by expressing key myogenic markers such as desmin, MEF2, MyHC2 and ACTN2. Stimulation with HGF and IGF-1 induces an upregulation of different myogenic markers, but probably is not essential for myogenic differentiation. IGFBPs play a role during myogenic differentiation, varying amongst culture and stimulation conditions. Three-dimensional cultivation of co-cultures enhances the myogenic differentiation capacity. PCL-collagen nanofibers especially represent a promising scaffold, mimicking the structure of skeletal muscle and inducing parallel alignment of MSCs and myoblasts. The results of this study represent important starting points for future studies and in vivo applications for the TE of skeletal muscle.

## Additional files

**Additional file 1:** Fluorescence microscopy of MyoD. Primary rat myoblasts after being passaged two times since isolation. A merge of DAPI (blue) and MyoD (red, with Alexa fluor 594 as secondary antibody) staining is shown. (PDF 472 kb)

**Additional file 2:** Fluorescence microscopy of MyHC2 after 7d. Single stainings of MyHC2 in MSC and Mb co-cultures under HGF stimulation for 7 d. (a) Nuclear staining with DAPI. (b) GFP-transduced MSC in green colour. (c) Staining for MyHC2 with Alexa fluor 594 as secondary antibody. Scale bars represent 20 μm. Magnification 400x. (PDF 2443 kb)

**Additional file 3:** Fluorescence microscopy of MyHC2 after 14d. Single stainings of MyHC2 in MSC and Mb co-cultures under HGF stimulation for 14 d. (a) Nuclear staining with DAPI. (b) GFP-transduced MSC in green colour. (c) Staining for MyHC2 with Alexa fluor 594 as secondary antibody. Scale bars represent 50 μm. Magnification 200x. (PDF 1334 kb)

**Additional file 4:** Fluorescence microscopy of MEF2. Single stainings of MEF2 in MSC and Mb co-cultures in control groups without HGF/IGF-1 after 14 d. (a) Nuclear staining with DAPI. (b) GFP-transduced MSC are in green colour. (c) Staining for MEF2 with Alexa fluor 594 as secondary antibody. Scale bars represent 50 μm. Magnification 200x. (PDF 792 kb)

**Additional file 5:** Life-cell tracking of MSC and myoblast co-cultures. GFP-transduced MSCs (green) and CM-Dil-myoblasts (red) were co-cultivated in basic differentiation medium over a period of 5 d. Double-labelled cells indicate fusion of myoblasts and MSCs. (MPG 5416 kb)

## Abbreviations

2D: Two-dimensional; 3D: Three-dimensional; ACTN2: Alpha-sarcomeric actinin; bFGF: Basic fibroblast growth factor; d: Days; DAPI: Diamidine-phenylindole-dihydrochloride; DES: Desmin; DHS: Donor horse serum; FACS: Fluorescence-activated cell sorting; GFP: Green Fluorescent Protein; h: Hours; HGF: Hepatocyte growth factor; IGF-1: Insulin-like growth factor-1; IGF-1.R: IGF-1 receptor; IGFBPs: IGF-binding proteins; Mb: Myoblasts; MEF2: Myogenic Enhancer Factor 2; MSC: Mesenchymal stem cells; MyHC2: Myosin heavy chain; MYOG: Myogenin; PCL: Poly-ε-caprolacton; RPL13a: Ribosomal protein L13a; RT-PCR: Real-time PCR; SEM: Scanning Electron Microscopy; TE: Tissue engineering

## Acknowledgements

The present work was performed in fulfillment of the requirements for obtaining the degree "Dr. med." for Ramona Witt.
We would like to thank Stefan Fleischer and Marie-Louise Gorkisch for their excellent technical support.

## Funding

This study was funded by the DFG (Deutsche Forschungsgemeinschaft, BE 4803/3-1), the Interdisciplinary Center for Clinical Research (IZKF, Faculty of Medicine Friedrich-Alexander University Erlangen-Nürnberg) and the ELAN Fonds of the University Hospital of Erlangen.
We acknowledge support by Deutsche Forschungsgemeinschaft and Friedrich-Alexander-Universität Erlangen-Nürnberg (FAU) within the funding programme Open Access Publishing.

## Authors' contributions

RW designed the study and performed all cell culture experiments and analysed the data. AW designed the study and performed flow cytometry, assisted the cell culture experiments and analysed the data. AMB participated in the immunocytochemistry. AC supported the cell experiments. DD carried out the electrospinning process. ARB supported the electrospinning. DWS participated in the electrospinning process. MH participated in the RT-PCR studies. C L performed MSC isolation and GFP-transduction. AA supported the cell experiments and the statistical analysis. REH participated in the immunocytochemistry and the coordination of the study. JPB designed the study, participated in the flow cytometry and cell culture studies and helped to draft the manuscript. All authors have seen and agreed to the final submitted version of the paper.

## Competing interests

The authors declare that they have no competing interest.

## Author details

[1]Department of Plastic and Hand Surgery and Laboratory for Tissue Engineering and Regenerative Medicine, University Hospital of Erlangen, Friedrich-Alexander University of Erlangen-Nürnberg (FAU), Krankenhausstraße 12, 91054 Erlangen, Germany. [2]Institute of Biomaterials, Department of Materials Science and Engineering, University of Erlangen-Nürnberg (FAU), Cauerstraße 6, 91058 Erlangen, Germany. [3]Institute of Polymer Materials, Department of Materials Science and Engineering, University of Erlangen- Nürnberg (FAU), Martensstrasse 7, 91058 Erlangen, Germany. [4]Interdisciplinary Clinic for Stem Cell Transplantation, University Cancer Center Hamburg (UCCH), 20246 Hamburg, Germany.

## References

1. Zanou N, Gailly P. Skeletal muscle hypertrophy and regeneration: interplay between the myogenic regulatory factors (MRFs) and insulin-like growth factors (IGFs) pathways. Cell Mol Life Sci. 2013;70(21):4117–30.
2. Megeney LA, Kablar B, Garrett K, Anderson JE, Rudnicki MA. Myod is required for myogenic stem cell function in adult skeletal muscle. Genes Dev. 1996;10(10):1173–83.
3. Mauro A. Satellite cell of skeletal muscle fibers. J Biophys Biochem Cytol. 1961;9:493–5.
4. Charge SBP, Rudnicki MA. Cellular and molecular regulation of muscle regeneration. Physiol Rev. 2004;89:209–38.
5. Cittadella Vigodarzere G, Mantero S. Skeletal muscle tissue engineering: strategies for volumetric constructs. Front Physiol. 2014;5:362.
6. Grogan BF, Hsu JR. Volumetric muscle loss. J Am Acad Orthop Surg. 2011;19 Suppl 1:S35–37.
7. Collins CA, Olsen I, Zammit PS, Heslop L, Petrie A, Partridge TA, Morgan JE. Stem cell function, self-renewal, and behavioral heterogeneity of cells from the adult muscle satellite cell niche. Cell. 2005;122(2):289–301.
8. Rando TA. The adult muscle stem cell comes of age. Nat Med. 2005;11(8):829–31.
9. Boonen KJ, Post MJ. The muscle stem cell niche: regulation of satellite cells during regeneration. Tissue Eng B Rev. 2008;14(4):419–31.
10. Machida S, Spangenburg EE, Booth FW. Primary rat muscle progenitor cells have decreased proliferation and myotube formation during passages. Cell Prolif. 2004;37(4):267–77.
11. Drost AC, Weng S, Feil G, Schafer J, Baumann S, Kanz L, Sievert KD, Stenzl A, Mohle R. In vitro myogenic differentiation of human bone marrow-derived mesenchymal stem cells as a potential treatment for urethral sphincter muscle repair. Ann N Y Acad Sci. 2009;1176:135–43.
12. Dezawa M, Ishikawa H, Itokazu Y, Yoshihara T, Hoshino M, Takeda S, Ide C, Nabeshima Y. Bone marrow stromal cells generate muscle cells and repair muscle degeneration. Science. 2005;309(5732):314–7.
13. Kulesza A, Burdzinska A, Szczepanska I, Zarychta-Wisniewska W, Pajak B, Bojarczuk K, Dybowski B, Paczek L. The mutual interactions between mesenchymal stem cells and myoblasts in an autologous co-culture model. PLoS One. 2016;11(8):e0161693.
14. Smolina NA, Davydova A, Shchukina IA, Karpushev AV, Malashicheva AB, Dmitrieva RI, Kostareva AA. Comparative assessment of different approaches for obtaining terminally differentiated muscle cells. Tsitologiia. 2014;56(4):291–9.
15. Gang EJ, Darabi R, Bosnakovski D, Xu Z, Kamm KE, Kyba M, Perlingeiro RC. Engraftment of mesenchymal stem cells into dystrophin-deficient mice is not accompanied by functional recovery. Exp Cell Res. 2009;315(15):2624–36.
16. Shudo Y, Miyagawa S, Ohkura H, Fukushima S, Saito A, Shiozaki M, Kawaguchi N, Matsuura N, Shimizu T, Okano T, et al. Addition of mesenchymal stem cells enhances the therapeutic effects of skeletal myoblast cell-sheet transplantation in a rat ischemic cardiomyopathy model. Tissue Eng A. 2014;20(3-4):728–39.
17. Wagers AJ, Conboy IM. Cellular and molecular signatures of muscle regeneration: current concepts and controversies in adult myogenesis. Cell. 2005;122(5):659 67.
18. Beier JP, Bitto FF, Lange C, Klumpp D, Arkudas A, Bleiziffer O, Boos AM, Horch RE, Kneser U. Myogenic differentiation of mesenchymal stem cells co-cultured with primary myoblasts. Cell Biol Int. 2011;35(4):397–406.
19. Miller KJ, Thaloor D, Matteson S, Pavlath GK. Hepatocyte growth factor affects satellite cell activation and differentiation in regenerating skeletal muscle. Am J Physiol Cell Physiol. 2000;278(1):C174–181.
20. Florini JR, Ewton DZ, Coolican SA. Growth hormone and the insulin-like growth factor system in myogenesis. Endocr Rev. 1996;17(5):481–517.
21. Kelley KM, Oh Y, Gargosky SE, Gucev Z, Matsumoto T, Hwa V, Ng L, Simpson DM, Rosenfeld RG. Insulin-like growth factor-binding proteins (IGFBPs) and their regulatory dynamics. Int J Biochem Cell Biol. 1996;28(6):619–37.
22. Duan C. Specifying the cellular responses to IGF signals: roles of IGF-binding proteins. J Endocrinol. 2002;175:41–54.
23. Duan C, Ren H, Gao S. Insulin-like growth factors (IGFs), IGF receptors, and IGF-binding proteins: roles in skeletal muscle growth and differentiation. Gen Comp Endocrinol. 2010;167(3):344–51.
24. Clemmons DR. Use of mutagenesis to probe IGF-binding protein structure/function relationships. Endocr Rev. 2001;22(6):800–17.
25. Heher P, Maleiner B, Pruller J, Teuschl AH, Kollmitzer J, Monforte X, Wolbank S, Redl H, Runzler D, Fuchs C. A novel bioreactor for the generation of highly aligned 3D skeletal muscle-like constructs through orientation of fibrin via application of static strain. Acta Biomater. 2015;24:251–65.
26. Choi JS, Lee SJ, Christ GJ, Atala A, Yoo JJ. The influence of electrospun aligned poly(epsilon-caprolactone)/collagen nanofiber meshes on the formation of self-aligned skeletal muscle myotubes. Biomaterials. 2008;29(19):2899–906.
27. Beier JP, Klumpp D, Rudisile M, Dersch R, Wendorff JH, Bleiziffer O, Arkudas A, Polykandriotis E, Horch RE, Kneser U. Collagen matrices from sponge to nano: new perspectives for tissue engineering of skeletal muscle. BMC Biotechnol. 2009;9:34.
28. McKinsey TA, Zhang CL, Olson EN. MEF2: a calcium-dependent regulator of cell division, differentiation and death. Trends Biochem Sci. 2002;27(1):40–7.
29. Jin W, Liu M, Peng J, Jiang S. Function analysis of Mef2c promoter in muscle differentiation. Biotechnol Appl Biochem. 2016:doi:10.1002/bab.1524.
30. Black BL, Olson EN. Transcriptional control of muscle development by myocyte enhancer factor-2 (MEF2) proteins. Annu Rev Cell Dev Biol. 1998;14:167–96.
31. Snyder CM, Rice AL, Estrella NL, Held A, Kandarian SC, Naya FJ. MEF2A regulates the Gtl2-Dio3 microRNA mega-cluster to modulate WNT signaling in skeletal muscle regeneration. Development. 2013;140(1):31–42.
32. Salucci S, Baldassarri V, Falcieri E, Burattini S. Alpha-Actinin involvement in Z-disk assembly during skeletal muscle C2C12 cells in vitro differentiation. Micron. 2015;68:47–53.
33. Sjoblom B, Salmazo A, Djinovic-Carugo K. Alpha-actinin structure and regulation. Cell Mol Life Sci. 2008;65(17):2688–701.
34. Hong J, Park JS, Lee H, Jeong J, Hyeon Yun H, Yun Kim H, Ko YG, Lee JH. Myosin heavy chain is stabilized by BCL-2 interacting cell death suppressor (BIS) in skeletal muscle. Exp Mol Med. 2016;48:e225.
35. Weintraub H, Davis R, Tapscott S, Thayer M, Krause M, Benezra R, Blackwell TK, Turner D, Rupp R, Hollenberg S, et al. The myoD gene family: nodal point during specification of the muscle cell lineage. Science. 1991;251(4995):761–6.
36. Lange C, Togel F, Ittrich H, Clayton F, Nolte-Ernsting C, Zander AR, Westenfelder C. Administered mesenchymal stem cells enhance recovery from ischemia/reperfusion-induced acute renal failure in rats. Kidney Int. 2005;68(4):1613–7.
37. Javazon EH, Colter DC, Schwarz EJ, Prockop DJ. Rat marrow stromal cells are more sensitive to plating density and expand more rapidly from single-cell-derived colonies than human marrow stromal cells. Stem Cells. 2001;19(3):219–25.
38. Walker N, Kahamba T, Woudberg N, Goetsch K, Niesler C. Dose-dependent modulation of myogenesis by HGF: implications for c-Met expression and downstream signalling pathways. Growth Factors (Chur, Switzerland). 2015; 33(3):229–41.
39. Yamada M, Tatsumi R, Yamanouchi K, Hosoyama T, Shiratsuchi S, Sato A, Mizunoya W, Ikeuchi Y, Furuse M, Allen RE. High concentrations of Hgf inhibit skeletal muscle satellite cell proliferation in vitro by inducing expression of myostatin: a possible mechanism for reestablishing satellite cell quiescence in vivo. Am J Physiol Cell Physiol. 2010;298(3):C465–476.

40. Gal-Levi R, Leshem Y, Aoki S, Nakamura T, Halevy O. Hepatocyte growth factor plays a dual role in regulating skeletal muscle satellite cell proliferation and differentiation. Biochim Biophys Acta. 1998;1402(1):39–51.

41. Navarro M, Barenton B, Garandel V, Schnekenburger J, Bernardi H. Insulin-like growth factor I (IGF-I) receptor overexpression abolishes the IGF requirement for differentiation and induces a ligand-dependent transformed phenotype in C2 inducible myoblasts. Endocrinology. 1997;138(12):5210–9.

42. Quinn LS, Ehsan M, Steinmetz B, Kaleko M. Ligand-dependent inhibition of myoblast differentiation by overexpression of the type-1 insulin-like growth factor receptor. J Cell Physiol. 1993;156(3):453–61.

43. Anastasi S, Giordano S, Sthandier O, Gambarotta G, Maione R, Comoglio P, Amati P. A natural hepatocyte growth factor/scatter factor autocrine loop in myoblast cells and the effect of the constitutive met kinase activation on myogenic differentiation. J Cell Biol. 1997;137(5):1057–68.

44. Gutierrez J, Cabrera D, Brandan E. Glypican-1 regulates myoblast response to HGF via met in a lipid raft-dependent mechanism: effect on migration of skeletal muscle precursor cells. Skelet Muscle. 2014;4(1):5.

45. Muguruma Y, Reyes M, Nakamura Y, Sato T, Matsuzawa H, Miyatake H, Akatsuka A, Itoh J, Yahata T, Ando K, et al. In vivo and in vitro differentiation of myocytes from human bone marrow-derived multipotent progenitor cells. Exp Hematol. 2003;31(12):1323–30.

46. Li Z, Gu TX, Zhang YH. Hepatocyte growth factor combined with insulin like growth factor-1 improves expression of GATA-4 in mesenchymal stem cells cocultured with cardiomyocytes. Chin Med J (Engl). 2008;121(4):336–40.

47. Kadam S, Patki S, Bhonde R. Human Fallopian tube as a novel source of multipotent stem cells with potential for islet neogenesis. J Stem Cells Regen Med. 2009;5(1):37–42.

48. Chen W, Xie M, Yang B, Bharadwaj S, Song L, Liu G, Yi S, Ye G, Atala A, Zhang Y. Skeletal myogenic differentiation of human urine-derived cells as a potential source for skeletal muscle regeneration. J Tissue Eng Regen Med. 2014;11(2):334–41.

49. Tasli PN, Dogan A, Demirci S, Sahin F. Myogenic and neurogenic differentiation of human tooth germ stem cells (hTGSCs) are regulated by pluronic block copolymers. Cytotechnology. 2016;68(2):319–29.

50. Weiss A, Leinwand LA. The mammalian myosin heavy chain gene family. Annu Rev Cell Dev Biol. 1996;12:417–39.

51. Karalaki M, Fili S, Philippou A, Koutsilieris M. Muscle regeneration: cellular and molecular events. In Vivo. 2009;23:779–96.

52. Syverud BC, VanDusen KW, Larkin LM. Effects of dexamethasone on satellite cells and tissue engineered skeletal muscle units. Tissue Eng A. 2016;22(5-6):480–9.

53. Singaravelu K, Padanilam BJ. In vitro differentiation of MSC into cells with a renal tubular epithelial-like phenotype. Ren Fail. 2009;31(6):492–502.

54. Alves AN, Ribeiro BG, Fernandes KP, Souza NH, Rocha LA, Nunes FD, Bussadori SK, Mesquita-Ferrari RA. Comparative effects of low-level laser therapy pre- and post-injury on mRNA expression of MyoD, myogenin, and IL-6 during the skeletal muscle repair. Lasers Med Sci. 2016;31(4):679–85.

55. Bentzinger CF, Wang YX, Rudnicki MA. Building muscle: molecular regulation of myogenesis. Cold Spring Harb Perspect Biol. 2012;4(2). doi:10.1101/cshperspect.a008342.

56. Sebastian S, Goulding L, Kuchipudi SV, Chang KC. Extended 2D myotube culture recapitulates postnatal fibre type plasticity. BMC Cell Biol. 2015;16:23.

57. Wright WE, Sassoon DA, Lin VK. Myogenin, a factor regulating myogenesis, has a domain homologous to MyoD. Cell. 1989;56(4):607–17.

58. Hinterberger TJ, Sassoon DA, Rhodes SJ, Konieczny SF. Expression of the muscle regulatory factor MRF4 during somite and skeletal myofiber development. Dev Biol. 1991;147(1):144–56.

59. Yamaguchi A, Sakuma K, Fujikawa T, Morita I. Expression of specific IGFBPs is associated with those of the proliferating and differentiating markers in regenerating rat plantaris muscle. J Physiol Sci. 2013;63(1):71–7.

60. Li M, Li Y, Lu L, Wang X, Gong Q, Duan C. Structural, gene expression, and functional analysis of the fugu (Takifugu rubripes) insulin-like growth factor binding protein-4 gene. Am J Physiol Regul Integr Comp Physiol. 2009;296(3):R558–566.

61. Salih DA, Tripathi G, Holding C, Szestak TA, Gonzalez MI, Carter EJ, Cobb LJ, Eisemann JE, Pell JM. Insulin-like growth factor-binding protein 5 (IGFBP5) compromises survival, growth, muscle development, and fertility in mice. Proc Natl Acad Sci U S A. 2004;101(12):4314–9.

62. Wang L, Wu Y, Guo B, Ma PX. Nanofiber Yarn/Hydrogel Core-Shell Scaffolds Mimicking Native Skeletal Muscle Tissue for Guiding 3D Myoblast Alignment, Elongation, and Differentiation. ACS Nano. 2015;9(9):9167–79.

63. Jana S, Leung M, Chang J, Zhang M. Effect of nano- and micro-scale topological features on alignment of muscle cells and commitment of myogenic differentiation. Biofabrication. 2014;6(3):035012.

64. Zhao W, Ju YM, Christ G, Atala A, Yoo JJ, Lee SJ. Diaphragmatic muscle reconstruction with an aligned electrospun poly(epsilon-caprolactone)/collagen hybrid scaffold. Biomaterials. 2013;34(33):8235–40.

65. Bitto FF, Klumpp D, Lange C, Boos AM, Arkudas A, Bleiziffer O, Horch RE, Kneser U, Beier JP. Myogenic differentiation of mesenchymal stem cells in a newly developed neurotised AV-Loop model. Biomed Res Int. 2013;2013:935046.

66. Horch RE, Beier JP, Kneser U, Arkudas A. Successful human long-term application of in situ bone tissue engineering. J Cell Mol Med. 2014;18(7):1478–85.

67. Horch RE, Lang W, Arkudas A, Taeger C, Kneser U, Schmitz M, Beier JP. Nutrient free flaps with vascular bypasses for extremity salvage in patients with chronic limb ischemia. J Cardiovasc Surg (Torino). 2014;55(2 Suppl 1):265–72.

# 11

# Intermittent parathyroid hormone (PTH) promotes cementogenesis and alleviates the catabolic effects of mechanical strain in cementoblasts

Yuyu Li[1,2], Zhiai Hu[1,2], Chenchen Zhou[1,2], Yang Xu[1,2], Li Huang[1,2], Xin Wang[1,2] and Shujuan Zou[1,2*] (iD)

## Abstract

**Background:** External root resorption, commonly starting from cementum, is a severe side effect of orthodontic treatment. In this pathological process and repairing course followed, cementoblasts play a significant role. Previous studies implicated that parathyroid hormone (PTH) could act on committed osteoblast precursors to promote differentiation, and inhibit apoptosis. But little was known about the role of PTH in cementoblasts. The purpose of this study was to investigate the effects of intermittent PTH on cementoblasts and its influence after mechanical strain treatment.

**Results:** Higher levels of cementogenesis- and differentiation-related biomarkers (bone sialoprotein (BSP), osteocalcin (OCN), Collagen type I (COL1) and Osterix (Osx)) were shown in 1–3 cycles of intermittent PTH treated groups than the control group. Additionally, intermittent PTH increased alkaline phosphatase (ALP) activity and mineralized nodules formation, as measured by ALP staining, quantitative ALP assay, Alizarin red S staining and quantitative calcium assay. The morphology of OCCM-30 cells changed after mechanical strain exertion. Expression of BSP, ALP, OCN, osteopontin (OPN) and Osx was restrained after 18 h mechanical strain. Furthermore, intermittent PTH significantly increased the expression of cementogenesis- and differentiation-related biomarkers in mechanical strain treated OCCM-30 cells.

**Conclusions:** Taken together, these data suggested that intermittent PTH promoted cementum formation through activating cementogenesis- and differentiation-related biomarkers, and attenuated the catabolic effects of mechanical strain in immortalized cementoblasts OCCM-30.

**Keywords:** OCCM-30 cell, Intermittent parathyroid hormone (PTH), Mechanical strain, Cementogenesis, Tooth root resorption

## Background

External root resorption is a pathological process, which tends to occur following multiple mechanical or chemical stimuli such as infection, trauma or orthodontic treatment and usually begins with the resorption of cementum. This condition may result in pain, swelling and even mobility of the tooth. Treatment alternatives are usually case-

dependent, and lack efficacy for the management of external root resorption [1]. Mechanical stimuli with different strain magnitude, frequency, rate or gradients has potent influences on modeling and remodeling of bone and associated cells and signaling pathways [2]. Similar to bone, cementum also undergoes considerable alteration under the influence of mechanical stimuli, and during this procedure, cementoblasts are strain-responsive and have the unique ability to transduce mechanical stimuli into biological events [3]. Moreover, it has been suggested that cementoblasts assist in the process of cementum repair

* Correspondence: drzsj@scu.edu.cn
[1]Department of Orthodontics, West China Hospital of Stomatology, Sichuan University, No. 14, 3rd Section, Renmin South Road, Chengdu 610041, China
[2]State Key Laboratory of Oral Diseases, National Clinical Research Center for Oral Diseases, West China Hospital of Stomatology, Sichuan University, No. 14, 3rd Section, Renmin South Road, Chengdu 610041, China

by forming either cellular intrinsic fiber cementum or acellular extrinsic fiber cementum [4].

Endogenous parathyroid hormone (PTH) is the primary regulator of calcium and phosphate metabolism in bone and kidney. Teriparatide/Forteo is a recombinant of human parathyroid hormone [5] that contains the 1–34 amino acid sequence of the complete PTH molecule (PTH (1–34)). The effects of PTH on bone homeostasis are dependent on the mode and dosage of administration. That is, intermittent low-dose PTH increases bone turnover with a greater stimulation of bone formation than bone resorption, whereas continuous high-dose PTH tends to cause catabolic outcomes [6]. Furthermore, the catabolic effects of continuous PTH on cementoblasts have been reported [7, 8]. Evidence has suggested that cementum and bone share many characteristics including gene expression profiles and cell morphologies. However, the effects of intermittent PTH on cementum still remain to be elucidated. We hypothesize that intermittent PTH could also be used to attenuate external root resorption by helping regenerate cementum.

In the present study, we investigated the biological changes of cementoblasts (OCCM-30) after a combined treatment of PTH and mechanical strain by four-point bending system (University of Electronic Science and Technology of China, Chengdu, China). It was shown that

intermittent PTH treatment directly regulated cementoblast behavior by detecting the expression of parathyroid hormone receptor type 1 (PTHR1) and some cementogenesis- and differentiation-related biomarkers using qPCR or western blot. Also, intermittent PTH enhanced the ALP activity and the formation of mineralized nodules of OCCM-30 cells. Furthermore, intermittent PTH alleviated the catabolic effect of mechanical strain via regulating the expression of BSP, ALP, OCN, COL1, Runt-related transcription factor 2 (Runx2) and Osx. These results provide the opportunity for evaluating the potential role of intermittent PTH in the cementum regeneration therapies.

## Results
### Intermittent PTH and mechanical strain regulated PTHR1 expression in OCCM-30 cells

It was reported that PTHR1 mRNA was expressed in cememtoblast SV-CM410 subclone [9]. To confirm if OCCM-30 was target cell for PTH, the expression of PTHR1 was examined by western blot (Fig. 1). PTHR1 protein was observed in OCCM-30 cells in the control group, and the upregulation of its expression was PTH-cycle dependent (Fig. 1a and Additional file 1: Figure S1). The increase amounted to 670% after 3 cycles of intermittent PTH, compared to the corresponding control group

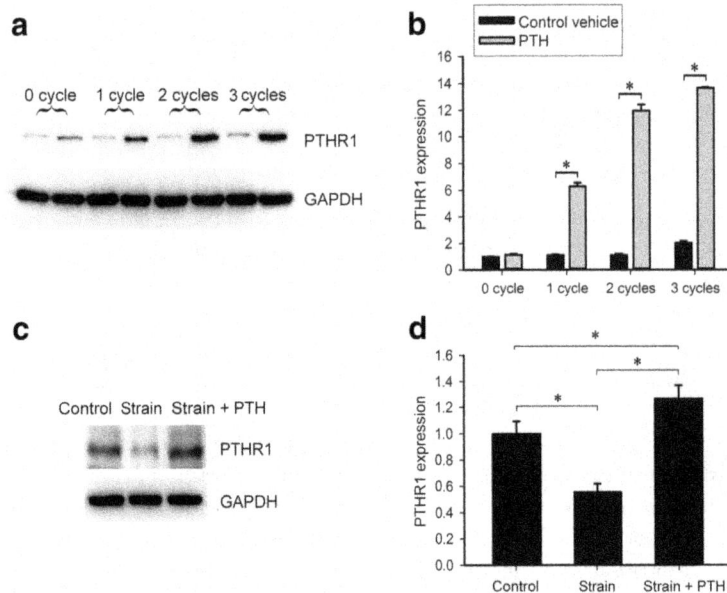

**Fig. 1** Intermittent PTH treatment and mechanical strain affected the protein level of PTHR1 in OCCM-30 cells. **a** PTH-cycle dependent changes in PTHR1 expression detected by western blot. In each cycle, for the first 6 h, the control group was treated with vehicle culture medium containing acetic acid, and the PTH group received PTH treatment. Then they were both cultured in fresh medium without acetic acid or PTH for 18 h. GAPDH was used as a loading control. **b** Densitometry of bands was performed and normalized to GAPDH. **c** Western blot analysis of PTHR1 and GAPDH levels in OCCM-30 cells in the presence of mechanical strain and intermittent PTH. The Strain + PTH group represented cells that accepted 18 h mechanical strain, followed by 3 cycles of intermittent PTH treatment. The strain group cells were treated with 18 h mechanical strain, followed by 3 cycles of vehicle culture medium. There was neither mechanical strain nor intermittent PTH treatment in the control group. GAPDH was used as an endogenous reference. The experiment was repeated 3 times and the representative result is shown. **d** The bands were quantified and normalized to GAPDH. The control group was set to 1. Bars represent means ± SD of triplicate measurements. * $p < 0.05$

(Fig. 1b and Additional file 2: Table S1, * $p < 0.05$). To investigate the effects of mechanical strain on OCCM-30, we used the SXG4201 four-point bending device (University of Electronic Science and Technology of China, Chengdu, China; Fig. 2a-c) to apply cyclic mechanical strain to the OCCM-30 cells. OCCM-30 cells were seeded onto the plates and were subjected to 2000 µε mechanical strain at 0.5 Hz (loading displacement is 1.12 mm) for 18 h. The deformation of the plates maintained in culture medium caused the attached cells to deform (Fig. 2d). The mechanical strain treatment protocol was all the same throughout the study. Control cultures grew under the same conditions but without the mechanical strain treatment. It was shown that PTHR1 protein expression was decreased by 45% after 18 h mechanical strain (Fig. 1c-d, Additional file 1: Figure S1 and Additional file 2: Table S2, * $p < 0.05$). Additionally, 3 cycles of intermittent PTH treatment after 18 h mechanical strain induced a 26% increase in PTHR1 expression, compared to the control group (Fig. 1c-d, Additional file 1: Figure S1 and Additional file 2: Table S2, * $p < 0.05$). These results suggested that intermittent PTH and mechanical strain had opposite effects on PTHR1 expression and possibly on cementogenic process.

### PTH promoted cementoblast mineralization in OCCM-30 cells

It was reported that PTH protected against periodontitis-associated bone loss through the regulation of osteoblast

activity [10]. We determined mineralization capacity of cementoblasts after intermittent PTH treatment by examining ALP activity and the formation of mineralized nodules. We performed ALP staining, quantitative ALP assay, Alizarin red S staining and quantitative calcium assay. Strong ALP staining was observed after 3 cycles of intermittent PTH treatment (Fig. 3a). As shown in Fig. 3b and Additional file 2: Table S3, intermittent PTH increased ALP activity in a time-dependent manner in OCCM-30 cells (* $p < 0.05$). Besides, after 10 days of culture, it was shown that 3 cycles of intermittent PTH enhanced the area of mineralized nodules compared to the untreated control (Fig. 3c). For quantitative calcium measurement, stained Alizarin red S was eluted, and the quantification of eluted dye showed enhancement in the intermittent PTH groups, reaching a 1.8-fold increase in the 3 cycles intermittent PTH group as compared to the control group (Fig. 3d and Additional file 2: Table S4, * $p < 0.05$). These results suggested that intermittent PTH could have a promotive role in cementoblast differentiation.

### PTH controlled the expression of cementogenesis-related proteins and regulation of cementoblast differentiation

Next, we investigated the expression of some mineralization-related proteins involving in cementogenesis and cementoblast differentiation after intermittent PTH treatment. BSP, OCN and COL1 are essential mineralization-related proteins

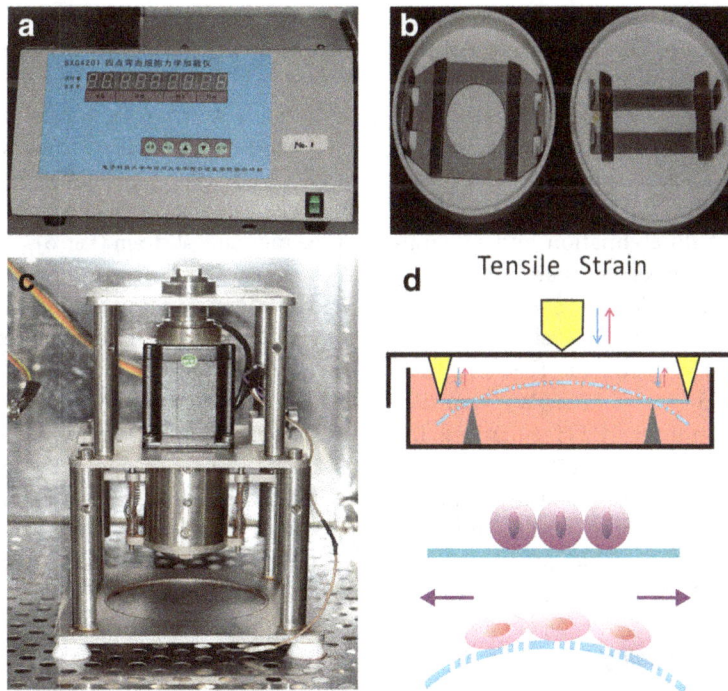

**Fig. 2** Components and schematic diagram of the uniaxial four-point bending system. The uniaxial four-point bending system is made up of **a** a digital control unit, **b** bending boxes and **c** an actuator. **d** Schematic diagram illustrates how the four-point bending unit works. When the bending box is compressed by the actuator, the cell culture plate is bended by four points to generate tensile strain

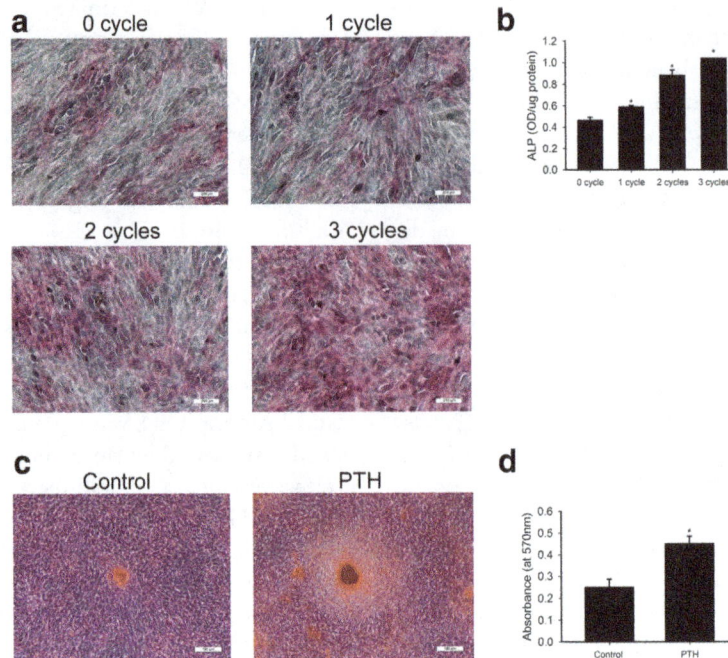

**Fig. 3** Intermittent PTH administration enhanced ALP activity and the formation of mineralized nodules. **a** Representative images of ALP staining of OCCM-30 cells after treatment with vehicle, 1 cycle, 2 cycles and 3 cycles of intermittent PTH. They together demonstrated a progressively enhanced ALP⁺ staining consistent with the increasing numbers of administration cycles. Scale bar: 200 μm. **b** ALP assay of 0, 1, 2 and 3 cycles of intermittent PTH treated groups. * indicates significant difference between 0 cycle group and other three groups ($p < 0.05$). **c** Microscopic findings of the Alizarin red S staining of the control group and the PTH group. The area of observable mineralized nodules in intermittent PTH treated group was obviously larger in comparison with that of the control group. Scale bar: 100 μm. **d** Alizarin red S extraction assay from the control group and the intermittent PTH treated group. * $p < 0.05$

in cementogenesis [11], and regarded as important biomarkers for root regeneration. The mRNA levels of BSP, OCN and COL1 were significantly increased after intermittent PTH treatment. 3 cycles of intermittent PTH induced 2.0-, 1.5- and 4.2-fold increase in BSP, OCN and COL1 expression, respectively (Fig. 4a-c and Additional file 2: Table S5, * $p < 0.05$). Osx is required for cementoblast differentiation and mineralized tissue formation. We observed that intermittent PTH also induced Osx mRNA expression, and the expression of Osx reached maximal levels after 3 cycles of intermittent PTH treatment (Fig. 4d and Additional file 2: Table S5, * $p < 0.05$). Western blot analysis showed that intermittent PTH enhanced the expression of BSP, OCN, COL1 and Osx proteins (Fig. 4e-i, Additional file 3: Figure S2 and Additional file 2: Table S6, * $p < 0.05$). These results indicated that BSP, OCN, COL1 and Osx could play important roles in intermittent PTH-induced cementogenesis.

## Mechanical strain induced morphological changes in OCCM-30 cells
Mechanical strain is a major regulator of cementum remodeling and resorption [12]. After mechanical strain treatment, the morphological changes of OCCM-30

cells were detected under an optical microscope. Untreated OCCM-30 cells exhibited irregular polygonal- or spindle-shaped morphologies, with the nucleus in the middle of the cytoplasm. And the cells were randomly oriented. However, mechanical strain induced a markedly altered morphology in OCCM-30 cells. Cells under mechanical strain condition were elongated and realigned, conforming to the direction of the mechanical strain (Fig. 5).

## Mechanical strain attenuated cementogenesis- and differentiation-related gene expression
To further investigate the effects of mechanical strain on cementogenesis in OCCM-30 cells, cememtogenesis-related genes BSP, ALP, OCN and OPN were examined by qPCR. Mechanical strain significantly inhibited cementogenesis-related genes expression in OCCM-30 cells. The expression of BSP decreased by 77% after exposed to mechanical strain of 2000 με for 18 h (Fig. 6a, * $p < 0.05$). Similarly, the expression of ALP, OCN and OPN in mechanical strain group also decreased by 62%, 66% and 74%, respectively (Fig. 6b-d, * $p < 0.05$). We also examined the expression of two transcription factors Runx2 and Osx that are important for cementogenic differentiation. Mechanical strain induced 10% decrease in Runx2 expression (Fig. 6e, $p > 0.05$). However, the expression of Osx decreased by 67% in

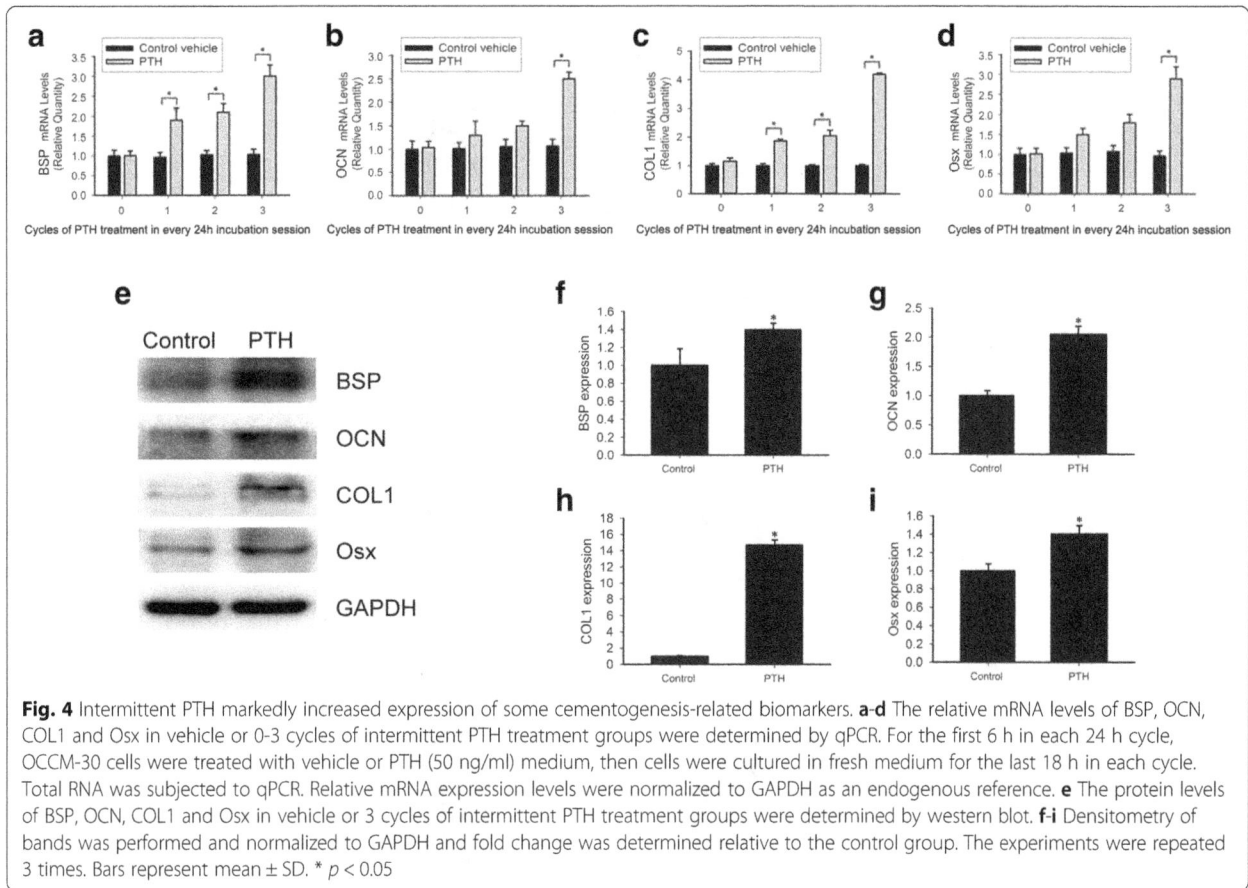

**Fig. 4** Intermittent PTH markedly increased expression of some cementogenesis-related biomarkers. **a-d** The relative mRNA levels of BSP, OCN, COL1 and Osx in vehicle or 0-3 cycles of intermittent PTH treatment groups were determined by qPCR. For the first 6 h in each 24 h cycle, OCCM-30 cells were treated with vehicle or PTH (50 ng/ml) medium, then cells were cultured in fresh medium for the last 18 h in each cycle. Total RNA was subjected to qPCR. Relative mRNA expression levels were normalized to GAPDH as an endogenous reference. **e** The protein levels of BSP, OCN, COL1 and Osx in vehicle or 3 cycles of intermittent PTH treatment groups were determined by western blot. **f-i** Densitometry of bands was performed and normalized to GAPDH and fold change was determined relative to the control group. The experiments were repeated 3 times. Bars represent mean ± SD. * $p < 0.05$

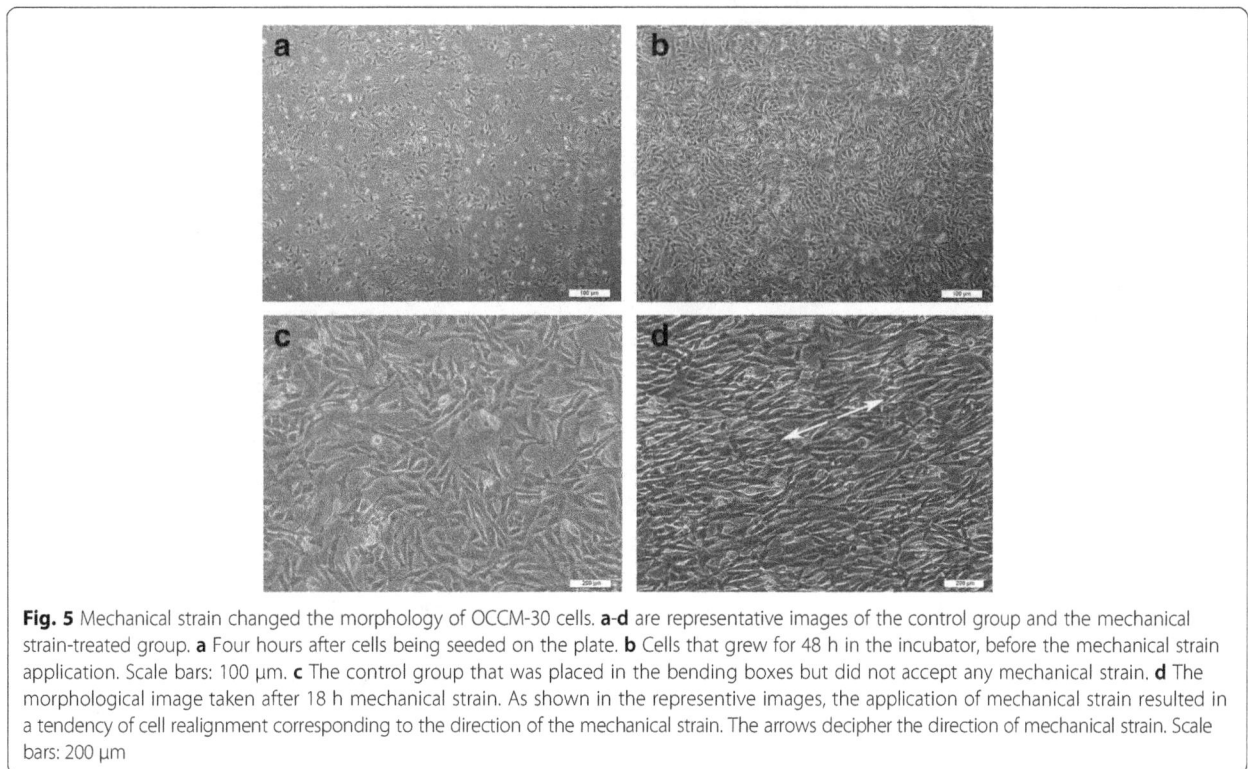

**Fig. 5** Mechanical strain changed the morphology of OCCM-30 cells. **a-d** are representative images of the control group and the mechanical strain-treated group. **a** Four hours after cells being seeded on the plate. **b** Cells that grew for 48 h in the incubator, before the mechanical strain application. Scale bars: 100 μm. **c** The control group that was placed in the bending boxes but did not accept any mechanical strain. **d** The morphological image taken after 18 h mechanical strain. As shown in the representative images, the application of mechanical strain resulted in a tendency of cell realignment corresponding to the direction of the mechanical strain. The arrows decipher the direction of mechanical strain. Scale bars: 200 μm

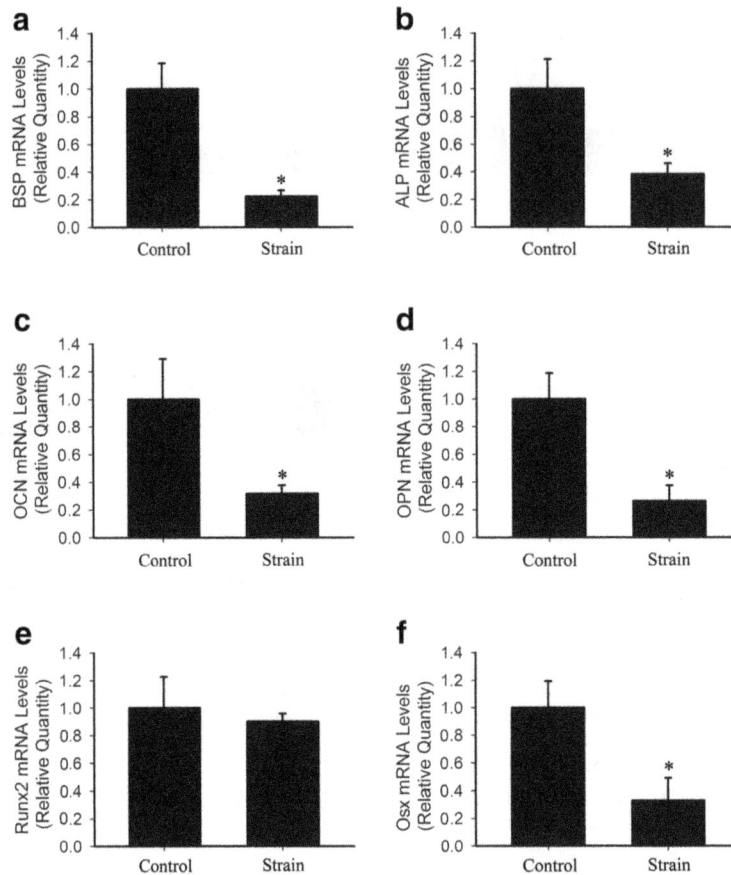

**Fig. 6** Mechanical strain impaired gene expression of certain cementogenesis-related biomarkers. **a-f** The relative mRNA levels of BSP, ALP, OCN, OPN, Runx2 and Osx in OCCM-30 cells in the presence or absence of 18 h mechanical strain. Total RNA was subjected to qPCR. Relative mRNA expression levels were normalized to GAPDH as an endogenous reference. The experiments were repeated 3 times. Bars represent means ± SD. * $p < 0.05$

mechanical strain group (Fig. 6f, * $p < 0.05$). Collectively, these findings suggested that mechanical strain might inhibit cementum formation via attenuating BSP, ALP, OCN, OPN and Osx expression. More detailed data are shown in Additional file 2: Table S7.

## PTH alleviated the inhibition of cementogenesis and cementoblast differentiation by mechanical strain

To verify whether or not the impairment of cementogenic ability induced by mechanical strain could be restored by intermittent PTH, OCCM-30 cells were subjected to mechanical strain of 2000 με for 18 h, followed by 3 cycles of intermittent PTH treatment. The expression of BSP, ALP, OCN and COL1 proteins decreased to 20%, 64%, 29% and 8% separately in mechanical strain-only group, which was consistent with our qPCR results that mechanical strain inhibited cementogenesis. However, intermittent PTH significantly upregulated these cementogenesis-related markers, the expression of which rebounded to 49%, 88%, 53% and 21%, respectively, in the strain + PTH group (Fig. 7b-e and Additional file 2: Table S8, * $p < 0.05$). These

opposite effects were also observed in differentiation-related proteins. The expression level of Runx2 and Osx in mechanical strain-only group was 47% and 36%, respectively, compared to the control group, while it was 90% and 79% in mechanical strain and PTH group (Fig. 7f-g and Additional file 2: Table S8, * $p < 0.05$). These results demonstrated that intermittent PTH suppressed the adverse effects of mechanical strain on cementum formation through upregulating important biomarkers involved in cementogenic differentiation and cementogenesis processes.

## Discussion

Cementum is a highly responsive mineralized tissue which plays an important role in the development of functional periodontal ligaments. Besides, cementoblastic activities of cementoblasts are thought to be critical in root resorption/repair [4]. Here we investigated the interactive effects of intermittent PTH and mechanical strain on cementogenesis in OCCM-30 cells. Intermittent PTH upregulated PTHR1 expression in OCCM-30 cells. Also, intermittent PTH elevated both mRNA and

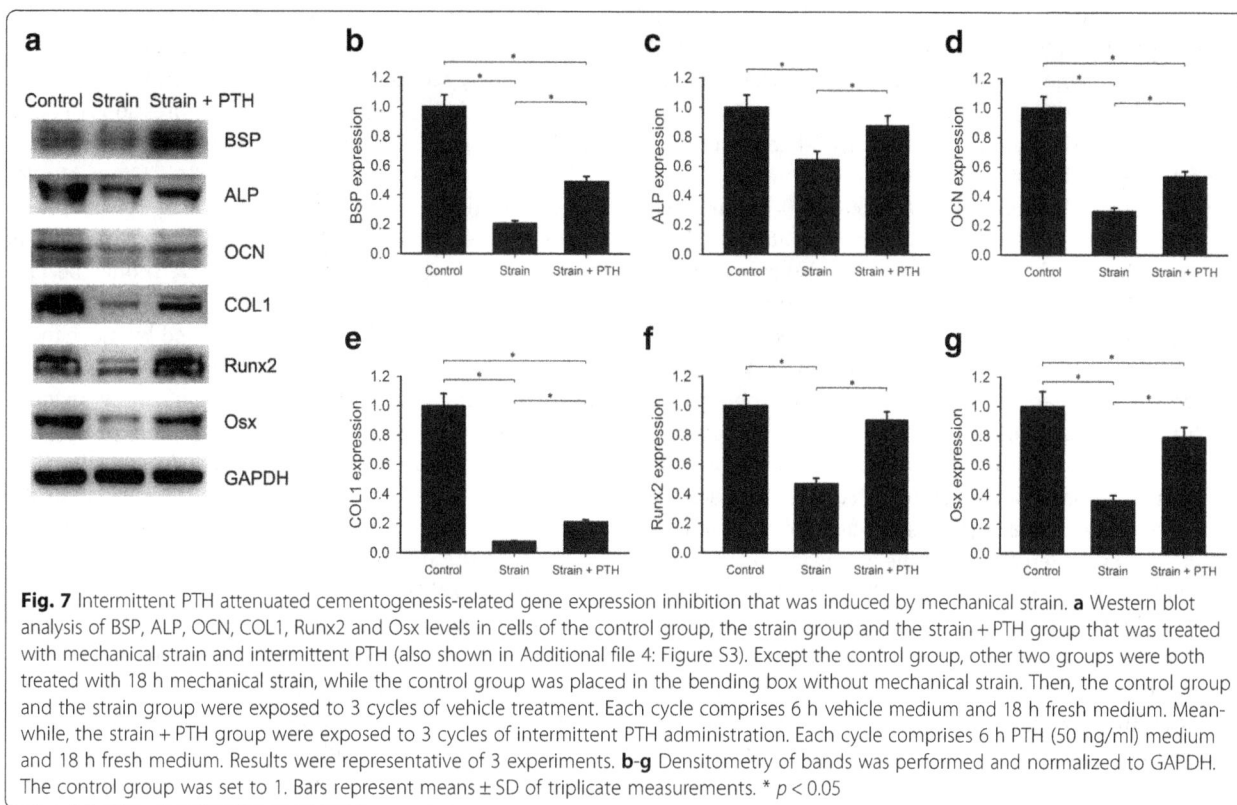

**Fig. 7** Intermittent PTH attenuated cementogenesis-related gene expression inhibition that was induced by mechanical strain. **a** Western blot analysis of BSP, ALP, OCN, COL1, Runx2 and Osx levels in cells of the control group, the strain group and the strain + PTH group that was treated with mechanical strain and intermittent PTH (also shown in Additional file 4: Figure S3). Except the control group, other two groups were both treated with 18 h mechanical strain, while the control group was placed in the bending box without mechanical strain. Then, the control group and the strain group were exposed to 3 cycles of vehicle treatment. Each cycle comprises 6 h vehicle medium and 18 h fresh medium. Meanwhile, the strain + PTH group were exposed to 3 cycles of intermittent PTH administration. Each cycle comprises 6 h PTH (50 ng/ml) medium and 18 h fresh medium. Results were representative of 3 experiments. **b-g** Densitometry of bands was performed and normalized to GAPDH. The control group was set to 1. Bars represent means ± SD of triplicate measurements. * $p < 0.05$

protein levels of BSP, OCN, COL1 and Osx. However, mechanical strain inhibited cementogenesis as evidenced by the decrease in cementogenesis- and differentiation-related genes expression. The results also demonstrated that intermittent PTH alleviated the mechanical strain-induced inhibition of cementoblast differentiation and cementogenesis based on changes in the expression of cementogenesis- and differentiation-related biomarkers.

PTHR1 [13] is a G protein-coupled receptor (GPCR) that transduces signals into cells via interacting with its two ligands PTH and PTH-related protein (PTHrP). Activation of PTHR1 in osteoblasts and chondrocytes modulated the rates of proliferation and apoptosis, as well as regulated a variety of signaling factors related to the histogenesis and remodeling of bone and cartilage [14]. Our data showed that PTHR1 was endogenously expressed in OCCM-30 cells. And intermittent PTH further upregulated the expression of PTHR1 in OCCM-30 cells. Based on the above, it was concluded that cementoblasts were the target cells of PTH. Along with the result that intermittent PTH activated the cementoblastic activities of cementoblasts, we presume that intermittent PTH can be a new method to promote the regeneration of cementum.

PTH is a major regulator of calcium homeostasis and consequently of bone metabolism. PTH exerts either anabolic or catabolic effects on bone [15, 16] depending upon the dosage and duration of administration. The anabolic effect requires intermittent exposure to low doses of PTH, while continuous exposure to high levels is associated with a catabolic effect [17]. In the present study, we showed that 3 cycles of intermittent PTH treatment enhanced the mineralization capacity of cementoblasts by examining ALP activity and the formation of mineralized nodules. This was further evidenced by quantitative ALP assay and quantitative calcium assay. The cementogenesis- and differentiation-related biomarkers BSP, OCN, COL1 and Osx were also stimulated by intermittent PTH. Previous in vivo studies showed that PTH administration increased cementum width in an osteoporotic rabbit model or improved the formation of newly formed cementum-like tissue in a rat model [18, 19]. Therefore, these findings together suggest that intermittent PTH can be a potential therapeutic agent for root resorption.

Mechanical loading include mechanical strain, compressive strain, fluid shear strain and vibration. Cells perceive and translate mechanical energy into biochemical responses via mechanotransduction pathways [20]. Mechanical energy regulates multiple cellular processes including adhesion, proliferation, differentiation and apoptosis. In order to simulate the action of orthodontic forces that mostly act in a continuous mode accompanying by interruption [21], we selected the four-point bending system to exert cyclic mechanical strain on OCCM-30 cells. Here we showed the regulatory potential of mechanical strain in the biological activities of cementoblasts. First, mechanical strain from the four-point bending device altered the morphology and alignment of cementoblasts. Further, molecular analysis showed that the expression levels of BSP, ALP, OCN, OPN, Runx2 and Osx decreased when OCCM-30 cells were exposed to mechanical strain. Consistently, a previous study showed that mechanical strain of 2000–4000 $\mu\varepsilon$ inhibited cell proliferation and BSP expression in cementoblasts [22]. A study of human dental pulp stem cells also revealed that mechanical strain suppressed the expression of osteogenesis-related genes bone morphogenetic protein 2 (BMP2), OCN and ALP, as well as odontogenic differentiation-related genes dentin sialophosphoprotein (DSPP), dentin sialoprotein (DSP) and BSP [23]. However, the effects of mechanical strain on cells are controversial. For example, a study demonstrated that after exposure of human intraoral mesenchymal stem and progenitor cells (MSPCs) to cyclic tensile strain, the osteogenic relative factors, such as osteonectin, BMP2, OPN and OCN were upregulated [24]. Considering the complexity of mechanical environment in dental roots under physiologic and pathologic conditions, the elements and mechanisms of mechanical strain that determine the conversion between anabolic and catabolic effects on cementum need further study.

Our study revealed that intermittent PTH increased the expression of BSP, OCN and COL1. However, BSP, ALP, OCN and OPN were inhibited by mechanical strain. BSP, ALP, OCN and OPN are commonly regarded as anabolic markers towards osteoinductivity. BSP is a mineralized, tissue-specific and non-collagenous protein. Its binding to collagen is thought to be important for the initiation of bone mineralization and formation [25]. OCN is a serum marker of osteoblastic bone formation and is believed to act in the bone matrix to regulate mineralization [26]. COL1 is the predominant type of protein that forms the extracellular matrix of bone. It determines the bending and compressive biomechanical properties of cortical bone, and is independent of bone mineral density [27]. Collagens also play vital roles during organ development, wound healing and tissue repair through interacting with growth factors and cytokines. ALP is needed for the normal mineralization during bone modeling and remodeling. It hydrolyzes inorganic pyrophosphate, a natural inhibitor of hydroxyapatite formation, to phosphate. Our research provides a way that PTH can be a potent method to regulate mineral activities in cementoblasts, no matter whether the cells are under the effects of mechanical strain. In our research, the changes of Runx2 and Osx were consistent with the expression of BSP either after intermittent PTH administration or mechanical strain. It was known that Runx2 and Osx were two key transcriptional factors in the differentiation and function of cementoblasts [28–31]. Some studies indicated that Runx2 was an upstream regulator of Osx [32]. Appropriate Runx2 regulatory function

required proper association with specific enhancers in the regulatory regions of target genes such as OCN, BSP and OPN, where it recruited other nuclear components to the transcriptional apparatus [33]. Overexpression of Osx greatly accelerated the formation of cellular cementum in a COL1-Osx transgenic mice model [34]. Other studies also demonstrated that Osx regulated cementoblast differentiation through activation of Dkk1 and inhibition of Wnt/$\beta$-catenin signaling pathway [28]. It should be emphasized that, in our data, intermittent PTH administration alleviated the catabolic effects induced by mechanical strain. Similarly, a previous study showed that transgenic mice expressing murine Dkk1 had significantly reduced bone mass, while daily PTH administration resulted in comparable increase in bone mass at all skeletal sites [35]. Taken together, we suppose that intermittent PTH has the potential to assist the prevention of tooth root resorption, especially in those cases that improper mechanical loading is a major causative factor.

However, several issues remain to be resolved before the application of intermittent PTH to clinical root resorption treatment. For example, orthodontically induced root resorption is a complex and sterile inflammatory process, including components such as forces, tooth roots, bone, surrounding matrix, cells, and certain biologic messengers. The actions of PTH in this complicated network have not yet been well established. Besides, There are a host of potential contributory factors associated with root resorption among orthodontic patients, including gender, ethnicity, age and genetic disposition of patients, root shape, initial resorption, type of dentition and malocclusion, as well as mechanics and duration of treatment [36]. In light of the complexity of this background, if PTH can be relied on to prevent root resorption regardless of these factors remains to be clarified. Moreover, whether systemic application or local injection of intermittent PTH would be proper for clinical treatment, and what would be the side effects still need further preclinical and clinical data to clarify.

## Conclusions

In summary, this study identified the potential of intermittent PTH to promote cementogenesis. It was shown that intermittent PTH treatment enhanced the mineralization capacity of cementoblasts. Our data also indicated that intermittent PTH improved the expression of cementogenesis- and differentiation-related biomarkers in OCCM-30 cells. Mechanical strain induced visibly morphological changes and suppressed some cementogenesis- and differentiation-related genes expression. Our data also demonstrated that intermittent PTH restrained the inhibition of cementogenesis and cementoblast differentiation by mechanical strain. Taken together, these findings suggest that intermittent PTH can be therapeutically exploited to improve prognosis of tooth root resorption.

# Methods

## Cell culture and reagents

The immortalized mouse cementoblast cell line OCCM-30 was a kind gift from Prof. Somerman and maintained as described previously [37–39]. Briefly, OCCM-30 cells were cultured in Dulbeco's modified Eagle medium (DMEM) with 10% fetal bovine serum, 100 U/ml penicillin G, and 100 µg/ml streptomycin (Gibco, Grand Island, NY, USA). OCCM-30 cells were incubated in a humidified chamber (5%$CO_2$/95% air) at 37 °C. Monolayer cells at 80% confluence were trypsinized and harvested for further study. PTH (1–34) (Bachem, Torrance, CA, USA) was dissolved in 0.1% acetic acid (containing 0.1% bovine serum albumin) according to the manufacturer's protocol. For intermittent PTH incubation [40], OCCM-30 cells were exposed to 50 ng/ml PTH for the first 6 h in each 24 h treatment session, while the vehicle culture medium of the control group also contained the same concentration of acetic acid but without any PTH. Then cells were all cultured in fresh medium without acetic acid or PTH during the remainder of the session. After 1-3 cycles of treatment, cells were collected and examined.

## Application of mechanical strain

OCCM-30 cells ($4 \times 10^5$ cells in 1 ml) were seeded onto the loading plates to form a $2 \times 2$ $cm^2$ square. The material of the loading plates was similar to that of cell culture dishes without any surface treatment. Cells were cultured in fresh medium for 48 h to allow cell adhesion, synchronized by 12 h serum starvation and prepared for cyclic uniaxial strain. Mechanical strain was imposed on OCCM-30 cells using the SXG4201 four-point bending device (University of Electronic Science and Technology of China, Chengdu, China; Fig. 2a-c), with the loading plate being maintained in culture medium during loading. The deformation of the plates caused the attached cells to deform (Fig. 2d). Control cultures grew under the same conditions but without the mechanical strain treatment protocol. Cells were subjected to 2000 µε mechanical strain at a frequency of 0.5 Hz (As strain is defined as the ratio of the change in length to the original length, 2000 µε mechanical strain means 2% changes of the cell length. Correspondingly, the loading displacement of our four-point bending device is 1.12 mm) [22] for 18 h. For the combined treatment of intermittent PTH and mechanical strain, cells were firstly exposed to 18 h mechanical strain in fresh medium, followed by 3 cycles of intermittent PTH treatment.

## Cellular morphological analysis

The morphological changes of OCCM-30 cells after mechanical strain treatment were examined with an Olympus IX70 microscope (Olympus, Tokyo, Japan) at 40× or 100× magnification. 10 fields on each bending plate were randomly selected and images were captured. Cell viability and appearance were analyzed by Image-Pro Plus 6.0 software.

## qPCR

Total RNA was extracted from OCCM-30 cells using Trizol (Invitrogen, Carlsbad, CA, USA). cDNA was synthesized from 1 µg of total RNA as a template, using PrimeScript™ RT reagent Kit with gDNA Eraser (Takara, Tokyo, Japan). qPCR was performed in 20 µL reaction mixtures containing QuantiFast SYBR Green PCR buffer (Qiagen, Valencia, CA, USA), gene-specific primers and cDNAs on a MyIQ thermocycler (Bio-Rad, Hercules, CA, USA). The primer pairs were listed (Table 1). Target gene expression was normalized to GAPDH and analyzed with the use of MyIQ software (Bio-Rad, Hercules, CA, USA).

## Western blot analysis

Cells were rinsed with PBS, trypsinized and collected by centrifugation at 4200 rpm and 4 °C for 5 min. Cell lysis was obtained by incubating in lysis containing RIPA, 1 mM PMSF and complete EDTA-free protease inhibitor cocktail (Roche, Mannheim, Germany). Equivalent amounts of protein (20–40 µg) were subjected to SDS-PAGE (Beyotime, Shanghai, China) and transferred onto a PVDF membrane (Millipore, Bedford, MA, USA). The membrane was blocked with 5% non-fat dried milk in TBST for 1 h and incubated with specific antibodies at 4 °C overnight. Proteins were detected using HRP-conjugated secondary antibodies and visualized using an enhanced chemiluminescence kit (Millipore, Bedford, MA, USA). The intensity of each band was quantified using Quantity One software (Bio-Rad, Hercules, CA, USA) after normalization to GAPDH. For immunoblotting, anti-PTHR1 antibody (Abcam, Cambridge, UK), anti-Runx2 antibody (Abcam, Cambridge, UK), anti-Osx antibody (Abcam, Cambridge, UK), anti-COL1 antibody (Abcam, Cambridge, UK), anti-OPN

**Table 1** Sequences of primers used in qPCR

| Gene | Forward primer sequence (5'–3') | Reverse primer sequence (5'–3') |
| --- | --- | --- |
| GAPDH | GACATCAAGAAGGTGGTGAAGC | GAAGGTGGAAGAGTGGGAGTT |
| ALP | CCAACTCTTTTGTGCCAGAGA | GGCTACATTGGTGTTGAGCTTTT |
| BSP | TAGGAGTTTCCAGGTTTCTGATGA | CTGCCCTTTCCGTTGTTGTC |
| COL1 | CTGGCGGTTCAGGTCCAAT | TTCCAGGCAATCCACGAGC |
| OCN | TGCTTGTGACGAGCTATCAG | GAGGACAGGGAGGATCAAGT |
| OPN | TAGGAGTTTCCAGGTTTCTGATGA | CTGCCCTTTCCGTTGTTGTC |
| Osterix | CTCACCAGGTCCAGGCAACA | GGAGCAAAGTCAGATGGGTAAGTAG |
| Runx2 | GGACGAGGCAAGAGTTTCACC | GAGGCGATCAGAGAACAAACTAGG |

antibody (Abcam, Cambridge, UK), anti-ALP antibody (Biorbyt, Cambridge, UK), anti-OCN antibody (Biorbyt, Cambridge, UK), and anti-BSP antibody (Biorbyt, Cambridge, UK) were used at 1:1000 dilution.

## ALP staining

ALP activity was performed using an ALP staining kit (Jiancheng, Nanjing, China) according to the manufacturer's protocol. Briefly, cells were fixed with 10% formalin for 5 min and rinsed in deionized water for 30s. The samples were stained with ALP substrate solution for 15 min at 37 °C in the dark. The stain was removed by washing with deionized water for 30s, and cells were counterstained with hematoxylin. ALP staining intensity was observed under an Olympus IX70 microscope (Olympus, Tokyo, Japan) and analyzed by Image-Pro Plus 6.0 software (Media Cybernetics, Silver Spring, MD, USA).

## Quantitative ALP assay

Corresponding to ALP staining, for quantitative ALP assay, OCCM-30 cells were also treated with 0 cycle, 1 cycle, 2 cycles and 3 cycles of intermittent PTH. The cells were then evaluated for ALP activity. Briefly, cells were rinsed three times with PBS, collected and lysed with ultrasound. The total protein content of these samples was measured by the bicinchoninic acid (BCA) method, using a protein assay kit (Beyotime, Shanghai, China). Then ALP was determined according to a quantitative ALP assay kit (Beyotime, Shanghai, China). The solution mixed by samples and ALP assay working solution was distributed at 100 µL per well on a 96-well plate, and was incubated for 30 min at 37 °C. The optical density (OD) was measured at 405 nm. The ALP activity was normalized to the total protein concentration and was calculated as OD per µg of the proteins.

## Alizarin red S staining and quantitative calcium assay

The Alizarin red S assay was used to determine the degree of mineralization in intermittent PTH and vehicle cultures. The mineralization solution contained ascorbic acid (50 µM), dexamethasone (100 nM), β-glycerophosphate (10 mM) (Sigma, St. Louis, MO, USA) and 10% fetal bovine serum (FBS) in DMEM. 50 ng/ml PTH was dissolved in 0.1% acetic acid in the PTH mineralization solution, while the vehicle mineralization solution contained 0.1% acetic acid without PTH. OCCM-30 cells were seeded in 6-well plates at a density of $1 \times 10^5$ cells per well. They were exposed to PTH or vehicle mineralization solution for 3 cycles. Then cells were cultured in mineralization solution without acetic acid or PTH for 7 days. After 10 days of culture, OCCM-30 cells were washed with PBS and fixed with 4% paraformaldehyde at room temperature for 15 min for subsequent Alizarin red S staining (2% (w/v) Alizarin red S

(Sigma-Aldrich, St. Louis, MO, USA) solution; pH 4.2), which was conducted (2 ml/each well) for 5 min at room temperature. Following staining, the cells were rinsed with deionized water. The formation of mineralized nodules was observed under an Olympus IX70 microscope (Olympus, Tokyo, Japan) and analyzed by Image-Pro Plus 6.0 software (Media Cybernetics, Silver Spring, MD, USA). For quantitative calcium measurement, 10% cetylpyridinium chloride (J&KCHEMICA, Beijing, China) solution was then added to each well for elution of the dye. After incubation at room temperature for 1 h with shaking, samples of the resulting solution were distributed on a 96-well plate. And absorbance was read at 570 nm.

## Statistical analysis

Data were presented as the mean ± standard deviation (SD). Statistical analyses were performed using ANOVA of factorial design (for factorial designed experiments) or one-way ANOVA (for three or more groups), or independent samples $t$ test (between two groups). Values of $p < 0.05$ were defined as statistical significance. All statistical analysis was determined by using SPSS 13.0 (SPSS).

## Additional files

> **Additional file 1: Figure S1.** Bands of Fig. 1 (western blot analysis). (TIF 254 kb)
>
> **Additional file 2: Table S1.** Densitometry for the bands (western blot analysis) of 0, 1, 2 and 3 cycles of intermittent PTH and the corresponding control groups in Fig. 1a-b. **Table S2.** Densitometry for the bands (western blot analysis) of the control group, the strain group and the strain + PTH group in Fig. 1c-d. **Table S3.** Quantitative analysis of ALP activity. Data indicated the levels of the control group, 1, 2 and 3 cycles of intermittent PTH groups respectively in Fig. 3b. **Table S4.** Data of quantitative calcium assay of the control group and 3 cycles of intermittent PTH group in Fig. 3d. **Table S5.** Data indicating the mRNA levels of BSP, OCN, COL1 and Osx of 0, 1, 2 and 3 cycles of intermittent PTH and the corresponding groups in Fig. 4a-d. **Table S6.** Densitometry for the bands (western blot analysis) of the control group and 3 cycles of intermittent PTH in Fig. 4e-i. **Table S7.** Data indicating the mRNA levels of BSP, ALP, OCN, OPN, Runx2 and Osx of the control group and the strain group after 18 h of mechanical strain treatment in Fig. 6a-f. **Table S8.** Densitometry for the bands (western blot analysis) of the control group, the strain group and the strain + PTH group in Fig. 7a-g. Data were presented as mean ± SD. (DOCX 21 kb)
>
> **Additional file 3: Figure S2.** Bands of Fig. 4 (western blot analysis). (TIF 1478 kb)
>
> **Additional file 4: Figure S3.** Bands of Fig. 7 (western blot analysis). (TIF 3040 kb)

## Abbreviations

ALP: Alkaline phosphatase; BMP2: Bone morphogenetic protein 2; BSP: Bone sialoprotein; COL1: Collagen type I; DSP: Dentin sialoprotein; DSPP: Dentin sialophosphoprotein; MSPC: Mesenchymal stem and progenitor cell; OCN: Osteocalcin; OPN: Osteopontin; Osx: Osterix; PTH: Parathyroid hormone; PTHR1: Parathyroid hormone receptor type 1; PTHrP: Parathyroid hormone-related protein; Runx2: Runt-related transcription factor 2

## Acknowledgments

This work was supported by the National Natural Science Foundation of China [grant number: 81470777].

## Funding
This work was financially supported by the National Natural Science Foundation of China [grant number: 81470777].

## Authors' contributions
SZ and YL conceived and designed the study. LY carried out the experiments and sample preparation. CZ and LH performed the image analysis. ZH, YX and XW performed the data processing and statistical analysis. YL wrote the manuscript and SZ edited it. All authors read and approved the final manuscript.

## Competing interests
The authors declare that they have no competing interests.

## References
1. Ahangari Z, Nasser M, Mahdian M, Fedorowicz Z, Marchesan MA. Interventions for the management of external root resorption. Cochrane Database Syst Rev. 2015;Cd008003. doi: 10.1002/14651858.CD008003.pub3.
2. Zernicke R, MacKay C, Lorincz C. Mechanisms of bone remodeling during weight-bearing exercise. Appl Physiol Nutr Metab. 2006;31:655–60.
3. Rego EB, Inubushi T, Kawazoe A, Miyauchi M, Tanaka E, Takata T, Tanne K. Effect of PGE2 induced by compressive and tensile stresses on cementoblast differentiation in vitro. Arch Oral Biol. 2011;56:1238–46.
4. Bosshardt DD. Are cementoblasts a subpopulation of osteoblasts or a unique phenotype? J Dent Res. 2005;84:390–406.
5. DEMPSTER DW, COSMAN F, PARISIEN M, SHEN V, LINDSAY R. Anabolic actions of parathyroid hormone on bone. Endocr Rev. 1993;14:690–709.
6. Tam CS, Heersche JN, Murray TM, Parsons JA. Parathyroid hormone stimulates the bone apposition rate independently of its resorptive action: differential effects of intermittent and continuous administration. Endocrinology. 1982;110:506–12.
7. Qin C, D'Souza R, Feng JQ. Dentin Matrix Protein 1 (DMP1): New and important roles for biomineralization and phosphate homeostasis. J Dent Res. 2007;86:1134–41.
8. Wang L, Tran AB, Nociti Jr FH, Thumbigere-Math V, Foster BL, Krieger CC, Kantovitz KR, Novince CM, Koh AJ, McCauley LK, Somerman MJ. PTH and vitamin D repress DMP1 in cementoblasts. J Dent Res. 2015;94:1408–16.
9. Ouyang H, McCauley LK, Berry JE, D'Errico JA, Strayhorn CL, Somerman MJ. Response of immortalized murine cementoblasts/periodontal ligament cells to parathyroid hormone and parathyroid hormone-related protein in vitro. Arch Oral Biol. 2000;45:293–303.
10. Barros SP, Silva MA, Somerman MJ, Nociti Jr FH. Parathyroid hormone protects against periodontitis-associated bone loss. J Dent Res. 2003;82:791–5.
11. Hakki SS, Bozkurt SB, Hakki EE, Belli S. Effects of mineral trioxide aggregate on cell survival, gene expression associated with mineralized tissues, and biomineralization of cementoblasts. J Endod. 2009;35:513–9.
12. Jang AT, Merkle AP, Fahey KP, Gansky SA, Ho SP. Multiscale biomechanical responses of adapted bone–periodontal ligament–tooth fibrous joints. Bone. 2015;81:196–207.
13. Kimura S, Yoshioka K. Parathyroid hormone and parathyroid hormone type-1 receptor accelerate myocyte differentiation. Sci Rep. 2014;4:5066.
14. Zhao Q, Brauer PR, Xiao L, McGuire MH, Yee JA. Expression of parathyroid hormone-related peptide (PTHrP) and its receptor (PTH1R) during the histogenesis of cartilage and bone in the chicken mandibular process. J Anat. 2002;201:137–51.
15. Liu Y-Q, Hong Z-L, Zhan L-B, Chu H-Y, Zhang X-Z, Li G-H. Wedelolactone enhances osteoblastogenesis by regulating Wnt/β-catenin signaling pathway but suppresses osteoclastogenesis by NF-κB/c-fos/NFATc1 pathway. Scientific Reports. 2016;6:32260.
16. Kasagi S, Chen W. TGF-beta1 on osteoimmunology and the bone component cells. Cell Biosci. 2013;3:4.
17. Locklin RM, Khosla S, Turner RT, Riggs BL. Mediators of the biphasic responses of bone to intermittent and continuously administered parathyroid hormone. J Cell Biochem. 2003;89:180–90.
18. Bellido M, Lugo L, Castaneda S, Roman-Blas JA, Rufian-Henares JA, Navarro-Alarcon M, Largo R, Herrero-Beaumont G. PTH increases jaw mineral density in a rabbit model of osteoporosis. J Dent Res. 2010;89:360–5.
19. Vasconcelos DF, Marques MR, Benatti BB, Barros SP, Nociti Jr FH, Novaes PD. Intermittent parathyroid hormone administration improves periodontal healing in rats. J Periodontol. 2014;85:721–8.
20. Ziegler N, Alonso A, Steinberg T, Woodnutt D, Kohl A, Müssig E, Schulz S, Tomakidi P. Mechano-transduction in periodontal ligament cells identifies activated states of MAP-kinases p42/44 and p38-stress kinase as a mechanism for MMP-13 expression. BMC Cell Biol. 2010;11:10.
21. Krishnan V, Davidovitch Z. Cellular, molecular, and tissue-level reactions to orthodontic force. Am J Orthod Dentofacial Orthop. 2006;129(4):469.e1–469.e32.
22. Huang L, Meng Y, Ren A, Han X, Bai D, Bao L. Response of cementoblast-like cells to mechanical tensile or compressive stress at physiological levels in vitro. Mol Biol Rep. 2009;36:1741–8.
23. Cai X, Zhang Y, Yang X, Grottkau BE, Lin Y. Uniaxial cyclic tensile stretch inhibits osteogenic and odontogenic differentiation of human dental pulp stem cells. J Tissue Eng Regen Med. 2011;5:347–53.
24. Lohberger B, Kaltenegger H, Stuendl N, Payer M, Rinner B, Leithner A. Effect of cyclic mechanical stimulation on the expression of osteogenesis genes in human intraoral mesenchymal stromal and progenitor cells. Biomed Res Int. 2014;2014:10.
25. Choi YJ, Lee JY, Chung C-P, Park YJ. Enhanced osteogenesis by collagen-binding peptide from bone sialoprotein in vitro and in vivo. J Biomed Mater Res A. 2013;101A:547–54.
26. Zoch ML, Clemens TL, Riddle RC. New insights into the biology of osteocalcin. Bone. 2016;82:42–9.
27. Garnero P, Borel O, Gineyts E, Duboeuf F, Solberg H, Bouxsein ML, Christiansen C, Delmas PD. Extracellular post-translational modifications of collagen are major determinants of biomechanical properties of fetal bovine cortical bone. Bone. 2006;38:300–9.
28. Cao Z, Liu R, Zhang H, Liao H, Zhang Y, Hinton RJ, Feng JQ. Osterix controls cementoblast differentiation through downregulation of Wnt-signaling via enhancing DKK1 expression. Int J Biol Sci. 2015;11:335–44.
29. Nemoto E, Sakisaka Y, Tsuchiya M, Tamura M, Nakamura T, Kanaya S, Shimonishi M, Shimauchi H. Wnt3a signaling induces murine dental follicle cells to differentiate into cementoblastic/osteoblastic cells via an osterix-dependent pathway. J Periodontal Res. 2016;51:164–74.
30. Pan K, Sun Q, Zhang J, Ge S, Li S, Zhao Y, Yang P. Multilineage differentiation of dental follicle cells and the roles of Runx2 over-expression in enhancing osteoblast/cementoblast-related gene expression in dental follicle cells. Cell Prolif. 2010;43:219–28.
31. Hirata A, Sugahara T, Nakamura H. Localization of Runx2, osterix, and osteopontin in tooth root formation in Rat molars. J Histochem Cytochem. 2009;57:397–403.
32. Nakashima K, de Crombrugghe B. Transcriptional mechanisms in osteoblast differentiation and bone formation. Trends Genet. 2003;19:458–66.
33. Li Y, Ge C, Long JP, Begun DL, Rodriguez JA, Goldstein SA, Franceschi RT. Biomechanical stimulation of osteoblast gene expression requires phosphorylation of the RUNX2 transcription factor. J Bone Miner Res. 2012;27:1263–74.
34. Cao Z, Zhang H, Zhou X, Han X, Ren Y, Gao T, Xiao Y, de Crombrugghe B, Somerman MJ, Feng JQ. Genetic evidence for the vital function of osterix in cementogenesis. J Bone Miner Res. 2012;27:1080–92.
35. Yao G-Q, Wu J-J, Troiano N, Insogna K. Targeted overexpression of Dkk1 in osteoblasts reduces bone mass but does not impair the anabolic response to intermittent PTH treatment in mice. J Bone Miner Metab. 2011;29:141–8.
36. Orthodontically Induced Inflammatory Root Resorption. Part II: the clinical aspects. The Angle Orthodontist. 2002;72:180–4.
37. Berry JE, Zhao M, Jin Q, Foster BL, Viswanathan H, Somerman MJ. Exploring the origins of cementoblasts and their trigger factors. Connect Tissue Res. 2003;44 Suppl 1:97–102.
38. D'Errico JA, Berry JE, Ouyang H, Strayhorn CL, Windle JJ, Somerman MJ. Employing a transgenic animal model to obtain cementoblasts in vitro. J Periodontol. 2000;71:63–72.
39. Ouyang H, McCauley LK, Berry JE, Saygin NE, Tokiyasu Y, Somerman MJ. Parathyroid hormone-related protein regulates extracellular matrix gene expression in cementoblasts and inhibits cementoblast-mediated mineralization in vitro. J Bone Miner Res. 2000;15:2140–53.
40. Ishizuya T, Yokose S, Hori M, Noda T, Suda T, Yoshiki S, Yamaguchi A. Parathyroid hormone exerts disparate effects on osteoblast differentiation depending on exposure time in rat osteoblastic cells. J Clin Investig. 1997;99:2961–70.

# 12

# Impact of the *MDM2* splice-variants *MDM2-A*, *MDM2-B* and *MDM2-C* on cytotoxic stress response in breast cancer cells

Johanna Huun[1], Liv B. Gansmo[1], Bård Mannsåker[1,2,4], Gjertrud Titlestad Iversen[2], Jan Inge Øvrebø[3,5], Per E. Lønning[1,2] and Stian Knappskog[1,2*]

## Abstract

**Background:** The murine double minute 2 (MDM2) is an oncogene and a negative regulator of the tumor suppressor protein p53. *MDM2* is known to be amplified in numerous human cancers, and upregulation of MDM2 is considered to be an alternative mechanism of p53 inactivation. The presence of many splice variants of *MDM2* has been observed in both normal tissues and malignant cells; however their impact and functional properties in response to chemotherapy treatment are not fully understood.

Here, we investigate the biological effects of three widely expressed alternatively spliced variants of *MDM2*; MDM2-A, MDM2-B and MDM2-C, both in unstressed MCF-7 breast cancer cells and in cells subjected to chemotherapy. We assessed protein stability, subcellular localization and induction of downstream genes known to be regulated by the MDM2-network, as well as impact on cellular endpoints, such as apoptosis, cell cycle arrest and senescence.

**Results:** We found both the splice variants MDM2-B and -C, to have a much longer half-life than MDM2 full-length (FL) protein after chemotherapy treatment indicating that, under stressed conditions, the regulation of degradation of these two variants differs from that of MDM2-FL. Interestingly, we observed all three splice variants to deviate from MDM2-FL protein with respect to subcellular distribution. Furthermore, while MDM2-A and -B induced the expression of the pro-apoptotic gene *PUMA*, this effect did not manifest in an increased level of apoptosis.

**Conclusion:** Although MDM2-B induced slight changes in the cell cycle profile, overall, we found the impact of the three *MDM2* splice variants on potential cellular endpoints upon doxorubicin treatment to be limited.

**Keywords:** MDM2, *MDM2* splice variants, MDM2-A, MDM2-B, MDM2-C, Breast cancer, Doxorubicin

## Background

The E3 ubiquitin ligase Murine Double Minute 2 (MDM2) is the key negative regulator of the p53 tumor suppressor protein. MDM2 binds and ubiquitinates p53, facilitating its proteasomal degradation [1–4]. p53, on the other hand, can induce transcription of *MDM2*, generating a negative feedback loop [5, 6]. Furthermore, amplification and overexpression of *MDM2* have been implicated in various types of cancer [1, 7, 8].

The *MDM2* gene consists of 12 exons encoding 491 amino acids [9]. MDM2 has a well characterized p53 binding domain at the N-terminal and a highly conserved RING domain at the C-terminus, responsible for the E3 ligase activity [10–13]. Additionally, MDM2 contains a well-defined nuclear localization signal (NLS), a nuclear export signal (NES) and a nucleolar localization signal (NoLS), responsible for MDM2 localization both in the nucleus and in the cytoplasm [14].

Two decades ago, the first alternatively spliced MDM2 transcript was identified in human tumors. To date 72 different *MDM2* splice variants have been identified in human cancer and normal tissue [9, 15–18]. The presence of *MDM2* splice variants has been observed in both normal tissues and malignant cells, yet their functional properties are not fully understood. Several studies have attempted to determine whether the splice variants

* Correspondence: stian.knappskog@uib.no
[1]Section of Oncology, Department of Clinical Science, University of Bergen, 5020 Bergen, Norway
[2]Department of Oncology, Haukeland University Hospital, Bergen, Norway
Full list of author information is available at the end of the article

contribute to tumor formation or if they are expressed as a consequence of cancer progression. However, the finding that expression of *MDM2* splice variants increase upon genotoxic stress suggests that they might have a potential role in the response to chemotherapy treatment [19].

So far, MDM2-A (ALT2), MDM2-B (ALT1) and MDM2-C (ALT3) are the three most commonly detected and extensively studied splice variants of *MDM2*. Each variant is observed in several types of cancer, including breast cancer [18, 20–22]. MDM2-A lacks exon 4–9, while -B and -C are lacking exons 4–11 and exons 5–9, respectively (Fig. 1a). Thus, MDM2-A, -B and -C all have the p53 binding site at the N-terminal spliced out, but the RING-finger binding domain is still

present at the C-terminal, making them capable of binding to MDM2-FL.

MDM2-A has been characterized as an activator of p53, inhibiting growth in a p53-dependent manner, and to cause a decrease in the transformation and tumorigenesis in vitro [23]. Contrasting this, the same variant has also been shown to induce increased expression levels of Cyclin D1 and E, hence, suggesting that this splice variant has a tumor promoting activity in vivo [24].

MDM2-B is the splice variant most commonly overexpressed in human tumors [9]. MDM2-B is known to interact with MDM2-FL and sequester it in the cytoplasm, leading to inhibition of the MDM2-FL-p53 interaction and thereby causing stabilization and transactivation of p53 and induction of cellular growth arrest

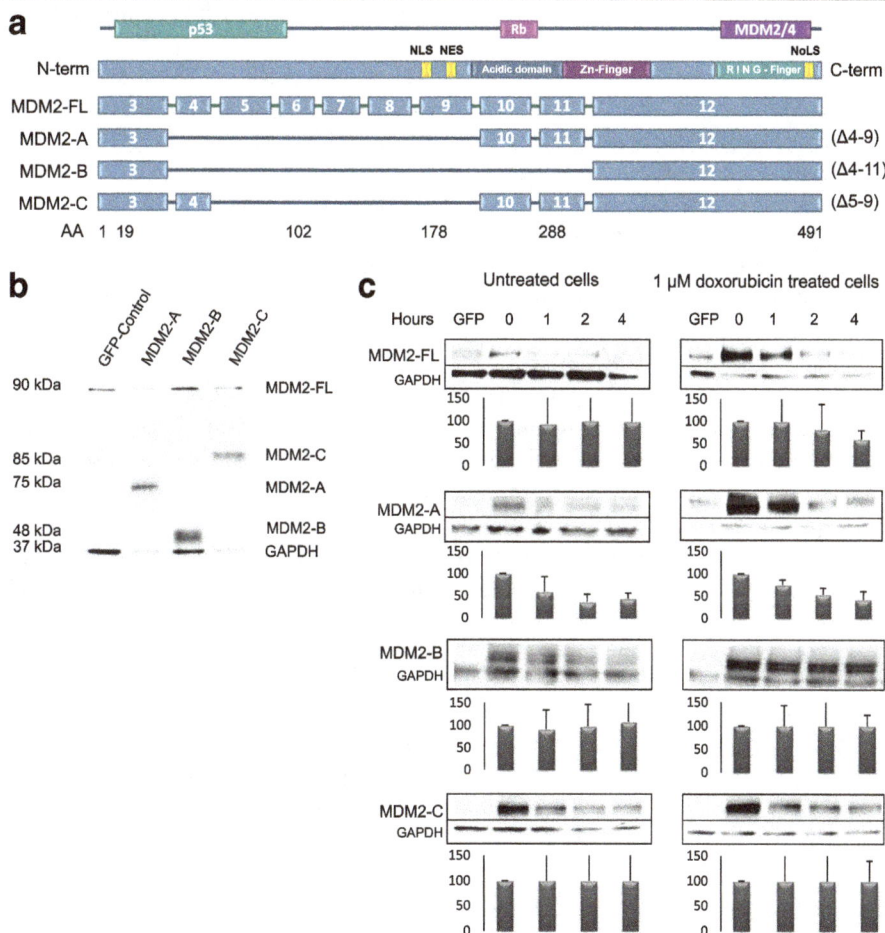

Fig. 1 (a) Schematic representation of MDM2 and the splice variants MDM2-A, -B and -C. MDM2 consists of 491 amino acids. Localization of the p53, pRb and MDM2/4 binding sites, NLS, NES, NoLS, the acidic domain, the Zn-finger domain and the RING-finger domain are indicated, as well as the exon distribution. (b) Validation of protein expression. MCF-7 breast cancer cells transfected with pCMV-GFP control (GFP-control), MDM2-A (75 kDa), -B (48 kDa) and -C (85 kDa) analyzed 24 h post transfection. GAPDH was used as loading control. Primary antibodies were Anti-MDM2 (N-20) Sc-813 (Santa Cruz) and GAPDH (SantaCruz). (c) Protein stability of the expressed splice variants. MCF-7 cells transfected with MDM2-FL, MDM2-A, -B and -C at 0, 1, 2 and 4 h post cycloheximide treatment, respectively. In addition to cycloheximide treatment, *left panel* shows cells without doxorubicin treatment, right panel shows cells treated with 1 μM doxorubicin for 24 h. Primary antibodies Anti-MDM2 (N-20) Sc-813 (Santa Cruz) and GAPDH (SantaCruz). Histograms under immunoblots represent averages of triplicate experiments and show levels of the MDM2-variants relative to GAPDH-levels for each sample

[22, 25–27]. In addition MDM2-B seems capable of inducing p53-independent cell growth [28]. Expression of MDM2-B is also shown to have tumor promoting activity by causing increased levels of Cyclin D1 and E in vivo [24].

MDM2-C is by far the least studied splice variant of the three, however -C is also known to bind MDM2-FL and has been shown to have an effect on cellular transformation independent of p53 [29].

In the present study, we aimed to investigate the potential roles of the three MDM2 splice variants MDM2-A, -B and -C in breast cancer cells in response to cytotoxic stress induced by chemotherapy. Thus, we conducted comprehensive molecular and cellular analyses in order to identify functions similar to, or differing from the well-established functions of the MDM2-FL protein.

## Methods

### Expression vectors

The sequences encoding MDM2-FL and the respective splice variants; MDM2-A, -B and -C were assembled from synthetic oligonucleotides and cloned into E.coli expression vectors (Geneart Life Technologies). *MDM2* encoding fragments were cut out using the BamHI and XhoI restriction sites. Following agarose gel purification the fragments were ligated into a pCMV eukaryotic expression vector (CMV-MCS-V5-6xHis-BGHpolyA in pCMV-cyto-EGFP-myc) using T4 DNA ligase. The utilized vector contained a sequence encoding an enhanced green fluorescent protein (eGFP) expressed from an independent CMV promoter region. Performing immunofluorescence, apoptosis and senescence analysis, a pcDNA3.1 V5-vector (TOPO) was used, providing a C-terminal V5-tag (Invitrogen). The plasmids were amplified in One Shot TOP10 Chemically Competent E.coli cells (Invitrogen) by Ampicillin selection, followed by colony PCR and purified using the QIAprep Spin Miniprep Kit (Qiagen). The constructed plasmids encoding MDM2-FL and splice variants were confirmed by sequencing using the BigDye1.1 system and Sanger sequencing prior large scale purification from E.coli by the HiSpeed plasmid maxi kit (Qiagen), according to the manufacturer's instructions. The resulting stock solutions of the plasmids were validated by sequencing prior to introduction to a eukaryotic cell system.

### Cell culture, transfection and treatment

Cells were fingerprinted with AmpFISTR Profiler and Cofiler plus (Applied Biosystems by Life technologies) before use. MCF-7 (HTB-22; ATCC) breast cancer cells were cultivated in EMEM (Eagle's minimum essential medium; ATCC), HCT116 *TP53+/+* (CCL-247; ATCC) and HCT116 *TP53−/−* colon cancer cells were cultivated

in McCoy's medium (ATCC). The media were supplemented with 10% FBS, 2% L-Glutamine and 2% Penicillin Streptavidin (Lonza). HCT116 *TP53−/−* were a generous gift from Dr. F. Bunz, B. Vogelstein & K. W. Kinzler at John Hopkins University and Howard Hughes Medical Institute, MD. Prioritizing high transfection efficacy, transfection was performed using 1.85 µg/ml plasmid and 1.7 µl/ml Lipofectamin-2000 (Invitrogen). Cells were treated with dimethyl sulfoxide (DMSO) as negative control (same final concentration as for the parallel cells treated with 1 µM doxorubicin in DMSO).

### Sorting cells by flow cytometry and GFP- transfection efficiency

The cells were sorted and harvested by Flow Cytometry (FACS Aria, BD Biosciences) based on GFP-expression 12 h post transfection, before they were seeded and added new growth medium for further analysis. The average transfection efficiency was found to be 33, 38 and 37% for the MDM2-A, -B and -C expressing vectors, respectively (data not shown).

### Western blot analysis

Cells were harvested with Trypsin EDTA (Lonza), lysed with IPH buffer with protease- inhibitor, debris removed and the proteins were denatured by boiling in SDS buffer and loaded on 12% SDS-polyacrylamide gels (Bio Rad) with PS11 protein ladder (GeneOn). Separated proteins were transferred onto 0.2 µM nitrocellulose membranes by turbo blotting for 7 min, 2.5A and 25 V using the Bio Rad system. Unspecific protein binding were blocked by incubation in 5% non-fat milk in TBS-Tween$_{0.05\%}$ for one hour at room temperature or overnight at 4 °C. Membranes were subsequently incubated in MDM2 specific antibody Sc-813 (SantaCruz) recognizing exon 3. Anti-Actin (Thermo Scientific) or Glyceraldehyde 3-phosphate dehydrogenase (GAPDH; SantaCruz) was used as loading control. Further, washing of membranes in TBS-Tween$_{0.05\%}$, proteins bound by the primary antibody were detected by HRP-conjugated secondary antibody (Sigma). Signals were detected using SuperSignal West Femto Chemiluminescent Substrate (Thermo Scientific) and the LAS 4000 imager (GE Healthcare).

### Protein stability analysis

Twenty four hours post transfection the cells were treated with DMSO or 1 µM doxorubicin for 24 h. The cells were then exposed to 100 µg/ml cycloheximide (Sigma) and harvested after 1, 2 and 4 h. Cells were washed (ice-cold PBS), de-attached (Trypsin-EDTA) and neutralized (EMEM) followed by centrifugation and subsequent washing of the generated cell pellet in cold PBS. Proteins were released by cell lysis for 10 min in IPH buffer with protease and phosphatase inhibitors. Cell debris was removed by centrifugation at 13 000 rpm for 1 min. Protein concentration was measured by absorption measurements at 280 nm using the NanoDrop2000 (Thermo Scientific). By varying the amount of cell lysate, equal amount of total protein was denatured in Sodium dodecyl (SDS) sample buffer and boiled at 95 °C for 5 min, followed by Western blotting.

### Subcellular localization by indirect immunofluorescence

Cells were grown on glass coverslips and transfected. Forty eight hours after transfection the cells were fixed for 15 min in 3.7% formaldehyde. Cells were then permeabilized with 0.1% Triton X-100 for 10 min and blocked with 1% BSA in PBS for 30 min. The coverslips were then incubated with MDM2 specific antibody Sc-813 (Santa Cruz) for 1 h followed by AlexaFluor 647 conjugated secondary antibody (Life Technologies) for 30 min. The cells were incubated 10 min in 0.1% Hoechst 33 342 (Molecular Probes, Life Technologies) in PBS, washed in 1 × PBS and mounted with Fluka Eucitt quick hardening mounting solution (Life Technologies). Localization of the different splice variants were determined independently by three investigators, blinded to each other's results, who analyzed >50 cells transfected with each of the *MDM2* splice variants.

### Quantitative PCR

Transfected and sorted cells were treated with 1 μM doxorubicin or DMSO (control) for 12 h. The total RNA was isolated using Trizol reagent (Life Technologies, Gaithersburg, MD) according to the manufacturer's instructions. Single stranded cDNA synthesis was performed using 500 ng total RNA, oligo-dT — and random hexamer primers (Sigma) with Transcriptor Reverse Transcriptase (Roche) in accordance with manufacturer's instructions. mRNA levels of MDM2 EXON3 (total MDM2), MDM2 3'UTR (endogenous MDM2) and RPLP2 (reference) were determined individually; EXON3 was amplified with the primers 5′-AACATGTCTGTACC-TACTGATGGTGC-3′ and 5′-CAGGGTCTCTTGTTC CGAAGC-3′ and the hydrolysis probe 6FAM-AACCAC CTCAC AGATTCC-BBQ. 3'UTR was amplified with the primers 5′-TGCTCCATCACCCATGCTAGA-3′ and 5′-TGGTGGTACATGCCTGTAATC-3′ and the hydrolysis probe 6FAM-TAGCTTGAACCCAGAAGGCGGA-BBQ, while *MDM4, RB1, E2F1, TP53, MTOR, CDK1A, ATM, PTEN, BCL2, APAF, GLB1* and *PUMA* by quantitative amplification reactions using custom made Realtime Ready plates (Roche; Configurator no: 100054567) using the LightCycler 480 instrument (Roche). LC480 Probes Master (Roche) was used as reaction mix. Reaction concentrations were 0.5 μM of each primer and 0.125 μM of each hydrolysis probe.

### Apoptosis assay: AnnexinV detection

Cells were transfected for 24 h, followed by treatment with DMSO or 1 μM doxorubicin for 24 h before they were trypsinated and washed in 1xPBS. Further, the cells were incubated 15 min at 37 °C in AnnexinV (Biotium) and Hoechst (Chemometec). The cells were washed once in AnnexinV buffer (Biotium) before they were re-suspended in AnnexinV buffer with 4% Propidium Iodid (PI; Chemometec) and analysed with the NucleoCounter 3000 (Chemometec). The analysis was repeated in three independent experiments.

### Cell cycle analysis

Transfected and treated cells were incubated 5 min at 37 °C in lysis buffer with Hoechst (Chemometec). Thereafter, cells received stabilizing buffer, before they were analyzed on a NucleoCounter 3000 (Chemometec) for DNA quantitation in three independent experiments.

### Cell proliferation assay by cell count

Exactly 30 000 transfected cells were seeded and counted with NucleoCounter 3000 after 24, 48 and 72 h respectively.

### Senescence assay: β-galactosidase staining

Cells were transfected for 24 h, followed by treatment with DMSO or 0.25 μM doxorubicin. After 7 days the cells were stained for β-galactosidase activity with the Senescence β-galactosidase Staining Kit (Cell Signaling) according to the manufacturer's instructions. The staining of the cells was performed by incubation in staining solution for 14 h in a cultivation incubator with humidified atmosphere, without $CO_2$ at 37 °C. Cells detected as blue were β-galactosidase positive cells versus un-colored, negative cells upon microscopic inspection (Nikon eclipse TS100). For all samples, presence of transfected plasmid was confirmed at the time of analysis by detection of GFP (manual inspection of cells under microscope).

### Statistics

Differences in subcellular localization were tested by Chi-square. Apoptosis, phase distribution in the cell cycle and senescence were tested by univariate analysis. All *p*-values are reported as two-sided.

## Results

In the present study, we examined the *MDM2* splice variants MDM2-A, -B and -C for their biological functions in breast cancer cells with and without chemotherapy treatment.

After generation of expression vectors for each of the three splice variants, we transfected MCF-7 cells and verified exogenous protein expression from the vectors by Western blot analyses: we detected MDM2-A, -B and -C at 75 kDa, 48 kDa and 85 kDa, respectively (Fig. 1b).

## Splice variant protein stability

The stability of the three *MDM2* splice variants expressed as proteins was determined subsequent to incubation with cycloheximide.

In untreated cells, while MDM2-A was degraded at a rate resembling MDM2-FL, and MDM2-B was degraded more rapidly than MDM2-FL, MDM2-C was degraded at a slightly slower rate. However, after doxorubicin treatment, MDM2-B in particular, but also -C was found stabilized as compared to MDM2-FL and MDM2-A (both MDM2-B and -C were stable after 4 h; Fig. 1c). Thus, our data strongly indicate that MDM2-B, and potentially also -C, are stabilized by different mechanisms than the MDM2-FL protein, upon cytotoxic stress.

## Subcellular localization

MDM2-FL is observed exclusively in the nucleus in the majority of cells (85%). In contrast, a different distribution pattern was observed for MDM2-A and -C; here, we observed 38 and 44% of the cells to harbor MDM2-A and -C in the nucleus only; 38 and 40% revealed cytoplasmic distribution only, while 24 and 16% respectively expressed MDM2-A and -C in both compartments ($p <$ 0.001 for both MDM2-A and -C comparing distribution to MDM2-FL). For MDM2-B, the subcellular distribution was even more different from MDM2-FL, with 19% of the cells revealing a nuclear localization only, 9% revealing localization in both compartments, while as many as 72% of the cells revealed exclusively cytoplasmic distribution of MDM2-B ($p <$ 0.001, compared to the MDM2-FL; Fig. 2).

## Feedback on expression of endogenous *MDM2*

We quantitated the levels of endogenous MDM2-FL mRNA upon overexpression of each of the *MDM2* splice variants.

Endogenous levels of MDM2 mRNA remained unchanged after upregulation of MDM2-A, both before and after treatment with doxorubicin (Fig. 3). While overexpression of MDM2-B lead to a notable 1.8 fold increase in the level of endogenous MDM2-FL ($p = 0.011$), overexpression of MDM2-C caused a similar but non-significant upregulation in untreated cells. Both splice variants in addition caused a non-significant upregulation of MDM2-FL in doxorubicin treated cells.

## Regulation of potential target genes

Further, we assessed the potential influence of the splice variants on expression levels of several genes known to be regulated by MDM2 / the MDM2 network: *MDM4, RB1, E2F1, TP53, MTOR, CDK1A, ATM, PTEN, BCL2, APAF, GLB1* and *PUMA*.

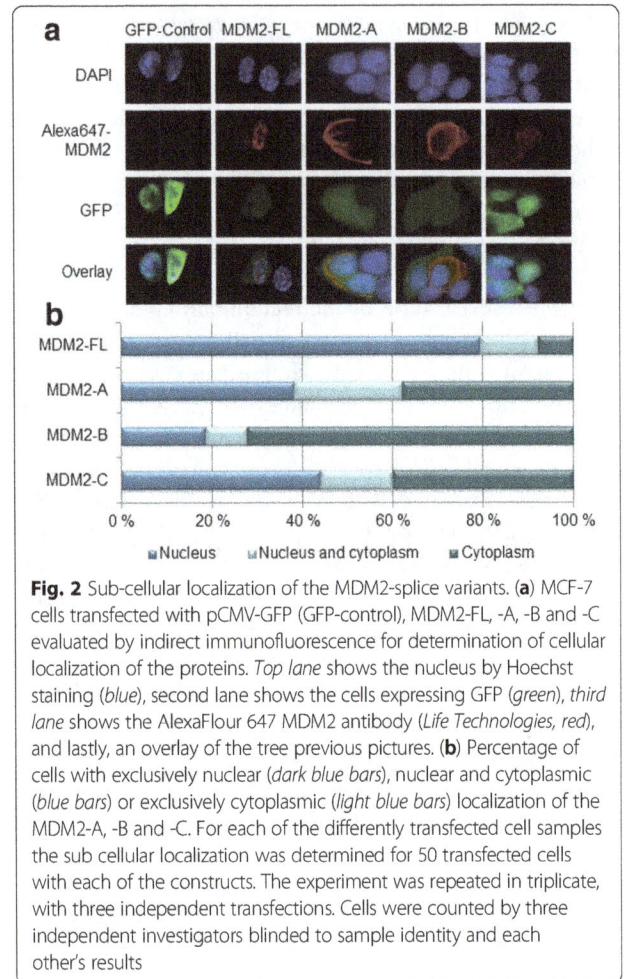

**Fig. 2** Sub-cellular localization of the MDM2-splice variants. (**a**) MCF-7 cells transfected with pCMV-GFP (GFP-control), MDM2-FL, -A, -B and -C evaluated by indirect immunofluorescence for determination of cellular localization of the proteins. *Top lane* shows the nucleus by Hoechst staining (*blue*), second lane shows the cells expressing GFP (*green*), *third lane* shows the AlexaFlour 647 MDM2 antibody (*Life Technologies, red*), and lastly, an overlay of the tree previous pictures. (**b**) Percentage of cells with exclusively nuclear (*dark blue bars*), nuclear and cytoplasmic (*blue bars*) or exclusively cytoplasmic (*light blue bars*) localization of the MDM2-A, -B and -C. For each of the differently transfected cell samples the sub cellular localization was determined for 50 transfected cells with each of the constructs. The experiment was repeated in triplicate, with three independent transfections. Cells were counted by three independent investigators blinded to sample identity and each other's results

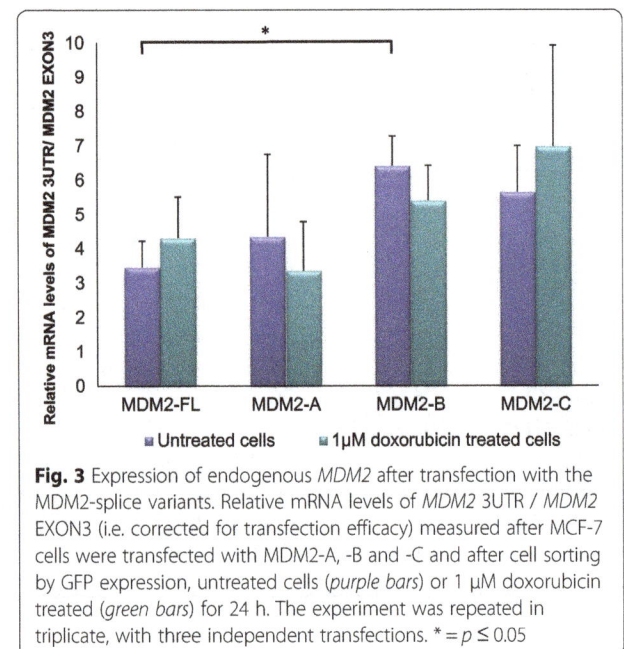

**Fig. 3** Expression of endogenous *MDM2* after transfection with the MDM2-splice variants. Relative mRNA levels of *MDM2* 3UTR / *MDM2* EXON3 (i.e. corrected for transfection efficacy) measured after MCF-7 cells were transfected with MDM2-A, -B and -C and after cell sorting by GFP expression, untreated cells (*purple bars*) or 1 μM doxorubicin treated (*green bars*) for 24 h. The experiment was repeated in triplicate, with three independent transfections. * = $p \leq 0.05$

While overexpression of the three splice variants had no effect or minor a influence on most of the examined genes only, we found overexpression of MDM2-A and -B under doxorubicin treatment both to lead to an approximately two-fold increase in the induction of the pro-apoptotic gene *PUMA*, as compared to the control cells ($p = 4.0 \times 10^{-4}$ and $p = 0.024$, respectively; Fig. 4). A similar effect was not observed for MDM2-FL or -C.

### Induction of apoptosis

None of the three splice variants had any influence on apoptosis assessed by AnnexinV, neither in untreated cells nor subsequent to doxorubicin exposure (Fig. 5). Since this was somewhat surprising, given the observed upregulation of *PUMA* in MDM2-A and -B overexpressing cells, we sought to validate this finding in additional cell lines. We therefore repeated these experiments in isogenic versions of HCT116 colon cancer cells, differing with respect to their *TP53* status only (one version being *TP53+/+* and the other one being *TP53−/−*). Neither overexpression of MDM2-A, -B or -C had any significant effect on apoptosis either in untreated cells or after exposure to doxorubicin in either cell line (Additional file 1).

### Impact on cell cycle progression

Addressing the effect of the *MDM2* splice variants on cell growth, we found no effect of overexpression of either MDM2-A, -B, or -C in untreated cells (Fig. 6a).

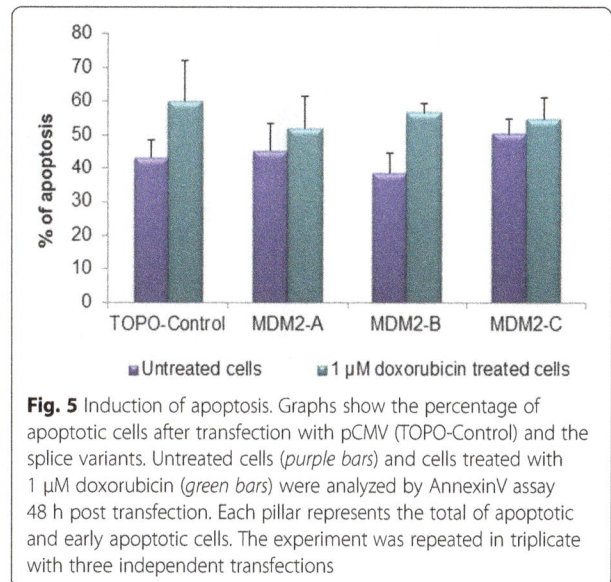

**Fig. 5** Induction of apoptosis. Graphs show the percentage of apoptotic cells after transfection with pCMV (TOPO-Control) and the splice variants. Untreated cells (*purple bars*) and cells treated with 1 µM doxorubicin (*green bars*) were analyzed by AnnexinV assay 48 h post transfection. Each pillar represents the total of apoptotic and early apoptotic cells. The experiment was repeated in triplicate with three independent transfections

In contrast, doxorubicin treatment of cells overexpressing MDM2-B increased the fraction of cells in G1/G-phase ($p = 0.044$) with a corresponding decrease in S-phase ($p = 0.022$) as compared to the doxorubicin treated control cells (Fig. 6b), indicating a potential role for this splice variant in cell cycle regulation of stressed cells. While similar effects were observed for both MDM2-A and -C, the differences did not reach statistical significance.

### Activation of senescence

In cells not exposed to doxorubicin, overexpression of MDM2-B and -C both lead to an increase in the number of senescent cells ($p = 0.004$ and $p = 0.012$, respectively), while overexpression of MDM2-A lead to a smaller, non-significant increase (Fig. 7). No such effect was observed in doxorubicin treated cells.

### Discussion

p53 is activated in response to genotoxic stress and contributes to various processes including apoptosis, cell cycle arrest and senescence [30]. MDM2 is the major regulator of p53, and overexpression of *MDM2* is considered to be a mechanism of direct p53 inactivation [31, 32]. While several splice variants of *MDM2* are identified in tumor- as well as normal cells [33], their functional effects are poorly understood. In the present study we investigated the functional effects of overexpressing the alternative spliced variants MDM2-A, -B and -C in response to chemotherapy treatment.

In line with previous findings [26, 29, 34], we found all three splice variants to be stably expressed at the protein level. Here, we show that upon cytotoxic stress, MDM2-B, and potentially also -C, might be stabilized by different mechanisms than the MDM2-FL protein, suggesting

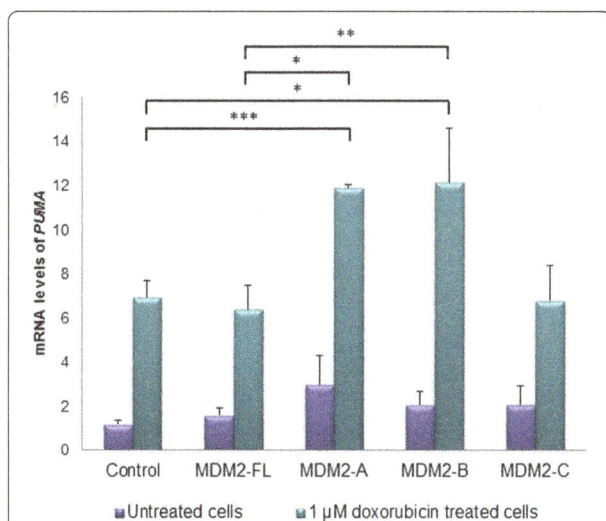

**Fig. 4** Expression of *PUMA* upon overexpression of the *MDM2* splice variants. Activation of *PUMA* after transfection with the splice variants by qPCR assay revealing relative mRNA levels of *PUMA*. MCF-7 cells transfected with pCMV-GFP (GFP-control), MDM2-A, -B and -C, sorted after 12 h based on GFP expression. Graphs show the untreated cells (*purple bars*) and the 24 h 1 µM doxorubicin (*green bars*). The experiment was repeated in triplicate, with three independent transfections. * = $p \leq 0.05$, ** = $p \leq 0.01$, *** = $p \leq 0.001$

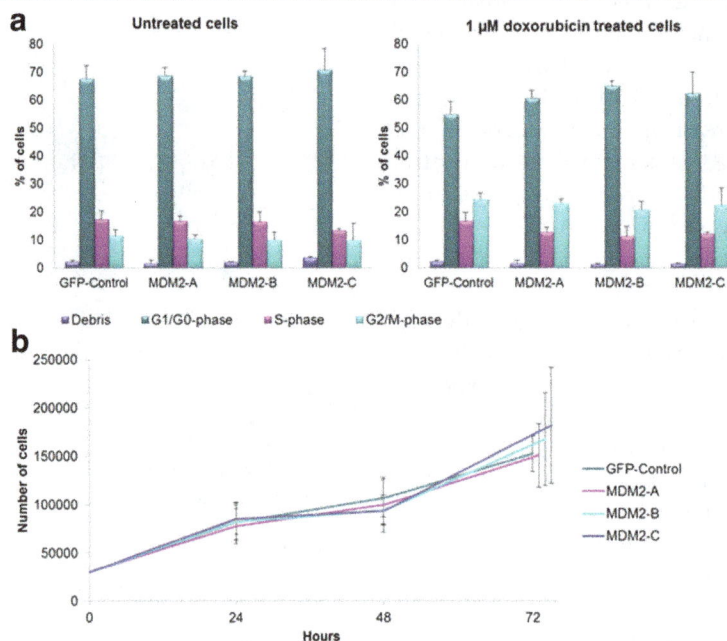

**Fig. 6** Cell cycle analysis. (**a**) MCF-7 breast cancer cells were transfected with pCMV-GFP (GFP-control), MDM2-A, -B and -C. DMSO control cells (*left graph*) or cells treated with 1 μM doxorubicin (*right graph*) for 24 h. The cells were analyzed by NucleoCounter-3000 for status of cell cycle progression. Bars represent cell debris (*purple bars*), cells in G1/G0-phase (*green bars*), cells in S-phase (*pink bars*) and cells in G2/M-phase (*blue bars*) respectively. (**b**) Cell proliferation analyzed by cell count. MCF-7 breast cancer cells transfected with pCMV-GFP (GFP-control), MDM2-A, -B and -C were counted to examine the cell proliferation 24, 48 and 72 h after transfection (data points at 72 h have been slightly shifted on the x-axis of the graph, for clarity). *Both* the experiments were repeated in triplicate, with three independent transfections

that a potential functional role of these splices might be related to cellular stress response.

Interestingly, MDM2-B was found to have a most distinct subcellular localization, differing not only from MDM2-FL, but also from MDM2-A and -C. We found a

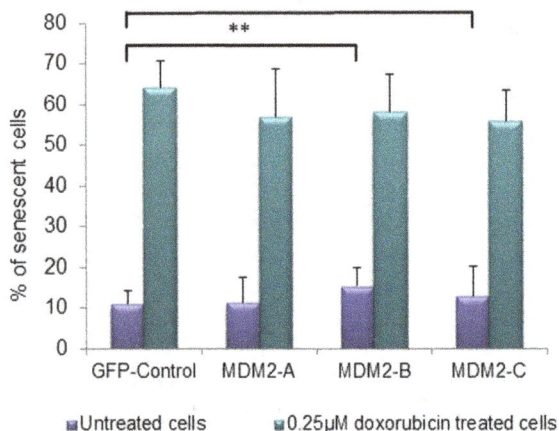

**Fig. 7** Induction of senescence. Graphs show the percentage of senescent cells after transfection with pCMV (TOPO-Control) and the splice variants, untreated (*purple bars*) or treated with 0.25 μM doxorubicin (*green bars*). Cells were analyzed by β-galactosidase assay 8 days post transfection. The experiment was repeated in triplicate with three independent transfections

large majority of MDM2-B exclusively located in the cytoplasm, possibly indicating a distinct function in the cell. This corresponds to previous studies, finding MDM2-B to be localized in the cytoplasm in H1299 cells (lung carcinoma) and NIH/3 T3 cells (mouse fibroblasts) [24]. There are, however, contradictory findings in mouse embryonic fibroblasts, showing exogenously expressed MDM2-B predominantly to be localize in the nucleus [35]. Regarding MDM2-C, this variant is previously found in both the nucleus and in the cytoplasm in MANCA (Burkitt lymphoma), T47D (breast cancer) and also in MCF-7 cells [29], in line with our observations in the present study. Notably, MDM2-A, -B and -C are all known to bind and co-localize with MDM2-FL in the nucleus [29, 35], but MDM2-B is also found to bind to MDM2-FL in the cytoplasm making MDM2-FL unable to enter the nucleus and inhibit p53 in unstressed MCF-7 cells [26].

Strikingly, we found significantly increased levels of endogenous MDM2-FL mRNA as a result of MDM2-B overexpression. This was not the case for MDM2-A and -C overexpression. Although the mechanism behind this upregulation remains unknown, we may speculate that, since we find MDM2-B mostly located in the cytoplasm, it could bind to the MDM2-FL there, making the MDM2-FL-protein unable to enter the nucleus and inhibit p53. This could in turn affect the feedback-loop between p53 and MDM2, and cause MDM2-FL levels to increase as p53 is accumulated in the nucleus, as previously shown by others [26].

Interestingly, *PUMA*, one of the major genes leading to activation of apoptosis, was significantly upregulated at the mRNA level, after MDM2-A and -B overexpression and doxorubicin treatment. Given that PUMA is a direct transcriptional target for p53, it is likely that the MDM2-splice variants' impact on PUMA levels is mediated via p53. Recent studies have indicated that MDM2-B has a functional role of in tumor cells as response to cellular stress (UV and Cisplatin) through activation of *p21* rather than *PUMA* [36], however, we found no change in *p21* level after overexpression of MDM2-B.

Based on our findings of *PUMA* upregulation that could indicate a role for MDM2-A and MDM2-B in apoptosis, together with previous findings indicating that MDM2-A may by associated with aging and senescence [34, 37], we assessed the impact of the splice variants on potential cellular endpoints upon doxorubicin treatment: apoptosis, growth/cell cycle progression and senescence. However, none of the three *MDM2* splice variants seemed to have any impact on activation of apoptosis or senescence, neither in untreated cells nor doxorubicin treated cells. Although this may seem contradictory to the finding of upregulated *PUMA*, notably, p53's functions as a transcription factor and as an inducer of apoptosis have, in several systems been shown to be independent [38, 39].

Regarding cell cycle distribution, we found that the cell cycle profile changed in the chemotherapy treated cells with up-regulated MDM2-B. This could potentially be explained by MDM2-B's ability to bind MDM2-FL in the cytoplasm (as discussed above), making MDM2-FL unable to enter the nucleus for inhibition of p53. In turn this could lead to accumulation of p53 that may trigger arrest of cells in the G1/G0-phase. Regarding MDM2-A, a previous studie has found this splice variant to inhibit growth in a p53-dependent manner [23], however, in the present study, we observed no change in cell cycle distribution as a result of overexpressing MDM2-A.

## Conclusions
We found the impact of the three *MDM2* splice variants on potential cellular endpoints upon chemotherapy treatment of breast cancer cell lines to be limited.

## Additional file

**Additional file 1:** Induction of apoptosis in HCT116 *TP53*+/+ and *TP53*–/–. Graphs show the percentage of apoptotic cells after transfection with pCMV (TOPO-Control), MDM2-FL and the splice variants, untreated (purple bars) or treated with 1 μM doxorubicin (green bars) analyzed by AnnexinV assay 24 h post transfection. Each pillar represents the total of apoptotic and early apoptotic cells. The experiment was repeated in triplicate with three independent transfections. (TIF 99 kb)

### Abbreviations
ATCC: American Type Culture Collection; BSA: Bovine serum albumin; DMSO: Dimethyl sulfoxide; EMEM: Eagle's minimum essential medium; FBS: Fetal bovine serum; GAPDH: Glyceraldehyde 3-phosphate dehydrogenase; GFP: Green fluorescent protein; HRP: Horse radish peroxidase; MDM2: Murine Double Minute 2; MDM4: Murine double minute 4; PI: Propidium Iodid; SDS: Sodium dodecyl; WT: Wild-type

### Acknowledgements
We thank Beryl Leirvaag, Elise de Faveri and Silje Bjørneklett for technical assistance and Einar Birkeland for scientific and technical advice.

### Funding
This work was performed in the Mohn Cancer Research Laboratory. The project was supported by grants from the Norwegian Cancer Society, the Norwegian Health Region West and the Bergen Medical Research Foundation.

### Authors' contributions
Designed experiments: JH, LG, BM, GI, SK. Performed the experiments: JH, LG, BM, GI, JIØ. Analyzed the data: JH, LG, GI, PEL, SK. Wrote the paper: JH, PEL, SK. All authors read and approved the final manuscript.

### Competing interests
The authors declare that they have no competing interests.

### Author details
[1]Section of Oncology, Department of Clinical Science, University of Bergen, 5020 Bergen, Norway. [2]Department of Oncology, Haukeland University Hospital, Bergen, Norway. [3]Department of Biology, University of Bergen, Bergen, Norway. [4]Present address: Department of Oncology and Palliative Medicine, Bodø, Norway. [5]Present address: Huntsman Cancer Institute, University of Utah Health Care, Salt Lake City, USA.

### References
1. Fakharzadeh SS, Trusko SP, George DL. Tumorigenic potential associated with enhanced expression of a gene that is amplified in a mouse tumor cell line. EMBO J. 1991;10(6):1565–9.
2. Haupt Y, Maya R, Kazaz A, Oren M. Mdm2 promotes the rapid degradation of p53. Nature. 1997;387(6630):296–9.
3. Honda R, Tanaka H, Yasuda H. Oncoprotein MDM2 is a ubiquitin ligase E3 for tumor suppressor p53. FEBS Lett. 1997;420(1):25–7.
4. Momand J, Zambetti GP, Olson DC, George D, Levine AJ. The mdm-2 oncogene product forms a complex with the p53 protein and inhibits p53-mediated transactivation. Cell. 1992;69(7):1237–45.
5. Wu X, Bayle JH, Olson D, Levine AJ. The p53-mdm-2 autoregulatory feedback loop. Genes Dev. 1993;7(7A):1126–32.
6. Zauberman A, Barak Y, Ragimov N, Levy N, Oren M. Sequence-specific DNA binding by p53: identification of target sites and lack of binding to p53-MDM2 complexes. EMBO J. 1993;12(7):2799–808.
7. Momand J, Jung D, Wilczynski S, Niland J. The MDM2 gene amplification database. Nucleic Acids Res. 1998;26(15):3453–9.
8. Michael D, Oren M. The p53 and Mdm2 families in cancer. Curr Opin Genet Dev. 2002;12(1):53–9.
9. Bartel F, Taubert H, Harris LC. Alternative and aberrant splicing of MDM2 mRNA in human cancer. Cancer Cell. 2002;2(1):9–15.
10. Chen J, Marechal V, Levine AJ. Mapping of the p53 and mdm-2 interaction domains. Mol Cell Biol. 1993;13(7):4107–14.
11. Freedman DA, Epstein CB, Roth JC, Levine AJ. A genetic approach to mapping the p53 binding site in the MDM2 protein. Mol Med. 1997;3(4):248–59.
12. Momand J, Villegas A, Belyi VA. The evolution of MDM2 family genes. Gene. 2011;486(1–2):23–30.
13. Freedman DA, Wu L, Levine AJ. Functions of the MDM2 oncoprotein. Cell Mol Life Sci. 1999;55(1):96–107.
14. Roth J, Dobbelstein M, Freedman DA, Shenk T, Levine AJ. Nucleo-cytoplasmic shuttling of the hdm2 oncoprotein regulates the levels of the p53 protein via a pathway used by the human immunodeficiency virus rev protein. EMBO J. 1998;17(2):554–64.

15. Arva NC, Talbott KE, Okoro DR, Brekman A, Qiu WG, Bargonetti J. Disruption of the p53-Mdm2 complex by Nutlin-3 reveals different cancer cell phenotypes. Ethn Dis. 2008;18(2 Suppl 2):S2. 1–8.

16. Bartl S, Ban J, Weninger H, Jug G, Kovar H. A small nuclear RNA, hdm365, is the major processing product of the human mdm2 gene. Nucleic Acids Res. 2003;31(4):1136–47.

17. Sam KK, Gan CP, Yee PS, Chong CE, Lim KP, Karen-Ng LP, Chang WS, Nathan S, Rahman ZA, Ismail SM, et al. Novel MDM2 splice variants identified from oral squamous cell carcinoma. Oral Oncol. 2012;48(11):1128–35.

18. Sigalas I, Calvert AH, Anderson JJ, Neal DE, Lunec J. Alternatively spliced mdm2 transcripts with loss of p53 binding domain sequences: transforming ability and frequent detection in human cancer. Nat Med. 1996;2(8):912–7.

19. Bartel F, Harris LC, Wurl P, Taubert H. MDM2 and its splice variant messenger RNAs: expression in tumors and down-regulation using antisense oligonucleotides. Mol Cancer Res. 2004;2(1):29–35.

20. Jeyaraj S, O'Brien DM, Chandler DS. MDM2 and MDM4 splicing: an integral part of the cancer spliceome. Front Biosci. 2009;14:2647–56.

21. Lukas J, Gao DQ, Keshmeshian M, Wen WH, Tsao-Wei D, Rosenberg S, Press MF. Alternative and aberrant messenger RNA splicing of the mdm2 oncogene in invasive breast cancer. Cancer Res. 2001;61(7):3212–9.

22. Chandler DS, Singh RK, Caldwell LC, Bitler JL, Lozano G. Genotoxic stress induces coordinately regulated alternative splicing of the p53 modulators MDM2 and MDM4. Cancer Res. 2006;66(19):9502–8.

23. Volk EL, Fan L, Schuster K, Rehg JE, Harris LC. The MDM2-a splice variant of MDM2 alters transformation in vitro and the tumor spectrum in both Arf- and p53-null models of tumorigenesis. Mol Cancer Res. 2009;7(6):863–9.

24. Sanchez-Aguilera A, Garcia JF, Sanchez-Beato M, Piris MA. Hodgkin's lymphoma cells express alternatively spliced forms of HDM2 with multiple effects on cell cycle control. Oncogene. 2006;25(18):2565–74.

25. Dias CS, Liu Y, Yau A, Westrick L, Evans SC. Regulation of hdm2 by stress-induced hdm2alt1 in tumor and nontumorigenic cell lines correlating with p53 stability. Cancer Res. 2006;66(19):9467–73.

26. Evans SC, Viswanathan M, Grier JD, Narayana M, El-Naggar AK, Lozano G. An alternatively spliced HDM2 product increases p53 activity by inhibiting HDM2. Oncogene. 2001;20(30):4041–9.

27. Zheng T, Wang J, Zhao Y, Zhang C, Lin M, Wang X, Yu H, Liu L, Feng Z, Hu W. Spliced MDM2 isoforms promote mutant p53 accumulation and gain-of-function in tumorigenesis. Nat Commun. 2013;4:2996.

28. Steinman HA, Burstein E, Lengner C, Gosselin J, Pihan G, Duckett CS, Jones SN. An alternative splice form of Mdm2 induces p53-independent cell growth and tumorigenesis. J Biol Chem. 2004;279(6):4877–86.

29. Okoro DR, Arva N, Gao C, Polotskaia A, Puente C, Rosso M, Bargonetti J. Endogenous human MDM2-C is highly expressed in human cancers and functions as a p53-independent growth activator. Plos One. 2013;8(10):e77643.

30. Vousden KH, Lu X. Live or let die: the cell's response to p53. Nat Rev Cancer. 2002;2(8):594–604.

31. Cummings J, Anderson L, Willmott N, Smyth JF. The molecular pharmacology of doxorubicin in vivo. Eur J Cancer. 1991;27(5):532–5.

32. Fahraeus R, Olivares-Illana V. MDM2's social network. Oncogene. 2014;33(35): 4365–76.

33. Bartel F, Pinkert D, Fiedler W, Kappler M, Wurl P, Schmidt H, Taubert H. Expression of alternatively and aberrantly spliced transcripts of the MDM2 mRNA is not tumor-specific. Int J Oncol. 2004;24(1):143–51.

34. Volk EL, Schuster K, Nemeth KM, Fan L, Harris LC. MDM2-A, a common Mdm2 splice variant, causes perinatal lethality, reduced longevity and enhanced senescence. Dis Model Mech. 2009;2(1–2):47–55.

35. Schuster K, Fan L, Harris LC. MDM2 splice variants predominantly localize to the nucleoplasm mediated by a COOH-terminal nuclear localization signal. Mol Cancer Res. 2007;5(4):403–12.

36. Jacob AG, Singh RK, Comiskey Jr DF, Rouhier MF, Mohammad F, Bebee TW, Chandler DS. Stress-induced alternative splice forms of MDM2 and MDMX modulate the p53-pathway in distinct ways. Plos One. 2014;9(8):e104444.

37. Campisi J. Cellular senescence as a tumor-suppressor mechanism. Trends Cell Biol. 2001;11(11):S27–31.

38. Moll UM, Wolff S, Speidel D, Deppert W. Transcription-independent pro-apoptotic functions of p53. Curr Opin Cell Biol. 2005;17(6):631–6.

39. Chipuk JE, Kuwana T, Bouchier-Hayes L, Droin NM, Newmeyer DD, Schuler M, Green DR. Direct activation of Bax by p53 mediates mitochondrial membrane permeabilization and apoptosis. Science. 2004;303(5660):1010–4.

# 13

# Molecular targets and signaling pathways regulated by nuclear translocation of syndecan-1

Tünde Szatmári[1*] (iD), Filip Mundt[2], Ashish Kumar-Singh[1], Lena Möbus[1], Rita Ötvös[1], Anders Hjerpe[1,2] and Katalin Dobra[1,2]

## Abstract

**Background:** The cell-surface heparan sulfate proteoglycan syndecan-1 is important for tumor cell proliferation, migration, and cell cycle regulation in a broad spectrum of malignancies. Syndecan-1, however, also translocates to the cell nucleus, where it might regulate various molecular functions.

**Results:** We used a fibrosarcoma model to dissect the functions of syndecan-1 related to the nucleus and separate them from functions related to the cell-surface. Nuclear translocation of syndecan-1 hampered the proliferation of fibrosarcoma cells compared to the mutant lacking nuclear localization signal. The growth inhibitory effect of nuclear syndecan-1 was accompanied by significant accumulation of cells in the G0/G1 phase, which indicated a possible G1/S phase arrest.

We implemented multiple, unsupervised global transcriptome and proteome profiling approaches and combined them with functional assays to disclose the molecular mechanisms that governed nuclear translocation and its related functions. We identified genes and pathways related to the nuclear compartment with network enrichment analysis of the transcriptome and proteome. The TGF-β pathway was activated by nuclear syndecan-1, and three genes were significantly altered with the deletion of nuclear localization signal: EGR-1 (early growth response 1), NEK11 (never-in-mitosis gene a-related kinase 11), and DOCK8 (dedicator of cytokinesis 8). These candidate genes were coupled to growth and cell-cycle regulation. Nuclear translocation of syndecan-1 influenced the activity of several other transcription factors, including E2F, NFκβ, and OCT-1. The transcripts and proteins affected by syndecan-1 showed a striking overlap in their corresponding biological processes. These processes were dominated by protein phosphorylation and post-translation modifications, indicative of alterations in intracellular signaling. In addition, we identified molecules involved in the known functions of syndecan-1, including extracellular matrix organization and transmembrane transport.

**Conclusion:** Collectively, abrogation of nuclear translocation of syndecan-1 resulted in a set of changes clustering in distinct patterns, which highlighted the functional importance of nuclear syndecan-1 in hampering cell proliferation and the cell cycle. This study emphasizes the importance of the localization of syndecan-1 when considering its effects on tumor cell fate.

**Keywords:** Syndecan-1, Nuclear localization, Transcriptomic, Proteomic, Pathway analysis, Proliferation, Cell cycle

* Correspondence: tunde.szatmari@ki.se
[1]Department of Laboratory Medicine, Division of Pathology, Karolinska Institutet, SE-14186 Stockholm, Sweden
Full list of author information is available at the end of the article

## Background

Syndecan-1 is a transmembrane heparan sulfate proteoglycan (HSPG), which carries heparan-sulfate (HS) and chondroitin-sulfate glycosaminoglycans on its ectodomain. Syndecan-1 acts as a co-receptor for growth factors, chemokines, and cytokines; thus, it regulates a multitude of cellular functions, including cell growth, proliferation, adhesion, and migration [1]. In these processes, the sub-cellular localization of syndecan-1 is critical [2]. Syndecan-1 is typically referred to as a cell-surface proteoglycan, but it can also be found in the stroma [3], and it can be shed into body fluids [4–6]. We have previously reported that syndecan-1 also translocates to the nucleus in a highly regulated manner by a tubulin-mediated transport mechanism [7]. In the nucleus, it co-localizes with FGF-2 and heparanase [8]. Although syndecan-1 has been detected in the nuclear compartment of various tumor types [7, 9, 10], the functions associated with nuclear translocation remain incompletely understood (for review, see [11–13]).

The presence and functions of HS in the nucleus have been studied extensively; however, less research has investigated the translocation of the core protein itself. The nuclear occurrence of HS [14, 15] was lately extended to include the whole syndecan-1 core protein [7, 16]. Other HSPGs, including syndecans-2 and -3 and glypican-1, were also identified in the nuclear compartments of various cell-types [17, 18]. The structural requirement for the nuclear HSPG translocation implies a nuclear localization signals (NLSs) found in the core proteins of several HSPGs. Syndecan-1 harbors the RMKKK motif in the juxta-membrane region of the cytoplasmic domain, which is the minimal, sufficient sequence required for nuclear localization [8]. Moreover, the MKKK sequence is essential for lipid raft-mediated endocytosis [19].

The nuclear HS has an anti-proliferative effect [15, 20], and the extent of growth inhibition depends on the cell confluence, the fine structure and the sulfation pattern of the nuclear HS. Moreover, the effect of nuclear HS differs between malignant and benign cells. The nuclear entry of HS depends on certain cell-cycle phases, and cell cycle progression is regulated by the amount of HS or HSPG in the nucleus [7, 21–25]. However, the exact mechanisms of action have not been established. Another well-studied function of HS is to shuttle heparin-binding growth factors and other macromolecules into the nucleus. These factors are internalized with HSPGs and they co-localize in the nucleus [8, 26–30].

Nuclear HS regulates gene expression through at least two mechanisms. First, it regulates the transcription machinery by inhibiting DNA topoisomerase; this activity prevents DNA relaxation, and the DNA remains inaccessible to transcription factors [31]. Moreover, HS directly inhibits transcription factors [32, 33], probably through direct interactions, because the DNA binding domains of some transcription factors contain high affinity heparin binding sequences [13]. Nuclear HS can also regulate gene expression by modulating the acetylation status of histone proteins. Both nuclear syndecan-1 [34] and HS chains [35] inhibit histone acetyltransferases. This activity can at least partly explain the anti-proliferative effects of HS.

Previously, we stably transfected fibrosarcoma cells with full-length syndecan-1 (FLs1) and a mutated syndecan-1 that lacked the RMKKK nuclear localization signal (NLSdel) motif in the juxtamembrane region of the cytoplasmic domain. We showed that FLs1 entered the nucleus normally, but deletion of the RMKKK motif abolished the nuclear translocation of this proteoglycan [25].

In the current study, we elucidated the functions of nuclear syndecan-1 on both transcriptomic and proteomic levels, combining the results to visualize the affected signaling patterns. With the same two fibrosarcoma cell-sub-lines (one transfected with FLs1 and the other with NLSdel), it was possible to separate the nuclear and cell-surface functions of syndecan-1. We demonstrated a differential impact of nuclear syndecan-1 on cell cycle progression, viability and apoptosis. The transcript of FLs1, translocating to the nucleus (but not the NLSdel mutant, with predominant membrane and cytosolic distribution), induced the accumulation of cells in G1/G0 phase and hampered the proliferation of fibrosarcoma cells. We delineated the molecular background of these changes, and we identified nuclear proteins and transcription factors responsible for these effects.

## Results

### Syndecan-1 level and its subcellular localization in different constructs

Syndecan-1 levels corresponded to 1.5- to 2-fold increase in the FLs1 and NLSdel transfected cell lines compared to controls (Additional file 1: Figure S1).

The subcellular localization of syndecan-1 was confined to the nuclear compartment in cells transfected with FLs1 and it was mainly cytoplasmic in the NLSdel and empty vector transfected cells (Additional file 2: Figure S2).

### Effects of nuclear syndecan-1 on cell proliferation and cell cycle progression

Proliferation was significantly altered in fibrosarcoma cells transfected with different syndecan-1 constructs. The doubling time of cells transfected with NLSdel was shorter (32.3 h) compared to cells that overexpressed FLs1 (38.9 h) and empty vector (41.09 h); Fig. 1). Consequently, cells with preserved nuclear localization (FLs1) had lower proliferation rate compared to the NLSdel mutant that displayed impaired nuclear localization.

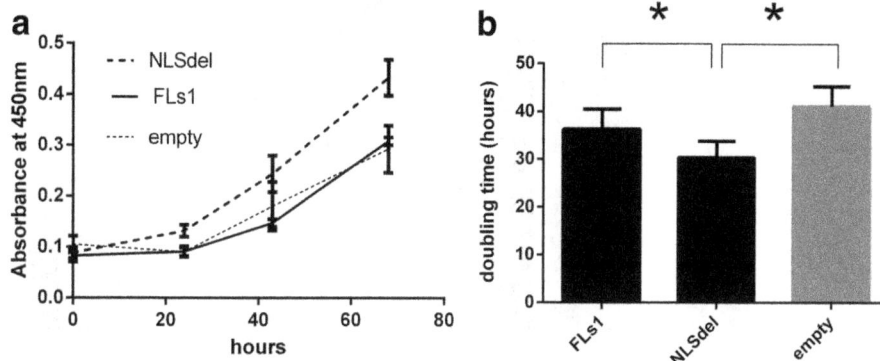

**Fig. 1** Nuclear translocation of syndecan-1 inhibits fibrosarcoma cell proliferation. **a** B6FS fibrosarcoma cells transfected with full-length (FLs1), syndecan-1 lacking the nuclear localization signal (NLSdel) and corresponding empty vector (empty) were seeded at a density of 3000 cells/well in 96 well plates and absorbance values were measured using WST-1 proliferation assay. Symbols represent the mean ± SD ($n = 3$). Cell proliferation increased when the nuclear translocation of syndecan-1 was impaired in the NLS deleted construct ($p = 0.018$). **b** Doubling time of cells transfected with different constructs shows that NLSdel grows faster while the difference between FLs1 and control is not significant. Bars represent the average of three independent experiments ±SD. *$p < 0.05$, based on the paired t-test

Cell cycle analysis showed that significantly ($p \leq 0.05$) fewer cells were in the G1 phase in NLSdel cells than in FLs1 cells; thus, cells transfected with NLSdel passed through the G1/S checkpoint more rapidly than cells transfected with FLs1 (Fig. 2).

Immunocytochemical stating with Ki-67 revealed very high proliferation index at 48 h after seeding, corresponding to 99% in all cell lines transfected with the three different constructs. The proportion of Ki-67 positive cells was 83% in empty vector, 94% and 96% in the full-length (FLs1) and NLSdel, respectively, after 72 h (Additional file 3: Figure S3).

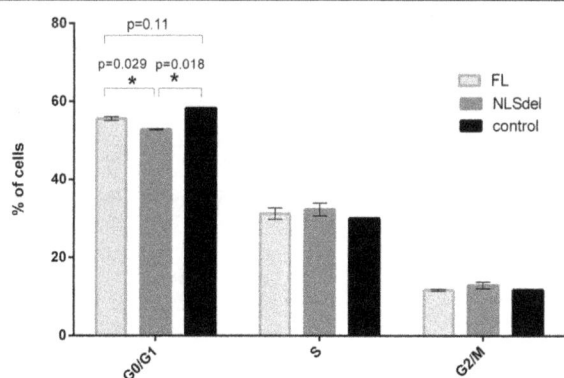

**Fig. 2** Nuclear translocation of syndecan-1 affects the cell cycle distribution of fibrosarcoma cells. Cells transfected with the full-length (FLs1) or syndecan-1 lacking the nuclear localization signal (NLSdel) and empty vector (empty) were assayed with propidium iodide staining followed by flow cytometry, at 48 h after cell seeding. Columns represent the mean percentage ± SD (n = 3) of cells in the indicated phase of the cell cycle. Cells accumulated in the G1 phase in the presence of nuclear syndecan-1 compared to the deletion mutant. *p values were calculated using the paired t-test

### Effects of nuclear translocation of syndecan-1 on the spontaneous apoptosis of fibrosarcoma cells

Nuclear translocation of syndecan-1 caused a small, but significant ($p \leq 0.05$) inhibition of spontaneous apoptosis at 48 and 72 h after cell seeding, compared to the apoptosis of cells transfected with NLSdel. Apoptosis of fibrosarcoma cells slightly increased in both samples over time. At 48 h after cell seeding, the fractions of apoptotic cells increased by $3.2 \pm 0.6\%$ in FLs1 cells and by $5.7 \pm 1.3\%$ in NLSdel cells (Fig. 3a). At 72 h after seeding, the fractions of apoptotic cells increased by 4.7 $\pm 2.3\%$ and $6.7 \pm 2.2\%$, respectively (Fig. 3b).

### Differential gene and protein expression in the presence and absence of nuclear syndecan-1

To provide new insight into the regulatory pathways governed by nuclear syndecan-1 we performed transcriptomic and proteomic screenings of the B6FS fibrosarcoma cells transfected with the three syndecan-1 constructs.

Nuclear translocation of syndecan-1 resulted in 20 differentially expressed genes compared to the deletion mutant unable to translocate to the nucleus. Of these, 2 genes were downregulated and 18 were upregulated (Table 1).

We successfully validated three significantly altered genes by RT-qPCR (Table 2a): early growth response 1 (EGR1), never in mitosis gene a-related kinase 11 (NEK11), and dedicator of cytokinesis 8 (DOCK8). The first two proteins encoded by these genes are localized to the nucleus, whereas DOCK8 is mostly cytosolic.

The overexpression of FLs1 resulted in the modulation of 119 genes compared to control cells transfected with empty vector. Of these, 63 genes were downregulated. Following validation, we found that several matrix and

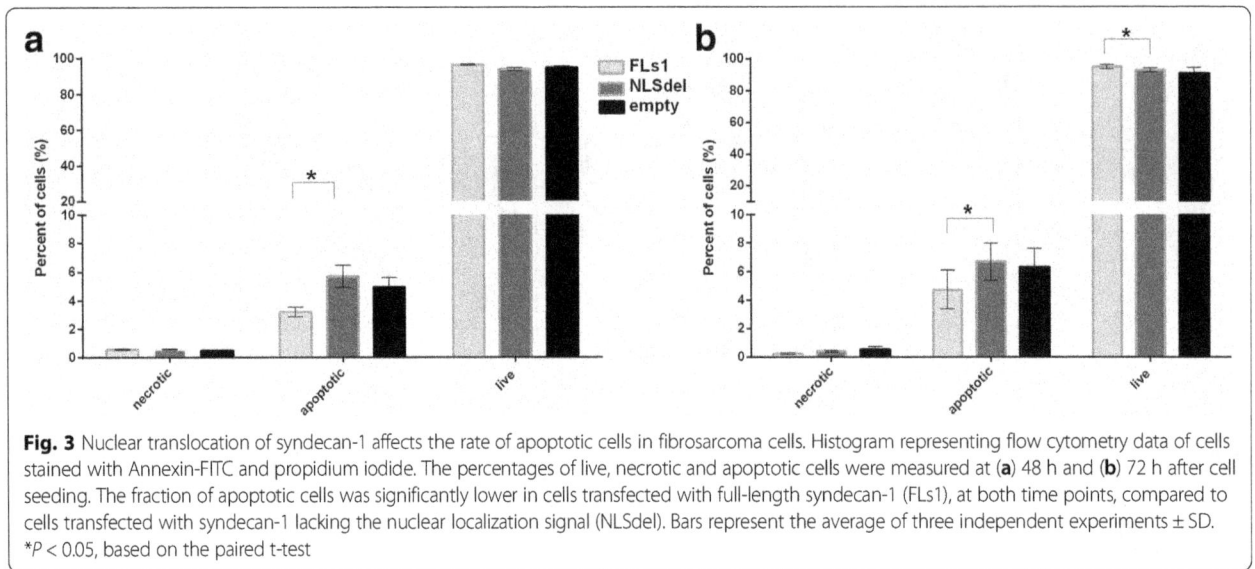

**Fig. 3** Nuclear translocation of syndecan-1 affects the rate of apoptotic cells in fibrosarcoma cells. Histogram representing flow cytometry data of cells stained with Annexin-FITC and propidium iodide. The percentages of live, necrotic and apoptotic cells were measured at (**a**) 48 h and (**b**) 72 h after cell seeding. The fraction of apoptotic cells was significantly lower in cells transfected with full-length syndecan-1 (FLs1), at both time points, compared to cells transfected with syndecan-1 lacking the nuclear localization signal (NLSdel). Bars represent the average of three independent experiments ± SD. *P < 0.05, based on the paired t-test

membrane related proteins (e.g., AREG, COL1A2, PCDH18, SERPINB4) showed altered expression; in addition, several intracellular and nuclear factors were affected that had roles in signaling and cell growth, including DACH1, ITGA8, and PIP5K1B (Table 2b). Compared to control cells (V), the overexpression of

NLSdel resulted in alterations in 42 genes, and all were downregulated. These genes encoded several secreted proteins, including COL19A1, FAP, IL2RB, SERPINA3, SERPINB4, and IL2RB (Table 2c).

With MS-based proteomics, we identified and quantified 8963 proteins across all samples. Changes in the

**Table 1** Differentially expressed genes in cells with nuclear syndecan-1 (FLs1) versus cells with syndecan-1 lacking the NLS signal (NLSdel)

| FLs1 vs NLSdel Symbol | Gene name | FC | q |
|---|---|---|---|
| EFCAB6 | EF-hand calcium binding domain 6 | 2.24 | 0.013 |
| CCKAR | cholecystokinin A receptor | 1.97 | 0.025 |
| EGR1 | early growth response 1 | 1.85 | 0.003 |
| CDCP1 | CUB domain containing protein 1 | 1.84 | 0.00 |
| ZNF676 | zinc finger protein 676 | 1.84 | 0.031 |
| NEK11 | NIMA (never in mitosis gene a- related kinase 11 | 1.77 | 0.001 |
| SLC16A4 | solute carrier family 16, member 4 (monocarboxylic acid transporter 5) | 1.75 | 0.013 |
| DOCK8 | dedicator of cytokinesis 8 | 1.72 | 0.00 |
| LMBRD2 | LMBR1 domain containing 2 | 1.68 | 0.037 |
| SNORA56 | small nucleolar RNA, H/ACA box 56 | 1.67 | 0.015 |
| LPCAT2 | lysophosphatidylcholine acyltransferase 2 | 1.6 | 0.017 |
| LSM14B | LSM14B, SCD6 homolog B (S. cerevisiae) | 1.58 | 0.047 |
| HBG1 | hemoglobin, gamma A | 1.57 | 0.01 |
| SNORA13 | small nucleolar RNA, H/ACA box 13 | 1.57 | 0.026 |
| SNORD116–6 | small nucleolar RNA, C/D box 116–6 | 1.57 | 0.00 |
| CCDC88C | coiled-coil domain containing 88C | 1.56 | 0.03 |
| EHF | ets homologous factor | 1.52 | 0.006 |
| PMFBP1 | polyamine modulated factor 1 binding protein 1 | 1.51 | 0.015 |
| PLA2G5 | phospholipase A2, group V | −1.59 | 0.002 |
| CYP3A7 | cytochrome P450, family 3, subfamily A, polypeptide 7 | −1.6 | 0.001 |

Differentially expressed genes identified by microarray analysis in fibrosarcoma cells with nuclear syndecan-1 (FLs1) versus syndecan-1 lacking the NLS signal (NLSdel). *FC* fold change, *q* false discovery rate

**Table 2** Differentially expressed genes from Affymetrix array, validated by qRT-PCR

| Comparison | Gene product | Microarray (FC) | qPCR (FC, relative expression ± SD, P-value) |
|---|---|---|---|
| a. FLs1 vs NLSdel | EGR1 | 1.85 | 1.39 ± 0.06 ($p = 0.0004$) |
| | NEK11 | 1.77 | 1.67 ± 0.41 ($p = 0.04$) |
| | DOCK8 | 1.72 | 2.13 ± 0.49 ($p = 0.08$) |
| b. FLs1 vs V | COL1A2 | 0.58 | 0.51 ± 0,07 ($p = 0.00001$) |
| | PCDH18 | 0.54 | 0.46 ± 0,04 ($p = 0.00001$) |
| | ITGA8 | 0.48 | 0.50 ± 0.22 ($p = 0.0019$) |
| | PIP5K1B | 0.47 | 0.28 ± 0.18 ($p = 0.0001$) |
| | DACH1 | 0.41 | 0.29 ± 0.09 ($p = 0.0001$) |
| | AREG | 0.40 | 0.28 ± 0.25 ($p = 0.0005$) |
| | SERPINB4 | 0.36 | 0.21 ± 0.13 ($p = 0.0005$) |
| c. NLSdel vs V | COL19A1 | 0.89 | 0.53 ± 0.37 ($p = 0.02$) |
| | FAP | 0.59 | 0.67 ± 0.07 ($p = 0.00003$) |
| | SERPINA3 | 0.48 | 0.59 ± 0.26 ($p = 0.009$) |
| | SERPINB4 | 0.37 | 0.66 ± 0.24 ($p = 0.03$) |
| | IL2RB | 0.33 | 0.55 ± 0.25 ($p = 0.02$) |

Quantitative real-time PCR analysis of gene expression in B6FS cells overexpressing full-length syndecan-1 (FLs1), cells overexpressing NLS deleted syndecan-1 (NLSdel) and control cells (V). A subset of genes fulfilling the criteria of >1.5 fold up- or downregulation and a false discovery rate (q) ≤0.05 were further analyzed by qPCR. The table shows the significantly altered genes by both microarray and qPCR. FC fold change, SD-standard deviation. qPCR was performed three times, each in triplicates

proteome were modest, but the NLSdel cells showed more pronounced proteomic changes than the FLs1-transfected cells (both normalized to mock-transfected controls). At most, 0.5% of the detected proteins showed expression changes that exceeded 1.5-fold in individual samples (Additional file 1: Figure S4a). All except two replicates showed good Pearson correlations ($r > 0.5$) with their respective cluster groups (Additional file 4: Figure S4b and Additional file 5: Figure S5).

When the two non-clustering samples were excluded, the number of proteins regulated differentially between the FLs1 and the NLSdel groups increased from 21 to 122 at 1.5-fold changes and from none to 40 at 2-fold changes. Of the initial 21 proteins that were differentially regulated, 15 remained detectable after excluding the two non-clustering samples (Additional file 6: File S1). Based on these initial findings, we decided to perform subsequent analyses on a two-by-two sample basis.

### Upstream and downstream signaling events regulated by nuclear translocation of syndecan-1

Because few transcripts were regulated, we have presented only proteomic results from the upstream and downstream regulatory analyses performed with the IPA, which compared FLs1 and NLSdel expression. Based on the pattern of differential protein expression, the IPA analyses indicated that TGF-β1, SMAD3, and RAC1 were activated. TGF-β1 was overexpressed in the FLs1 sample compared to NLSdel sample (1.5fold change, q < 0.05; Fig. 4a). The only regulator predicted to be inhibited by nuclear syndecan-1 was the estrogen receptor group (Fig. 4b). The proteomic data patterns predicted a consistent

downstream biological effect, where cell death was activated and cell proliferation was inhibited (Fig. 4c).

Although we found no overlap between the differentially regulated transcripts and proteins, our analysis of associated GO terms showed considerable overlap in biological processes. The networks generated by this method contained both the differentially expressed genes and their binding partners, and the lack of NLS was associated with several cellular functions. These data indicated a strong effect on protein modifications, particularly protein phosphorylation, transcription regulation, and apoptosis (Fig. 5). The GSEA analysis performed on the transcriptome of NLSdel versus Fls1 identified 114 pathways that were significantly enriched, following nuclear translocation of syndecan-1, and 51 pathways were identified by analyzing the proteome dataset. The overlap between the two analyses contained 12 pathways (Fig. 6), which depicted the common effects of syndecan-1 nuclear translocation on mRNA and protein levels. In the mRNA dataset, most of the significantly enriched pathways belonged to the categories of cell cycle regulation (13 pathways), DNA synthesis and transcription (10 pathways), and immune responses (9 pathways) (Table 3). In contrast, the top enriched pathways in the proteome dataset were related to cell adhesion and cell membrane transport; these functions were previously associated with syndecan-1. Interestingly, the proteome dataset also indicated the enrichment of pathways related to the immune system. Several pathways related to TGF-β were also enriched in the proteome dataset (Table 4 and Additional file 7: File S2).

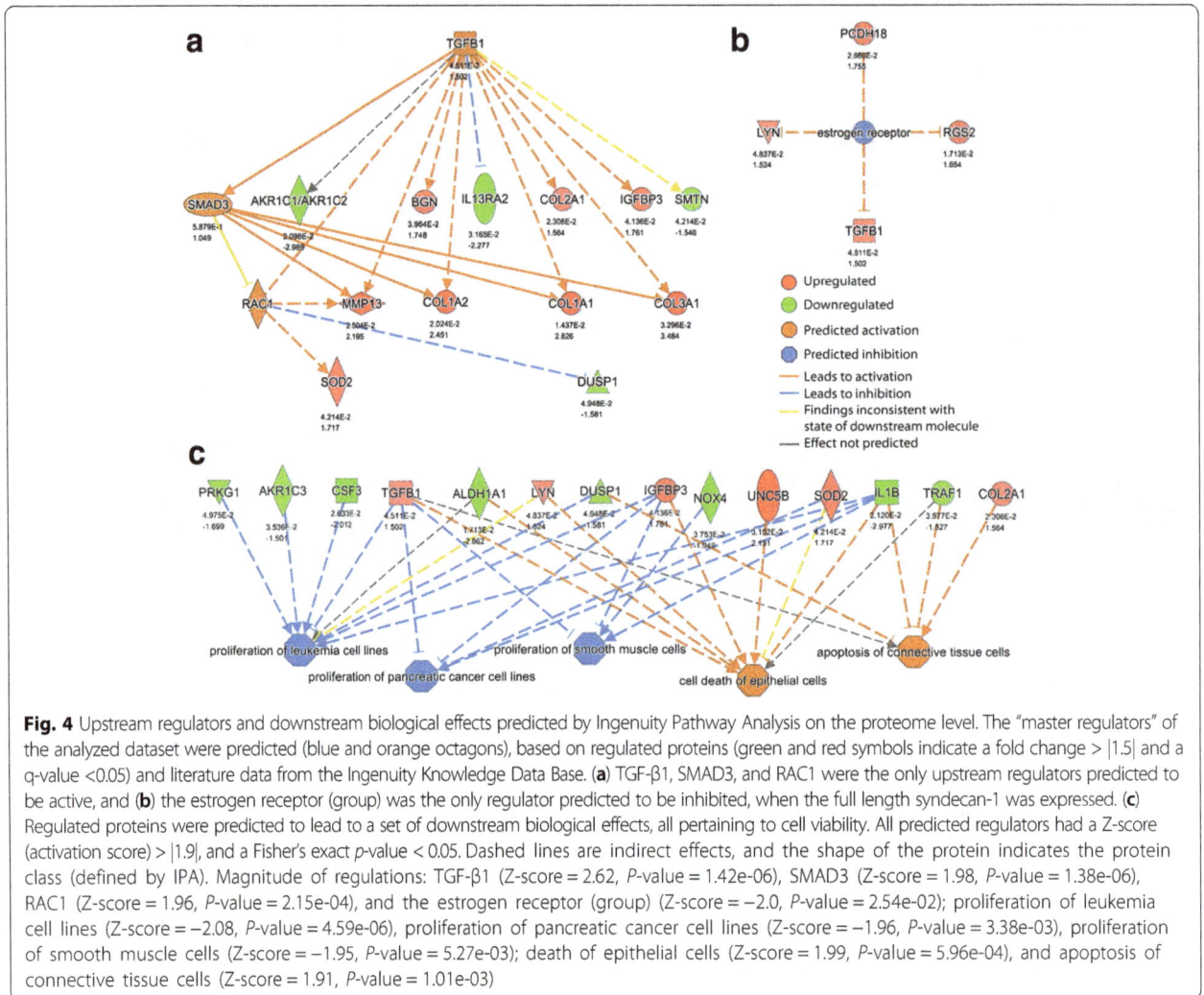

**Fig. 4** Upstream regulators and downstream biological effects predicted by Ingenuity Pathway Analysis on the proteome level. The "master regulators" of the analyzed dataset were predicted (blue and orange octagons), based on regulated proteins (green and red symbols indicate a fold change > |1.5| and a q-value <0.05) and literature data from the Ingenuity Knowledge Data Base. (**a**) TGF-β1, SMAD3, and RAC1 were the only upstream regulators predicted to be active, and (**b**) the estrogen receptor (group) was the only regulator predicted to be inhibited, when the full length syndecan-1 was expressed. (**c**) Regulated proteins were predicted to lead to a set of downstream biological effects, all pertaining to cell viability. All predicted regulators had a Z-score (activation score) > |1.9|, and a Fisher's exact p-value < 0.05. Dashed lines are indirect effects, and the shape of the protein indicates the protein class (defined by IPA). Magnitude of regulations: TGF-β1 (Z-score = 2.62, P-value = 1.42e-06), SMAD3 (Z-score = 1.98, P-value = 1.38e-06), RAC1 (Z-score = 1.96, P-value = 2.15e-04), and the estrogen receptor (group) (Z-score = −2.0, P-value = 2.54e-02); proliferation of leukemia cell lines (Z-score = −2.08, P-value = 4.59e-06), proliferation of pancreatic cancer cell lines (Z-score = −1.96, P-value = 3.38e-03), proliferation of smooth muscle cells (Z-score = −1.95, P-value = 5.27e-03); death of epithelial cells (Z-score = 1.99, P-value = 5.96e-04), and apoptosis of connective tissue cells (Z-score = 1.91, P-value = 1.01e-03).

## Sub-cellular localization of differentially regulated transcripts and proteins

Next, we studied the possible subcellular co-localization of the network components. When we compared the FLs1 and NLSdel transcriptome datasets, the proteins encoded by the most enriched genes (i.e., enriched GO cellular components) were nuclear. In contrast, when we compared either the FLs1 versus V or the NLSdel versus V datasets, the cytoplasm and membrane-related components and processes were enriched (Fig. 7).

## Transcription factors regulating cell growth are affected by nuclear translocation of syndecan-1

To elucidate the early events leading to these complex changes, we profiled the activity of cell-growth-related transcription factors that were affected by the nuclear translocation of syndecan-1. We extracted the nuclear proteins from both FLs1 and NLSdel transfected cells and hybridized them with a panel of consensus sequences that represented transcription factors with important roles in cell growth. The transcription factors with altered expression that were identified in repeated experiments are shown in Fig. 8.

The nuclear translocation of syndecan-1 significantly activated the nuclear factor kappa-light-chain-enhancer of activated B cells (NFκβ), E2F transcription factor 1 (E2F-1), and EGR. The activities of the POU domain, class 2, transcription factor 1 (OCT-1), paired box 3 (Pax-3), and Specificity Protein 1 (Sp1) were also altered in all three experiments, but the activation levels did not reach significance.

## Discussion

Syndecan-1 is critically involved in tumor cell proliferation and migration in a wide range of malignancies. The effects of syndecan-1 are tissue-dependent and largely vary in tumors of different origin [24, 25, 36–39]. The sub-cellular localization of syndecan-1 is important in this context. Syndecan-1 can be anchored to the cell

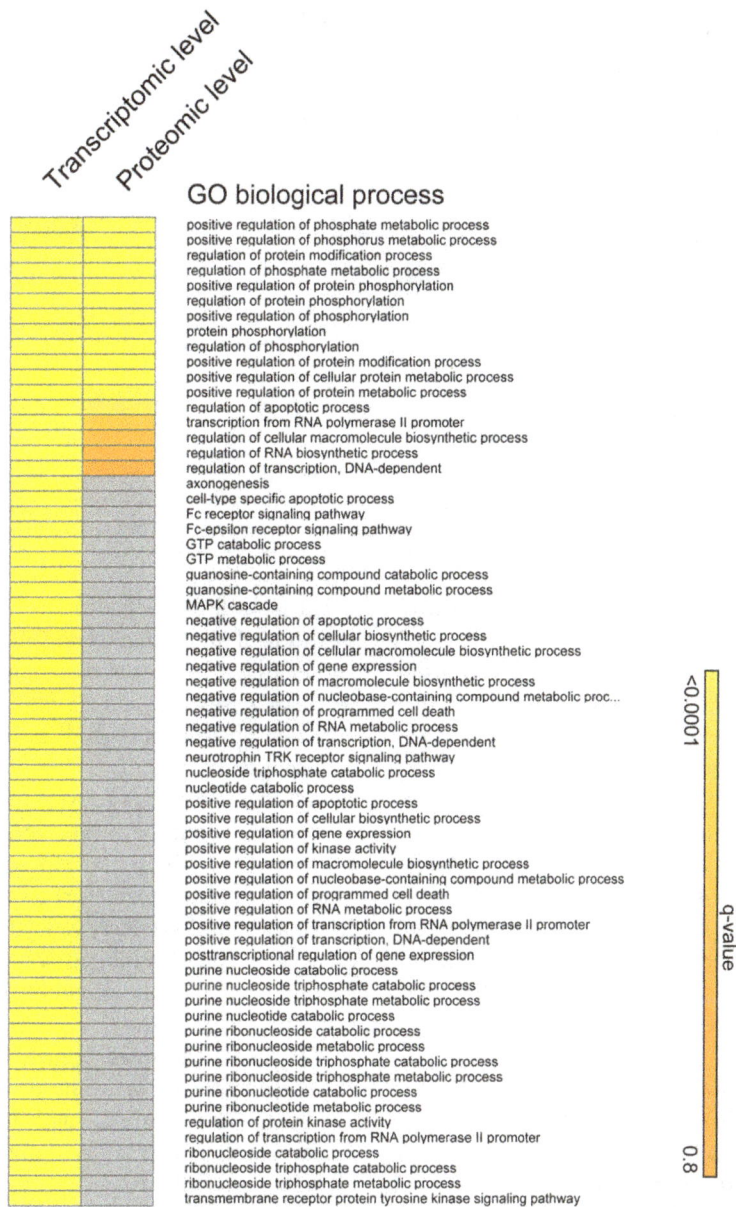

**Fig. 5** Network enrichment analysis highlights the overlap between transcriptomic and proteomic data. The most regulated transcripts or proteins (fold changes > |1.5| and q-values <0.05 between cells transfected with either syndecan-1 or syndecan-1 without a nuclear localization signal) were used separately to assess *GO biological processes* (Funcoup 3.0). The results were diagramed with the Gene-E program. Grey cells represent missing values

membrane, translocated to the nucleus or shed and these locations have profound influences on its functions [40, 41]. Nikolova et al. found that the soluble syndecan-1 affects proliferation and invasiveness of breast cancer cells associated to a molecular signature including downregulation of TIMP-1, alteration in levels of uPAR, the Rho family of small guanosine triphosphatases and of integrins [41].

To test the functions and molecular pathways related to nuclear localization, we separately studied the functions of syndecan-1 in the nucleus and on the cell surface by employing a fibrosarcoma model, with preserved and impaired nuclear localization [25]. With this model system, we studied the functions regulated by the nuclear translocation of syndecan-1, with focus on cell growth. We combined transcriptomic and proteomic approaches to map the molecular mechanisms governing these functions on a global scale. The fact that most of the differentially expressed genes found in the omics screenings, were validated mainly by qPCR and only a

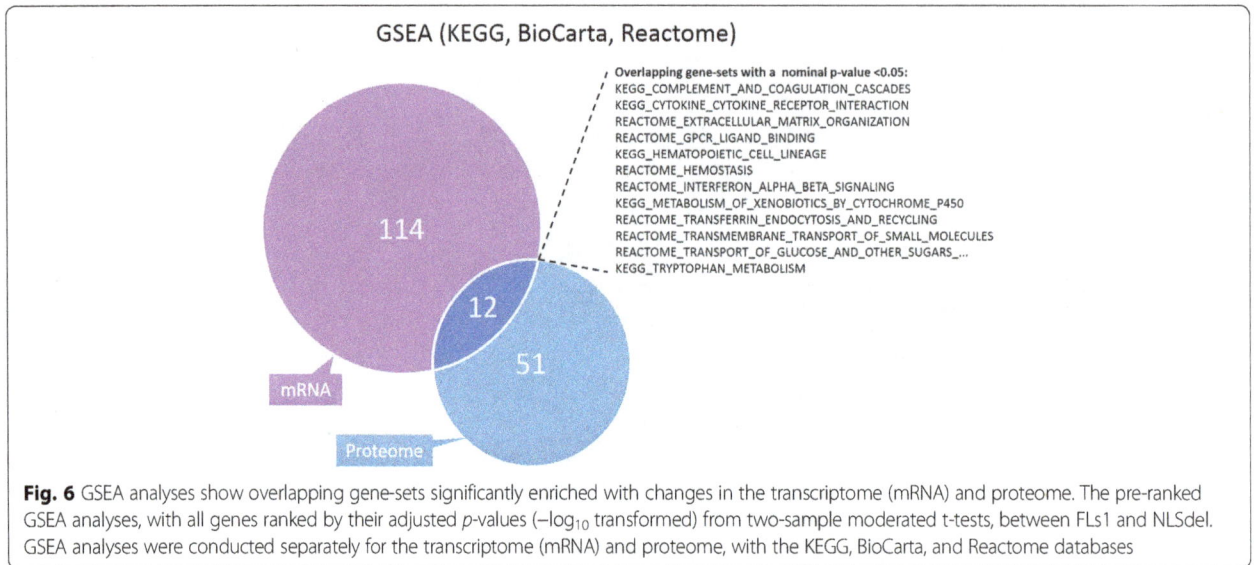

**Fig. 6** GSEA analyses show overlapping gene-sets significantly enriched with changes in the transcriptome (mRNA) and proteome. The pre-ranked GSEA analyses, with all genes ranked by their adjusted $p$-values ($-\log_{10}$ transformed) from two-sample moderated t-tests, between FLs1 and NLSdel. GSEA analyses were conducted separately for the transcriptome (mRNA) and proteome, with the KEGG, BioCarta, and Reactome databases

few of them by a transcription factor array might constitute a limitation.

Previously, we have shown that syndecan-1 translocated to the nucleus in a regulated manner [7]. Here we demonstrate for the first time that the nuclear translocation of syndecan-1 has anti-proliferative effects; as cells with abolished nuclear localization proliferated at a significantly higher rate than those transfected with the full-length syndecan-1. Moreover, cells with nuclear syndecan-1 accumulated in the G0/G1 phase of the cell cycle at a higher extent compared to those with impaired nuclear localization. Ki-67 staining did not show significant differences among different constructs, all having a high proliferation rate in vitro.

There are evidences supporting both anti- and proapoptotic effects of syndecan-1 in different cell types. In myeloma syndecan-1 inhibited apoptosis [42] while its knock-down resulted in increase of apoptosis in endometrial cells [43], myeloma [44] and urothelial carcinoma cells [45]. Interestingly, in our experimental setting, the growth inhibitory effect of nuclear syndecan-1 compared to the cells with abolished nuclear translocation was accompanied by inhibition of spontaneous apoptosis, indicating that these two mechanisms might be interlinked. Similarly, Cortes et al. observed that overexpression of cell surface syndecan-1 in hepatocytes was associated with increased cell proliferation and apoptosis [46]. On the other hand, proliferation might be induced in the neighborhood of apoptotic cells as a compensatory mechanism, where, although apoptosis is initiated, the effector caspases are inhibited, and thus, the living cells constantly emit mitogenic signals, which stimulate the surrounding cells to proliferate [47, 48].

The subcellular localization of syndecan-1 elicited a plethora of molecular changes which were categorized and analyzed by means of extensive bioinformatics. Network analyses pointed predominantly toward altered genes and pathways related to the nuclear compartment. In accordance with our earlier data [22, 49, 50], we found that syndecan-1 overexpression altered TGF-β-related signaling pathways and cell cycle regulation. Moreover, the TGF-β pathway was predicted by bioinformatics as master regulator associated with the nuclear translocation of syndecan-1 in this setting. In mesothelioma cells TGF-β inhibited the nuclear translocation of syndecan-1, and this inhibition hampered the proliferation of the cells [51].

We identified three genes that were significantly enhanced by the nuclear translocation of syndecan-1: EGR-1, NEK11, and DOCK8, suggesting that these genes are responsive to the nuclear translocation of syndecan-1. However, it remains to be determined whether they are direct targets or mediators of syndecan-1 effect in the nucleus.

EGR-1 is a transcription factor activated by a wide variety of extracellular stimuli and apoptotic signals. NEK11 is a DNA damage-response protein. Both proteins are localized in the nucleus and play multiple roles in the cell cycle. NEK11 kinase activity directly phosphorylates CDC25A; thus, it is required for DNA damage-induced G2/M arrest [52]. NEK11 in turn is dependent on the cell cycle; its highest expression occurs in the G2/M phase [53], but its activation through an association with Nek2A is enhanced in G1/S-arrested cells [54]. Similar to NEK11, EGR-1 is important in cell cycle progression: it regulates the G0/G1 transition [55], and it activates cyclin D2 [56]. Thus, it plays a role in the G1/S transition, and increases entry into the S/G2-phase. NEK11 is primarily associated with DNA

**Table 3** Categories of top enriched pathways in transcriptomic datasets according to Gene Set Enrichment Analysis (GSEA)

| Cell Cycle | SIZE | p-val | rank |
|---|---|---|---|
| REACTOME_G1_S_TRANSITION | 100 | 0.004 | 20 |
| REACTOME_MITOTIC_G1_G1_S_PHASES | 124 | 0.01 | 30 |
| REACTOME_M_G1_TRANSITION | 72 | 0.01 | 31 |
| REACTOME_S_PHASE | 100 | 0.011 | 37 |
| REACTOME_CELL_CYCLE_CHECKPOINTS | 105 | 0.015 | 43 |
| REACTOME_FGFR_LIGAND_BINDING_AND_ACTIVATION | 22 | 0.022 | 62 |
| REACTOME_CELL_CYCLE | 384 | 0.022 | 64 |
| REACTOME_CYCLIN_E_ASSOCIATED_EVENTS_DURING_G1_S_TRANSITION_ | 62 | 0.027 | 78 |
| REACTOME_G2_M_CHECKPOINTS | 35 | 0.036 | 96 |
| REACTOME_MITOTIC_M_M_G1_PHASES | 162 | 0.04 | 109 |
| REACTOME_CDK_MEDIATED_PHOSPHORYLATION_AND_REMOVAL_OF_CDC6 | 46 | 0.041 | 111 |
| REACTOME_P53_DEPENDENT_G1_DNA_DAMAGE_RESPONSE | 53 | 0.044 | 114 |
| BIOCARTA_EIF_PATHWAY | 16 | 0.046 | 117 |
| DNA synthesis and transcription | SIZE | p-val | rank |
| REACTOME_RNA_POL_I_PROMOTER_OPENING | 55 | ≤0.001 | 1 |
| REACTOME_SYNTHESIS_OF_DNA | 84 | 0.003 | 16 |
| REACTOME_RNA_POL_I_TRANSCRIPTION | 79 | 0.009 | 29 |
| REACTOME_DNA_STRAND_ELONGATION | 30 | 0.013 | 42 |
| KEGG_DNA_REPLICATION | 36 | 0.015 | 44 |
| REACTOME_DNA_REPLICATION | 182 | 0.023 | 67 |
| REACTOME_RNA_POLI_RNA_POLIII_AND_MITOCH._TRANSCRIPTION | 112 | 0.025 | 70 |
| REACTOME_ACTIVATION_OF_THE_PRE_REPLICATIVE_COMPLEX | 24 | 0.026 | 73 |
| REACTOME_ASSEMBLY_OF_THE_PRE_REPLICATIVE_COMPLEX | 57 | 0.026 | 75 |
| REACTOME_TRANSCRIPTION | 189 | 0.035 | 92 |
| Immune System | SIZE | p-val | rank |
| REACTOME_INTERFERON_SIGNALING | 151 | 0.017 | 47 |
| BIOCARTA_CYTOKINE_PATHWAY | 21 | 0.025 | 71 |
| BIOCARTA_TH1TH2_PATHWAY | 18 | 0.026 | 72 |
| REACTOME_DOWNSTREAM_SIGNALING_EVENTS_OF_B-CELL_RECEPTOR_BCR | 92 | 0.03 | 83 |
| BIOCARTA_NKCELLS_PATHWAY | 20 | 0.032 | 86 |
| REACTOME_ANTIVIRAL_MECHANISM_BY_IFN_STIMULATED_GENES | 64 | 0.034 | 90 |
| REACTOME_CYTOKINE_SIGNALING_IN_IMMUNE_SYSTEM | 260 | 0.037 | 99 |
| BIOCARTA_DC_PATHWAY | 22 | 0.044 | 115 |
| REACTOME_SIGNALING_BY_THE_B_CELL_RECEPTOR_BCR | 121 | 0.049 | 125 |

The GSEA analysis was performed on the transcriptome of NLSdel versus Fls1. The significantly enriched pathways belong to three categories: cell cycle regulation, DNA synthesis and transcription, and immune responses

replication and damage, stress-responses, and drug resistance [53, 57]; it is activated by DNA-damaging agents and DNA replication inhibitors [52, 53]. In our experimental setting, nuclear syndecan-1 activated NEK11, and thus, inhibited cell proliferation by causing cell cycle arrest.

In some tumor types, EGR-1 promotes growth and induces resistance to apoptosis [56]. In other tumor types, it can promote apoptosis [58] and significantly suppress tumor growth [59]. In HT1080 fibrosarcoma cells, tumor suppression is associated with inhibition of p53-dependent apoptosis [60]. EGR-1 regulates multiple tumor suppressors, in addition to p53, including TGF-β and PTEN [61]. Moreover, there is a complex relationship between TGF-β and EGR-1. In kidney [62] and colon cancer cells, EGR-1 induced TGF-β1 to suppress growth and tumorigenicity. In HT1080 fibrosarcoma cells, TGF-β induction was associated with increased adhesion [63].

**Table 4** Categories of top enriched pathways identified by Gene Set Enrichment Analysis (GSEA) of proteomic dataset

| Functions related to syndecan-1 | SIZE | p-val | rank |
|---|---|---|---|
| REACTOME_COLLAGEN_FORMATION | 28 | ≤0.001 | 1 |
| REACTOME_NCAM1_INTERACTIONS | 16 | ≤0.001 | 2 |
| REACTOME_EXTRACELLULAR_MATRIX_ORGANIZATION | 38 | ≤0.001 | 3 |
| KEGG_ECM_RECEPTOR_INTERACTION | 51 | ≤0.001 | 4 |
| REACTOME_KERATAN_SULFATE_KERATIN_METABOLISM | 16 | 0.002 | 10 |
| REACTOME_CHONDROITIN_SULFATE_DERMATAN_SULFATE_METABOLISM | 29 | 0.001 | 16 |
| REACTOME_INTEGRIN_CELL_SURFACE_INTERACTIONS | 47 | ≤0.001 | 17 |
| KEGG_CELL_ADHESION_MOLECULES_CAMS | 47 | ≤0.001 | |
| REACTOME_GLYCOSAMINOGLYCAN_METABOLISM | 57 | 0.001 | 28 |
| KEGG_FOCAL_ADHESION | 137 | ≤0.001 | 34 |
| REACTOME_TRANSMEMBRANE_TRANSPORT_OF_SMALL_MOLECULES | 154 | 0.001 | 41 |
| REACTOME_CELL_JUNCTION_ORGANIZATION | 36 | 0.032 | 42 |
| REACTOME_HEPARAN_SULFATE_HEPARIN_HS_GAG_METABOLISM | 27 | 0.045 | 56 |
| TGF β | SIZE | p-val | rank |
| REACTOME_DOWNREGULATION_OF_TGF_BETA_RECEPTOR_SIGNALING | 17 | 0.014 | 26 |
| REACTOME_TGF_BETA_RECEPTOR_SIGNALING_ACTIVATES_SMADS | 19 | 0.018 | 27 |
| KEGG_TGF_BETA_SIGNALING_PATHWAY | 44 | 0.019 | 45 |
| Immune System | SIZE | p-val | rank |
| KEGG_HEMATOPOIETIC_CELL_LINEAGE | 25 | 0.001 | 12 |
| REACTOME_INTERFERON_GAMMA_SIGNALING | 32 | 0.001 | 15 |
| REACTOME_IMMUNOREG_INTERACTIONS_BETWEEN_A_LYMPHOID_ AND_A_NON_LYMPHOID_CELL | 16 | 0.005 | 18 |
| BIOCARTA_IL1R_PATHWAY | 24 | 0.001 | 23 |
| REACTOME_IL1_SIGNALING | 27 | 0.014 | 33 |
| KEGG_NATURAL_KILLER_CELL_MEDIATED_CYTOTOXICITY | 58 | 0.023 | 63 |

The GSEA analysis was performed on the proteomic dataset of NLSdel versus Fls1. Most of the top enriched pathways belong to categories already associated with syndecan-1. In addition, functions related to TGF β signaling and immune regulation were significantly enriched

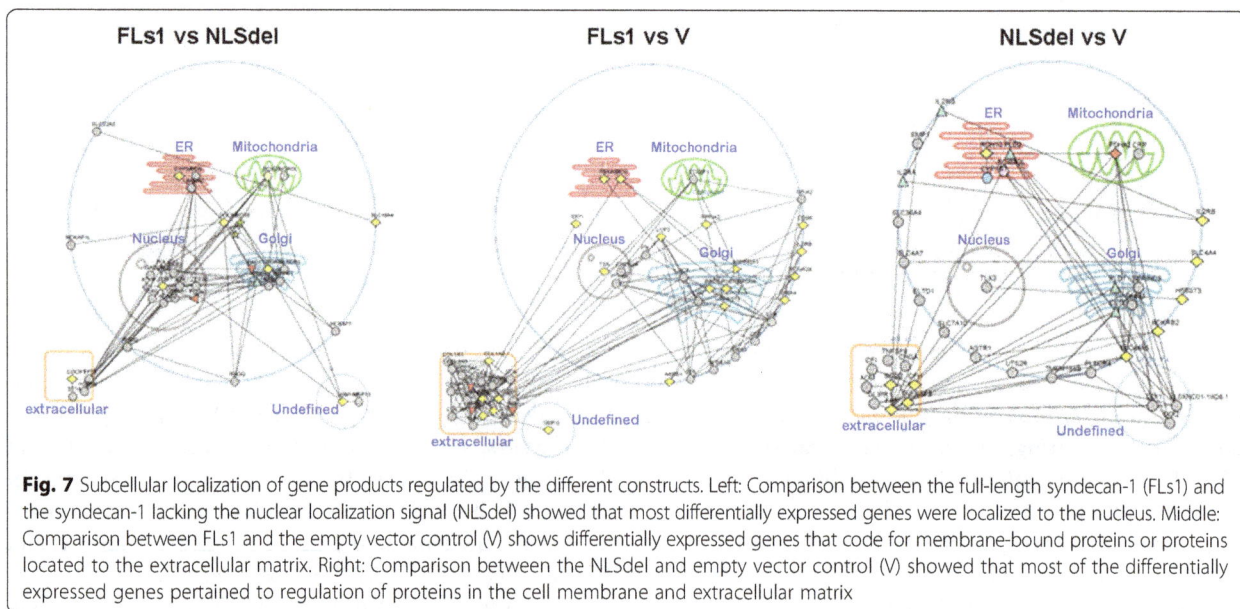

**Fig. 7** Subcellular localization of gene products regulated by the different constructs. Left: Comparison between the full-length syndecan-1 (FLs1) and the syndecan-1 lacking the nuclear localization signal (NLSdel) showed that most differentially expressed genes were localized to the nucleus. Middle: Comparison between FLs1 and the empty vector control (V) shows differentially expressed genes that code for membrane-bound proteins or proteins located to the extracellular matrix. Right: Comparison between the NLSdel and empty vector control (V) showed that most of the differentially expressed genes pertained to regulation of proteins in the cell membrane and extracellular matrix

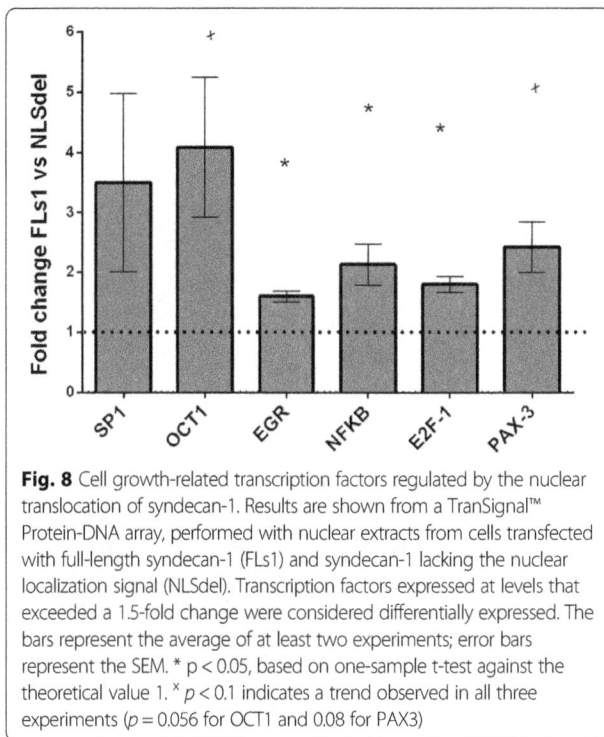

**Fig. 8** Cell growth-related transcription factors regulated by the nuclear translocation of syndecan-1. Results are shown from a TranSignal™ Protein-DNA array, performed with nuclear extracts from cells transfected with full-length syndecan-1 (FLs1) and syndecan-1 lacking the nuclear localization signal (NLSdel). Transcription factors expressed at levels that exceeded a 1.5-fold change were considered differentially expressed. The bars represent the average of at least two experiments; error bars represent the SEM. * $p < 0.05$, based on one-sample t-test against the theoretical value 1. ˣ $p < 0.1$ indicates a trend observed in all three experiments ($p = 0.056$ for OCT1 and 0.08 for PAX3)

Interestingly, in non-small lung cancer, EGR-1 counteracted the TGF-β-induced epithelial-to-mesenchymal transition [64]. In turn, EGR-1 was identified as a TGF-β target [65, 66]. In our study, nuclear syndecan-1 induced EGR-1 expression, which was associated with activating the TGF-β pathway and a slight inhibition of apoptosis. These features could be interconnected. The induction of EGR-1 expression by nuclear syndecan-1 was a very consistent result throughout our experiments. Our assessment of the activation of cell growth-related transcription factors showed that the magnitude of EGR-1 activation was in concordance to its upregulation at the RNA level, based on the Affymetrix array and the qPCR results.

DOCK8 is a member of a guanine nucleotide exchange factors family, involved in regulating cell morphology and intracellular signaling. It interacts with the Rho GTPase Cdc42 [67], and acts as a guanine nucleotide exchange factor [68]; thus, its deficiency might lead to impaired tumor immune surveillance. DOCK8 also participates in regulating tumor cell invasion [69] and metastatic processes [70]. Its expression was reduced in lung cancer [71] and altered in gliomas [72].

Our data also point toward several other transcription factors that were differentially regulated by nuclear syndecan-1. Some of these factors could directly activate the basic transcription machinery, like SP1 and E2F-1 [73]. SP1 induces apoptosis and inhibits cell cycle progression. E2F-1 is required for entry into the S1 phase [74]. Syndecan-1 is regulated by SP1, as the promoter of the syndecan-1 gene contains a SP1 binding site [75].

EGR-1 also interacts with other transcription factors and can compete with SP1, activate NFκβ and AP-1 [76], and in turn, it is activated by NFκβ and by E2F-1 [77].

Among these transcription factors, only NFκβ was previously associated with syndecan-1. One study showed reduced cellular NFκβ levels, when syndecan-1 was silenced [78]. NFκβ is considered a positive mediator of cell growth [79]. Its growth-promoting effects are typically associated with the inhibition of apoptosis. However, it was also demonstrated that environmental signals determine whether NFκβ induction leads to apoptosis or survival [80]. The dependency upon environmental factors is valid for most transcription factors as they may play context-dependent and dual roles in mediating cell growth and apoptosis. For example, E2F-1 knockout mice developed tumors [81], despite the fact that E2F-1 is crucial for progression through the S-phase. This result could be explained by the fact that E2F-1 is also important in apoptosis [82]. NFκβ stimulates proliferation in some environments [83], but leads to apoptosis in others [82]. OCT-1 could mediate growth arrest in some tissue types [84]; in other settings, low OCT-1 levels activated IFN-g and had pro-proliferative effects, but high OCT-1 levels had pro-apoptotic effects [85].

The effect of nuclear syndecan-1 seem to be consistent with these observations: by co-activating several transcription factors, nuclear syndecan-1 initiates a series of molecular events, which ultimately lead to the inhibition of both proliferation and apoptosis (Fig. 9). Our findings are consistent with recently described processes, where first, transcription factors activate proliferation-related genes at relatively low levels; then later, as the transcription factors accumulate, apoptosis-related genes are activated. Other factors, like histone-modifying genes or microRNAs could also affect the timing of this process. This theory is based on observations that the same transcription factors are involved in both proliferation and apoptosis [82, 86].

Although the deletion of the syndecan-1 nuclear localization signal caused only modest net changes in the transcriptome and proteome, the effects were considerable, bearing in mind that only a pentameric amino acid sequence was removed. All the elicited changes clustered in distinct patterns, which indicated the functional importance of nuclear syndecan-1. The overlap between the differentially regulated transcripts and proteins were limited, which may be related to the kinetics of nuclear translocation and transcription. When the cells were trypsinized and seeded, they lost all syndecan-1 proteins, in both the cell membrane and nucleus; however, at 48 h after seeding, this proteoglycan could be detected in the nucleus [7]. At this time point, some syndecan-1-regulated transcripts may not have completed translation. However, when we evaluated the

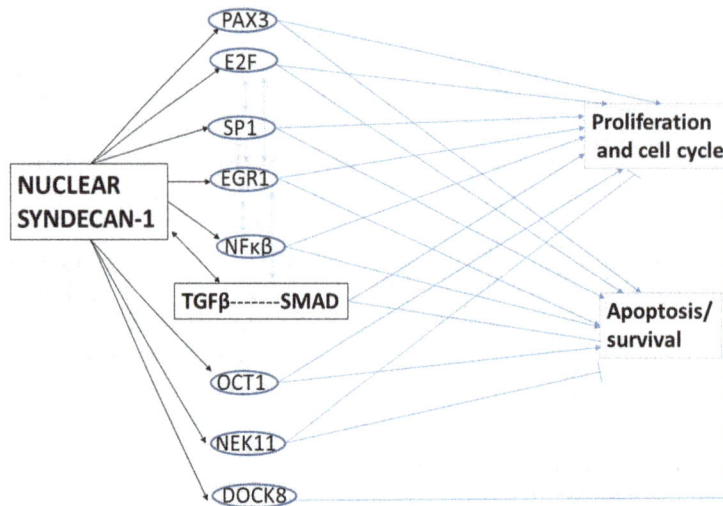

**Fig. 9** Regulatory network elicited by nuclear translocation of syndecan-1 leading to inhibition of proliferation and survival. Data are based on differentially expressed genes obtained by microarray and transcription factor array on fibrosarcoma cells with preserved and impaired nuclear translocation of syndecan-1. Nuclear syndecan-1 activates several transcription factors (ovals) and induces immediate early genes. Transcription factors activate (arrows) or inhibit (bars) their targets, and in addition, they trigger the TGF-β pathway (rectangle). In turn, TGF-β can also induce EGR-1 and other transcription factors, which can provide a feedback loop. The arrows in blue represent the current knowledge about the role of target genes in governing different processes. The functional outcome of nuclear syndecan-1 signaling is the measured inhibition of proliferation and cell survival

effects of the regulated transcripts and proteins in GO terms, we found an overlap in the biological processes affected by syndecan-1. These processes were dominated by phosphorylation and other post-translation protein modifications, which indicated alterations in intracellular signaling. Additionally, we found overlapping enrichments in several KEGG and Reactome pathways related to extracellular matrix organization, transmembrane transport, and endocytosis. These pathways played roles in functions previously associated with syndecan-1. Thus, the effects on the proteome were related to several known functions of syndecan-1. In contrast, the effects on transcription were linked to gene expression and cell cycle control.

With the IPA, we dissected the proteomic changes in more detail by linking significantly regulated proteins to certain *regulators*; then, by inferring literature-based directionality, we could determine whether these regulators were predicted to be activated or inhibited. We found a clear pattern of significantly regulated proteins that were related to an active TGF-β1/SMAD3/RAC1 axis. The connection between syndecan-1 and TGF-β1 was reported previously [49, 87, 88], and this pathway represented the only regulators significantly activated in our dataset. Importantly, the outcomes of the TGF- β-mediated signaling events were fine-tuned and highly dependent on the spatial distribution and the sub-cellular localization of various members of the signaling cascade. Independent studies have confirmed the inhibitory role of HS on TGF-β1 signaling; it facilitated lipid raft/caveolae-mediated endocytosis and rapid degradation [89].

## Conclusion

We showed that nuclear syndecan-1 inhibited proliferation and cell cycle progression in fibrosarcoma cells. The global characterization of the transcriptome and proteome related to nuclear syndecan-1 indicated that these effects were delicately regulated by multiple actors in related signaling pathways, where TGF-β1 seemed to play a central role. The nuclear ligands of syndecan-1 and the subsequent signaling pathways should be further elucidated to clarify our understanding of the importance of this HSPG in the nucleus. Our study results suggest that EGR1, NEK11, and several other transcription factors such as NFκβ and E2F-1 are syndecan-1 targets in the nucleus.

## Methods

### Cell characteristics and culture conditions

We used subtypes of a human fibrosarcoma cell line (B6FS) [90] that had low endogenous syndecan-1 levels, transfected with three different constructs: 1.) a plasmid carrying the full-length syndecan-1 gene (FLs1); 2.) the same plasmid carrying syndecan-1, but lacking the RMKKK nuclear localization signal (NLSdel), and 3.) the empty vector (V) as a control (Fig. 10). For detailed description of these plasmids and cell transfection see [25]. The stably transfected cells were cultured under selective pressure with Geneticin (G418, Roche Diagnostics GmbH, Mannheim, Germany). Previously, we showed that after transfection of FLs1, syndecan-1 was detected in the nucleus, whereas the nuclear translocation of

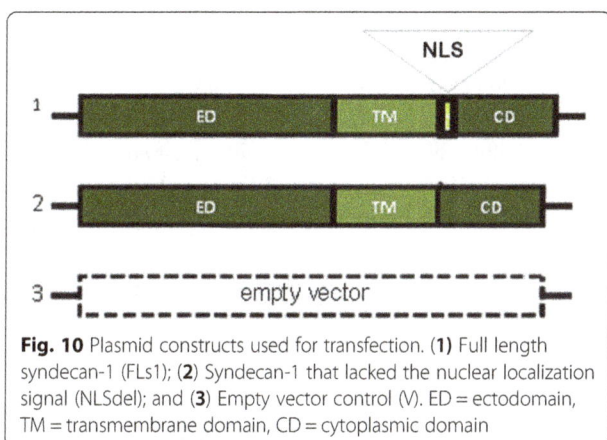

**Fig. 10** Plasmid constructs used for transfection. (**1**) Full length syndecan-1 (FLs1); (**2**) Syndecan-1 that lacked the nuclear localization signal (NLSdel); and (**3**) Empty vector control (V). ED = ectodomain, TM = transmembrane domain, CD = cytoplasmic domain

syndecan-1 was hampered in NLSdel [25]. We cultured these cells in RPMI 1640-GlutamaxTM-I medium (72,400, Gibco) supplemented with 10% fetal bovine serum (FBS), under standard incubation conditions, in humidified 5% ($v$/v) CO2 at 37 °C.

We carefully controlled the experimental conditions to obtain similar levels of syndecan-1 expression throughout the experiments. We regularly verified the syndecan-1 levels by fluorescence activated cell sorting (FACS) and western blotting prior to RNA extractions and mass spectrometry. This verification ensured that the differences detected were related to the presence or lack of nuclear syndecan-1 and not to differences in syndecan-1 expression levels.

## Fluorescence activated cell sorting (FACS)
For FACS analyses, cells were detached with enzyme-free Cell Dissociation Buffer (Gibco, 13,151–014) for 15 min, and when necessary, cells were scraped from the plate. Cells were collected, counted, and fixed in 2% buffered formaldehyde. We then incubated the cells with antibodies against the ectodomain of syndecan-1 (MCA658) for 15 min at 4 °C. After washing, cells were stained with Alexa 488-conjugated goat anti-mouse secondary antibody (Molecular Probes, A-11001) for 15 min at room temperature (RT), in the dark. Subsequent experiments were performed when the syndecan-1 levels in both FLs1 and NLSdel cells ranged between 1.5- and 2-fold above the levels in control cells.

## Western blotting
For western blotting sub-confluent cells were dissociated with 0.5% Trypsin-EDTA and washed twice with PBS. Lysis was achieved by incubation for 15 min in buffer containing: 50 mM Tris-HCl, pH 8.0, with 150 mM sodium chloride, 1.0% Igepal CA-630 (NP-40), 0.5% sodium deoxycholate, 0.1% sodium dodecyl sulfate and protease inhibitor (Thermo scientific). Lysed cells were spun at 16000 x g for 5 min at 4 °C, and the supernatant was

collected and mixed with 2X Laemmli loading buffer with 2-mercaptoethanol. Samples were separated by SDS-polyacrylamide gel electrophoresis. Transfer was performed to a PVDF membrane using the trans-blot turbo transfer system (Bio-Rad). The membrane was blocked for 1 h in 0.5% milk and incubated overnight at 4 °C with primary antibodies for Syndecan-1 (C-20) (Santa Cruz Biotechnology cat.nr. SC-7099) diluted 1:200, and monoclonal Anti-Actin (Clone AC-40) (Sigma-Aldrich, cat.nr. A4700) diluted 1:500. Following washes the membrane was incubated with secondary antibodies (Rabbit Anti-Goat IgG, F (ab')$_2$ Fragment Specific, Peroxidase Conjugated (Thermo scientific) and ECL™ Anti-mouse IgG, Horseradish peroxidase linked F(ab')$_2$ fragment (from sheep) (GE Healthcare) at 1:5000 dilution for 1 h at room temperature. For chemiluminescent detection, chemiluminescent HRP Substrate (Advansta, cat.nr. K-12043-D10) was added and the membrane was incubated for 1 min. The Odyssey Imaging System (LI-COR) was used to develop the membrane and the relative expression of syndecan-1 was normalized to actin as loading control using the ImageJ software.

## Immunocytochemical staining and subcellular localization of the newly synthesized syndecan-1
The subcellular localization and the level of syndecan-1 was further verified using immunocytochemical analysis and subsequent fluorescent microscopy, as described previously 25]. Cells were seeded on to POLYSINE coated microscopy slides (Menzel-Gläser, Braunschweig, Germany). After 48 h cells were fixed in 3% paraformaldehyde followed by permeabilization with 0.1% Triton X-100 (Sigma, Steinheim, Germany); non-specific binding was blocked with 3% goat serum (Dako A/S, Glostrup, Denmark) for 30 min. Mouse anti Human CD138 monoclonal antibody (MCA-681) Serotec LTD, Kidlington, Oxford, England) and mouse IgG1 (Dako A/S, Glostrup, Denmark)as negative control was used to stain syndecan-1, followed by incubation with Alexa 488 goat anti-mouse F(ab')2 fragment of IgG (H + L), (Molecular Probes, Leiden, The Netherlands, A11017). Samples were then counterstained with 1 mg/L bisbenzimide H33342 (Fluka, Steinheim, Germany). Detailed visualization was performed using Nikon microphot-FXA EPI-FL3 fluorescence microscope.

## Cell proliferation assay
Different densities of FLs1 and NLSdel cells (2000, 3000, or 4000 cells/well) were seeded on 96-well plates. Cell proliferation was measured with the Cell Proliferation Reagent, WST-1 (Roche Diagnostics Scandinavia AB, Bromma, Sweden) at different time points, according to the manufacturer's instructions. Briefly, cells were incubated with 1/10 ($v$/v) WST1 reagent for 2 h at 37 °C. Samples were analyzed with a Spectramax spectrophotometer at 450 nm with background subtraction at

630 nm. Three independent experiments were performed, each in triplicate. The paired Student's t test was applied to determine statistical significance, with GraphPad Prism software. Doubling time was calculated from the logarithmic phase of the growth curve [91].

## Immunocytochemical detection of Ki-67

As an additional measure of proliferation, we used immunocytochemistry to label proliferating cells with the proliferation marker Ki-67. For this purpose, cytospin preparations of cells transfected with the three different constructs were done on SuperFrost Plus glass slides (Thermo Fisher Scientific Inc., Waltham, MA, USA). Cells were fixed in H2O with 25% ethanol, 25% methanol and 3% polyethylene glycol (PEG). Prior to staining, PEG was extracted by decreasing concentrations of ethanol in H2O. For epitope retrieval slides were incubated at 100 °C for 5 min in a citrate buffer pH 6.0 (Bond Epitope Retrieval Solution 1, Leica Microsystems GmbH). Endogenous peroxidase activity was abolished with 3% hydrogen peroxide in H2O. Slides were then incubated for 30 min with primary antibody (Dako M7240) diluted 1:200 in BOND Primary Antibody Diluent (Leica Microsystems GmbH). Secondary IgG was added and incubated for 15 min and detected with the Bond Polymer Refine Detection kit (Leica Microsystems GmbH); as described in the manufacturer's protocol. Following 15 min incubation with a poly-HRP, bound antibodies were visualized by Diaminobenzidine. Cell nuclei were counterstain with hematoxylin. The immunostaining was performed in a Leica BOND-III automated IHC with relevant controls.

For each cell line, random microscopic fields have been photo documented and evaluated. For each construct, at least 200 cells were counted. The presence or absence of nuclear reactivity to Ki67 was recorded and the percentage of Ki-67 positive cells was related to the total number of cells.

## Cell cycle analysis

FLs1and NLSdel cells were grown for 48 h; then, they were harvested, fixed in 1 mL of 70% cold ethanol, and incubated at 4 °C overnight. Cells were washed in PBS, resuspended in 500 μL staining solution containing 50 μg/mL propidium iodide (Sigma-Aldrich, MO, USA) and 100 μg/mL RNAse A (Sigma-Aldrich, MO, USA), and incubated for 30 min at 37 °C. The cell cycle distribution was measured for 10,000 cells in each sample with a FACSCalibur cytometer (Becton Dickinson, CA, USA). Results were analyzed with ModFit LT software (Verity Software House, ME, USA). Statistical significance was tested with the paired t test in GraphPad software.

## Measurement of spontaneous apoptosis

We detected apoptosis by performing FACS analysis with the FITC Annexin V Apoptosis detection kit (BD Pharmingen), according to the manufacturer's instructions. Briefly, cells were trypsinized, washed with PBS, and resuspended in Binding Buffer with Annexin V-FITC and propidium iodide. Then, cells were incubated for 15 min in the dark, followed by FACS analysis. The apoptosis was measured at 48 and 72 h after cell seeding. Three independent experiments were performed for both time points. Statistical significance was assessed with the paired t test, in GraphPad software.

## Transcriptomic and proteomic data generation
### RNA isolation

At 48 h after seeding, we isolated total RNA from fibrosarcoma cells transfected with FLs1, NLSdel, or control V with the High Pure RNA Isolation Kit (Roche Diagnostics GmbH Mannheim, Germany), in accordance to the manufacturer's protocol. Three biological replicates were used for each construct. The yield and purity of the RNA were determined by measuring the UV absorbance at 260 and 280 nm with a NanoDrop spectrophotometer (NanoDrop Technologies Inc.).

### Affymetrix gene expression array

To disclose the molecular mechanisms underlying syndecan-1 nuclear translocation, we performed microarray analysis on cells that overexpressed FLs1 and NLSdel at similar levels. RNAs isolated from the cells were subjected to microarray analysis with the GeneChip Human Gene 1.1 ST Array (Affymetrix Inc., Santa Clara, CA, USA), which covered the whole transcript. Target synthesis and hybridization was performed in the Affymetrix core facility (Novum, Karolinska Institutet, Stockholm, Sweden). The raw data has been deposited in the MIAME compliant database Gene Expression Omnibus (accession number GSE81504). Image analysis and data pre-processing was performed with the Affymetrix Gene Chip Command Console. For data processing, we performed background correction with the PM-GCBG method (subtracting the GC-content specific background); data normalization with the Global Median method; and raw intensity value summarizations with PLIER (Probe Logarithmic Intensity Error). For each sample, the analysis generated a signal that represented the relative measure of transcript abundance. Individual signals that exceeded a value of 10 were considered for further analysis.

### Preparation of cells for mass spectrometry-based proteomics

Cells were grown in 75 cm$^2$ culture dishes for 48 h in triplicate. Then, cells were lysed with 4% SDS, 25 mM HEPES, and 1 mM DTT, on ice. Cell lysates were heated to 95 °C for 5 min, followed by 1 min sonication, and 15 min centrifugation at 14,000 g. Proteins were reduced, alkylated, and digested to peptides according to an adapted FASP protocol [92]. Individual samples were labeled with TMT-10plex isobaric labels (Thermo Fischer Scientific, San Jose, CA, USA), according to the manufacturer's instructions. Briefly, 80 μg of peptides from each sample was combined with a designated TMT reagent, and labeling was performed at room temperature for 3 h. Labeling controls were preformed to guarantee >99% labeling of primary amines. Then, samples were combined (i.e., a total of 800 μg) and cleaned on a SCX column (Phenomenex, Torrance, CA, USA).

### High resolution isoelectric focusing

We used isoelectric focusing to fractionate our TMT-10plexes, and thereby reduce the complexity of the proteome. Specifically, we applied the recently developed, high resolution, isoelectric focusing method (HiRIEF) [93]), with an immobilized pH gradient of 3.7 to 4.9 (kindly provided by GE healthcare, Uppsala, Sweden). The TMT pooled sample (390 μg) was applied to the HiRIEF strip and run on an Ettan IPGphor (GE Healthcare) until at least 100 kVh had been reached (around 24 h). The fractionated sample was extracted from the gel strip in an automated manner, to yield 72 individual fractions. These fractions were then injected separately on a Q Exactive mass spectrometer (see section 2.6.5). This procedure was previously described in more detail [93].

### nanoLC-MS/MS analysis

Peptides were separated with an online 3000 RSLCnano system. Samples were trapped on an Acclaim PepMap nanotrap column (C18, 3 μm, 100 Å, 75 μm × 20 mm), and separated on an Acclaim PepMap RSLC column (C18, 2 μm, 100 Å, 75 μm × 50 cm; Thermo scientific). Next, HiRIEF-fractionated peptides were separated on a gradient of A (5% DMSO, 0.1% Formic acid; FA) combined with B (90% Acetonitrile; ACN, 5% DMSO, 0.1% FA), where B ranged from 3% to 37%. Samples were run for 50 min at a flowrate of 0.25 μL/min. The Q Exactive instrument (Thermo Fischer Scientific, San Jose, CA, USA) was operated in a data-dependent manner, where the top 5 precursors were selected for HCD fragmentation and MS/MS. The survey scan was performed at 70,000 resolution over a range of 300–1600 $m/z$, with a maximum injection time of 100 ms and target of $1 \times 10^6$ ions. HCD fragmentation spectra were generated with a maximum ion injection time of 150 ms and an AGC of $1 \times 10^5$. Then, fragmentation was performed at 30% normalized collision energy, with 35,000 resolution. Precursors were isolated with a width of 2 $m/z$ and placed on the exclusion list for 70 s. For 4-h gradients, we used a top 10 method, with a survey scan over the range of 400–1600 $m/z$ and a maximum injection of

140 ms. Single and unassigned charge states were rejected from precursor selection.

## Data analysis and bioinformatics
### Affymetrix data analysis
We performed a differential gene expression analysis, based on Affymetrix data, with the OCplus package provided in R software *(http://www.R-project.org/)*. [94] We conducted three pairwise comparisons, including FLs1 versus NLSdel, FLs1 versus V, and NLSdel versus V. We compared signals between samples with paired t-tests. The p-values were converted to false discovery rates (q-values) with a multiple-testing correction. A threshold of q ≤ 0.05 was applied, and differentially expressed genes were ranked by the fold-change (i. e., the ratio of expression values between a sample and a control). Thus, a syndecan-1 modulated sample was compared to its corresponding control. A transcript was considered significantly up- or down-regulated, when the fold change exceeded |1.5|. Probeset IDs were converted to HUGO gene symbols to denote the genes. We performed network enrichment analysis with Funcoup 3.0 network of functional coupling (http://funcoup.sbc.su.se) [95], in two different ways. In the first approach, we applied the functional analysis on the previously established, differentially expressed genes, for each pair of data. This method was suited to disclosing the possible involvement of differentially expressed genes in various cellular functions and to map their distribution to different cellular compartments. In the second approach, in addition to a differential analysis of the fold-change, we performed a global network analysis of functional coupling to reveal the involvement of genes with specific biological functions, which were apparent when syndecan-1 was overexpressed with or without the NLS. This approach allowed investigation of functional relationships between differentially expressed genes, particularly when summarizing small changes in many related genes. It also highlighted differentially expressed genes that might be direct binding partners of syndecan-1, based on currently available data from the literature available in the curated resources, Gene Ontology (GO), Reactome, and KEGG.

### Peptide identification, protein identification, and data analysis
We used the Proteome discoverer 1.4 with Sequest HT and percolator search algorithms to construct the proteome. The precursor mass tolerance was set to 10 ppm, and to 0.02 Da for fragments. We set oxidized methionine as a dynamic modification, and we set carbamidomethylation of cysteines, TMT10 on the N-terminus, and lysines as fixed modifications. Spectra were matched to the Uniprot human

database (downloaded 20,140,203), limited to a positive false-discovery rate (FDR) of 1%. The FDR was determined by searching against a decoy database of similar size with reversed sequences. All TMT10 quantifications were median-centered for each sample. FLs1 and NLSdel samples were normalized to the empty control V. A moderated t-test was performed to determine the number of proteins that were significantly changed between FLs1 and NLSdel samples. P-values were adjusted with the Benjamini-Hochberg correction (q-values). The moderated t-test was performed in the R software environment (version 3.1.2). As a quality control, samples were clustered (one minus the Pearson coefficient) with the Gene-E software platform (http://software.broadinstitute.org/morpheus/) [96]. The vast majority of the proteome remained unperturbed. This feature made clustering of samples on the global scale very susceptible to background fluctuations (technical or biological), and two of the replicates did not cluster together with their respective groups. Therefore, we decided to exclude these two samples from the final analyses and proceeded with the remaining four samples for subsequent analyses.

### Bioinformatic analyses of the proteome
Proteins with fold-changes that exceeded |1.5| and with adjusted $p$-values (two-sample moderated t-test) < 0.05 were considered for analysis with Ingenuity Pathway Analysis (IPA, version 23,814,503; QIAGEN). Only findings with experimental observations in human cell lines or tissues were considered. Upstream and downstream *(disease and function)* analyses were performed with data from the IPA Knowledge Data-Base, which predicted the activation or inhibition of regulators or downstream biological effects [97]. These predictions were reported, and considered significant, when they had a Z-score > 1.9 for activation and <1.9 for inhibition. A Fisher's exact p-value ≤0.05 was taken to indicate a significant overlap with upstream regulation or downstream biological effects. The upstream and downstream effects were discerned from the pattern of identified proteins, and the degree of consistency between the observed levels and those reported in the published scientific literature. Furthermore, we analyzed differentially regulated transcripts and proteins with Funcoup 3.0 to assess the overlap between findings in transcriptome and proteome spaces, based on the GO terms. These analyses were based on genes and proteins that were differentially expressed between FLs1 and NLSdel samples, with a fold-change that exceeded |1.5| and a q-value <0.05. Additionally, we performed a gene set enrichment analysis (GSEA; http://software.broadinstitute.org/gsea/index.jsp) [98] with a pre-ranked test, where all gene names were ranked by their adjusted $p$-values ($-\log_{10}$ transformed) from two-sample, moderated t-tests,

between FLs1 and NLSdel. GSEA analyses were conducted separately for the transcriptome and proteome, with the KEGG, BioCarta, and Reactome databases.

## Validation and functional assays
### RT-qPCR
We validated the Affymetrix results with real-time quantitative polymerase chain reaction (RT-qPCR) assays. cDNA synthesis was performed by reverse transcribing 2 mg RNA with a First-Strand cDNA Synthesis Kit (Amersham Pharmacia Biotech., Little Chalfont, Buckinghamshire, England). We used the same RNAs that were used for the Affymetrix analysis. We performed RT-qPCR with the Platinum SybrGreen qPCR SuperMix-UDG kit (Invitrogen) and DNA-polymerase, with a set of sense/antisense primers (CyberGene AB, Sweden).

The primers were designed based on gene sequences from GeneBank (NCBI), with the exception of GAPDH

**Table 5** Primer sequences used for RT-PCR validation

| GENE | PRIMERS (5'TO 3'ORIENTATION): FORWARD/REVERSE |
|---|---|
| GAPDH | ACATCATCCCTGCCTCTACTGG/ AGTGGGTGTCGCTGTTGAAGTC [99] |
| SDC1 | TCTGACAACTTCTCCGGCTC/CCACTTCTGGCAGGACTACA [100] |
| DOCK8 | AGTGCCGAGGACTTTGAGAA/ ATTCTGTTGCCCAGGTGTTC |
| EGR1 | TGACCGCAGAGTCTTTTCCT/ TGGGTTGGTCATGCTCACTA |
| NEK11 | AGAGGATGCCACATCTGACC/ GAAGTGCAACCCAGGACATT |
| ZNF676 | CTGGTCTTCCTGGGTATTGC/ TTGCTCTGGCCAAAACTCTT |
| CA9 | TAAGCAGCTCCACACCCTCT/ TCTCATCTGCACAAGGAACG |
| COL19A1 | GTGGTTTCTGTGGCAGGTTT/AGTCTGCCTCCTCGCAATTA |
| DACH1 | GTGGAAAACACCCCTCAGAA/ CTTGTTCCACATTGCACACC |
| EGR2 | CCTCCTTATTCTGGCTGTGC// CTGGGATCATTGGGAAGAGA |
| FAP | CTTGTCCTGGCTTCAGCTTC/ AGGTGGCAACTCCAAATACG |
| HS6ST3 | GGCTCACTGAGTTCCAGAGG/ TCTAGCTGCTTGGTGTGGTG |
| IL2RB | GCTGATCAACTGCAGGAACA/ TGTCCCTCTCCAGCACTTCT |
| PIP5KIB | CCAGGAATGGAAGGATGAGA/ AATTGTGGTTGCCAAGGAAG |
| SERPINA3 | CCAACGTGGACTTCGCTTTC/CTCTTGGCATCCTCCGTGAA |
| SERPINB4 | TCAGTGAAGCCAACACCAAG/ TGTTGCAGCTTTTTCTGTGG |
| TNRFS9 | CACTCTGTTGCTGGTCCTCA/ CACAGGTCCTTTGTCCACCT |
| VCAM1 | CAGACAGGAAGTCCCTGGAA/ TTCTTGCAGCTTTGTGGATG |
| ADAMTS5 | CCCAGCCTGGACACATTACT/ TTCCCCTGAGCATTTTCAC |
| AREG | TGGATTGGACCTCAATGACA/ AGCCAGGTATTTGTGGTTCG |
| CDK20 | ATGGCTAAGGTGGCATTGTC/ CGCTCATCCTGAGGGAGTAG |
| COL1A2 | CCTGGTAATCCTGGAGCAAA/ TTACCGCTCTCTCCTTTGGA |
| CXCL1 | AGGGAATTCACCCCAAGAAC/ CACCAGTGAGCTTCCTCCTC |
| ITGA8 | CACATTCTGGTGGACTGTGG/ AATCCCTTGTTGTTGCGTTC |
| MMP10 | GGCTCTTTCACTCAGCCAAC/ GGCTCTTTCACTCAGCCAAC |
| PCDH18 | AGCATCTGCAGCTTTTCCAT/ AGGGAATTTTCCCCAACATC |
| SULT1B1 | GGTTATCCCATGACCTGTGC/CCAGGGAGAGTCATTTCCAA |

[99] and syndecan-1 [100]. The primer sequences are shown in Table 5. All PCR reactions were performed with an iCycler machine (CFX96TM Real Time PCR Detection System, BioRAD Hercules, CA, USA), in triplicate, with a total volume of 10 µL/well, and a primer concentration of 200 nM. We performed the analyses with Bio-Rad CFX Manager Software 2.0 (BioRad Laboratories 2008). Data were analyzed with the $2^{-\Delta\Delta Ct}$ method. Each target was normalized to GAPDH, as the reference gene, and the fold-change in expression was measured for each target with respect to the corresponding controls. The data are expressed as the mean of at least three independent experiments.

### Nuclear extraction and transcription-factor array analysis

We prepared nuclear extracts from FLs1 or NLSdel cells, which contained activated transcription factors related to cell proliferation. Extracts were prepared with the Active Motif nuclear extraction kit (Rixensart, Belgium, cat. no. 40010). Cells were collected and resuspended in hypotonic buffer, which contained detergents. The cytoplasmic fraction was removed, and cell nuclei were lysed and solubilized in a lysis buffer, which contained protease inhibitors and 10 mM DTT. The protein concentrations in nuclear extracts were measured with the bicinchoninic acid (BCA) assay (Thermo Scientific, IL, USA, cat. no. 23225) at an optical density of 562 nm. The activity of cell growth-related transcription factors was profiled with TranSignal™ Cell Growth Protein/DNA Arrays (Affymetrix Inc., Panomics). 3 µg of nuclear extracts were preincubated with a set of biotin-labeled DNA binding oligonucleotides (TranSignal Probe Mix) to allow the formation of DNA/protein complexes; then, the protein/DNA complexes were separated from the free probes with spin column separation. The probes in the complexes were extracted and hybridized to the TranSignal Array membrane in an overnight incubation at 42 °C. The array was spotted (in duplicate, and at two dilutions) with consensus sequences that corresponded to 20 different transcription factors, which were known key players in cell growth and differentiation. We detected the hybridized signals with HRP-based chemiluminescence detection. The membranes were exposed to a chemiluminescence imaging system (FluorChem™ SP, Alpha Innotech, USA) for 5–10 min. Different signals corresponded to differently activated transcription factors from the nuclear extracts. Results were quantified with the ImageJ 1.47, open-source image analysis program. We calculated the ratio of data collected from FLs1 cells versus those collected from NLSdel cells. Three independent experiments were performed. The threshold for significance was a 1.5-fold change for each experiment.

## Additional files

Additional file 1: Figure S1. Syndecan-1 protein level following transfection with the full-length syndecan-1 (FLs1), nuclear localization signal deleted syndecan-1 (NLSdel) and empty vector control (EV). (a) Representative histogram of syndecan-1 protein level detected by Fluorescence Activated Cell Sorting (FACS) analysis. Dotted line represents the IgG control, green line corresponds to empty vector and the blue and red line to the full-length syndecan-1 (FLs1) and nuclear localization signal deleted syndecan-1 (NLSdel), respectively. (b) Quantitative syndecan-1 protein level by FACS analysis corresponding to three independent experiments. Error bars represent standard error of the mean (SEM). * denotes statistically significant differences. (c) Relative syndecan-1 levels measured by western blotting, using actin as loading control. (JPEG 175 kb)

Additional file 2: Figure S2. Immunocytochemical staining and subcellular localization of the newly synthesized syndecan-1. Panels (a, d and g) represent empty vector, (b, e and g) represent the nuclear localization signal deleted syndecan-1 (NLSdel) and panels (c, f and i) represent full-length syndecan-1 (FLs1) transfected cells. Green staining (a-c) shows syndecan-1, blue color shows (d-f) nuclear staining (Bisbenzimide H33342). Panels (g-i) show overlay of syndecan-1 and the nuclear staining. Immunoreactivity for syndecan-1 is observed mainly in the cell membrane and cytoplasm. In FLs1 syndecan-1 is localized also in the cell nucleus. The amount of total syndecan-1 is lower in empty vector than in the other two constructs. (TIFF 38 kb)

Additional file 3: Figure S3. Ki-67 proliferation index of the full length syndecan-1 (FLs1); nuclear localization signal deleted syndecan-1 (NLSdel); and Empty vector control (EV). Black bars represent the proportion of Ki-67 positive cells at 48 and gray bars at 72 h, respectively. (TIFF 624 kb)

Additional file 4: Figure S4. (a) At the level of the global proteome, the amplitudes of changes are small; less than 0.5% of the proteins showed >1.5-fold changes in regulation for each replicate. (b) Clustering of one minus the Pearson coefficient, in both columns (samples/replicates) and rows (proteins), shows that two of the replicates had patterns distinct from their respective groups (FL rep3 and NLSdel rep1). However, common features can be discerned between the remaining samples in the groups. (TIFF 523 kb)

Additional file 5: Figure S5. Moderated F-test results show proteins that are significantly regulated (Benjamini-Hochberg corrected p-value <0.05; red dots) between the full-length syndecan-1 group (FL) and the group with a syndecan-1 that lacked the nuclear localization signal (NLSdel). Numbers represent Pearson r correlations. The replicates, FL rep3 and NLSdel rep1, show discrepancies in protein expression. However, the other samples show good correlations (r > 0.50). (JPEG 748 kb)

Additional file 6: File S1. Differentially regulated proteins between the FLs1 and the NLSdel groups Sheet: "All samples" regards the two sample moderated t-test using all samples in each group. Sheet: "2 vs 2 samples" regards the same two sample moderated t-test analysis but excluding two replicates with low Pearson correlation. id = Uniprot accession numbers (XLSX 1981 kb)

Additional file 7: File S2. List of pathways enriched following nuclear translocation of syndecan-1, identified by GSEA analysis The GSEA analysis performed on the transcriptomic dataset of NLSdel versus Fls1 (sheet "mRNA") and the proteome dataset (sheet "Proteome") identified several enriched pathways. (XLSX 141 kb)

### Abbreviations

DOCK8: dedicator of cytokinesis 8; EGR-1: early growth response 1; FLs1: cells transfected with full-length syndecan-1; GO: Gene ontology; HS: heparan sulfate; HSPG: heparan sulfate proteoglycan; NEK11: never-in-mitosis gene a-related kinase 11; NLS: nuclear localization signal; NLSdel: cells transfected with syndecan-1 lacking the nuclear localization signal; V: cells transfected with empty vector

### Acknowledgements

The authors thank Prof. Janne Lehtiö and Henrik J. Johansson, Department of Oncology-Pathology, Cancer Proteomics, Mass spectrometry, Science for Life Laboratory, Karolinska Institutet, for assistance with the proteomic analysis; Andrey Alexeyenko for his help with the Affymetrix data analysis, Åsa-Lena

Dackland for skillful assistance with the FACS analyses and San Francisco Edit for language review.

## Funding
This work was supported by AFA Insurance, the Swedish Research Council (K2012-99X-21,999-01-3), and the Swedish Cancer Foundation (130491). The funding body did not have any role in the design of the study and collection, analysis, and interpretation of data and in writing the manuscript.

## Authors' contributions
TS contributed to conceiving and design of the experiments, performed the microarray experiments, analysed the data and wrote the paper. FM contributed to conceiving and design of the experiments, performed and analysed the proteomic data and contributed to paper writing. A K_S performed experiments, analysed the data, revised the paper. LM performed experiments and revised the paper. RO performed experiments and revised the paper. AH contributed to conceiving and design of the experiments and revised the paper. KD conceived the study and participated in overall experimental design, data interpretation, presentation, project administration, wrote and revised the paper. All authors have read and approved the final version of the manuscript.

## Competing interests
The authors declare that they have no competing interests.

## Author details
[1]Department of Laboratory Medicine, Division of Pathology, Karolinska Institutet, SE-14186 Stockholm, Sweden. [2]Division of Clinical Pathology/Cytology, Karolinska University Laboratory, Karolinska University Hospital, SE-14186 Stockholm, Sweden.

## References
1. Bernfield M, Gotte M, Park PW, Reizes O, Fitzgerald ML, Lincecum J, Zako M. Functions of cell surface heparan sulfate proteoglycans. Annu Rev Biochem. 1999;68:729–77.
2. Szatmari T, Otvos R, Hjerpe A, Dobra K. Syndecan-1 in cancer: implications for cell signaling, differentiation, and prognostication. Dis Markers. 2015; 2015:796052.
3. Yang N, Mosher R, Seo S, Beebe D, Friedl A. Syndecan-1 in breast cancer stroma fibroblasts regulates extracellular matrix fiber organization and carcinoma cell motility. Am J Pathol. 2011;178(1):325–35.
4. Yanagishita M, Hascall VC. Cell surface heparan sulfate proteoglycans. J Biol Chem. 1992;267(14):9451–4.
5. Seidel C, Sundan A, Hjorth M, Turesson I, Dahl IM, Abildgaard N, Waage A, Borset M. Serum syndecan-1: a new independent prognostic marker in multiple myeloma. Blood. 2000;95(2):388–92.
6. Mundt F, Heidari-Hamedani G, Nilsonne G, Metintas M, Hjerpe A, Dobra K. Diagnostic and prognostic value of soluble syndecan-1 in pleural malignancies. Biomed Res Int. 2014;2014:419853.
7. Brockstedt U, Dobra K, Nurminen M, Hjerpe A. Immunoreactivity to cell surface syndecans in cytoplasm and nucleus: tubulin-dependent rearrangements. Exp Cell Res. 2002;274(2):235–45.
8. Zong F, Fthenou E, Wolmer N, Hollosi P, Kovalszky I, Szilak L, Mogler C, Nilsonne G, Tzanakakis G, Dobra K. Syndecan-1 and FGF-2, but not FGF receptor-1, share a common transport route and co-localize with heparanase in the nuclei of mesenchymal tumor cells. PLoS One. 2009;4(10):e7346.
9. Chen L, Sanderson RD. Heparanase regulates levels of syndecan-1 in the nucleus. PLoS One. 2009;4(3):e4947.
10. Stewart MD, Ramani VC, Sanderson RD. Shed syndecan-1 translocates to the nucleus of cells delivering growth factors and inhibiting histone acetylation: a novel mechanism of tumor-host cross-talk. J Biol Chem. 2015;290(2):941–9.
11. Kovalszky I, Hjerpe A, Dobra K. Nuclear translocation of heparan sulfate proteoglycans and their functional significance. Biochim Biophys Acta. 2014;1840(8):2491–7.
12. Iozzo RV, Schaefer L. Proteoglycan form and function: a comprehensive nomenclature of proteoglycans. Matrix Biol. 2015;42:11–55.
13. Stewart MD, Sanderson RD. Heparan sulfate in the nucleus and its control of cellular functions. Matrix Biol. 2014;35:56–9.
14. Bhavanandan VP, Davidson EA. Mucopolysaccharides associated with nuclei of cultured mammalian cells. Proc Natl Acad Sci U S A. 1975;72(6):2032–6.
15. Margolis RK, Crockett CP, Kiang WL, Margolis RU. Glycosaminoglycans and glycoproteins associated with rat brain nuclei. Biochim Biophys Acta. 1976;451(2):465–9.
16. Richardson TP, Trinkaus-Randall V, Nugent MA. Regulation of heparan sulfate proteoglycan nuclear localization by fibronectin. J Cell Sci. 2001;114(Pt 9):1613–23.
17. Leadbeater WE, Gonzalez AM, Logaras N, Berry M, Turnbull JE, Logan A. Intracellular trafficking in neurones and glia of fibroblast growth factor-2, fibroblast growth factor receptor 1 and heparan sulphate proteoglycans in the injured adult rat cerebral cortex. J Neurochem. 2006;96(4):1189–200.
18. Liang Y, Haring M, Roughley PJ, Margolis RK, Margolis RU. Glypican and biglycan in the nuclei of neurons and glioma cells: presence of functional nuclear localization signals and dynamic changes in glypican during the cell cycle. J Cell Biol. 1997;139(4):851–64.
19. Chen K, Williams KJ. Molecular mediators for raft-dependent endocytosis of syndecan-1, a highly conserved, multifunctional receptor. J Biol Chem. 2013;288(20):13988–99.
20. Ishihara M, Fedarko NS, Conrad HE. Transport of heparan sulfate into the nuclei of hepatocytes. J Biol Chem. 1986;261(29):13575–80.
21. Fedarko NS, Ishihara M, Conrad HE. Control of cell division in hepatoma cells by exogenous heparan sulfate proteoglycan. J Cell Physiol. 1989;139(2):287–94.
22. Cheng F, Petersson P, Arroyo-Yanguas Y, Westergren-Thorsson G. Differences in the uptake and nuclear localization of anti-proliferative heparan sulfate between human lung fibroblasts and human lung carcinoma cells. J Cell Biochem. 2001;83(4):597–606.
23. Roghani M, Moscatelli D. Basic fibroblast growth factor is internalized through both receptor-mediated and heparan sulfate-mediated mechanisms. J Biol Chem. 1992;267(31):22156–62.
24. Zong F, Fthenou E, Castro J, Peterfia B, Kovalszky I, Szilak L, Tzanakakis G, Dobra K. Effect of syndecan-1 overexpression on mesenchymal tumour cell proliferation with focus on different functional domains. Cell Prolif. 2010;43(1):29–40.
25. Zong F, Fthenou E, Mundt F, Szatmari T, Kovalszky I, Szilak L, Brodin D, Tzanakakis G, Hjerpe A, Dobra K. Specific syndecan-1 domains regulate mesenchymal tumor cell adhesion, motility and migration. PLoS One. 2011;6(6):e14816.
26. Hsia E, Richardson TP, Nugent MA. Nuclear localization of basic fibroblast growth factor is mediated by heparan sulfate proteoglycans through protein kinase C signaling. J Cell Biochem. 2003;88(6):1214–25.
27. Quarto N, Amalric F. Heparan sulfate proteoglycans as transducers of FGF-2 signalling. J Cell Sci. 1994;107(Pt 11):3201–12.
28. Amalric F, Bouche G, Bonnet H, Brethenou P, Roman AM, Truchet I, Quarto N. Fibroblast growth factor-2 (FGF-2) in the nucleus: translocation process and targets. Biochem Pharmacol. 1994;47(1):111–5.
29. Wittrup A, Zhang SH, ten Dam GB, van Kuppevelt TH, Bengtson P, Johansson M, Welch J, Morgelin M, Belting M. ScFv antibody-induced translocation of cell-surface heparan sulfate proteoglycan to endocytic vesicles: evidence for heparan sulfate epitope specificity and role of both syndecan and glypican. J Biol Chem. 2009;284(47):32959–67.
30. Christianson HC, Belting M. Heparan sulfate proteoglycan as a cell-surface endocytosis receptor. Matrix Biol. 2014;35:51–5.
31. Kovalszky I, Dudas J, Olah-Nagy J, Pogany G, Tovary J, Timar J, Kopper L, Jeney A, Iozzo RV. Inhibition of DNA topoisomerase I activity by heparan sulfate and modulation by basic fibroblast growth factor. Mol Cell Biochem. 1998;183(1–2):11–23.
32. Busch SJ, Martin GA, Barnhart RL, Mano M, Cardin AD, Jackson RL. Trans-repressor activity of nuclear glycosaminoglycans on Fos and Jun/AP-1 oncoprotein-mediated transcription. J Cell Biol. 1992;116(1):31–42.
33. Dudas J, Ramadori G, Knittel T, Neubauer K, Raddatz D, Egedy K, Kovalszky I. Effect of heparin and liver heparan sulphate on interaction of HepG2-derived transcription factors and their cis-acting elements: altered potential of hepatocellular carcinoma heparan sulphate. Biochem J. 2000;350(Pt 1):245–51.
34. Purushothaman A, Hurst DR, Pisano C, Mizumoto S, Sugahara K, Sanderson RD: Heparanase-mediated loss of nuclear syndecan-1 enhances histone acetyltransferase (HAT) activity to promote expression of genes that drive an aggressive tumor phenotype. J Cell Chem 2011;286(35):30377–83
35. Buczek-Thomas JA, Hsia E, Rich CB, Foster JA, Nugent MA. Inhibition of histone acetyltransferase by glycosaminoglycans. J Cell Biochem. 2008;105(1):108–20.

36. Hirabayashi K, Numa F, Suminami Y, Murakami A, Murakami T, Kato H. Altered proliferative and metastatic potential associated with increased expression of syndecan-1. Tumour Biol. 1998;19(6):454–63.

37. Choi DS, Kim JH, Ryu HS, Kim HC, Han JH, Lee JS, Min CK. Syndecan-1, a key regulator of cell viability in endometrial cancer. Int J Cancer. 2007;121(4):741–50.

38. Mali M, Elenius K, Miettinen HM, Jalkanen M. Inhibition of basic fibroblast growth factor-induced growth promotion by overexpression of syndecan-1. J Biol Chem. 1993;268(32):24215–22.

39. Leppa S, Mali M, Miettinen HM, Jalkanen M. Syndecan expression regulates cell morphology and growth of mouse mammary epithelial tumor cells. Proc Natl Acad Sci U S A. 1992;89(3):932–6.

40. Garusi E, Rossi S, Perris R. Antithetic roles of proteoglycans in cancer. Cell Mol Life Sci. 2012;69(4):553–79.

41. Nikolova V, Koo CY, Ibrahim SA, Wang Z, Spillmann D, Dreier R, Kelsch R, Fischgrabe J, Smollich M, Rossi LH, et al. Differential roles for membrane-bound and soluble syndecan-1 (CD138) in breast cancer progression. Carcinogenesis. 2009;30(3):397–407.

42. Beauvais DM, Jung O, Yang Y, Sanderson RD, Rapraeger AC. Syndecan-1 (CD138) suppresses apoptosis in multiple myeloma by activating IGF1 receptor: prevention by SynstatinIGF1R inhibits tumor growth. Cancer Res. 2016;76(17):4981–93.

43. Boeddeker SJ, Baston-Buest DM, Altergot-Ahmad O, Kruessel JS, Hess AP. Syndecan-1 knockdown in endometrial epithelial cells alters their apoptotic protein profile and enhances the inducibility of apoptosis. Mol Hum Reprod. 2014;20(6):567–78.

44. Khotskaya YB, Dai Y, Ritchie JP, MacLeod V, Yang Y, Zinn K, Sanderson RD. Syndecan-1 is required for robust growth, vascularization, and metastasis of myeloma tumors in vivo. J Biol Chem. 2009;284(38):26085–95.

45. Shimada K, Nakamura M, De Velasco MA, Tanaka M, Ouji Y, Miyake M, Fujimoto K, Hirao K, Konishi N. Role of syndecan-1 (CD138) in cell survival of human urothelial carcinoma. Cancer Sci. 2010;101(1):155–60.

46. Cortes V, Amigo L, Donoso K, Valencia I, Quinones V, Zanlungo S, Brandan E, Rigotti A. Adenovirus-mediated hepatic syndecan-1 overexpression induces hepatocyte proliferation and hyperlipidaemia in mice. Liver Int. 2007;27(4):569–81.

47. Ryoo HD, Bergmann A. The role of apoptosis-induced proliferation for regeneration and cancer. Cold Spring Harb Perspect Biol. 2012;4(8):a008797.

48. Fan Y, Bergmann A. Apoptosis-induced compensatory proliferation. The cell is dead. Long live the cell! Trends Cell Biol. 2008;18(10):467–73.

49. Szatmari T, Mundt F, Heidari-Hamedani G, Zong F, Ferolla E, Alexeyenko A, Hjerpe A, Dobra K. Novel genes and pathways modulated by syndecan-1: implications for the proliferation and cell-cycle regulation of malignant mesothelioma cells. PLoS One. 2012;7(10):e48091.

50. Fedarko NS, Conrad HE. A unique heparan sulfate in the nuclei of hepatocytes: structural changes with the growth state of the cells. J Cell Biol. 1986;102(2):587–99.

51. Dobra K, Nurminen M, Hjerpe A. Growth factors regulate the expression profile of their syndecan co-receptors and the differentiation of mesothelioma cells. Anticancer Res. 2003;23(3B):2435–44.

52. Melixetian M, Klein DK, Sorensen CS, Helin K. NEK11 regulates CDC25A degradation and the IR-induced G2/M checkpoint. Nat Cell Biol. 2009;11(10):1247–53.

53. Noguchi K, Fukazawa H, Murakami Y, Uehara Y. Nek11, a new member of the NIMA family of kinases, involved in DNA replication and genotoxic stress responses. J Biol Chem. 2002;277(42):39655–65.

54. Fry AM, O'Regan L, Sabir SR, Bayliss R. Cell cycle regulation by the NEK family of protein kinases. J Cell Sci. 2012;125(Pt 19):4423–33.

55. Molnar G, Crozat A, Pardee AB. The immediate-early gene Egr-1 regulates the activity of the thymidine kinase promoter at the G0-to-G1 transition of the cell cycle. Mol Cell Biol. 1994;14(8):5242–8.

56. Virolle T, Krones-Herzig A, Baron V, De Gregorio G, Adamson ED, Mercola D. Egr1 promotes growth and survival of prostate cancer cells. Identification of novel Egr1 target genes. J Biol Chem. 2003;278(14):11802–10.

57. Liu X, Gao Y, Lu Y, Zhang J, Li L, Yin F. Downregulation of NEK11 is associated with drug resistance in ovarian cancer. Int J Oncol. 2014;45(3):1266–74.

58. Liu C, Rangnekar VM, Adamson E, Mercola D. Suppression of growth and transformation and induction of apoptosis by EGR-1. Cancer Gene Ther. 1998;5(1):3–28.

59. Huang RP, Fan Y, de Belle I, Niemeyer C, Gottardis MM, Mercola D, Adamson ED. Decreased Egr-1 expression in human, mouse and rat mammary cells and tissues correlates with tumor formation. Int J Cancer. 1997;72(1):102 9.

60. de Belle I, Huang RP, Fan Y, Liu C, Mercola D, Adamson ED. p53 and Egr-1 additively suppress transformed growth in HT1080 cells but Egr-1 counteracts p53-dependent apoptosis. Oncogene. 1999;18(24):3633–42.

61. Baron V, Adamson ED, Calogero A, Ragona G, Mercola D. The transcription factor Egr1 is a direct regulator of multiple tumor suppressors including TGFbeta1, PTEN, p53, and fibronectin. Cancer Gene Ther. 2006;13(2):115–24.

62. Dey BR, Sukhatme VP, Roberts AB, Sporn MB, Rauscher FJ 3rd, Kim SJ. Repression of the transforming growth factor-beta 1 gene by the Wilms' tumor suppressor WT1 gene product. Mol Endocrinol. 1994;8(5):595–602.

63. Liu C, Yao J, de Belle I, Huang RP, Adamson E, Mercola D. The transcription factor EGR-1 suppresses transformation of human fibrosarcoma HT1080 cells by coordinated induction of transforming growth factor-beta1, fibronectin, and plasminogen activator inhibitor-1. J Biol Chem. 1999;274(7):4400–11.

64. Shan LN, Song YG, Su D, Liu YL, Shi XB, Lu SJ. Early growth response Protein-1 involves in transforming growth factor-beta1 induced epithelial-Mesenchymal transition and inhibits migration of non-small-cell lung cancer cells. Asian Pac J Cancer Prev. 2015;16(9):4137–42.

65. Chen SJ, Ning H, Ishida W, Sodin-Semrl S, Takagawa S, Mori Y, Varga J. The early-immediate gene EGR-1 is induced by transforming growth factor-beta and mediates stimulation of collagen gene expression. J Biol Chem. 2006;281(30):21183–97.

66. Bhattacharyya S, Chen SJ, Wu M, Warner-Blankenship M, Ning H, Lakos G, Mori Y, Chang E, Nihijima C, Takehara K, et al. Smad-independent transforming growth factor-beta regulation of early growth response-1 and sustained expression in fibrosis: implications for scleroderma. Am J Pathol. 2008;173(4):1085–99.

67. Ruusala A, Aspenstrom P. Isolation and characterisation of DOCK8, a member of the DOCK180-related regulators of cell morphology. FEBS Lett. 2004;572(1–3):159–66.

68. Cote JF, Vuori K. Identification of an evolutionarily conserved superfamily of DOCK180-related proteins with guanine nucleotide exchange activity. J Cell Sci. 2002;115(Pt 24):4901–13.

69. Zhang Q, Dove CG, Hor JL, Murdock HM, Strauss-Albee DM, Garcia JA, Mandl JN, Grodick RA, Jing H, Chandler-Brown DB, et al. DOCK8 regulates lymphocyte shape integrity for skin antiviral immunity. J Exp Med. 2014;211(13):2549–66.

70. Wang SJ, Cui HY, Liu YM, Zhao P, Zhang Y, Fu ZG, Chen ZN, Jiang JL. CD147 promotes Src-dependent activation of Rac1 signaling through STAT3/DOCK8 during the motility of hepatocellular carcinoma cells. Oncotarget. 2015;6(1):243–57.

71. Takahashi K, Kohno T, Ajima R, Sasaki H, Minna JD, Fujiwara T, Tanaka N, Yokota J. Homozygous deletion and reduced expression of the DOCK8 gene in human lung cancer. Int J Oncol. 2006;28(2):321–8.

72. Idbaih A, Carvalho Silva R, Criniere E, Marie Y, Carpentier C, Boisselier B, Taillibert S, Rousseau A, Mokhtari K, Ducray F, et al. Genomic changes in progression of low-grade gliomas. J Neuro-Oncol. 2008;90(2):133–40.

73. Karlseder J, Rotheneder H, Wintersberger E. Interaction of Sp1 with the growth- and cell cycle-regulated transcription factor E2F. Mol Cell Biol. 1996;16(4):1659–67.

74. Grinstein E, Jundt F, Weinert I, Wernet P, Royer HD. Sp1 as G1 cell cycle phase specific transcription factor in epithelial cells. Oncogene. 2002;21(10):1485–92.

75. Vihinen T, Maatta A, Jaakkola P, Auvinen P, Jalkanen M. Functional characterization of mouse syndecan-1 promoter. J Biol Chem. 1996;271(21):12532–41.

76. Parra E, Ferreira J, Ortega A. Overexpression of EGR-1 modulates the activity of NF-kappaB and AP-1 in prostate carcinoma PC-3 and LNCaP cell lines. Int J Oncol. 2011;39(2):345–52.

77. Zheng C, Ren Z, Wang H, Zhang W, Kalvakolanu DV, Tian Z, Xiao W. E2F1 induces tumor cell survival via nuclear factor-kappaB-dependent induction of EGR1 transcription in prostate cancer cells. Cancer Res. 2009;69(6):2324–31.

78. Ibrahim SA, Hassan H, Vilardo L, Kumar SK, Kumar AV, Kelsch R, Schneider C, Kiesel L, Eich HT, Zucchi I, et al. Syndecan-1 (CD138) modulates triple-negative breast cancer stem cell properties via regulation of LRP-6 and IL-6-mediated STAT3 signaling. PLoS One. 2013;8(12):e85737.

79. Joyce D, Albanese C, Steer J, Fu M, Bouzahzah B, Pestell RG. NF-kappaB and cell-cycle regulation: the cyclin connection. Cytokine Growth Factor Rev. 2001;12(1):73–90.

80. Kaltschmidt B, Kaltschmidt C, Hofmann TG, Hehner SP, Droge W, Schmitz ML. The pro- or anti-apoptotic function of NF-kappaB is determined by the nature of the apoptotic stimulus. Eur J Biochem. 2000;267(12):3828–35.

82. Garcia M, Mauro JA, Ramsamooj M, Blanck G. Tumor suppressor genes are larger than apoptosis-effector genes and have more regions of active chromatin: connection to a stochastic paradigm for sequential gene expression programs. Cell Cycle. 2015;14(15):2494–500.

83. Bargou RC, Emmerich F, Krappmann D, Bommert K, Mapara MY, Arnold W, Royer HD, Grinstein E, Greiner A, Scheidereit C, et al. Constitutive nuclear factor-kappaB-RelA activation is required for proliferation and survival of Hodgkin's disease tumor cells. J Clin Invest. 1997;100(12):2961–9.

84. Dalvai M, Schubart K, Besson A, Matthias P. Oct1 is required for mTOR-induced G1 cell cycle arrest via the control of p27(Kip1) expression. Cell Cycle. 2010;9(19):3933–44.

85. Szekeres K, Koul R, Mauro J, Lloyd M, Johnson J, Blanck G. An Oct-1-based, feed-forward mechanism of apoptosis inhibited by co-culture with Raji B-cells: towards a model of the cancer cell/B-cell microenvironment. Exp Mol Pathol. 2014;97(3):585–9.

86. Mauro JA, Blanck G. Functionally distinct gene classes as bigger or smaller transcription factor traps: a possible stochastic component to sequential gene expression programs in cancer. Gene. 2014;536(2):398–406.

87. Hayashida K, Johnston DR, Goldberger O, Park PW. Syndecan-1 expression in epithelial cells is induced by transforming growth factor beta through a PKA-dependent pathway. J Biol Chem. 2006;281(34):24365–74.

88. Go K, Ishino T, Nakashimo Y, Miyahara N, Ookubo T, Takeno S, Hirakawa K. Analysis of syndecan-1 and TGF-beta expression in the nasal mucosa and nasal polyps. Auris Nasus Larynx. 2010;37(4):427–35.

89. Chen CL, Huang SS, Huang JS. Cellular heparan sulfate negatively modulates transforming growth factor-beta1 (TGF-beta1) responsiveness in epithelial cells. J Biol Chem. 2006;281(17):11506–14.

90. Thurzo V, Popovic M, Matoska J, Blasko M, Grofova M, Lizonova A, Steno M. Human neoplastic cells in tissue culture: two established cell lines derived from giant cell tumor and fibrosarcoma. Neoplasma. 1976;23(6):577–87.

91. Doubling Time Computing [ http://www.doubling-time.com/compute.php ].

92. Branca RM, Orre LM, Johansson HJ, Granholm V, Huss M, Perez-Bercoff A, Forshed J, Kall L, Lehtio J. HiRIEF LC-MS enables deep proteome coverage and unbiased proteogenomics. Nat Methods. 2014;11(1):59–62.

93. Dessau RB, Pipper CB. "R"–project for statistical computing. Ugeskr Laeger. 2008;170(5):328–30.

94. R: A language and environment for statistical computing. [ http://www.R-project.org/].

95. Schmitt T, Ogris C, Sonnhammer EL. FunCoup 3.0: database of genome-wide functional coupling networks. Nucleic Acids Res. 2014;42(Database issue):D380–8.

96. GENE.E: Interact with GENE-E from R. R package version 1.14.0 [http://software.broadinstitute.org/morpheus/].

97. Kramer A, Green J, Pollard J Jr, Tugendreich S. Causal analysis approaches in ingenuity pathway analysis. Bioinformatics. 2014;30(4):523–30.

98. Subramanian A, Tamayo P, Mootha VK, Mukherjee S, Ebert BL, Gillette MA, Paulovich A, Pomeroy SL, Golub TR, Lander ES, et al. Gene set enrichment analysis: a knowledge-based approach for interpreting genome-wide expression profiles. Proc Natl Acad Sci U S A. 2005;102(43):15545–50.

99. Takada Y, Shinkai F, Kondo S, Yamamoto S, Tsuboi H, Korenaga R, Ando J. Fluid shear stress increases the expression of thrombomodulin by cultured human endothelial cells. Biochem Biophys Res Commun. 1994;205(2):1345–52.

100. Ridley RC, Xiao H, Hata H, Woodliff J, Epstein J, Sanderson RD. Expression of syndecan regulates human myeloma plasma cell adhesion to type I collagen. Blood. 1993;81(3):767–74.

# 14

# IL-27 inhibits the TGF-β1-induced epithelial-mesenchymal transition in alveolar epithelial cells

Zhaoxing Dong[1*], Wenlin Tai[2], Wen Lei[1], Yin Wang[1], ZhenKun Li[1] and Tao Zhang[1*]

## Abstracts

**Background:** IL-27 is a multifunctional cytokine that has both pro-inflammatory and anti-inflammatory functions. Although IL-27 has been shown to potently inhibit lung fibrosis, the detailed mechanism of IL-27 in this process is poorly understood. Epithelial–mesenchymal transition (EMT) is one of the key mechanisms involved in pulmonary fibrosis. We assessed the effects of IL-27 on TGF-β1-induced EMT in alveolar epithelial cells.

**Methods:** A549 cells (a human AEC cell line) were incubated with TGF-β1, IL-27, or both TGF-β1 and IL-27, and changes in E-cadherin, β-catenin, vimentin and a-SMA levels were measured using real-time PCR, western blotting and fluorescence microscopy. The related proteins in the JAK/STAT and TGF-β/Smad signalling pathways were examined by western blot.

**Results:** IL-27 increased the expression of epithelial phenotypic markers, including E-cadherin and β-catenin, and inhibited mesenchymal phenotypic markers, including vimentin and a-SMA in A549 cells. Moreover, TGF-β1-induced EMT was attenuated by IL-27. Furthermore, we found that TGF-β1 activated the phosphorylation of JAK1, STAT1, STAT3, STAT5, Smad1, Smad3 and Smad5, and IL-27 partially inhibited these changes in this process. When cells were treated with the STAT3 specific inhibitor wp1006 and the Smad3 specific inhibitor SIS3, the inhibition of EMT by IL-27 was significantly strengthened.

**Conclusion:** Our results suggest that IL-27 attenuates epithelial–mesenchymal transition in alveolar epithelial cells in the absence or presence of TGF-β1 through the JAK/STAT and TGF-β/Smad signalling pathways.

**Keywords:** Interleukin 27, Epithelial–mesenchymal transition, Alveolar epithelial cells, Signalling pathway

## Background

Pulmonary fibrosis is characterized by the destruction of lung tissue architecture and the formation of fibrous foci. Some studies suggested that pulmonary fibrotic diseases included a three phase model of wound repair-injury, inflammation and repair [1]. A subsequent hypothesis suggested that epithelial injury and impaired wound repair, without preceding inflammation, were the aetiology of fibrosis [2]. Mounting evidence suggests that one possible mechanism of fibrotic disease pathogenesis involves alveolar epithelial cell (AEC)-derived fibroblasts through epithelial–mesenchymal transition (EMT) [3, 4].

Although research has made advances in unveiling the molecular mechanism of pulmonary fibrosis, current treatments for idiopathic pulmonary fibrosis show poor efficacy and do not prevent or reverse the disease progression [5].

IL-27 is a heterodimeric cytokines that includes EB virus-induced gene 3 (EBI3) and P28 (IL-27p28) and plays an important role in T cell differentiation. IL-27, by inhibiting the expression of the RORγt master transcription factor, prevented the development of proinflammatory Th17 cells and inhibited the production of IL-17A and IL-17 F in naive T cells [6]. The IL-27 receptor is made up of gp130 and WSX-1, and associates with cytoplasmic protein kinases, such as JAKs (Janus Activated Kinases) that mediate cytokine signalling [7]. The JAK/STAT signalling pathway was initially identified as a

* Correspondence: dongkm@hotmail.com; ZT6958@sina.com.cn
Wenlin Tai is the Co-first author.
[1]Department of Respiratory, The 2nd Affiliated Hospital of Kunming Medical University, Dianmian Road 374, Kunming, Yunnan 650101, China
Full list of author information is available at the end of the article

critical pathway for normal cellular processes but has also been implicated in pulmonary fibrosis [8]. Our previous work demonstrated that IL-27 might inhibit Th17 cell differentiation and the secretion of related inflammation factors in a bleomycin-induced pulmonary fibrosis model [9].

AECs are important target cells that can directly promote lung fibrosis by acquiring a mesenchymal phenotype through EMT. EMT is genetically characterized by a decreased expression of epithelial cell-associated genes (E-cadherin) and increased expression of mesenchymal cell-associated genes, such as α-smooth muscle actin (α-SMA) [10, 11]. Recent studies demonstrated the role of EMT in pulmonary fibrosis [3, 12]. The TGF-β/Smad signalling pathway is required for both EMT and fibrosis in a variety of organs [13].

Currently, the role of IL-27 in idiopathic pulmonary fibrosis is not clearly defined. Shen [14] found that IL-27 might be involved in DM and PM pathogenesis. Moreover, higher levels of IL-27 were measured in patients with interstitial lung disease (ILD). Given these results, we hypothesized that IL-27 may be involved in lung fibrosis. We previously established that IL-27 is involved in pulmonary fibrosis in a bleomycin-induced mouse model, but the specific mechanism was not determined. In this study, we identified potential molecular mechanisms of the effects of IL-27 on pulmonary fibrosis. We found that treatment of A549 cells with IL-27 inhibited EMT-related changes and attenuated the effects of TGF-β1.

## Methods

### Cell culture

A549 cells were purchased from the Kunming Animal Institute and cultured in complete medium containing Dulbecco's Modified Eagle's Medium (DMEM) with high levels of glucose and L-glutamine supplemented with 10 % (v/v) foetal bovine serum (FBS) and 1 % (v/v) antibiotic/antimycotic agents (all from Invitrogen Canada, Inc., Burlington, ON, Canada) and maintained in 5 % $CO_2$ at 37 °C. All procedures were performed in accordance with the Declaration of Helsinki of the World Medical Association. Additionally, the protocols were approved by the IRB/Ethics Committee of Kunming Medical University. For IL-27 and TGF-β1 cytokines used in this research were recombinant mouse cytokines from eBiosience(San Diego, California, USA).

### MTT assay

Cell viability was measured in a quantitative colorimetric MTT assay (Beyotime, Nantong, China). Briefly, cells were seeded in 96-well plates ($6 \times 10^3$ cells/well) and maintained in growth media for 24 h with 5 % $CO_2$ at 37 °C. When the cells reached 60 % confluence, they were treated with different concentrations of IL-27 or TGF-β1 for 48 h. Next, 10 μl of the MTT solution (5 mg/ml) was added to each well, and the cells were incubated for another 4 h at 37 °C. After formazan crystals formed, the MTT medium was aspirated and replaced with 150 μl of dimethyl sulfoxide (DMSO) (Beyotime, Nantong, China) to solubilize the crystals. Then, the plates were shaken for 5 min. The absorbance of each well was recorded using a microplate spectrophotometer at 570 nm. Relative cellular growth was determined by the ratio of the average absorbance of treated cells versus the average absorbance of control cells. Cell viability was calculated as the ratio of the optical densities.

### Real-time quantitative RT-PCR

RNA was obtained from cultured fibroblasts using TRIzol Reagent (TaKaRa, Japan) according to the manufacturer's protocol. RNA was then reverse transcribed using a Prime Script RT Reagent Kit (TaKaRa). The total RNA (1 μg) from each tissue sample was reverse-transcribed to cDNA as follows: 8 μl of 5X Prime Script Buffer (for real-time); 2 μl of Prime Script RT Enzyme Mix; 0.1 nmol oligo(dT) primer; 0.2 nM random hexamers; 2 μg of total RNA; and RNase-free deionized water to a final volume of 40 μl. The reverse transcription proceeded for 15 min at 37 °C and 5 s at 85 °C. The specific primers were designed using Primer Premier 5.0. All primers were synthesized by Sangon Biotechnology. An ABI 7300 Real-Time PCR System (ABI, USA) was used for RT-PCR amplification and detection. RT-PCR reactions were prepared in triplicates in 20-μl reaction volumes as follows: 10 μl of 2X SYBR Premix Ex Taq II, 0.4 μM of forward and reverse primers, 2 μl of cDNA template, and 6.4 μl of RNase-free water. Master Mix without cDNA template was used as a negative control. RT-PCR cycling conditions were used as suggested in the SYBR Premix Ex Taq II Kit instructions (TaKaRa, Japan). Melting curves were evaluated to ensure the specificity of the PCR products in the SYBR Green reactions. Relative mRNA levels of the target genes were normalized to β-actin mRNA. The following oligonucleotide primers specific for human genes were used. a-SMA, 5′-CGGGA-CATCAAGGAGAAACT (sense) and 5′-CCCAT-CAGGCAACTCGTAA-3′(antisense); E-cadherin, 5′-ATGCTGAGGATGATTGAGGTGGGT-3′(sense) and 5′-CAAATGTGTTCAGCTCAGCCAGCA-3 (antisense);β-catenin, 5′-TGCAGTTCGCCTTCACTATG-GACT-3′(sense) and 5′-GATTTGCGGGACAAAGGG CAAGAT-3′ (antisense); Vimentin, 5′-AGAACC TGC AGGAGGCAGAAGAAT-3′(sense) and 5′-TTCCAT TTCACGCATCTGGCGTTC -3′(antisense); β-actin, 5′-TGACGTGGACATCCGCAAAG-3′ (sense) and CT GGAA GGTGGACAGCGAGG-3′(antisense).

## Western blot

Total protein concentration was measured using a BCA Protein Assay Kit (Beyotime, ShangHai, China). For western blotting, 30 μg of protein was loaded into each lane of 10 % SDS PAGE gels and was followed by electrophoresis and protein transfer to PVDF membranes (Millipore). After the transfer, the membranes were blocked with 5 % BSA in PBST. Immunoblots were probed with primary antibody at 4 °C overnight followed by secondary antibodies (Proteintech, 1:5000 dilution) for 30 min at room temperature. After extensive washing, the membranes were incubated in ECL reagent (Pierce, Thermo Co., Ltd, USA) for HRP detection and then exposed to autoradiography film (Bio-Rad, Co., Ltd, USA) for band visualization. β-Actin was used as a loading control. The relative amounts of various proteins were analysed, and the results were quantified using Image J software.

## Immunofluorescence staining and fluorescence microscopy

Cells were grown in 6-well glass-bottomed dishes. After the cells were treated, they were fixed in 4 % paraformaldehyde for 30 min and then permeabilized with 0.2 % Triton X-100 in PBS. Non-specific binding sites were blocked with normal goat serum (Sigma-Aldrich, USA) diluted in 0.1 % Triton-X-100 in PBS for 2 h. Then, the cells were incubated overnight at 4 ° C with primary antibodies at a 1:200 dilution in blocking buffer. Primary antibodies were purchased from Proteintech Technology Company. The next day, the cells were incubated with appropriate fluorescein-conjugated secondary antibodies. DAPI was used to stain nuclei before acquiring images. The images were acquired using a fluorescence microscope (Olympus, Tokyo, Japan); the green or red fluorescence indicated positive antibody expression, and the blue fluorescence was nuclear DAPI labelling. The labelled fields of each section were analysed to produce a mean optical density value (MOD), which represents the strength of the staining signals as measured per positive pixel.

## Statistical analysis

Data are presented as the mean response ± S.E.M ($\bar{x} \pm s$) and analysed using GraphPad Prism 5.0 software to compare mean values between groups in a one-way ANOVA and Tukey's test, $*P < 0.05$, $**P < 0.01$, $***P < 0.001$.

# Results

## IL-27 affects A549 cells in a concentration-dependent manner

To explore the mechanism of IL-27 in alveolar epithelial A549 cells, we first examined the effects of different concentrations of IL-27 (10 to 100 ng/ml) on the proliferation of A549 cells in an MTT assay. We found that the lowest effective concentration of IL-27 was 20 ng/ml, and 100 ng/ml was the best concentration (Fig. 1a). To ascertain the effects of various concentration of IL-27 on EMT-related changes in A549 cells, we examined the epithelial phenotype markers E-cadherin and the mesenchymal phenotypic marker of vimentin by western blot. IL-27 increased E-cadherin protein levels and decreased vimentin protein levels in a concentration-dependent manner (Fig. 1b). Using the same method, we ascertained that the lowest effective concentration of TGF-β1 was 40 ng/ml. And treatment with TGF-β1 at the concentration of 40 ng/ml for 48 h, there is almost one fold increase in the expression of E-cadherin and one fold decrease in vimentin protein levels. Thus, we used 100 ng/ml of IL-27 and 40 ng/ml of TGF-β1 for subsequent experiments.

## IL-27 inhibiting EMT in A549 cells

To confirm the effect of IL-27 in alveolar epithelial cells, we first evaluated the transcription levels of E-cadherin, β-catenin, vimentin and a-SMA using real-time PCR. We found that IL-27 could enhance basal E-cadherin and β-catenin expression and decrease basal vimentin and a-SMA expression. TGF-β1 treatment of A549 cells for 48 h significantly reduced basal expression of E-cadherin and β-catenin and increased the basal expression of vimentin and a-SMA. E-cadherin and β-catenin levels in the presence of 40 ng/ml of TGF-β1 were increased by 100 ng/ml of IL-27, but vimentin and a-SMA levels were inhibited (Fig. 2a). To further examine the role of IL-27 in inhibiting EMT in alveolar epithelial cells, we performed western blots and immunofluorescence labelling to observe the protein expression level and obtained the same results as above (Fig. 2b, c). As the results shown, after exposure to IL-27 and TGF-β1 for 48 h, there is almost one fold increase in E-cadherin and β-catenin protein level and 0.5 fold decrease in vimentin and a-SMA expression compared to cells treated with TGF-β1 alone.

## IL-27 inhibits TGF-β1-induced EMT by inactivating the JAK/STAT signalling pathway

To further explore the molecular mechanisms of IL-27 in EMT, we examined the expression of the JAK/STAT signalling pathway. After exposure to cytokines for 1 h, we found that TGF-β1-induced phosphorylation of JAK1, STAT1, STAT3 and STAT5 in AECs. When cells were treated concomitantly with IL-27 and TGF-β1, activation of JAK1, STAT1, STAT3 and STAT5 was partially inhibited, indicating that IL-27 modulates JAK1, STAT1, STAT3, and STAT5 phosphorylation (Fig. 3a).

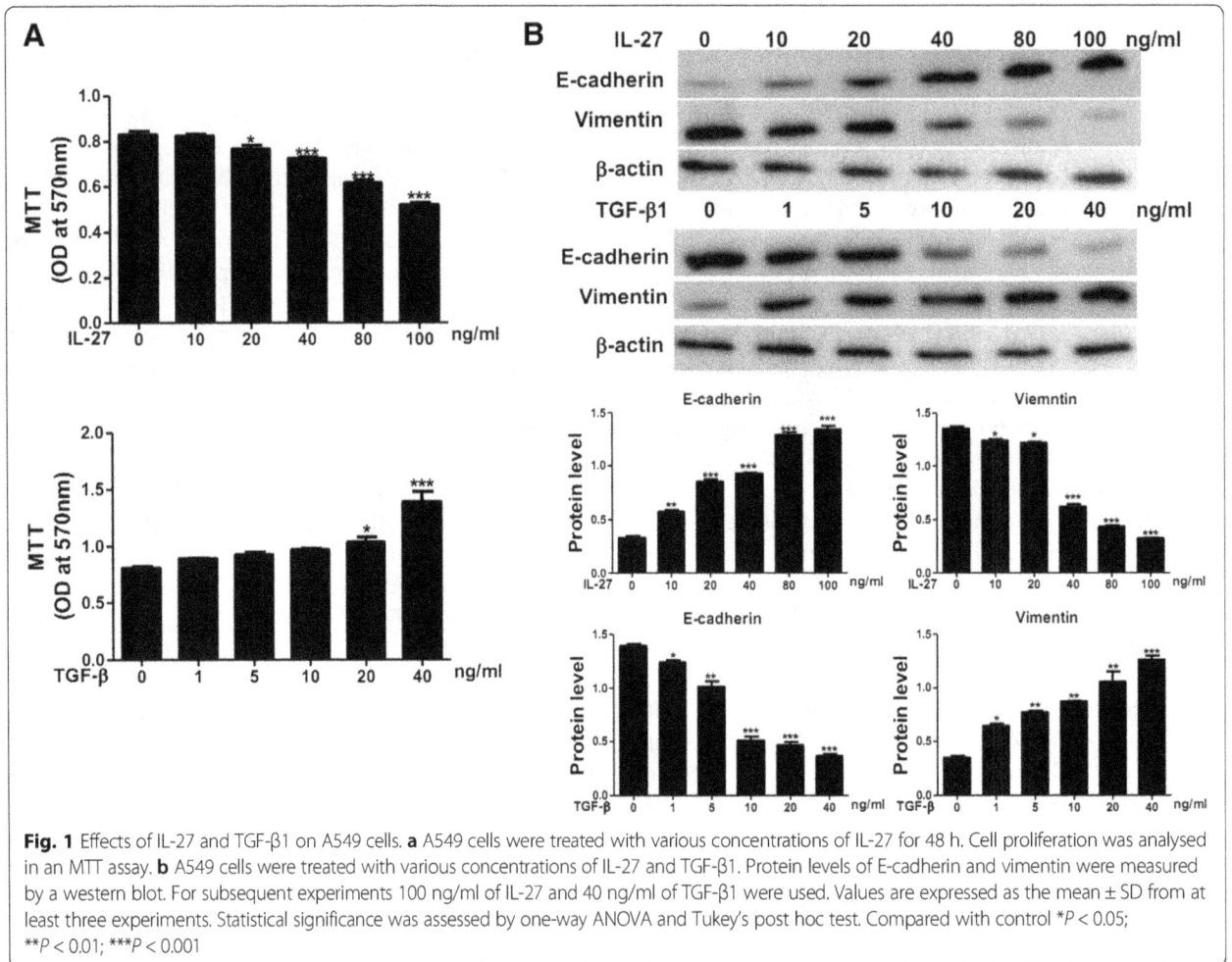

**Fig. 1** Effects of IL-27 and TGF-β1 on A549 cells. **a** A549 cells were treated with various concentrations of IL-27 for 48 h. Cell proliferation was analysed in an MTT assay. **b** A549 cells were treated with various concentrations of IL-27 and TGF-β1. Protein levels of E-cadherin and vimentin were measured by a western blot. For subsequent experiments 100 ng/ml of IL-27 and 40 ng/ml of TGF-β1 were used. Values are expressed as the mean ± SD from at least three experiments. Statistical significance was assessed by one-way ANOVA and Tukey's post hoc test. Compared with control *$P < 0.05$; **$P < 0.01$; ***$P < 0.001$

Furthermore, simultaneously treated AECs with the STAT3 specific inhibitor WP1006 could boost the inhibition effect of IL-27 on the mRNA and protein expression level of mesenchymal phenotypic markers vimentin and a-SMA and enhanced the expression of epithelial phenotypic markers E-cadherin and β-catenin compared with those treated with IL-27 alone (Fig. 3b, c). These above results indicated that the IL-27 inhibited EMT process partially through inactivating JAK/STAT signalling pathway.

### IL-27 affects TGF-β1-mediated phosphorylation of Smad1, Smad3 and Smad5

TGF-β/Smad is a classical signalling pathway involved in TGF-β1-induced EMT. To determine whether IL-27 affects TGF-β1-induced EMT through the TGF-β/Smad signalling pathway, we investigated the expression of major proteins in the TGF-β/Smad signalling pathway after treatment with IL-27 and/ TGF-β for 1 h. The phosphorylation of Smad1, Smad3, Smad5 and the expression of TGF-βR1 were inhibited by IL-27 in the absence or presence of TGF-β1 (Fig. 4a). Moreover, we

found that IL-27 could induce the expression of smad6 and smad7 and knocking down the expression of smad6 and smad7 could partially impair the function of IL-27 in reducing the phosphorylation of smad1/3/5- (Additional file 1: Figure S1). Consistently, addition of the Smad3 specific inhibitor SIS3 weakened the effect of TGF-β1-induced EMT but strengthened the effect of IL-27 as indicated in the mRNA and protein expression level of mesenchymal phenotypic markers vimentin and a-SMA and epithelial phenotypic markers E-cadherin and β-catenin (Fig. 4b, c).

### Discussion

To further explore the mechanism of IL-27 in pulmonary fibrosis, we examined the role of IL-27 in EMT of A549 cells. We found that IL-27 could inhibit the EMT-related changes in A549 cells. Some studies have reported that TGF-β1 is a key regulator of EMT in pulmonary fibrosis; TGF-β1-induced EMT might be central to the process of collagen production and fibrosis. In fact, adding TGF-beta to epithelial cells in culture is a convenient way to induce EMT in various epithelial cells

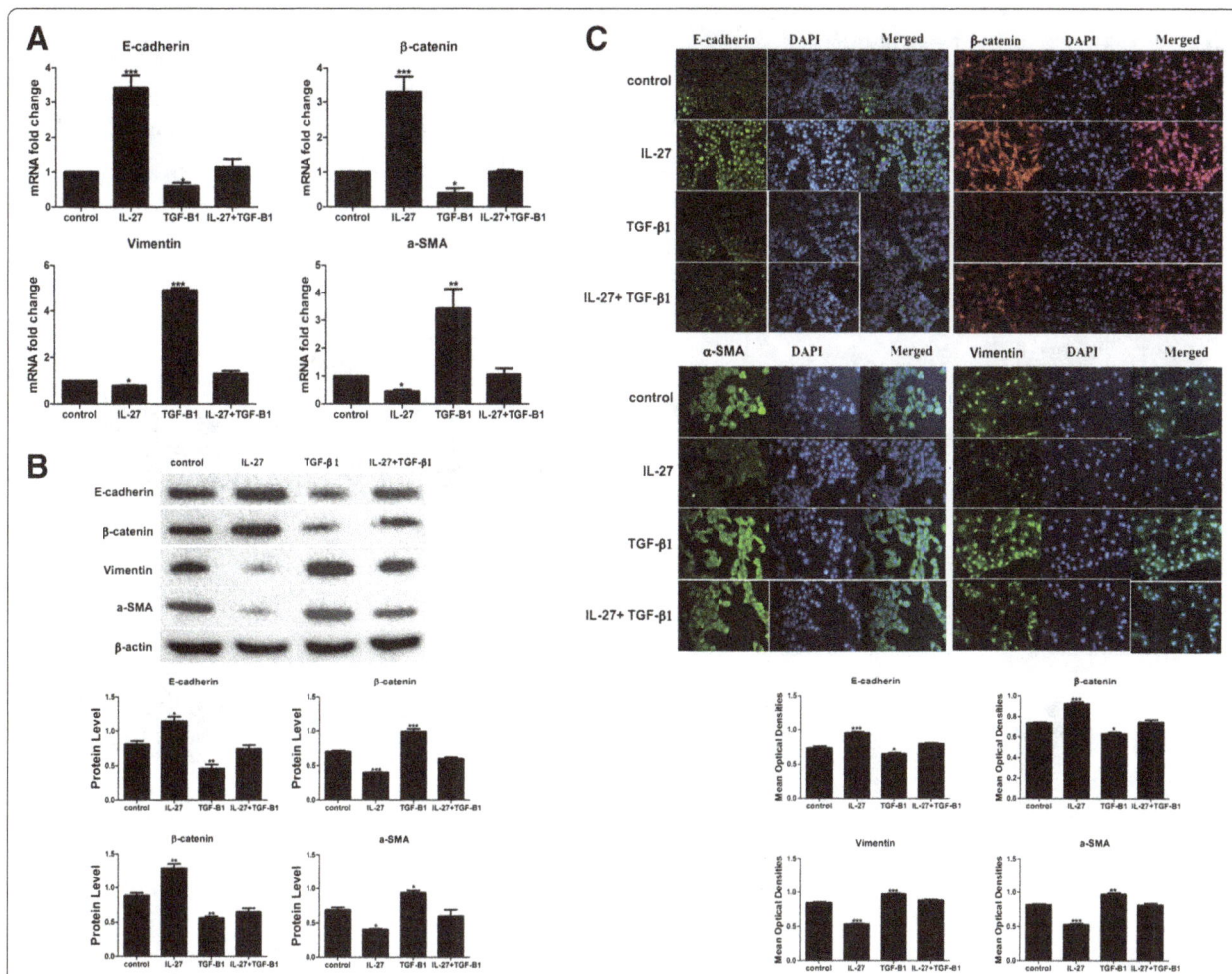

**Fig. 2** IL-27 attenuated TGF-β1-induced EMT in A549 cells. A549 cells were treated with 40 ng/ml of TGF-β1 or/and 100 ng/ml of IL-27 for 48 h. **a, b** The transcription levels of epithelial phenotypic markers E-cadherin and β-catenin and mesenchymal phenotypic markers vimentin and a-SMA were measured by real-time PCR and a western blot. **c** Fluorescence micrographs of E-cadherin, β-catenin, vimentin and α-SMA in AECs at 200X. Scale bars = 100 μm. Green or red indicates the protein of interest, and blue indicates the cell nucleus. Quantification of E-cadherin, β-catenin, vimentin and α-SMA in AECs was carried out using Image-Pro Plus 6.0 software. Mean optical densities were measured. All data are shown as the mean ± SD from at least three experiments. Statistical significance was assessed by one-way ANOVA and Tukey's post hoc test. $*P < 0.05$; $**P < 0.01$; $***P < 0.001$

[11, 13–15]. Our results demonstrated that IL-27 co-cultured with TGF-β1 in AEC attenuated the EMT.

The regulation of TGF-β1-induced EMT is complex because TGF-β1 signalling occurs through different pathways, including Smads, mitogen-activated protein kinase (MAPK), and phosphatidylinositol 3-kinase (PI3K) pathways. Kolosova and colleagues demonstrated that both Smad2 and Smad3 were important for TGF-β1 function in cultured pulmonary epithelial cells [16]. Kasai proposed that the signalling pathway involved in alveolar EMT was likely to be a Smad2-dependent pathway. Our results also verified that TGF-β1 activated the phosphorylation of Smad1, Smad3 and Smad5, and IL-27 weakened the phosphorylated levels of these proteins in the process of inhibiting TGF-β1-induced EMT. Moreover, the Smad3 specific inhibitor SIS3 significantly strengthened the role of IL-27. These

results suggest that IL-27 partially suppressed the TGF-β1-induced EMT in AEC through the TGF-β1/Smad signalling pathway. And inhibitory SMADs (SMAD6/7) as negative regulator of smad are involved in regulation of TGF-β pathway [17–19]. Our further results (Additional file 1: Figure S1) showed that IL-27 could induce the expression of smad6 and 7. And knocking down the expression of smad6 and smad7 by siRNA could partially impair the function of IL-27 in reducing the phosphorylation of smad1/3/5. These results suggest that IL-27 inhibiting the EMT partially through increase the function of inhibitory smads.

IL-27 is a member of the IL-12 family of cytokines and activates the JAK/STAT signal transduction pathway in a context-dependent manner [20]. In natives T cells, IL-27 induces T-Bet and IL-12Rbeta2 through Stat1 and Stat3 [21]. George proposed that IL-27

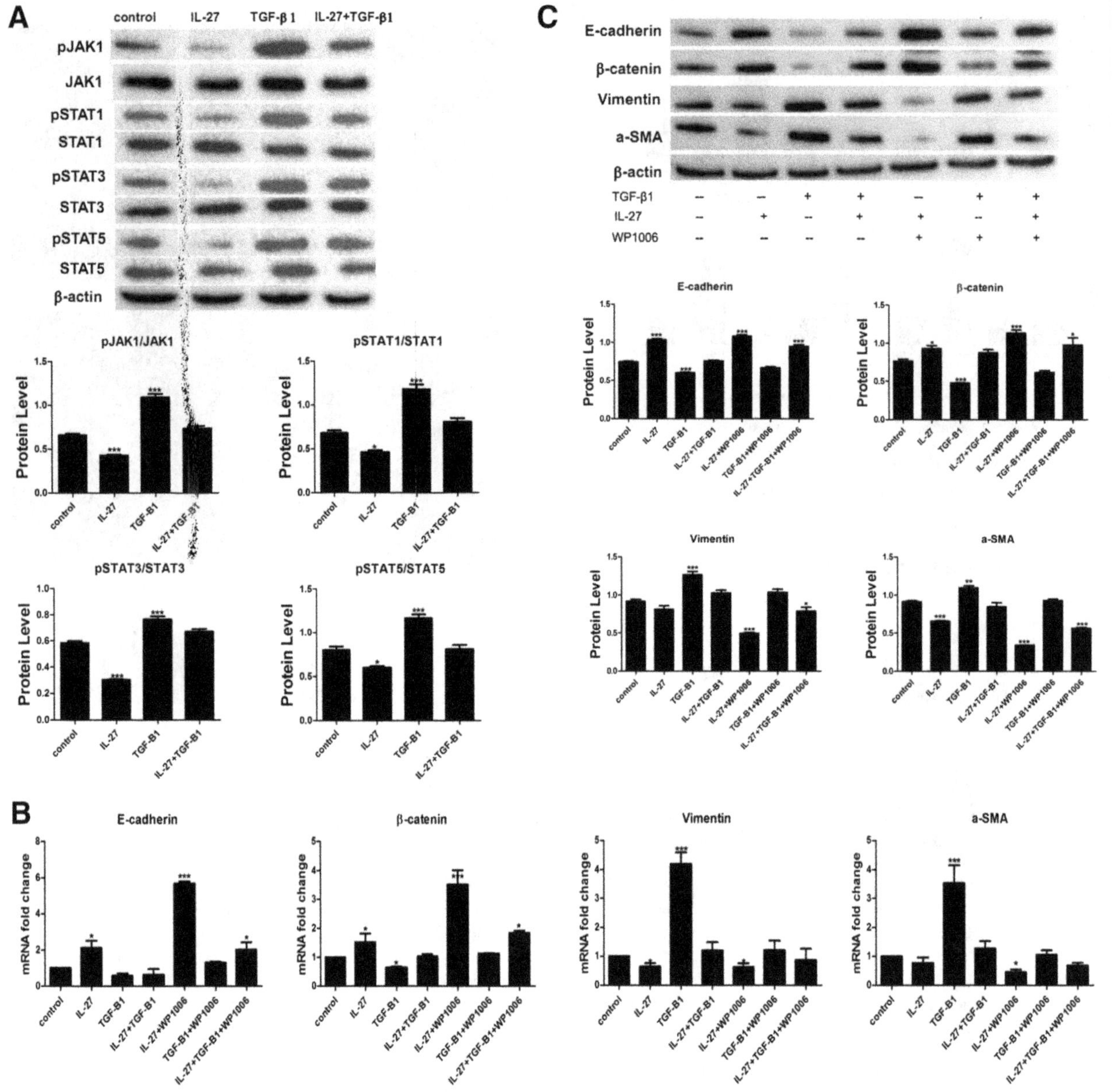

**Fig. 3** IL-27 inactivated the JAK/STAT signalling pathway in AEC. **a** JAK1, STAT1, STAT3 and STAT5 phosphorylated levels were measured by western blot after treatment with IL-27/and TGF-β1 for 1 h. **b, c** After exposure to a STAT3 specific inhibitor, E-cadherin, β-catenin, vimentin and α–SMA in AECs were measured by real-time PCR and western blot. All data are shown as the mean ± SD ($n = 3$). Statistical significance was assessed by one-way ANOVA and Tukey's post hoc test. *$P < 0.05$; **$P < 0.01$; ***$P < 0.001$

activates a STAT1-dominant pattern of signalling in human monocytes [22]. Kachroo demonstrated that IL-27 attenuated EMT and the production of pro-angiogenic factors in a STAT1-dominant pathway in human non-small cell lung cancer [23]. Yoshimoto proposed that IL-27 had an antiproliferative effect on melanomas through WSX-1/STAT1 signalling [24]. The aberrant activation of Stat occurs in many cancers [25]. Our results found that IL-27 could decrease the phosphorylation of STAT1, STAT3 and STAT5

during EMT. This is consistent with research done by Ko et al which previously validated that activation of STAT3 could lead to decreased expression of E-cadherin [26]. Our results suggest that the JAK/STAT signalling pathway might be a key molecular mechanism of IL-27 activity in pulmonary fibrosis. Moreover, our results demonstrated that IL-27 attenuated the phosphorylated levels of JAK1, STAT1, STAT3 and STAT5 during EMT. However, we did not determine which STAT subtype is most

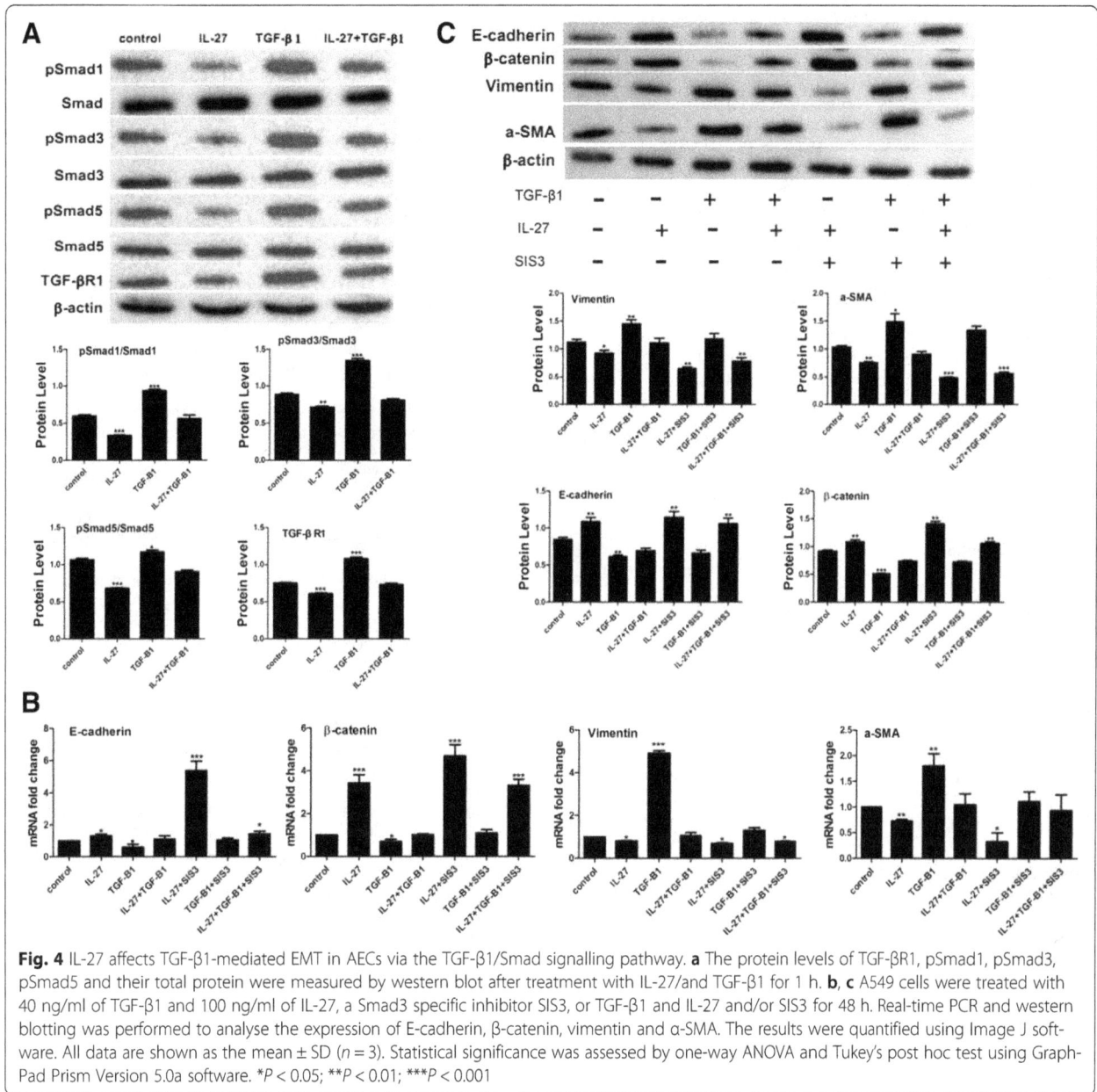

**Fig. 4** IL-27 affects TGF-β1-mediated EMT in AECs via the TGF-β1/Smad signalling pathway. **a** The protein levels of TGF-βR1, pSmad1, pSmad3, pSmad5 and their total protein were measured by western blot after treatment with IL-27/and TGF-β1 for 1 h. **b, c** A549 cells were treated with 40 ng/ml of TGF-β1 and 100 ng/ml of IL-27, a Smad3 specific inhibitor SIS3, or TGF-β1 and IL-27 and/or SIS3 for 48 h. Real-time PCR and western blotting was performed to analyse the expression of E-cadherin, β-catenin, vimentin and α-SMA. The results were quantified using Image J software. All data are shown as the mean ± SD ($n = 3$). Statistical significance was assessed by one-way ANOVA and Tukey's post hoc test using Graph-Pad Prism Version 5.0a software. *$P < 0.05$; **$P < 0.01$; ***$P < 0.001$

predominant. And a further step needs to take to uncover why IL-27 could decrease the phosphorylation of STAT.

## Conclusion

In summary, the present study is the first report of the effects of IL-27 on EMT-related changes in A549 cells. Here, we show that IL-27 could inhibit TGF-β1-induced EMT and inactivate Smad and STAT signal transduction pathways in AECs. Our results increase the current understanding of IL-27 in EMT and identify new potential targets for therapeutic intervention of pulmonary fibrosis.

## Additional file

**Additional file 1: Figure S1.** IL-27 affects TGF-β1-mediated EMT in AECs partially through inhibitory smads. A: The protein levels of smad6 and smad7, TGF-βR1, pSmad1, pSmad3, pSmad5 and their total protein were measured by western blot after treatment with IL-27/and TGF-β1 for 1 h. And statistically expression level of p-smad1, p-smad3 and p-smad5 was shown on the right, the relative expression level was compared with control and the change between TGF-β1 group and TGF-β1 + IL-27 group was also analysed. B: A549 cells were either left as control or transfected with siRNA specific to smad6 and smad7 for 6 h prior treated with 40 ng/ml of TGF-β1 and/or 100 ng/ml of IL-27. And the expression level of smad6 and smad7, TGF-βR1, pSmad1, pSmad3, pSmad5 and their total protein were measured by western blot in 4 different groups. And statistically expression level of p-smad1, p-smad3 and p-smad5 was shown on the right, the relative expression level was compared with control and the expression level between TGF-β1 + siRNA group and TGF-β1 + IL-27+ siRNA group was also analysed. All data are shown as the mean ± SD (n = 3). Statistical significance was assessed by one-way ANOVA and Tukey's post hoc test using GraphPad Prism Version 5.0a software. *means compared with control group; # means TGF-β1 group compared with TGF-β1 + IL-27 group. *$P < 0.05$; **$P < 0.01$; ***$P < 0.001$; ##, $P < 0.01$; ###, $P < 0.001$, ns, no significant difference. (JPG 381 kb)

## Abbreviations
AECs: alveolar epithelial cells; EMT: epithelial–mesenchymal transition; IL-27: interleukin-27; JAK: Janus protein tyrosine kinase; MOD: mean optical density; QPCR: real-time quantitative PCR; STAT: signal transducer and activator of transcription; TGF-β1: transforming growth factor-β1.

## Competing interests
The authors declare that they have no competing interests.

## Authors' contributions
ZD and TZ designed the study and the experiments. WL and YW carried out the PCR, and western blots. ZL and TZ performed immunofluorescence labelling and fluorescence microscopy. WT participated in the design of the study and performed the statistical analysis. ZD and TZ drafted the manuscript. All authors read and approved the final manuscript.

## Acknowledgements
This study was supported by the research grant 81360015 from the National Natural Science Foundation of China, the research grant 2013FB049 from the Natural Science Foundation of Technology Department in Yunnan province, and the research grant 2012Z082 from the Key Project of Education Department in Yunnan province.

## Author details
<sup>1</sup>Department of Respiratory, The 2nd Affiliated Hospital of Kunming Medical University, Dianmian Road 374, Kunming, Yunnan 650101, China. <sup>2</sup>Department of Clinical Laboratory, Yunnan Molecular Diagnostic Center, The 2nd Affiliated Hospital of Kunming Medical University, Dianmian Road, Kunming, Yunnan, China.

## References
1. Wilson MS, Wynn TA. Pulmonary fibrosis: pathogenesis, etiology and regulation. Mucosal Immunol. 2009;2(2):103–21.
2. Strieter RM, Mehrad B. New mechanisms of pulmonary fibrosis. Chest. 2009; 136(5):1364–70.
3. Willis BC, Liebler JM, Luby-Phelps K, et al. Induction of epithelial-mesenchymal transition in alveolar epithelial cells by transforming growth factor-beta1: potential role in idiopathic pulmonary fibrosis. Am J Pathol. 2005;166(5):1321–32.
4. Kim KK, Kugler MC, Wolters PJ, et al. Alveolar epithelial cell mesenchymal transition develops in vivo during pulmonary fibrosis and is regulated by the extracellular matrix. Proc Natl Acad Sci U S A. 2006;103(35):13180–5.
5. Raghu G, Collard HR, Egan JJ, et al. An official ATS/ERS/JRS/ALAT statement: idiopathic pulmonary fibrosis: evidence-based guidelines for diagnosis and management. Am J Respir Crit Care Med. 2011;183(6):788–824.
6. Diveu C, McGeachy MJ, Boniface K, et al. IL-27 blocks RORc expression to inhibit lineage commitment of Th17 cells. J Immunol. 2009;182(9):5748–56.
7. Villarino AV, Huang E, Hunter CA. Understanding the pro- and anti-inflammatory properties of IL-27. J Immunol. 2004;173(2):715–20.
8. Knight D, Mutsaers SE, Prele CM. STAT3 in tissue fibrosis: is there a role in the lung. Pulm Pharmacol Ther. 2011;24(2):193–8.
9. Dong Z, Lu X, Yang Y, et al. IL-27 alleviates the bleomycin-induced pulmonary fibrosis by regulating the Th17 cell differentiation. BMC Pulm Med. 2015;15:13.
10. Kalluri R, Neilson EG. Epithelial-mesenchymal transition and its implications for fibrosis. J Clin Invest. 2003;112(12):1776–84.
11. Willis BC, Borok Z. TGF-beta-induced EMT: mechanisms and implications for fibrotic lung disease. Am J Physiol Lung Cell Mol Physiol. 2007;293(3): L525–34.
12. Hisatomi K, Mukae H, Sakamoto N, et al. Pirfenidone inhibits TGF-beta1-induced over-expression of collagen type I and heat shock protein 47 in A549 cells. BMC Pulm Med. 2012;12:24.
13. Xu J, Lamouille S, Derynck R. TGF-beta-induced epithelial to mesenchymal transition. Cell Res. 2009;19(2):156–72.
14. Shen H, Xia L, Lu J. Pilot study of interleukin-27 in pathogenesis of dermatomyositis and polymyositis: associated with interstitial lung diseases. Cytokine. 2012;60(2):334–7.
15. Kasai H, Allen JT, Mason RM, Kamimura T, Zhang Z. TGF-beta1 induces human alveolar epithelial to mesenchymal cell transition (EMT). Respir Res. 2005;6:56.
16. Kolosova I, Nethery D, Kern JA. Role of Smad2/3 and p38 MAP kinase in TGF-beta1-induced epithelial-mesenchymal transition of pulmonary epithelial cells. J Cell Physiol. 2011;226(5):1248–54.
17. Meng XM, Tang PM, Li J1, Lan HY. TGF-β/Smad signaling in renal fibrosis. Front Physiol. 2015;6:82.
18. Jung SM, Lee JH, Park J, et al. Smad6 inhibits non-canonical TGF-β1 signalling by recruiting the deubiquitinase A20 to TRAF6. Nat Commun. 2013;4:2562.
19. Li Q. Inhibitory SMADs: potential regulators of ovarian function. Biol Reprod. 2015;92(2):50.
20. Pflanz S, Hibbert L, Mattson J, et al. WSX-1 and glycoprotein 130 constitute a signal-transducing receptor for IL-27. J Immunol. 2004;172(4):2225–31.
21. Hibbert L, Pflanz S, De Waal MR, Kastelein RA. IL-27 and IFN-alpha signal via Stat1 and Stat3 and induce T-Bet and IL-12Rbeta2 in naive T cells. J Interferon Cytokine Res. 2003;23(9):513–22.
22. Kalliolias GD, Ivashkiv LB. IL-27 activates human monocytes via STAT1 and suppresses IL-10 production but the inflammatory functions of IL-27 are abrogated by TLRs and p38. J Immunol. 2008;180(9):6325–33.
23. Kachroo P, Lee MH, Zhang L, et al. IL-27 inhibits epithelial-mesenchymal transition and angiogenic factor production in a STAT1-dominant pathway in human non-small cell lung cancer. J Exp Clin Cancer Res. 2013;32:97.
24. Yoshimoto T, Morishima N, Mizoguchi I, et al. Antiproliferative activity of IL-27 on melanoma. J Immunol. 2008;180(10):6527–35.
25. Yue P, Turkson J. Targeting STAT3 in cancer: How successful are we? Expert Opin Investig Drugs. 2009;18(1):45–56.
26. Ko HS, Choi SK, Kang HK, Kim HS, Jeon JH, Park IY, et al. Oncostatin M stimulates cell migration and proliferation by down-regulating E-cadherin in HTR8/SVneo cell line through STAT3 activation. Reprod Biol Endocrinol. 2013;11:93.

# 15

# Spatiotemporal patterns of sortilin and SorCS2 localization during organ development

Simon Boggild[1,2*], Simon Molgaard[1,2], Simon Glerup[2] and Jens Randel Nyengaard[1,3]

## Abstract

**Background:** Sortilin and SorCS2 are part of the Vps10p receptor family. They have both been studied in nervous tissue with several important functions revealed, while their expression and possible functions in developing peripheral tissue remain poorly understood. Here we deliver a thorough characterization of the prenatal localization of sortilin and SorCS2 in mouse peripheral tissue.

**Results:** Sortilin is highly expressed in epithelial tissues of the developing lung, nasal cavity, kidney, pancreas, salivary gland and developing intrahepatic bile ducts. Furthermore tissues such as the thyroid gland, developing cartilage and ossifying bone also show high expression of sortilin together with cell types such as megakaryocytes in the liver. SorCS2 is primarily expressed in mesodermally derived tissues such as striated muscle, adipose tissue, ossifying bone and general connective tissue throughout the body, as well as in lung epithelia. Furthermore, the adrenal gland and liver show high expression of SorCS2 in embryos 13.5 days old.

**Conclusions:** The possible functions relating to the expression patterns of Sortilin and SorCS2 in development are numerous and hopefully this paper will help to generate new hypotheses to further our understanding of the Vps10p receptor family.

**Keywords:** Sortilin, SorCS2, Development, Localization, Expression, Vps10p

## Background

Sortilin and SorCS2 are members of the Vps10p-domain-containing receptor family [1, 2]. Sortilin is a multi-ligand receptor shown to function at the level of the Trans Golgi Network (TGN), plasma membrane and lysosomes in sorting a variety of molecules [3]. Functions of sortilin have been investigated in the central and peripheral nervous system (CNS and PNS) and include trafficking and signaling of neurotrophins (NTs) and their receptors [4, 5]. In particular, the interaction between sortilin, the pro-versions of nerve growth factor (pro-NGF), brain derived neurotrophic factor (pro-BDNF), NT-3, the NT receptor p75[NTR] and the resulting apoptotic signals have been well examined [6–10]. SorCS2 has been studied considerably less but focus has also been on nervous structures. A recently published paper from our group has established that SorCS2 exists in CNS and PNS in single- and two-chain forms that regulate dopaminergic axon guidance and peripheral sensory neuron apoptosis, respectively, and interact with pro-BDNF/p75[NTR] [2]. Interestingly, SorCS2 has also been shown to be essential for pro-NGF mediated growth cone collapse [11]. Variations in the sortilin gene have been implicated in the processing of amyloid precursor protein (APP) and essential tremor and variations in the SorCS2 gene have been shown to occur more frequently in bipolar disorder, schizophrenia and ADHD suggesting both sortilin and SorCS2 as important for normal CNS function [12–17]. In addition, recent findings have indicated that sortilin and SorCS2 may be associated with increased risk of Alzheimer's disease [18]. Research into the physiological role of sortilin outside the nervous system has found sortilin to be involved in trafficking of glucose transporter type 4 vesicles [19] and expressed in several types of cancer cells [20–22].

* Correspondence: simon.boeggild.hansen@post.au.dk
[1]MIND Centre, Stereology and Electron Microscopy Laboratory, Aarhus University, 8000 C Aarhus, Denmark
[2]MIND Centre, Department of Biomedicine, Aarhus University, Ole Worms Allé 3, 8000 C Aarhus, Denmark
Full list of author information is available at the end of the article

Furthermore, GWAS studies have found an allele of the sortilin gene locus that increases hepatic sortilin expression and confers a protective effect against cardiovascular disease [23, 24]. Subsequent studies of this have produced results suggesting that increased expression of hepatic sortilin reduces levels of low density lipoprotein (LDL)-cholesterol in blood, and thereby atherosclerosis [25]. Surprisingly, lack of sortilin also leads to decreased LDL-cholesterol levels in blood through decreased VLDL secretion [26]. To complicate matters further, sortilin has been identified as a facilitator of proprotein convertase subtilisin/kexin type 9 (PCSK9) secretion, a protein that reduces LDL-receptor expression [19, 27]. Consequently, the exact role of sortilin in cholesterol metabolism remains a matter of debate. Sortilin and SorCS2 are expressed dissimilarly in the CNS, but seem to have partly overlapping functions, with both being able to transmit apoptotic signals via pro-BDNF and pro-NGF [2, 7]. This relationship of structural and in part functional similarities combined with complementary expression patterns might be found outside the CNS as well. Furthermore, the functions of SorCS2 and sortilin with these neurotrophins are relevant for neurodevelopment and conceivably so for the maturation of other cell systems. Thus, it is relevant to study sortilin and SorCS2 together in non-neuronal tissues. While there is data concerning the expression of sortilin and SorCS2 mRNA during development, the focus has so far been on the nervous system. Here, we deliver a thorough description of the spatiotemporal localization of sortilin and SorCS2 in organs and peripheral tissue of the developing mouse embryo, hopefully enabling new hypotheses toward their physiological roles to be created and tested.

## Methods
### Animals
Animals used for this study were C57j/bl6bom mice. Sortilin(−/−) and SorCS2(−/−) mice were bred on C57j/bl6bom background using homologous recombination [4]. All animals were bred and housed at the Animal Facility at Aarhus University in accordance with the Danish Animal Protection Act (12/09/2015). A maximum of five mice per plastic cage (42 x 25 x 15 cm) were housed with water ad libitum and fed standard chow (Altromin #1324). Cages contained nesting material, bedding, a metal tunnel and a wooden stick and were cleaned every week. Animals were kept under pathogen-free conditions with a 12-h light/12-h dark schedule. Experiments were approved by the Danish Animal Experiments Inspectorate (Permit 2011/561−119). For the expression analysis three embryos from two different mothers were used for each age point.

### Tissue
Fetal tissue was obtained through timed-pregnant C57j/bl6bom mice, on embryonic day (E) 13.5, 15.5 and 17.5. Midday on the day of vaginal plug detection was determined as E0.5. Pregnant mice were euthanized by cervical dislocation, the uterus removed and transferred to ice cold phosphate buffered saline (PBS). Embryos were dissected out and fixed in 4 % paraformaldehyde for 1–2 days. After cryoprotection in 30 % sucrose, embryos were covered in Tissue-Tek (Sakura, 4583) and immersed in liquid nitrogen. Sagittal and horizontal sections were cut at 10 μm using a cryostat (Microm HM 500 OM) and stored at −20 °C until use. Adult SorCS2(−/−) and sortilin(−/−) mice have no gross abnormalities of their organs. Therefore a presumption that SorCS2 and sortilin do not play a crucial role in the early organogenesis was made. Age points E13.5, E15.5 and E17.5 were thus selected as the different organs here are relatively easy to identify and important maturational processes take place.

### Immunohistochemistry
Antigen-retrieval was performed by microwaving the tissue for 2 × 5 min at 540 W in Target Retrieval Solution (DAKO, s1699) using a 1:10 dilution in PBS. Sections were cooled to room temperature, followed by 3x10 min wash in PBS. To permeabilize the cells and block endogenous reactivity, sections were incubated with PBS containing 1 % bovine serum albumin (BSA; Sigma, A4503) and 0.3 % Triton X-100 (Sigma, T8787) for 30 min. After a 10 min washing step, primary antibodies were added in a 50 nM Tris-based (TB; 6.06 g Tris, 8.77 g NaCl in 1 L $H_2O$, pH 7.4) buffer solution containing 1 % BSA and sections placed overnight at 4 °C. The next day, sections were left at room temperature for 1 h followed by 3 × 10 min of wash in PBS. Secondary antibodies were added in TB and sections left in the dark for 4 h at room temperature before washing 3 × 10 min. Nuclei were visualized by incubating with Hoechst stain using a 1:10,000 dilution in PBS for 10 min. Sections were mounted using fluorescent mounting medium (DAKO, s3023) and kept in the dark at 4 °C. Primary antibodies used were goat α-sortilin (R&D systems, AF2934,1:100), sheep α-SorCS2 (R&D systems, AF4237, 1:50), rabbit α-TTF-1 (Santa Cruz, H-190,1:250), rabbit α-calretinin (Milipore, ab5054, 1:250), rabbit α-Calbindin (Milipore, ab1778, 1:50) and rabbit α-Tyrosine Hydroxylase (Milipore, AB152, 1:250). Reviews of all antibodies used have been made public on the antibody rating site http://www.pabmabs.com. These extra primary antibodies were selected during the study to elaborate on observed sortilin and SorCS2 expression. Secondary antibodies used were donkey

α-sheep 488, donkey α-goat 488 and donkey α-rabbit 565 (Life Technologies, A-11015, A-11055, A-10042, 1:300).

Images were captured using a Zeiss LSM780 laser-scanning confocal microscope controlled by ZEN 2011 software. Specificity of sortilin and SorCS2 antibodies were ensured using sortilin(−/−) and SorCS2(−/−) tissue [2, 5]. The expression level and localization of sortilin and SorCS2 was determined visually.

## Results
### Sortilin
#### Respiratory tract

**Nose/upper airways** At all three ages E13.5, E15.5 and E17.5, moderately strong sortilin immunoreactivity (IR) was found in the epithelium of the nasal cavity. The expression was localized to the apical part of the epithelial cells and showed a granular pattern. At E15.5, cells in the epithelium marked by the calcium-binding protein calretinin, a marker for maturing olfactory cells [28], showed sortilin IR as well (Fig. 1a, b). In the respiratory part of the pharynx as well as the larynx and trachea, sortilin IR was again found in the apical part of the epithelial cells, albeit at a weaker level. This pattern was repeated through E13.5-E17.5. Lastly, strong expression of sortilin was found in epithelial glands of the nasal cavity at E17.5 (Fig. 1c).

**Lung/Lower airways** Sortilin was found highly expressed in the developing lung, and showed distinct spatial distribution throughout all three age points. At E13.5, in the pseudo glandular lung stage, sortilin reactivity was found to be pronounced in the epithelium of the entire developing bronchial system as marked by Thyroid Transcription Factor 1 (TTF-1), a marker for developing bronchial cells [29]. Expression had a granular pattern and was most prominent in the apical part of the epithelial cells (Fig. 1d, *d'*). The mesenchyme showed very sparse reactivity. At E15.5, in the early canalicular phase, the pattern was similar, albeit with a lower expression of sortilin in more proximal bronchi. At E17.5 in the beginning of the saccular phase the pattern of sortilin had changed somewhat. In the more distal branches of the tubular system sortilin was highly expressed in the apical vesicular pattern previously described. However, in the parts of the airways resembling developing alveoli, sortilin reactivity was very sparse (Fig. 1e, f). Furthermore, the mesenchyme/stroma began to show distinct sortilin IR. Along the borders of the prospective alveoli and bronchioli, solitary cells with a round soma approximately 10–12 μm in diameter and strong granular expression of sortilin were found. These cells resembled macrophages.

A final note on lungs, preliminary examination of two Sortilin(−/−) embryos give the impression that at E15.5 the lung epithelium have fewer branches per lung area and have a larger lumen compared to (two) WT mice. A comparison is shown in Fig. 2 but this must be checked in many more mice before a statistically relevant conclusion can be drawn.

In short summary, sortilin IR in the developing respiratory system was largely confined to the apical parts of epithelia, most pronounced in the roof of the nasal cavity and in the presumptive more immature stages/parts of the developing lower airways.

#### Digestive tract
##### Mouth and esophagus
At E13.5 the epithelium of the mouth and tongue both showed sortilin IR at a weak level, stronger in the latter (Fig. 7a, b). By E15.5 this expression had diminished, apart from a few epithelial cells dispersed in the posterior palate, and by E17.5, no sortilin-IR could be detected. In the esophagus, weak sortilin IR in the epithelium was found consistently at all the three age points.

#### Submandibular gland
Sortilin showed a distinct expression pattern in the submandibular gland. Expression was confined to the apical part of the epithelial cells of acini as well as ducts in a granular manner at E13.5. This persisted throughout the course of E15.5-E17.5, as the gland grew larger and more complex (Fig. 3b).

#### Gut
At E13.5 sortilin IR was found in midgut, most pronounced in the epithelium, as well as in a circular band with a thickness of 2–3 cells in the periphery of the gut wall. In the epithelial cells, sortilin IR was located apically in a granular pattern (v 3a). At E15.5 sortilin IR of the gut had decreased. In the developing small intestine, expression of sortilin was present weakly in the epithelium, and strongly in a subset of cells herein. In these cells sortilin was often located basally to the nucleus in resemblance of enteroendocrine cells, and sometimes in a small cellular process extending along the basal epithelial border (Fig. 3a, *b'* ,*c'*). In the large intestine, sortilin IR was noted in the epithelium at a moderate level (Fig. 3c). The same pattern of expression was apparent at E17.5 with the addition of expression in the ventricular epithelium.

#### Liver
Throughout all three stages investigated, similarities in sortilin IR were found in the liver. Large cells with intense granular expression of sortilin were found across the organ. These cells were up to 25 μm in diameter

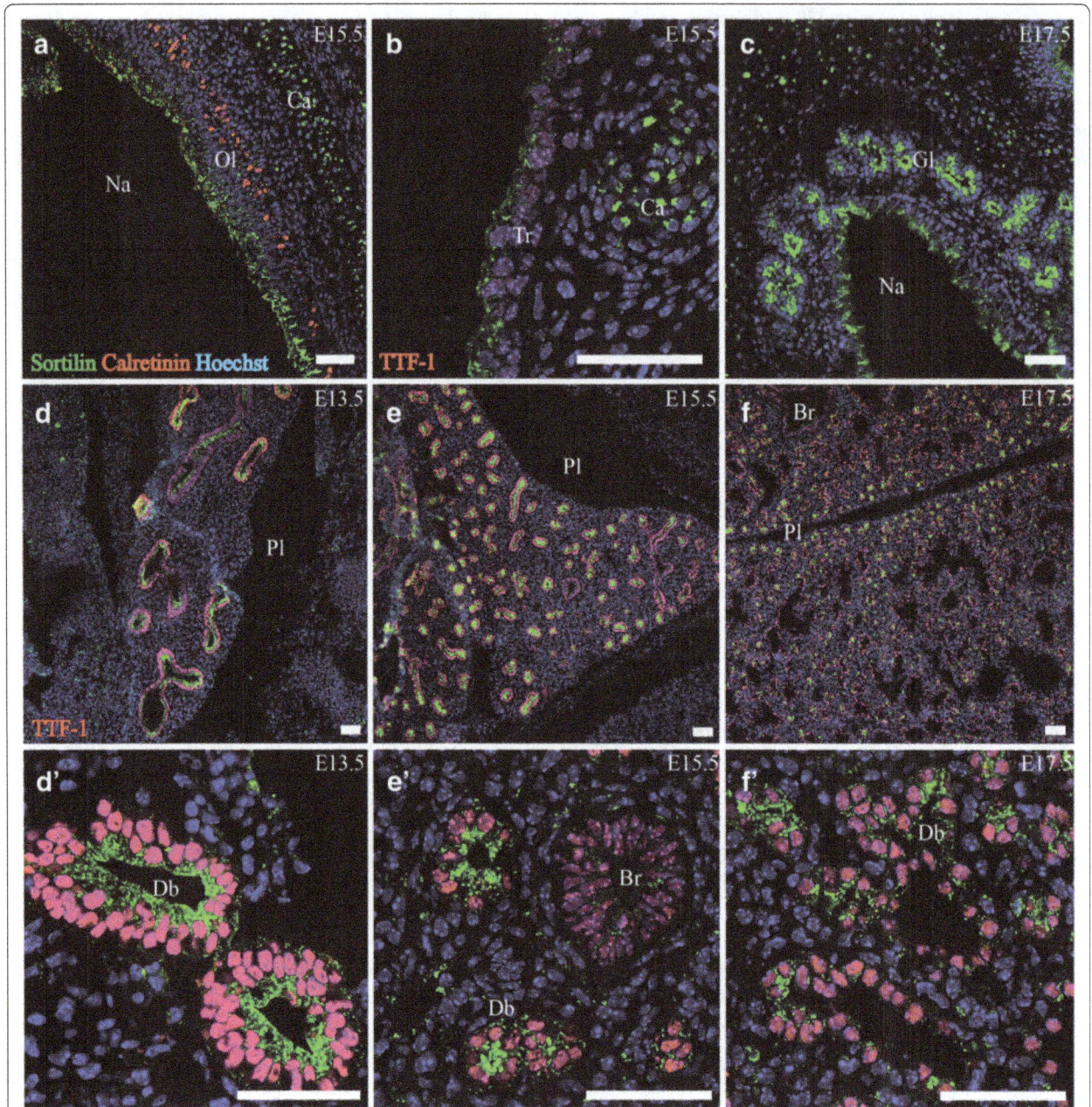

**Fig. 1** Sortilin expression in embryonic respiratory tract. Immunofluorescence was performed on sagittal sections (10 μm) of embryos on embryonic day E13.5 (**d**), E15.5 (**a**, **b**, **d**, **e**) and E17.5 (**c**, **f**). Sortilin localizes prominently to the epithelium and glands of the nasal cavity (**a**, **c**) and is also found in the tracheal epithelium (**b**). Furthermore, sortilin is highly expressed in the apical part of the epithelium of the branching bronchial network at all three time points investigated. The expression is confined primarily to distal branches and is diminished in proximal more mature bronchi (**d**, **e**). Green channel is sortilin, red channel is calretinin (**a**) or TTF-1 and blue channel is Hoechst nuclear stain. Scale bars = 50 μm. Br, bronchus; Ca, cartilage; Db, distal branch; Gl, epithelial gland; Na, nasal cavity; Ol, olfactory epithelium; Pl pleuric cavity; Tr, tracheal epithelium

with a large lobed nucleus resembling megakaryocytes. Additionally, a strong granular expression of sortilin was present in cells harboring a nucleus with several lobes, resembling neutrophil granulocytes, as well as in cells with a bended nucleus resembling eosinophils or monocytes. Specifically for E13.5, sortilin IR was found in cells with an elongated nucleus and cytoplasm that was often surrounding a single cell or a few identical looking blood cells of various lineages. At E15.5 and E17.5, clusters of sortilin IR cells in the near vicinity of large vessel-like structures were present. These clusters often had a bi-layered structure surrounding a small lumen, fitting the morphology of developing intrahepatic bile ducts [30]. Liver findings are shown in Fig. 4.

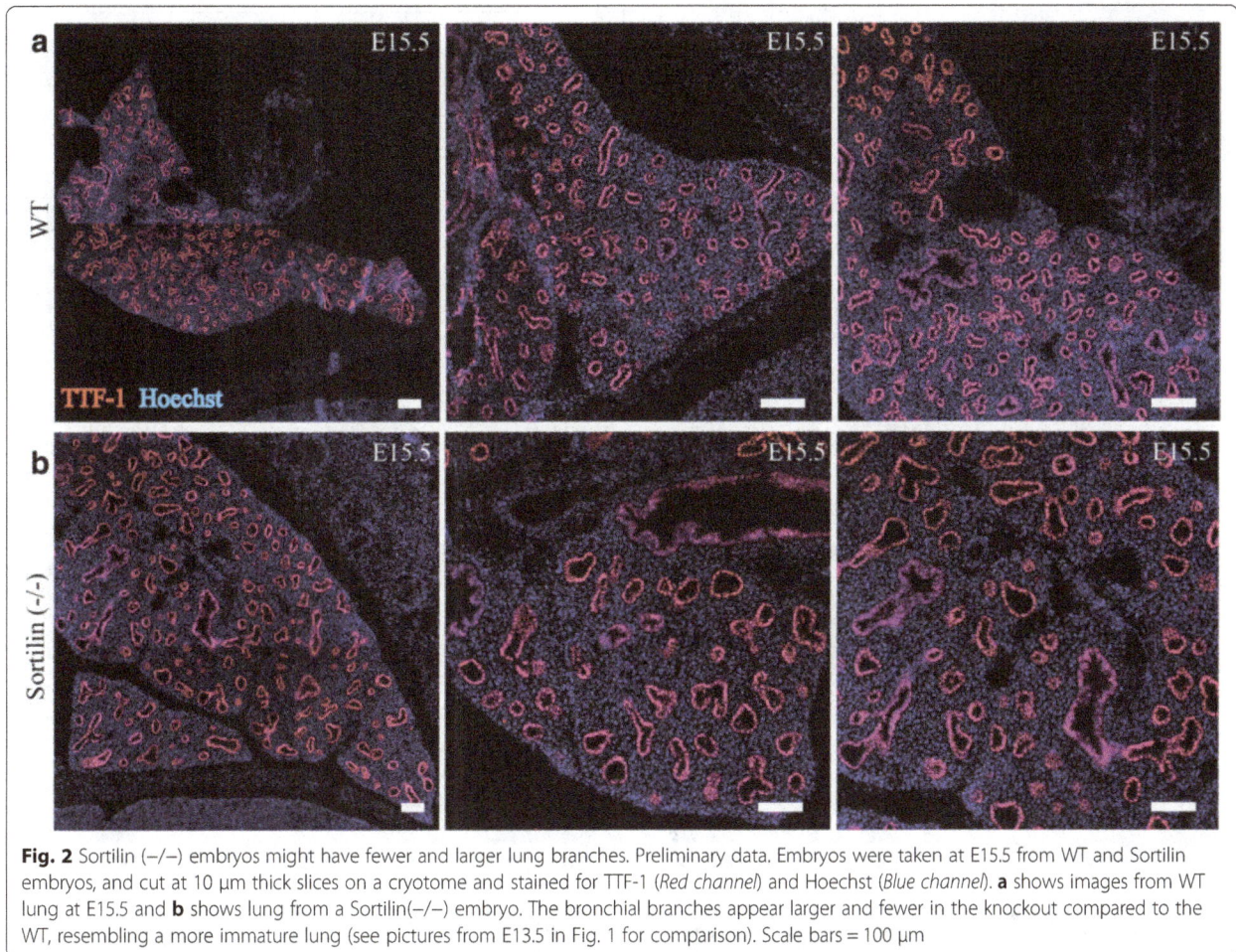

**Fig. 2** Sortilin (−/−) embryos might have fewer and larger lung branches. Preliminary data. Embryos were taken at E15.5 from WT and Sortilin embryos, and cut at 10 μm thick slices on a cryotome and stained for TTF-1 (*Red channel*) and Hoechst (*Blue channel*). **a** shows images from WT lung at E15.5 and **b** shows lung from a Sortilin(−/−) embryo. The bronchial branches appear larger and fewer in the knockout compared to the WT, resembling a more immature lung (see pictures from E13.5 in Fig. 1 for comparison). Scale bars = 100 μm

## Pancreas

In the pancreas, sortilin IR was localized to the parenchyma of the exocrine part. At E13.5, the epithelium of the ductal system showed high sortilin IR, primarily in the apical part and of granular appearance (Fig. 5). This was also true for E15.5 and E17.5, where the exocrine acini also showed granular sortilin IR in the apical parts of the acinar cells, although not as marked as the ductal cells.

In short summary, sortilin IR in the digestive system was found at the highest levels in the epithelium of the submandibular gland acini and ducts, and the secretory ducts of pancreas, but was also highly expressed in the liver in granulocytes, megakaryocytes and developing bile ducts. Lastly sortilin IR was found in cells resembling enteroendocrine cells of the gut as well as in epithelia of the ventricle and large intestine.

## Urinary system
### Kidney

Sortilin was found expressed in the tubular system inside the metanephros at all three age points. At E13.5 sortilin IR was pronounced in cells expressing Calbindin-d28k, a marker for ureteric branches [31], and developing nephrons (Fig. 6). Furthermore, metanephric tissue surrounding the ureteric buds showed expression of sortilin. At E15.5 sortilin was differentially expressed in the epithelial tubular system, with sparse reactivity in the mesenchyme. Most ureteric branches, now developing into collecting ducts, showed a similar expression pattern to that of E13.5. Some nephric tubules showed a strong granular pattern of sortilin IR, and these had a high cuboidal-cylindrical epithelium with sortilin predominantly located in the apical part. Other tubules with a lower, flattened cuboidal epithelium showed weaker sortilin IR. In developing renal corpuscles, weak-moderate sortilin IR was found along the parietal layer of Bowman's capsule (not shown), as well as in a few cells dispersed in the glomerulus. At E17.5 the pattern was similar to E15.5.

## Heart and vascular system
### Heart

Expression of sortilin in the heart showed a similar pattern throughout E13.5, E15.5 and E17.5. Very sparse reactivity was found in the ventricle, but contrary to this, a

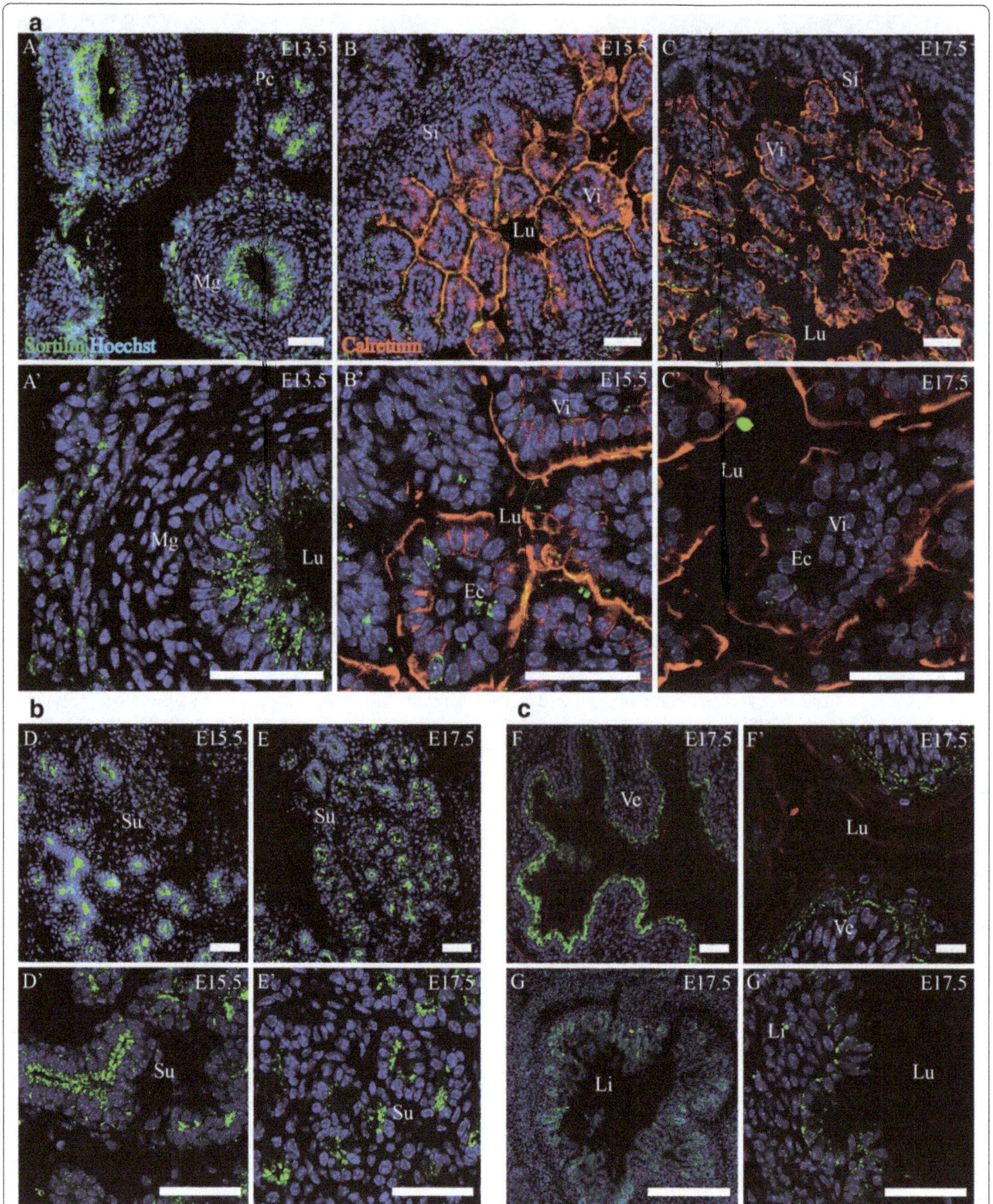

**Fig. 3** Expression of Sortilin in the developing alimental canal and submandibular gland. Immunohistofluorescence was performed on sagittal sections (10 μm) of embryos on embryonic day E13.5, E15.5 and E17.5. **a**. Sortilin is expressed in the epithelium of the midgut at E13.5 (*a, a'*) but is not observed in the small intestine from E15.5 and onwards, apart from a small subset of epithelial cells (*b-c'*). **b**. Sortilin is highly expressed in the epithelium of the developing submandibular gland and is primarily apically located (*d-e'*). **c**. Sortilin is moderately expressed in the epithelium of the developing ventricle (*f, f'*) and the large intestine at E17.5 (*g, g'*). Green channel is sortilin, red channel is calretinin and blue channel is Hoechst nuclear stain. Scale bars = 50 μm. Ec, presumptive enteroendocrine cell; Li, large intestine; Lu, lumen; Mg, midgut; Pc, pancreas; Si, small intestine; Su, submandibular gland branches; Ve, ventricle; Vi, intestinal villus

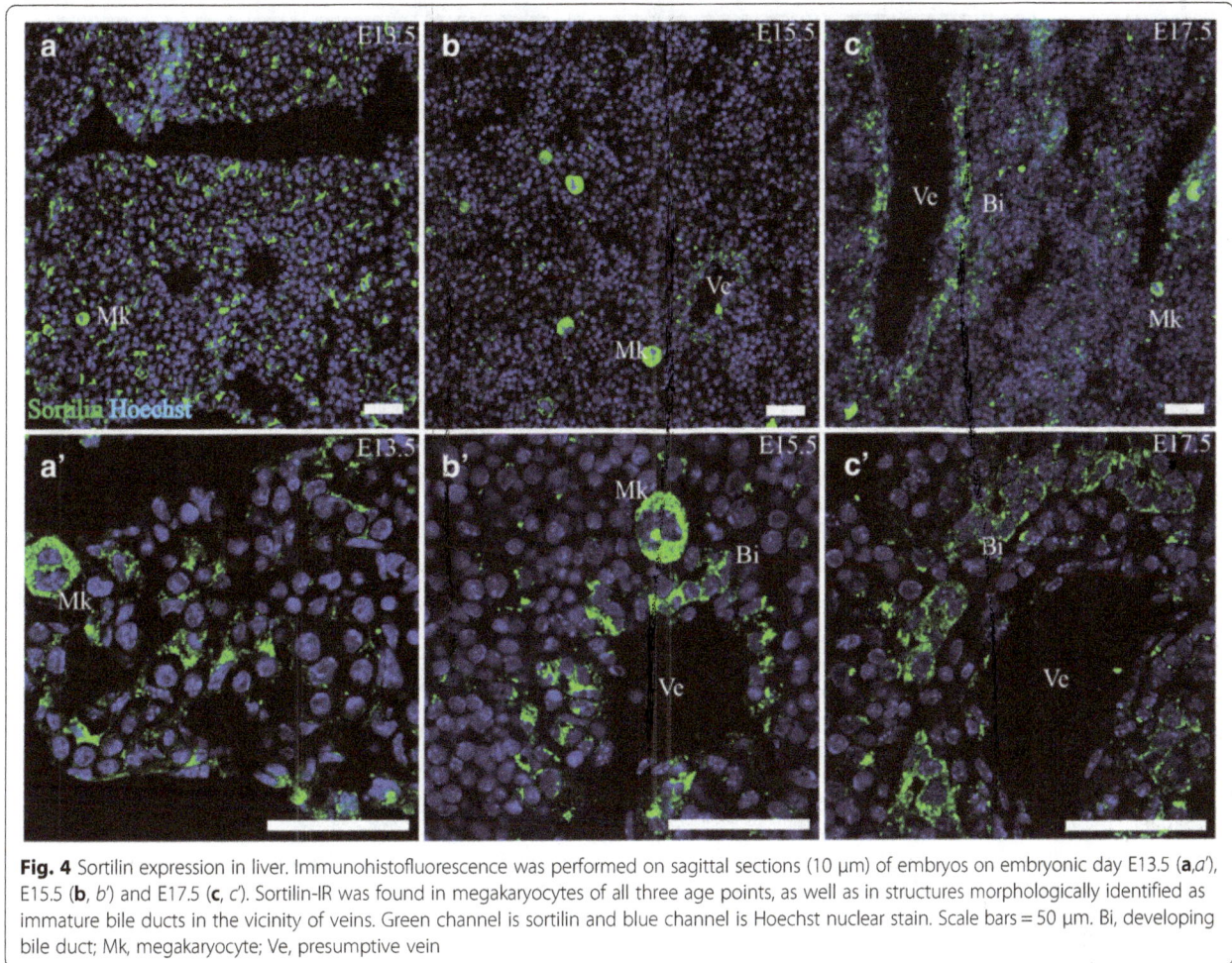

**Fig. 4** Sortilin expression in liver. Immunohistofluorescence was performed on sagittal sections (10 μm) of embryos on embryonic day E13.5 (**a**,*a'*), E15.5 (**b**, *b'*) and E17.5 (**c**, *c'*). Sortilin-IR was found in megakaryocytes of all three age points, as well as in structures morphologically identified as immature bile ducts in the vicinity of veins. Green channel is sortilin and blue channel is Hoechst nuclear stain. Scale bars = 50 μm. Bi, developing bile duct; Mk, megakaryocyte; Ve, presumptive vein

moderately strong expression was apparent in the atria (Fig. 7d, e). This spanned the wall as well as the valves.

### Large blood vessels

In the vascular system, the arteries showed no expression of sortilin while the veins showed moderate to strong expression at all three age points (Fig. 7c, f).

## Musculoskeletal system

### Cartilage/Bone

Sortilin IR was generally detected at high levels in the entire skeletal primordium as well as cartilage. In structures undergoing intramembranous ossification e.g. skull, sortilin was expressed in the entire structure. In structures undergoing endochondral ossification e.g. costae, sortilin expression was generally confined to cells in the center of the structure, an area with more space between nuclei compared to the surrounding concentric cell layers (Fig. 8a-c).

At E15.5, sortilin IR was pronounced in the cartilage anlagen of the long bones, facial bones and vertebrae. Furthermore sortilin was strongly expressed in costae,

cartilage of the larynx, trachea and the skull. Expression was confined to the central cells as described for E13.5. Furthermore, areas undergoing ossification in the spine showed strong sortilin expression in a centered pattern (Fig. 8d, e).

At E17.5, cartilage of the trachea and larynx as well as costae showed similar sortilin reactivity to earlier age points. In the ossifying spine, the pattern of expression had changed somewhat. Sortilin IR cells were no longer found in the center, but rather in the posterodorsal and anteroventral parts of the vertebrae, confined to a cluster of cells: some cells with a diameter of ~10 μm, and other large cells at ~15–20 μm in diameter (Fig. 8f). These cells are possibly osteoblasts and osteoclasts, respectively, where both showed a strong granular pattern of sortilin expression dispersed throughout the cytoplasm.

### Skin

The skin showed little sortilin IR at E13.5, but this increased to moderate levels at E15.5 and E17.5. Expression was located in the apical part of the epidermis (Fig. 7g-i).

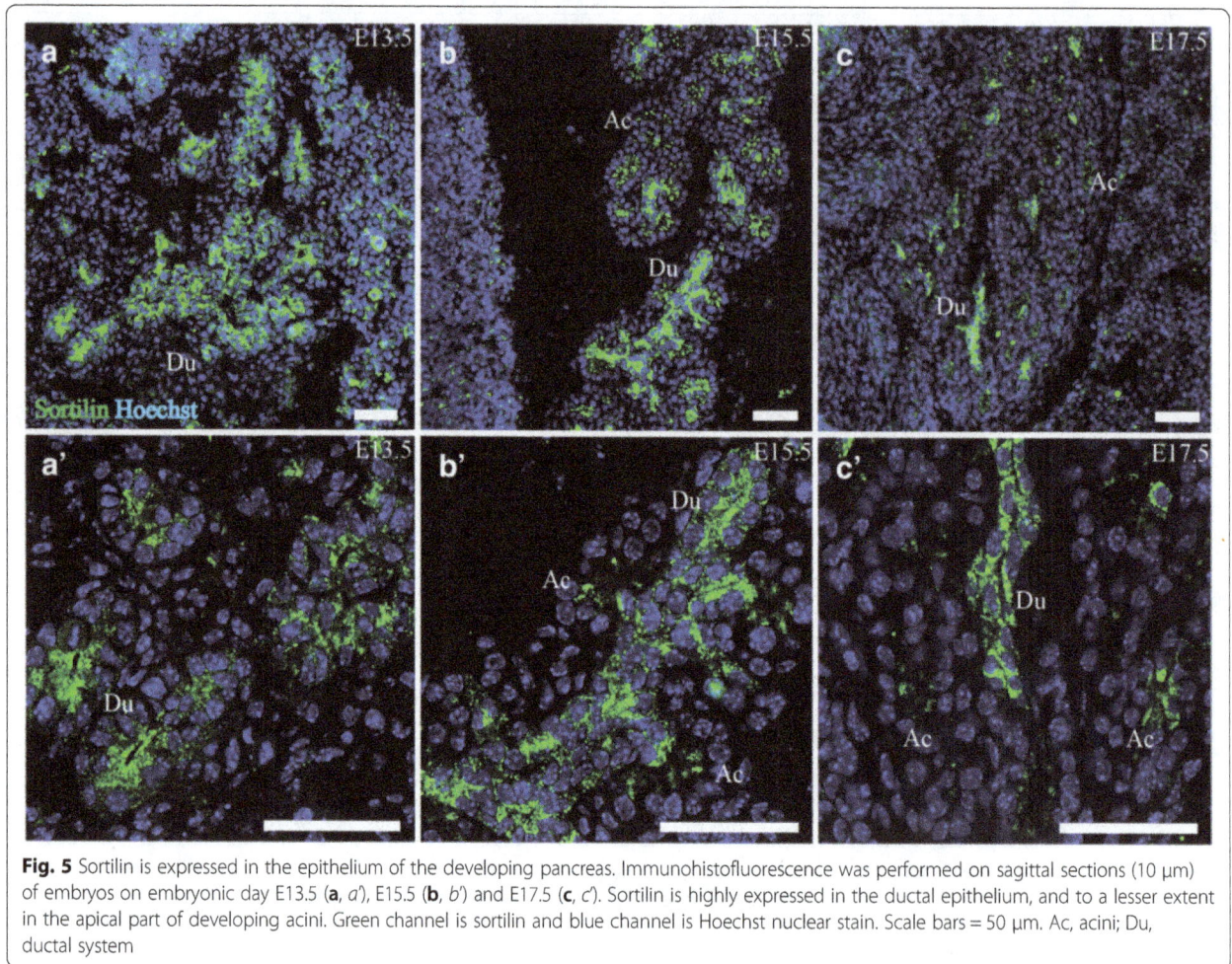

**Fig. 5** Sortilin is expressed in the epithelium of the developing pancreas. Immunohistofluorescence was performed on sagittal sections (10 μm) of embryos on embryonic day E13.5 (**a**, *a'*), E15.5 (**b**, *b'*) and E17.5 (**c**, *c'*). Sortilin is highly expressed in the ductal epithelium, and to a lesser extent in the apical part of developing acini. Green channel is sortilin and blue channel is Hoechst nuclear stain. Scale bars = 50 μm. Ac, acini; Du, ductal system

### Adipose tissue

At E13.5 and E15.5, adipose tissue showed no expression of sortilin. By E17.5, the brown (multilobular) fat deposits in the neck and back region showed strong sortilin IR, with a peak along the edges of deposits (Fig. 8g, h)

In summary, sortilin expression was found pronounced in skeletal structures undergoing both endochondral and intramembranous ossification, as well as in cartilage of e.g. larynx. Furthermore brown fat deposits showed strong sortilin IR by E17.5.

### Endocrine system
### Thyroid gland

At E13.5, the thyroid gland showed strong sortilin IR in cells expressing TTF-1 (Fig. 9). By E15.5, folliculogenesis has commenced, and TTF-1 IR cells have begun to arrange in small groups with sortilin IR in the part of the cells facing the center. At E17.5, small follicles are clearly visible, with strong granular sortilin expression in the apical cytoplasm of TTF-1 IR cells facing the emerging lumen.

### Adrenal glands

In the developing adrenal glands, expression of sortilin was noted at E13.5 in cells staining positive for tyrosine hydroxylase (TH). Furthermore, cells surrounding the TH positive cells in the adrenal glands also showed sortilin IR. This expression diminished to non-detectable levels at E15.5 and E17.5.

### SorCS2
### Respiratory tract

**Lung** SorCS2 was found expressed at high levels in the apical parts of developing bronchi analogous to the expression of sortilin, throughout E13.5, E15.5 and E17.5, with the exception that SorCS2 showed a less pronounced granular pattern. The visceral blade of the pleura also stained clearly for SorCS2 (Fig. 10a-c).

### Digestive tract

**Gut** Moderate expression of SorCS2 was found in the connective tissue of the lamina propria and

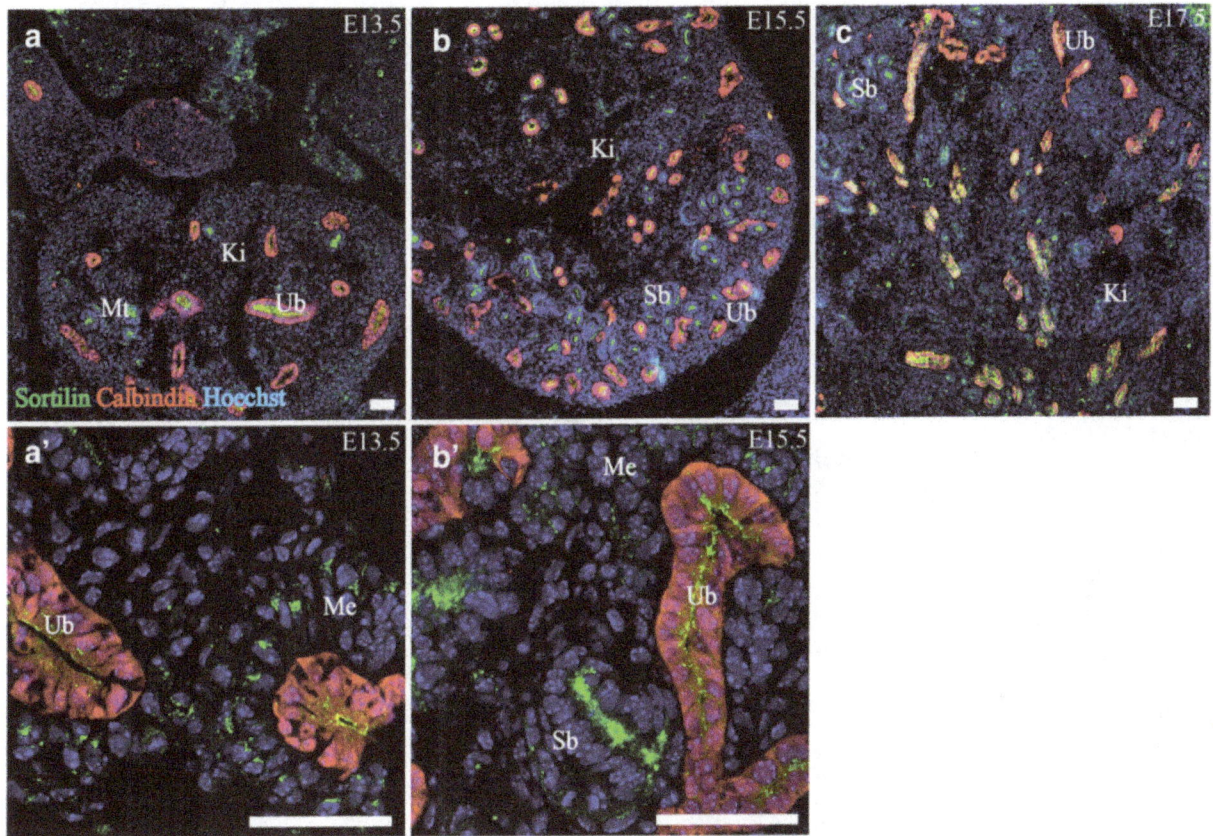

**Fig. 6** Sortilin is expressed in the tubular system of the developing metanephros. Immunohistofluorescence was performed on sagittal sections (10 μm) of embryos on embryonic day E13.5 (**a**, *a'*), E15.5 (**b**, *b'*) and E17.5 (**c**). Sortilin shows marked apical expression in ureteric branches and a portion of the metanephric tubular system. This expression becomes more prominent from E13.5-E15.5 and remains stable at E17.5. Green channel is sortilin, red channel is Calbindin-D28K and blue channel is Hoechst nuclear stain. Scale bars = 50 μm. Ki, kidney; Me, metanephric mesenchyme; Sb, S-shaped body Ub, ureteric branch

submucosal layer of the developing gut at E15.5 and E17.5 as well as the submesothelial connective tissue (Fig. 10j-l).

**Liver** The liver showed strong expression of SorCS2 at E13.5. The cells expressing SorCS2 were spherical with a round nucleus and sparse cytoplasm around 10 μm in diameter and SorCS2 was located perinuclear in a granular pattern. The cells were often located together in small groups interspersed by areas of cells without expression of SorCS2 (Fig. 10d). By E15.5 and E17.5 this pattern of expression had diminished to very low levels. At E15.5 and E17.5 expression of SorCS2 is noted in vessel-like structures corresponding to central veins (Fig. 10e, f).

### Urinary system

**Kidney** Throughout all three time points there was a moderate expression of SorCS2 in the periphery of the developing kidney, corresponding to the cap mesenchyme/cortical interstitium. The level of this expression remained fairly constant throughout the three time points investigated (Fig. 10g-i).

### Heart and vascular system

**Heart** The ventricles and atria of the developing heart both showed weak levels of expression of SorCS2 at E13.5. This expression was diminished at E15.5 and was non-detectable at E17.5 (Fig. 11a, b).

**Vessels** Prospective small arterioles and medium-large arteries showed prominent expression of SorCS2 in both the systemic and pulmonic circulation. SorCS2-IR localized to concentric cell layers in the vessel wall disparate from the endothelium (Fig. 11c–e). Larger veins also showed faint expression of SorCS2 in the same pattern.

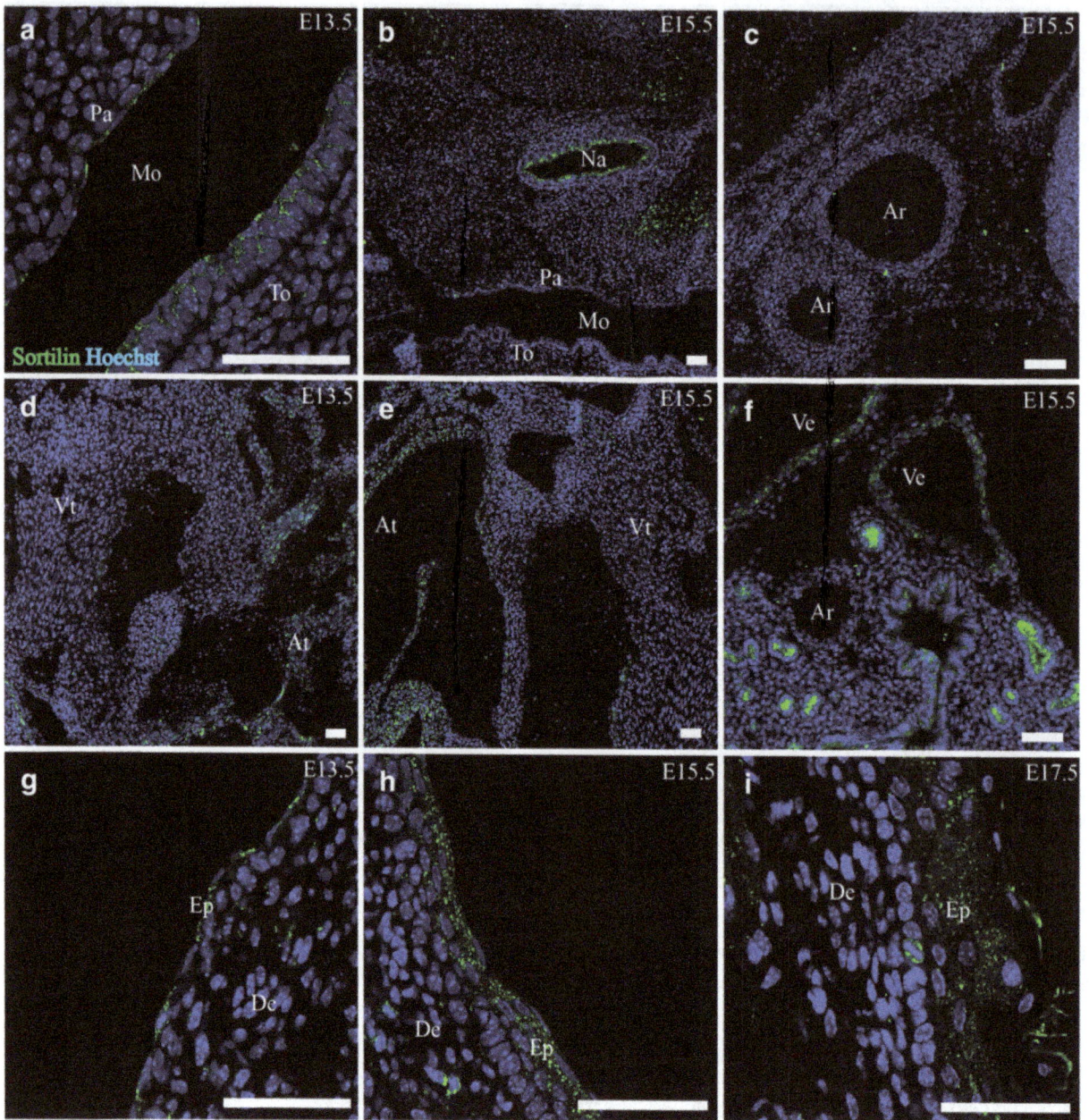

**Fig. 7** Sortilin expression in mouth, heart, vascular system and skin. Immunohistofluorescence was performed on sagittal sections (10 μm) of embryos on embryonic day E13.5, E15.5 and E17.5. Sortilin is found in the epithelium of the tongue and palate at E13.5 but is gone by E15.5 (**a,b**). In the cardiovascular system, sortilin is found in the atria and large veins (**d-f**) but not in the arteries (**c**), Finally, sortilin is found moderately expressed in the epidermal layers of the skin most prominently at E15.5 (**g-i**), Green channel is sortilin, and blue channel is Hoechst nuclear stain. Scale bars = 50 μm. Ar, arteries; At, atria of heart; De, dermis; Ep, epidermis; Mo, mouth cavity; Na, nasal cavity; Pa, epithelium of palate; To, tongue; Ve, veins, Vt, ventricle of heart

*Musculoskeletal system*

**Cartilage/Bone** SorCS2 was prominently expressed in ossifying bone such as the mandible and vertebrae at E15.5 and E17.5. The expression localized mostly to the outer 1/3 of the ossifying area. Furthermore, SorCS2 was prominently expressed in a concentric layer of

mesenchyme surrounding the costae, sternum, clavicle and skull (Fig. 12d, e).

**Striated muscle** Throughout E13.5, E15.5 and E17.5 SorCS2 was found expressed at high levels ubiquitously in striated muscle. The expression located mainly in satellite cells closely associated with muscle cells but also

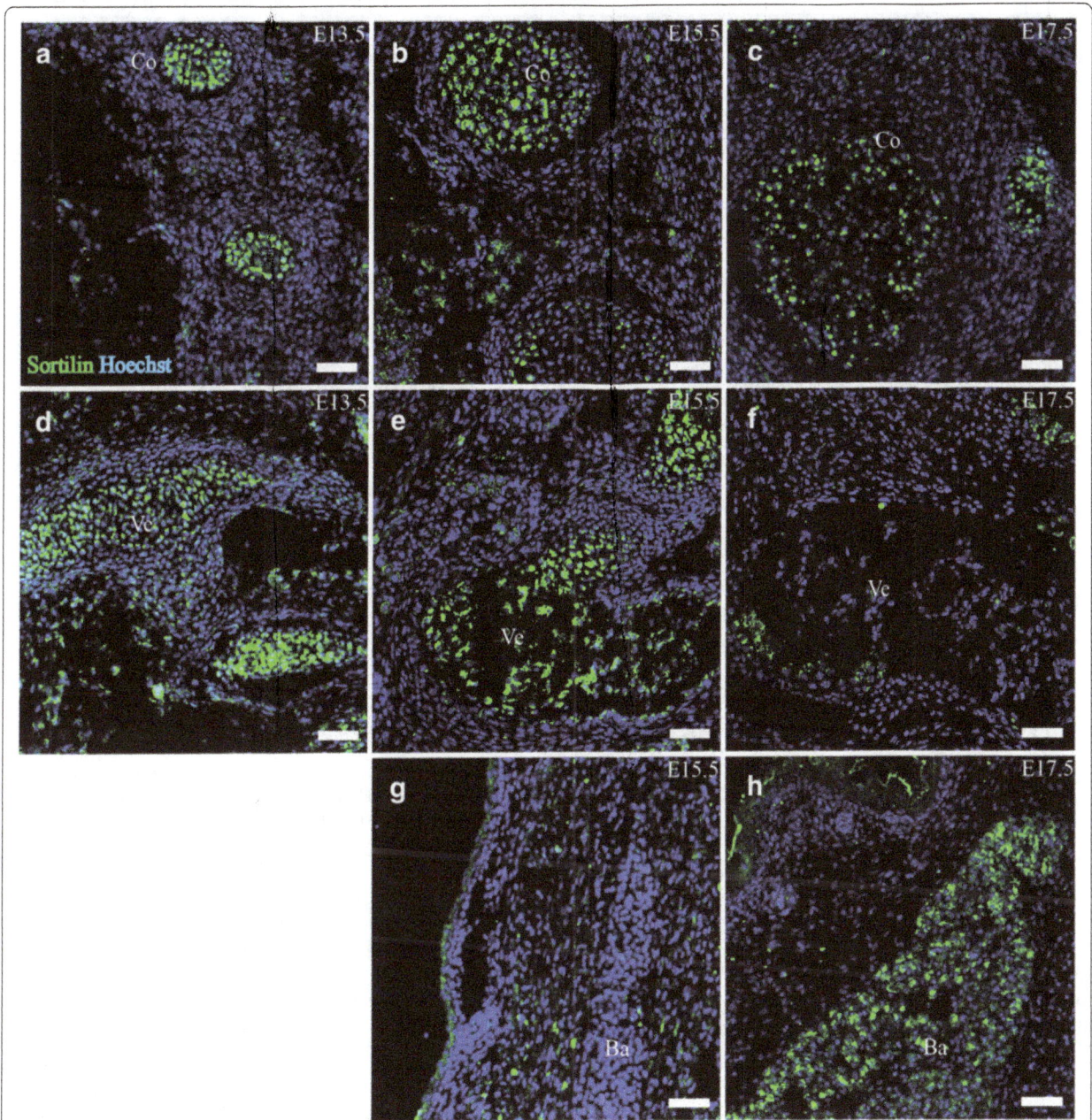

**Fig. 8** Sortilin is expressed in costae, vertebrae and brown adipose tissue. Immunohistofluorescence was performed on sagittal sections (10 μm) of embryos on embryonic day E13.5, E15.5 and E17.5. Sortilin show marked central expression in developing costae (**a-c**), as well as in the chondral primordium and later ossifying vertebrae (**d-f**). Sortilin expression is absent from brown adipose tissue until E17.5 where it is prominent (**g,h**). Green channel is sortilin and blue channel is Hoechst nuclear stain. Scale bars = 50 μm. Ba, brown adipose tissue; Co, costae; Ve, vertebrae

directly to one of the repeating bands of sarcomeres in muscle cells. The sarcomere part diminished at E17.5 (Fig. 12a-c).

**Skin** The connective tissue of the dermal layers of the skin had pronounced expression of SorCS2 at E15.5 and E17.5 (Fig. 11f, i).

**Adipose tissue** SorCS2 was found expressed abundantly in the brown multilobular adipose tissue of the neck at all three age points (Fig. 12h, i).

*Endocrine system*

**Adrenal glands** Expression of SorCS2 was prominent in the developing adrenal gland at E13.5, confined to cells

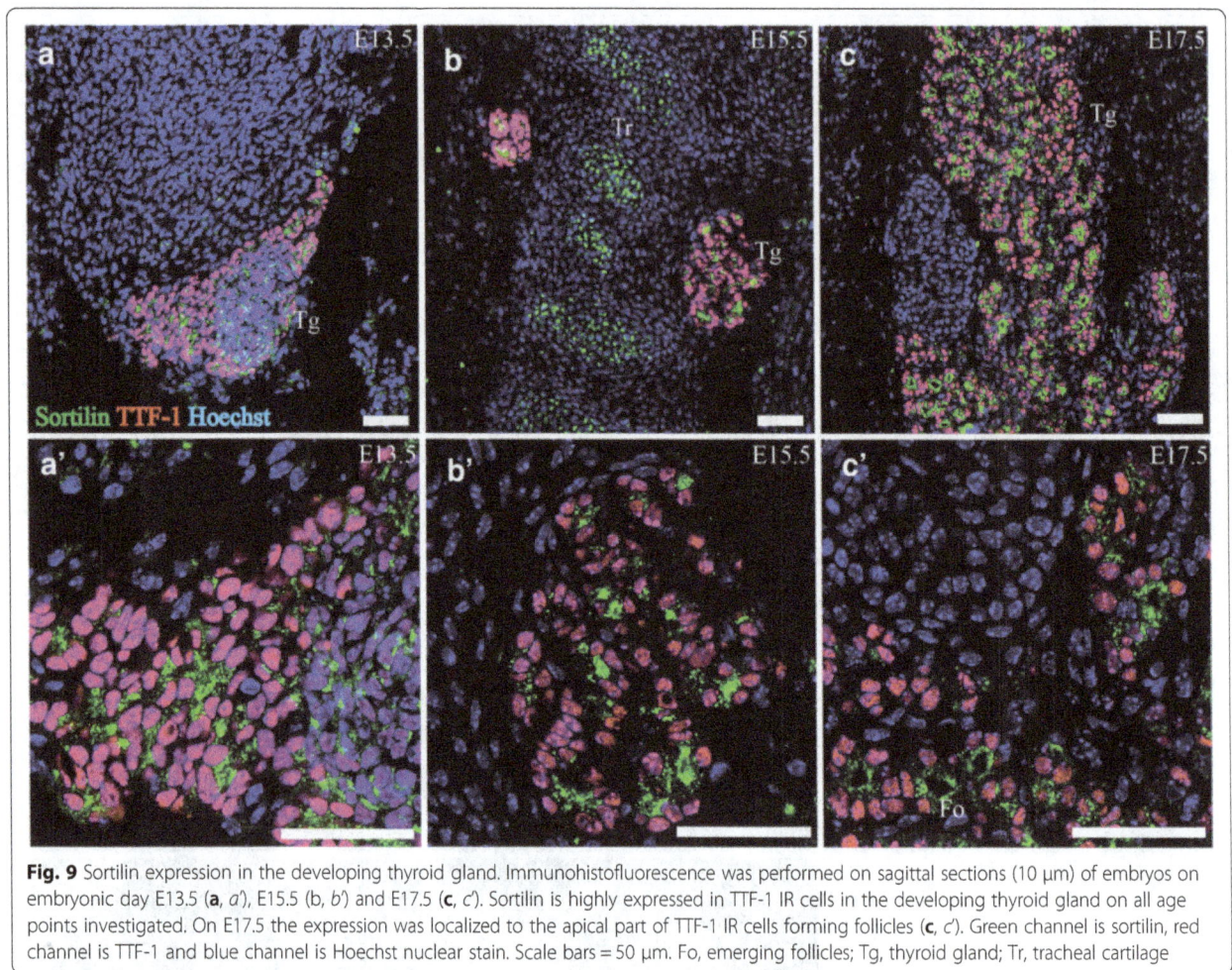

**Fig. 9** Sortilin expression in the developing thyroid gland. Immunohistofluorescence was performed on sagittal sections (10 μm) of embryos on embryonic day E13.5 (**a**, *a'*), E15.5 (**b**, *b'*) and E17.5 (**c**, *c'*). Sortilin is highly expressed in TTF-1 IR cells in the developing thyroid gland on all age points investigated. On E17.5 the expression was localized to the apical part of TTF-1 IR cells forming follicles (**c**, *c'*). Green channel is sortilin, red channel is TTF-1 and blue channel is Hoechst nuclear stain. Scale bars = 50 μm. Fo, emerging follicles; Tg, thyroid gland; Tr, tracheal cartilage

in close proximity to TH+ chromaffin progenitor cells. This corresponds to the steroidogenic cell population. Furthermore expression of SorCS2 was present in the peripheral mesenchyme. By E15.5 the expression in the steroidogenic cell population had diminished to low levels and was undetectable at E17.5 (Fig. 11g, h).

## Discussion

In the CNS, sortilin and SorCS2 show a complementary expression with regards to cell type and subcellular distribution [2]. This disparate expression pattern might indicate overlapping roles for sortilin and SorCS2 in the CNS, in addition to their unique functions. Indeed, the ability of both proteins to mediate apoptotic signals in the CNS could mean that the some of the differences in the amino-acid sequence of sortilin and SorCS2 help suit them to mediate similar effects in dissimilar cell types. As such it was interesting to see if these largely complementary expression patterns were also found in non-neuronal tissues of the developing embryo. This indeed seemed to be the case, as sortilin was mainly found in epithelial tissues, whereas SorCS2 predominated in mesodermally derived connective tissues.

## Sortilin in the developing embryo

Sortilin has been shown to bind multiple ligands, and to have physiological functions beyond the nervous system. Possible roles for sortilin in the development of the embryo have not been examined. We show sortilin to be abundantly expressed in the developing embryo in a surprisingly large number of organs and tissues. The most consistent finding is that sortilin is located to the epithelial tissues, and more specifically in the apical part of these. Many of these epithelial structures undergo significant branching in the period examined, notably lung, kidney, pancreas and the submandibular gland. The branching process is conceivably highly regulated and involves a variety of intrinsic and extrinsic mechanisms that work together to ensure the proper architecture of the given organ [32]. Among these mechanisms are several signaling molecules of well-known growth factor systems including members of the fibroblast growth factor family and bone morphogenetic proteins. During

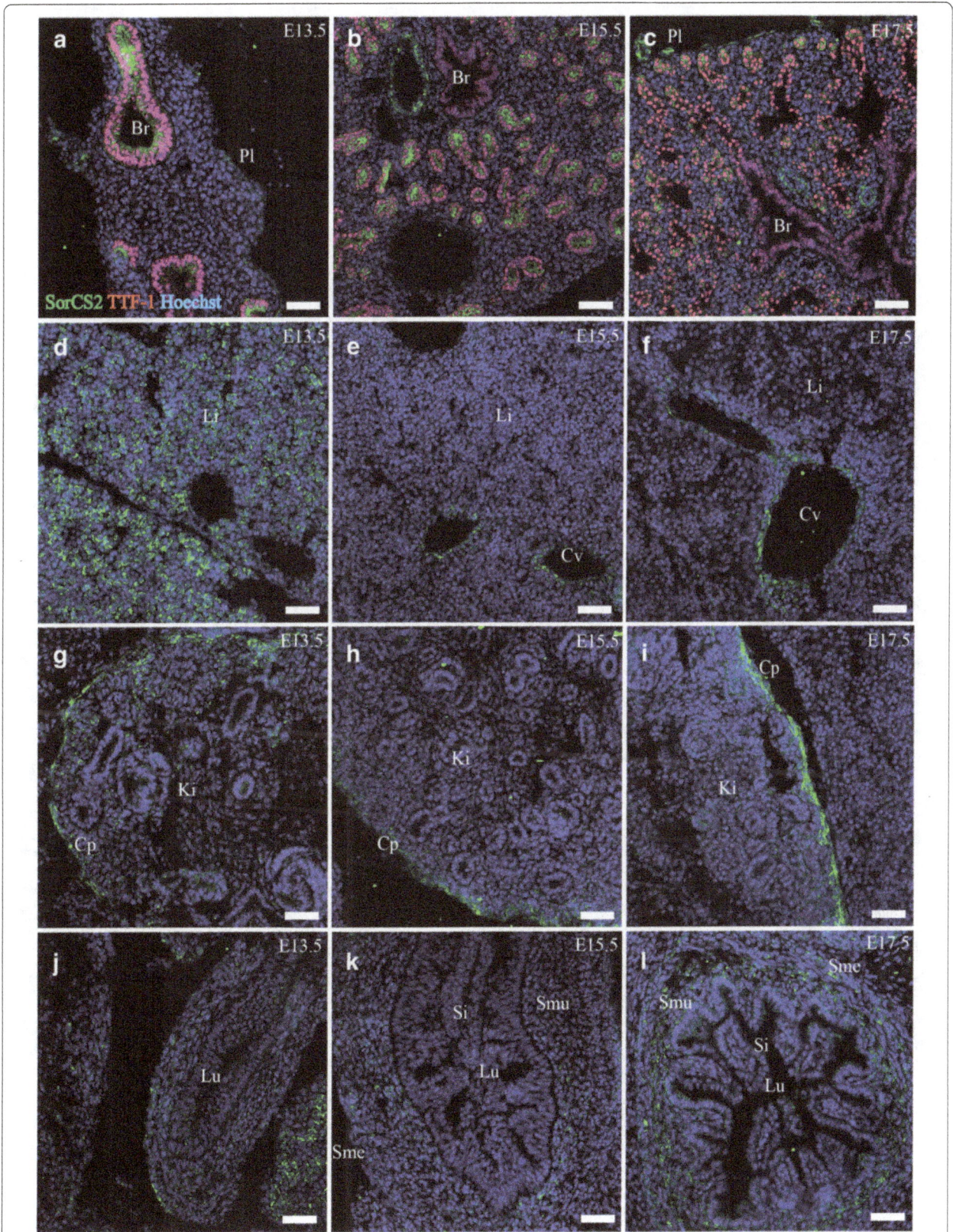

**Fig. 10** (See legend on next page.)

(See figure on previous page.)

**Fig. 10** SorCS2 expression in embryonic internal organs. Immunohistofluorescence was performed on sagittal sections (10 μm) of embryos on embryonic day E13.5, E15.5 and E17.5. SorCS2 is found prominently in the epithelium of the developing lung, and also the visceral pleural layer (**a-c**) It is highly expressed in the liver at E13.5 (**d**), but diminishes at later time points and instead localizes to central veins (**e,f**). SorCS2 is also expressed in the cap mesenchyme of the kidney (**g-i**), and in the connective tissue of the gut (**j-l**). Green channel is SorCS2, red channel is TTF-1 and blue channel is Hoechst nuclear stain. Scale bars = 50 μm. Br, bronchus; Cp, Cap mesenchyme; Cv, central vein; Ki, kidney; Li, liver; Lu, lumen; Pl pleuric cavity; Si, small Intestine; SMe, submesothelial connective tissue; SMu, submucosal connective tissue

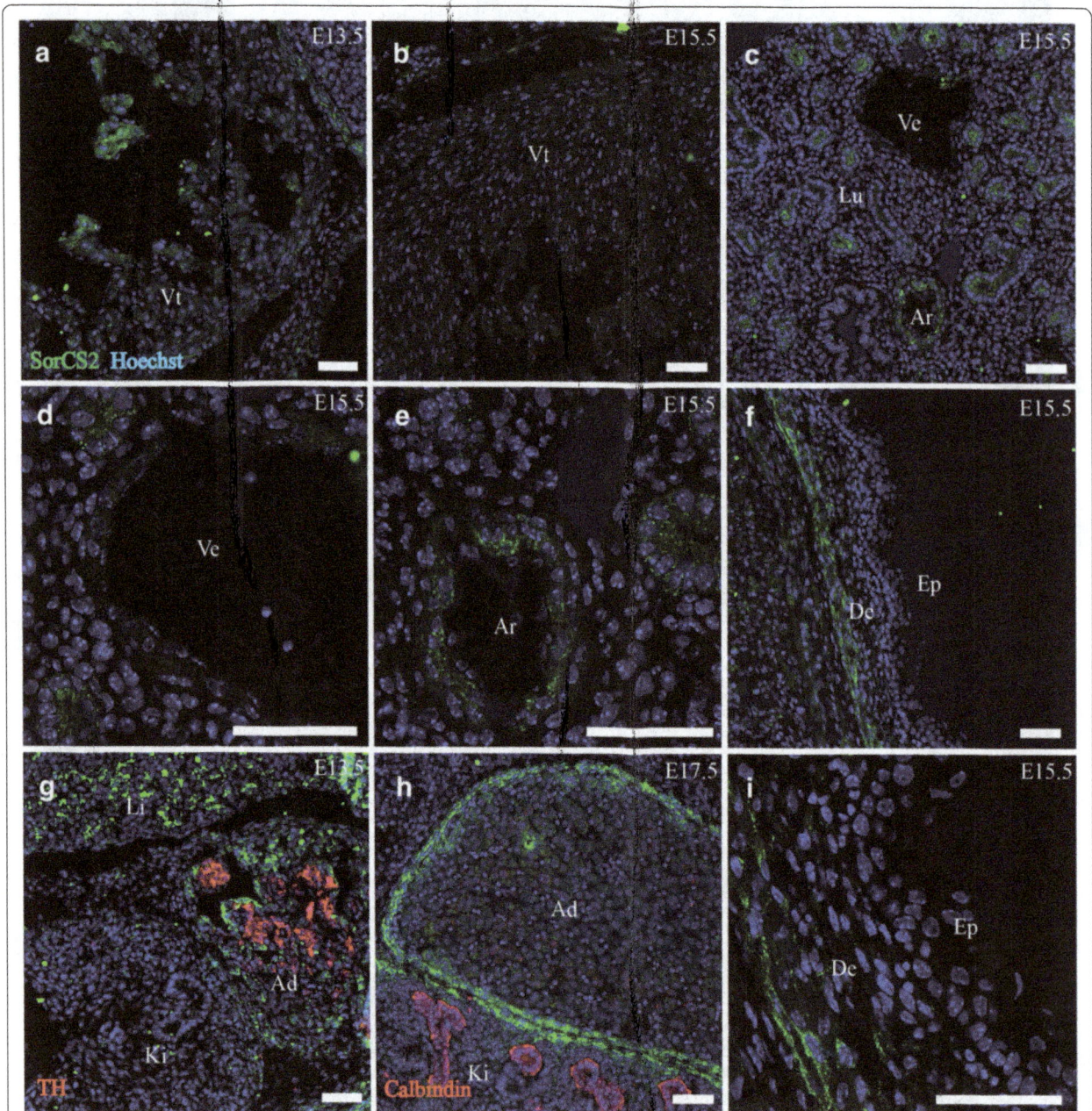

**Fig. 11** SorCS2 expression in cardiovascular system, skin and adrenal gland. Immunohistofluorescence was performed on sagittal sections (10 μm) of embryos on embryonic day E13.5, E15.5 and E17.5. SorCS2 is weakly expressed in the developing ventricle in a subset of cells (**a, b**). Medium-large arteries show prominent SorCS2 IR while veins show weak IR (**c-e**). In the skin SorCS2 is found localized to the deep dermal layers (**f, i**). Lastly, the adrenal gland shows pronounced expression of SorCS2 in a large portion of the cells (**g**) but by E17.5 this localization has changed to the periphery of the gland, analogous to the kidney (**h**). Green channel is SorCS2, red channel is TH (**g**) or calbindin (**h**) and blue channel is Hoechst nuclear stain. Scale bars = 50 μm. Ad, adrenal gland; Ar, arteries; De, dermis; Ep, epidermis; Ki, kidney; Lu, lung; Ve, vein; Vt, ventricle of heart

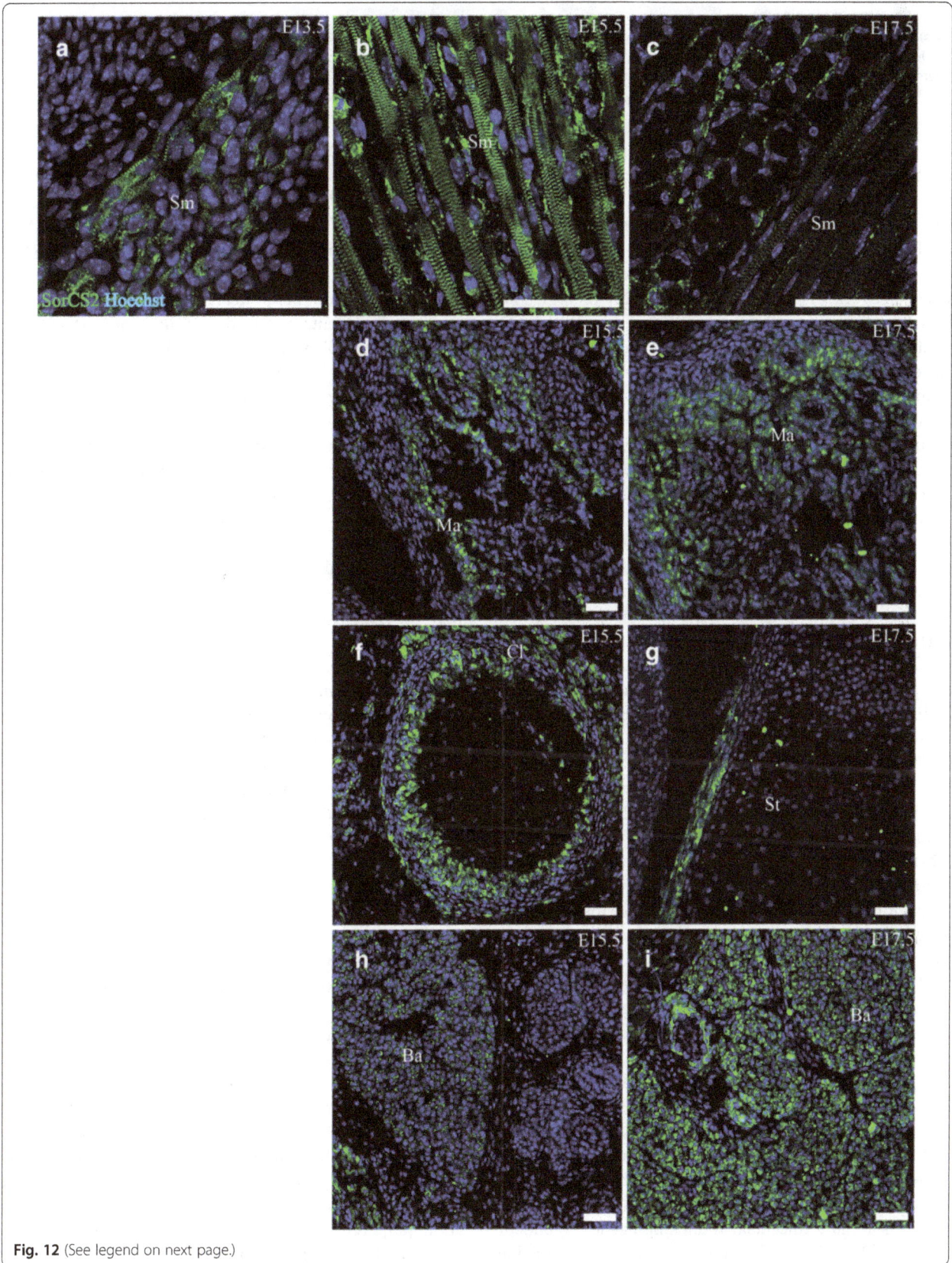

**Fig. 12** (See legend on next page.)

epithelial differentiation, there is a substantial amount of receptor regulation, where epithelia lose and acquire sensitivity to different extrinsic signals as they differentiate into more mature cell types. Sortilin is a multi-ligand receptor shown to function at the level of the trans Golgi network (TGN) in sorting a variety of molecules between different cellular compartments [3]. Sortilin could participate in the trafficking and regulation of various receptors or effector molecules involved in the branching process of these organs. The preliminary data from the lung of a sortilin(−/−) embryo could indicate a reduced level of branching, with fewer and larger bronchial branches compared to WT, however this needs to be confirmed with a more thorough investigation. Interestingly, sortilin has been shown to interact with transforming growth factor beta (TGFβ) and increase its degradation [33]. A study using a constitutively active form of TGFβ-1 in fetal mice showed arrest of lung development in the pseudo glandular stage, and lack of sortilin could hypothetically result in increased TGFβ signaling leading to delayed lung maturation [34]. The temporal shift in sortilin expression noted in the lung is in line with a hypothesis of sortilin playing a role in the branching process of this organ.

The developing thyroid gland showed strong sortilin expression through all three age points. Interestingly sortilin has been shown to bind thyroglobulin, the precursor protein of thyroid hormones T3 and T4 [35]. Thyroglobulin is secreted from the apical side of thyroid epithelia, the tyrosine residues are modified after which thyroglobulin is taken up again by endocytosis and proteolysed in lysosomes. Despite this, later investigations have showed that the levels of TSH and thyroglobulin as well as the morphology of the thyroid gland are normal in sortilin(−/−) mice. The only difference found was reduced T4 levels in male sortilin(−/−) mice [36]. Whether this is due to redundancy of sortilin in thyroid gland function, or a question of experimental circumstances remain to be discovered.

Another tissue with prominent sortilin IR is cartilage and bone. Both types of tissues produce and secrete large amounts of extracellular proteins regulated by extracellular signals, and sortilin could be involved in the substantial intracellular trafficking needed to effectuate this [37, 38]. Furthermore, sortilin is expressed in several other tissues ranging from blood cells to brown adipose tissue.

Earlier studies have examined the expression of sortilin and SorCS2. Petersen et al. investigated the expression of sortilin in adult human tissues by northern blotting [39]. They found expression of two different mRNA transcripts of 8 and 3.5 kb. The 8 kb transcript was found in high amounts in heart and skeletal muscle, thyroid gland, placenta and testis and in minor amounts in kidney, colon, liver and lymphoid organs. Hermans-Borgmeyer et al. have investigated the expression of sortilin mRNA in mouse embryological development with focus on the central nervous system. In the peripheral organs they found the sortilin mRNA transcripts most abundantly expressed in lung and kidney, but also in heart and skeletal muscle [40]. Sarret et al. investigated the distribution of Sortilin mRNA and protein in rat central nervous system. They found that sortilin predominantly localized in cell somata as well as in the proximal dendrites of neurons [41]. Generally, this is well in line with our findings with the exception of Petersen et al. who investigated adult human tissue. While the expression of sortilin in adult human vs fetal mouse tissue can be genuinely different there may be other explanations as well. The localization of mRNA transcripts and the protein sortilin are not necessarily the same, which could also be the reason why Hermans-Borgmeyer et al. find expression of sortilin in skeletal muscle in contrast to us. Furthermore, Petersen et al. do not state whether they had other human tissues available so it is possible that they would find expression also in lung, pancreas, salivary glands and so forth.

## SorCS2

Contrary to sortilin, SorCS2 was found expressed prominently in mesodermally derived structures such as adipose tissue, striated muscle tissue and developing bone. This could indicate a function for SorCS2 in these tissues. A genome wide association study found that a single nucleotide polymorphism in the locus of the SorCS2 gene was significantly associated with circulating levels of insulin-like growth factor binding protein 3 (IGFBP-3) in a human population [42]. IGFBP-3 binds and regulates the bioavailability of IGF-1 in plasma, and overexpression in a mouse model showed growth retardation [43]. IGF-1 is a primary effector of growth hormone, and stimulates growth of most tissues including muscle, adipose tissue and bone [44]. Thus, it is interesting to investigate whether SorCS2(−/−) show signs of reduced growth. Furthermore, SorCS2 was found expressed in several connective tissues, such as the dermis, submucosal and submesothelial layers of the gut. Finally SorCS2 was also found in the epithelium of the bronchial system, which differs somewhat from the general localization.

Rezgaoui et al. have investigated the expression of SorCS2 in mouse embryonic tissue by in situ hybridization in CD-1 mice. They found SorCS2 transcripts localized to facial mesenchyme, skeletal and smooth muscle, cartilaginous tissue and lung epithelial tissue. Furthermore, they also found weak levels of expression in heart muscle and olfactory epithelium. This corresponds well with our findings, apart from the SorCS2 protein we see in adipose tissue. This could be due to methodological differences as mentioned above, or to differences in gene expression between CD-1 and C57J/bl6bom mice.

The complementary expression pattern of sortilin and SorCS2 in some body parts raises the question whether this also translates to complementary functions in these tissues. This seems to be the case for dorsal root ganglion neurons and Schwann cells, which require sortilin and SorCS2 respectively, for $p75^{NTR}$ mediated apoptosis [2]. SorCS2 differs from sortilin in that it does not have complete cytoplasmic tail sorting motifs needed to convey transport between the TGN and lysosomes [45]. Thus it is possible that SorCS2 does not share functional similarities in this aspect, but rather in functions on the cell surface. It should be noted that the complementary pattern described is not complete, as there are several organs that contain sortilin but not SorCS2 e.g. pancreas, submandibular gland, and organs that contain both proteins in the same cells (lung).

## Conclusions

The developing mouse embryo shows an extensive and distinct expression of sortilin and SorCS2 in a large variety of tissues, with several dissimilar physiological roles possible. It is our hope that this structural characterization will encourage new studies of the functions of sortilin and SorCS2 in organ development.

#### Competing interests
The authors declare no competing interests.

#### Authors' contributions
SB carried out the experiments, drafted and revised the manuscript. SM, SG and JN helped conceiving the study, revised the manuscript and interpreted the data. All authors read and approved the final manuscript.

#### Acknowledgements
MIND Centre is supported by Lundbeck Foundation, Centre for Stochastic Geometry and Advanced Bioimaging is supported by Villum Foundation and Simon Boggild is supported by Aarhus University. Simon Glerup is supported by a Sapere Aude Starting Grant from the Danish Council for Independent Research. The authors would like to thank Helene M. Andersen, Benedicte Vestergaard and Anja Aagaard Danneskjold Pedersen for excellent technical assistance

#### Author details
[1]MIND Centre, Stereology and Electron Microscopy Laboratory, Aarhus University, 8000 C Aarhus, Denmark. [2]MIND Centre, Department of Biomedicine, Aarhus University, Ole Worms Allé 3, 8000 C Aarhus, Denmark. [3]Centre for Stochastic Geometry and Advanced Bioimaging, Aarhus University, 8000 C Aarhus, Denmark.

#### References
1. Willnow TE, Petersen CM, Nykjaer A. VPS10P-domain receptors - regulators of neuronal viability and function. Nat Rev Neurosci. 2008;9:899–909.
2. Glerup S, Olsen D, Vaegter CB, Gustafsen C, Sjoegaard SS, Hermey G, Kjolby M, Molgaard S, Ulrichsen M, Boggild S, Skeldal S, Fjorback AN, Nyengaard JR, Jacobsen J, Bender D, Bjarkam CR, Sorensen ES, Fuchtbauer EM, Eichele G, Madsen P, Willnow TE, Petersen CM, Nykjaer A. SorCS2 Regulates Dopaminergic Wiring and Is Processed into an Apoptotic Two-Chain Receptor in Peripheral Glia. Neuron. 2014;82:1074–87.
3. Nielsen MS, Madsen P, Christensen EI, Nykjaer A, GLIEMANN J, Kasper D, Pohlmann R, Petersen CM. The sortilin cytoplasmic tail conveys Golgi-endosome transport and binds the VHS domain of the GGA2 sorting protein. EMBO J. 2001;20:2180–90.
4. Jansen P, Giehl K, Nyengaard JR, Teng K, Lioubinski O, Sjoegaard SS, Breiderhoff T, Gotthardt M, Lin F, Eilers A, Petersen CM, Lewin GR, Hempstead BL, Willnow TE, Nykjaer A. Roles for the pro-neurotrophin receptor sortilin in neuronal development, aging and brain injury. Nat Neurosci. 2007;10:1449–57.
5. Vaegter CB, Jansen P, Fjorback AW, Glerup S, Skeldal S, Kjolby M, Richner M, Erdmann B, Nyengaard JR, Tessarollo L, Lewin GR, Willnow TE, Chao MV, Nykjaer A. Sortilin associates with Trk receptors to enhance anterograde transport and neurotrophin signaling. Nat Neurosci. 2011;14:54–61.
6. Nykjaer A, Lee R, Teng KK, Jansen P, Madsen P, Nielsen MS, Jacobsen C, Kliemannel M, Schwarz E, Willnow TE, Hempstead BL, Petersen CM. Sortilin is essential for proNGF-induced neuronal cell death. Nature. 2004;427:843–8.
7. Nykjaer A, Willnow TE. Sortilin: a receptor to regulate neuronal viability and function. Trends Neurosci. 2012;35:261.
8. Teng HK, Teng KK, Lee R, Wright S, Tevar S, Almeida RD, Kermani P, Torkin R, Chen ZY, Lee FS, Kraemer RT, Nykjaer A, Hempstead BL. ProBDNF induces neuronal apoptosis via activation of a receptor complex of p75NTR and sortilin. J Neurosci Off J Soc Neurosci. 2005;25:5455–63.
9. Tauris J, Gustafsen C, Christensen EI, Jansen P, Nykjaer A, Nyengaard JR, Teng KK, Schwarz E, Ovesen T, Madsen P, Petersen CM. Proneurotrophin-3 may induce Sortilin-dependent death in inner ear neurons. Eur J Neurosci. 2011;33:622–31.
10. Skeldal S, Sykes AM, Glerup S, Matusica D, Palstra N, Autio H, Boskovic Z, Madsen P, Castren E, Nykjaer A, Coulson EJ. Mapping of the interaction site between sortilin and the p75 neurotrophin receptor reveals a regulatory role for the sortilin intracellular domain in p75 neurotrophin receptor shedding and apoptosis. J Biol Chem. 2012;287:43798–809.
11. Deinhardt K, Kim T, Spellman DS, Mains RE, Eipper BA, Neubert TA, Chao MV, Hempstead BL. Neuronal growth cone retraction relies on proneurotrophin receptor signaling through Rac. Sci Signal. 2011;4:ra82.
12. Gustafsen C, Glerup S, Pallesen LT, Olsen D, Andersen OM, Nykjaer A, Madsen P, Petersen CM. Sortilin and SorLA display distinct roles in processing and trafficking of amyloid precursor protein. J Neurosci Off J Soc Neurosci. 2013;33:64–71.
13. Baum AE, Akula N, Cabanero M, Cardona I, Corona W, Klemens B, Schulze TG, Cichon S, Rietschel M, Nothen MM, Georgi A, Schumacher J, Schwarz M, Abou Jamra R, Hofels S, Propping P, Satagopan J, Detera-Wadleigh SD, Hardy J, Mcmahon FJ. A genome-wide association study implicates diacylglycerol kinase eta (DGKH) and several other genes in the etiology of bipolar disorder. Mol Psychiatry. 2008;13:197–207.
14. Christoforou A, Mcghee KA, Morris SW, Thomson PA, Anderson S, Mclean A, Torrance HS, LE Hellard S, Pickard BS, Stclair D, Muir WJ, Blackwood DH, Porteous DJ, Evans KL. Convergence of linkage, association and GWAS findings for a candidate region for bipolar disorder and schizophrenia on chromosome 4p. Mol Psychiatry. 2011;16:240–2.
15. Ollila HM, Soronen P, Silander K, Palo OM, Kieseppa T, Kaunisto MA, Lonnqvist J, Peltonen L, Partonen T, Paunio T. Findings from bipolar disorder genome-wide association studies replicate in a Finnish bipolar family-cohort. Mol Psychiatry. 2009;14:351–3.

16. Alemany S, Ribases M, Vilor-Tejedor N, Bustamante M, Sanchez-Mora C, Bosch R, Richarte V, Cormand B, Casas M, Ramos-Quiroga JA and Sunyer J. New suggestive genetic loci and biological pathways for attention function in adult attention-deficit/hyperactivity disorder. Am. J. Med. Genet. B Neuropsychiatr. Genet. 2015;168:459–70.

17. Sanchez E, Bergareche A, Krebs CE, Gorostidi A, Makarov V, Ruiz-Martinez J, Chorny A, Lopez de Munain A, Marti-Masso JF, Paisan-Ruiz C. SORT1 Mutation Resulting in Sortilin Deficiency and p75NTR Upregulation in a Family With Essential Tremor. ASN neuro. 2015; 7, 10.1177/1759091415598290. Print 2015 Jul.

18. Reitz C, Tosto G, Vardarajan B, Rogaeva E, Ghani M, Rogers RS, Conrad C, Haines JL, Pericak-Vance MA, Fallin MD, Foroud T, Farrer LA, Schellenberg GD, George-Hyslop PS, Mayeux R, Alzheimer's Disease Genetics Consortium (ADGC). Independent and epistatic effects of variants in VPS10-d receptors on Alzheimer disease risk and processing of the amyloid precursor protein (APP). Transl Psychiatry. 2013;3, e256.

19. Huang G, Buckler-Pena D, Nauta T, Singh M, Asmar A, Shi J, Kim JY, Kandror KV. Insulin responsiveness of glucose transporter 4 in 3T3-L1 cells depends on the presence of sortilin. Mol Biol Cell. 2013;24:3115–22.

20. Mijatovic T, Gailly P, Mathieu V, de Neve N, Yeaton P, Kiss R, Decaestecker C. Neurotensin is a versatile modulator of in vitro human pancreatic ductal adenocarcinoma cell (PDAC) migration. Cell Oncol. 2007;29:315–26.

21. Roselli S, Pundavela J, Demont Y, Faulkner S, Keene S, Attia J, Jiang CC, Zhang XD, Walker MM, Hondermarck H. Sortilin is associated with breast cancer aggressiveness and contributes to tumor cell adhesion and invasion. Oncotarget. 2015;6:10473–86.

22. Xiong J, Zhou L, Yang M, Lim Y, Zhu YH, Fu DL, Li ZW, Zhong JH, Xiao ZC, Zhou XF. ProBDNF and its receptors are upregulated in glioma and inhibit the growth of glioma cells in vitro. Neuro Oncol. 2013;15:990–1007.

23. Kathiresan S, Melander O, Guiducci C, Surti A, Burtt NP, Rieder MJ, Cooper GM, Roos C, Voight BF, Havulinna AS, Wahlstrand B, Hedner T, Corella D, Tai ES, Ordovas JM, Berglund G, Vartiainen E, Jousilahti P, Hedblad B, Taskinen MR, Newton-Cheh C, Salomaa V, Peltonen L, Groop L, Altshuler DM, Orho-Melander M. Six new loci associated with blood low-density lipoprotein cholesterol, high-density lipoprotein cholesterol or triglycerides in humans. Nat Genet. 2008;40:189–97.

24. Musunuru K, Strong A, Frank-Kamenetsky M, Lee NE, Ahfeldt T, Sachs KV, Li X, Li H, Kuperwasser N, Ruda VM, Pirruccello JP, Muchmore B, Prokunina-Olsson L, Hall JL, Schadt EE, Morales CR, Lund-Katz S, Phillips MC, Wong J, Cantley W, Racie T, Ejebe KG, Orho-Melander M, Melander O, Koteliansky V, Fitzgerald K, Krauss RM, Cowan CA, Kathiresan S, Rader DJ. From noncoding variant to phenotype via SORT1 at the 1p13 cholesterol locus. Nature. 2010; 466:714–9.

25. Strong A, Ding Q, Edmondson AC, Millar JS, Sachs KV, Li X, Kumaravel A, Wang MY, Ai D, Guo L, Alexander ET, Nguyen D, Lund-Katz S, Phillips MC, Morales CR, Tall AR, Kathiresan S, Fisher EA, Musunuru K, Rader DJ. Hepatic sortilin regulates both apolipoprotein B secretion and LDL catabolism. J Clin Invest. 2012;122:2807–16.

26. Kjolby M, Andersen OM, Breiderhoff T, Fjorback AW, Pedersen KM, Madsen P, Jansen P, Heeren J, Willnow TE, Nykjaer A. Sort1, encoded by the cardiovascular risk locus 1p13.3, is a regulator of hepatic lipoprotein export. Cell Metab. 2010;12:213–23.

27. Gustafsen C, Kjolby M, Nyegaard M, Mattheisen M, Lundhede J, Buttenschon H, Mors O, Bentzon JF, Madsen P, Nykjaer A, Glerup S. The hypercholesterolemia-risk gene SORT1 facilitates PCSK9 secretion. Cell Metab. 2014;19:310–8.

28. Wei H, Lang MF, Jiang X. Calretinin is expressed in the intermediate cells during olfactory receptor neuron development. Neurosci Lett. 2013;542:42–6.

29. Costa RH, Kalinichenko VV, Lim L. Transcription factors in mouse lung development and function. American journal of physiologyLung cellular and molecular physiology. 2001;280:L823–38.

30. Vestentoft PS. Development and molecular composition of the hepatic progenitor cell niche. Danish medical journal. 2013;60:B4640.

31. Liu L, Dunn ST, Christakos S, Hanson-Painton O, Bourdeau JE. Calbindin-D28k gene expression in the developing mouse kidney. Kidney Int. 1993;44: 322–30.

32. Iber D, Menshykau D. The control of branching morphogenesis. Open biology. 2013;3:130088.

33. Kwon S, Christian JL. Sortilin associates with transforming growth factor-beta family proteins to enhance lysosome-mediated degradation. J Biol Chem. 2011;286:21876–85.

34. Zeng X, Gray M, Stahlman MT, Whitsett JA. TGF-beta1 perturbs vascular development and inhibits epithelial differentiation in fetal lung in vivo. Dev Dyn. 2001;221:289–301.

35. Botta R, Lisi S, Pinchera A, Giorgi F, Marcocci C, Taddei AR, Fausto AM, Bernardini N, Ippolito C, Mattii L, Persani L, de Filippis T, Calebiro D, Madsen P, Petersen CM, Marino M. Sortilin is a putative postendocytic receptor of thyroglobulin. Endocrinology. 2009;150:509–18.

36. Lisi S, Madsen P, Botta R, Petersen CM, Nykjaer A, Latrofa F, Vitti P, Marino M. Absence of a Thyroid Phenotype in Sortilin Deficient Mice. Endocr. Pract. 2015; 21:981–5.

37. Chen ZY, Ieraci A, Teng H, Dall H, Meng CX, Herrera DG, Nykjaer A, Hempstead BL, Lee FS. Sortilin controls intracellular sorting of brain-derived neurotrophic factor to the regulated secretory pathway. J Neurosci Off J Soc Neurosci. 2005; 25:6156–66.

38. Hu F, Padukkavidana T, Vaegter CB, Brady OA, Zheng Y, Mackenzie IR, Feldman HH, Nykjaer A, Strittmatter SM. Sortilin-mediated endocytosis determines levels of the frontotemporal dementia protein, progranulin. Neuron. 2010;68:654–67.

39. Petersen CM, Nielsen MS, Nykjaer A, Jacobsen L, Tommerup N, Rasmussen HH, Roigaard H, Gliemann J, Madsen P, Moestrup SK. Molecular identification of a novel candidate sorting receptor purified from human brain by receptor-associated protein affinity chromatography. J Biol Chem. 1997;272:3599–605.

40. Hermans-Borgmeyer I, Hermey G, Nykjaer A, Schaller C. Expression of the 100-kDa neurotensin receptor sortilin during mouse embryonal development. Brain Res Mol Brain Res. 1999;65:216–9.

41. Sarret P, Krzywkowski P, Segal L, Nielsen MS, Petersen CM, Mazella J, Stroh T, Beaudet A. Distribution of NTS3 receptor/sortilin mRNA and protein in the rat central nervous system. J Comp Neurol. 2003;461:483–505.

42. Kaplan RC, Petersen AK, Chen MH, Teumer A, Glazer NL, Doring A, Lam CS, Friedrich N, Newman A, Muller M, Yang Q, Homuth G, Cappola A, Klopp N, Smith H, Ernst F, Psaty BM, Wichmann HE, Sawyer DB, Biffar R, Rotter JI, Gieger C, Sullivan LS, Volzke H, Rice K, Spyroglou A, Kroemer HK, Ida Chen YD, Manolopoulou J, Nauck M, Strickler HD, Goodarzi MO, Reincke M, Pollak MN, Bidlingmaier M, Vasan RS, Wallaschofski H. A genome-wide association study identifies novel loci associated with circulating IGF-I and IGFBP-3. Hum Mol Genet. 2011;20:1241–51.

43. Modric T, Silha JV, Shi Z, Gui Y, Suwanichkul A, Durham SK, Powell DR, Murphy LJ. Phenotypic manifestations of insulin-like growth factor-binding protein-3 overexpression in transgenic mice. Endocrinology. 2001;142:1958–67.

44. Semple RK, Bolander JR FF. The growth hormone axis. In: Baynes JW, Dominiczak MH, editors. Medical Biochemistry. Elsevier, China: Mosby; 2009. p. 545–546,547.

45. Hermey G, Riedel IB, Hampe W, Schaller HC, Hermans-Borgmeyer I. Identification and characterization of SorCS, a third member of a novel receptor family. Biochem Biophys Res Commun. 1999;266:347–51.

# 16

# Altered cellular localization and hemichannel activities of KID syndrome associated connexin26 I30N and D50Y mutations

Hande Aypek, Veysel Bay and Gülistan Meşe[*]

## Abstract

**Background:** Gap junctions facilitate exchange of small molecules between adjacent cells, serving a crucial function for the maintenance of cellular homeostasis. Mutations in connexins, the basic unit of gap junctions, are associated with several human hereditary disorders. For example, mutations in connexin26 (Cx26) cause both non-syndromic deafness and syndromic deafness associated with skin abnormalities such as keratitis-ichthyosis-deafness (KID) syndrome. These mutations can alter the formation and function of gap junction channels through different mechanisms, and in turn interfere with various cellular processes leading to distinct disorders. The KID associated Cx26 mutations were mostly shown to result in elevated hemichannel activities. However, the effects of these aberrant hemichannels on cellular processes are recently being deciphered. Here, we assessed the effect of two Cx26 mutations associated with KID syndrome, Cx26I30N and D50Y, on protein biosynthesis and channel function in N2A and HeLa cells.

**Results:** Immunostaining experiments showed that Cx26I30N and D50Y failed to form gap junction plaques at cell-cell contact sites. Further, these mutations resulted in the retention of Cx26 protein in the Golgi apparatus. Examination of hemichannel function by fluorescent dye uptake assays revealed that cells with Cx26I30N and D50Y mutations had increased dye uptake compared to Cx26WT (wild-type) containing cells, indicating abnormal hemichannel activities. Cells with mutant proteins had elevated intracellular calcium levels compared to Cx26WT transfected cells, which were abolished by a hemichannel blocker, carbenoxolone (CBX), as measured by Fluo-3 AM loading and flow cytometry.

**Conclusions:** Here, we demonstrated that Cx26I30N and D50Y mutations resulted in the formation of aberrant hemichannels that might result in elevated intracellular calcium levels, a process which may contribute to the hyperproliferative epidermal phenotypes of KID syndrome.

**Keywords:** Connexin26, Hemichannels, Keratitis-ichthyosis-deafness, Intracellular calcium

## Background

Gap junctions facilitate the intercellular communication between adjacent cells by allowing the exchange of small molecules that play roles in the regulation of many cellular events including proliferation, differentiation and cellular homeostasis [1, 2]. Gap junction biosynthesis starts in the ER-Golgi network by the oligomerization of six connexin subunits into hemichannels, known as connexons. Then, these connexons are transported to the plasma membrane where they can either align with other connexons from neighboring cells to complete the formation of intercellular channels, or function individually on the plasma membrane as hemichannels that can mediate the exchange of materials between the cell interior and the extracellular environment [3–8].

Gap junction channels and hemichannels are important modulators of tissue homeostasis as evidenced by the association of mutations in connexin genes with several hereditary disorders. For example, mutations in

---
* Correspondence: gulistanmese@iyte.edu.tr
Department of Molecular Biology and Genetics, Izmir Institute of Technology, Urla, Izmir, Turkey

connexin 26 (Cx26), a member of the connexin gene family, are the leading cause of non-syndromic hearing loss (NSHL) [9]. To date, more than 100 NSHL mutations throughout the Cx26 gene (GJB2) have been identified, and functional characterization of these mutations suggested that the majority are loss-of-function mutations that can lead to protein truncation, altered trafficking, misfolding of Cx26 protein or altered channel permeability [3, 10–12]. In addition to NSHL, Cx26 mutations were linked to syndromic deafness associated with various skin pathologies including keratitis-ichthyosis-deafness (KID) syndrome, palmoplantar keratoderma (PPK) and Vohwinkel syndrome (VS) [13]. Generation of diverse epidermal phenotypes during aforementioned skin disorders driven by different Cx26 mutations imply that associated mutations may have unique properties affecting distinct cellular machinery [3, 10, 14].

KID syndrome is a rare congenital genetic disorder with phenotypes of thickening of the skin, keratisis, scaly skin (ichthyosis) and deafness [15, 16]. The Cx26A40V was the first KID syndrome associated mutation that was shown to cause increased hemichannel activities and cell death in mRNA injected *Xenopus* oocytes [17]. Other studies on different KID syndrome mutations also suggested the involvement of altered hemichannel activities and cell death in both Xenopus oocytes and mammalian cell lines as a common mechanism for this disorder [18–22]. However, the molecular mechanisms that lead to cell death, and skin phenotypes due to active hemichannels are poorly understood. For that, we characterized two more KID syndrome mutations, Cx26I30N and D50Y, in order to understand whether they affect the protein biosynthetic pathway, hemichannel activities similar to previously characterized KID syndrome mutations and intracellular calcium levels that play essential roles in cellular processes,

especially in keratinocyte proliferation, differentiation and migration [23–25].

## Results

### Protein localization of Cx26I30N and D50Y mutant proteins

To examine the effects of Cx26I30N and D50Y KID syndrome associated mutations on protein synthesis and localization, gap junctional communication deficient cell line, HeLa, were transiently transfected with pIRES2EGFP2 Cx26WT (wild-type), I30N and D50Y constructs. 24 h after transfection, the protein synthesis and localization was determined by immunofluorescent staining of transfected cells (Fig. 1). Immunofluorescent staining with Cx26 specific antibody demonstrated that cells expressing Cx26WT, I30N and D50Y constructs were able to synthesize Cx26 proteins (Fig. 1). Further, cells transfected with Cx26WT targeted proteins to the plasma membrane where they formed gap junction plaques at the cell-to-cell junctions between adjacent cells (Fig. 1, white arrow head). On the other hand, in spite of positive protein synthesis in cells with Cx26I30N or D50Y constructs, no gap junctional plaques were observed between neighboring cells expressing mutant constructs (Fig. 1, red arrow heads). This suggested that Cx26I30N and D50Y proteins failed to form gap junction plaques between adjacent cells.

### Effect of Cx26I30N and D50Y mutations on the Cx26 protein trafficking

In order to determine the location of Cx26 proteins within cells, co-labelling of Cx26 protein with golgin-97, a Golgi apparatus marker, or Cx26 protein with Wheat germ agglutinin (WGA), for the plasma membrane staining, were performed (Fig. 2). As observed in images in Fig. 2a, there was no overlap between Cx26 and golgin-97 protein signals in Cx26WT expressing cells

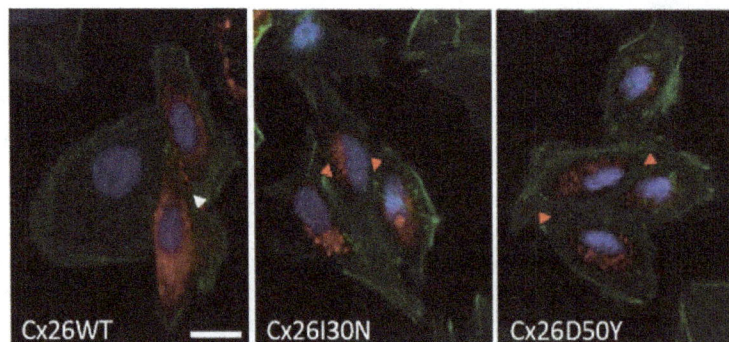

**Fig. 1** Effect of Cx26I30N and D50Y mutations on protein expression and localization. Merged images of Cx26WT, I30N and D50Y transfected HeLa cells that were co-stained with phalloidin for actin (*green*) and Cx26 antibody (*red*). *Blue* is for DAPI staining of the nucleus. *White arrow* head shows the gap junction plaques formed between adjacent Cx26WT expressing cells. Red arrow heads point the cell-to-cell contact sites between neighboring cells of Cx26I30N and D50Y. Cells with only green and blue signals indicate untransfected cells. Scale bar 10 μm

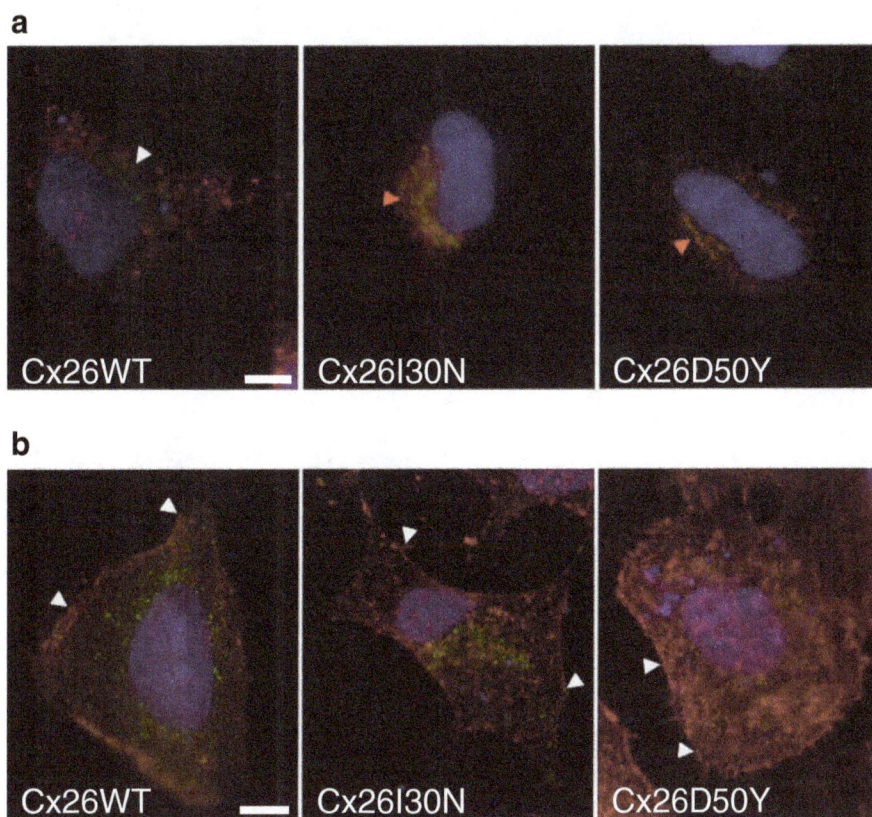

**Fig. 2** Effect of Cx26I30N and D50Y mutations on protein trafficking. **a** Merged images of Cx26WT, I30N and D50Y transfected cells that were co-stained with Cx26 (*red*) and the Golgi apparatus marker, golgin-97 (*green*) antibodies. Blue is for DAPI staining of the nucleus. Arrow heads point the location of the Golgi apparatus. Yellow signals in Cx26I30N and D50Y images show the colocalization of Cx26 and golgin-97. Scale bar 10 μm. **b** Merged images of Cx26WT, I30N and D50Y transfected cells that were co-stained with Cx26 (*green*) antibody and Rhodamine labeled wheat germ agglutinin (WGA). Blue is for DAPI staining of the nucleus. Yellow signals shown by arrow heads point out the colocalization of Cx26 and WGA. Scale bar 10 μm

while Cx26I30N or D50Y mutant proteins were widely co-localized with golgin-97 in transfected cells as evidenced by the presence of yellow signals (Fig. 2a, red arrow heads). Analysis of Costes' colocalization values through image processing revealed that 2 out of 25 (8 %) images for Cx26WT had significant colocalization for Cx26 protein and golgin-97 (p > 95 %). On the other hand, 19 out of 33 images (58 %) for Cx26I30N and 13 out of 26 (50 %) for Cx26D50Y had significant colocalization, ratios that were both significantly ($p < 0.01$) higher compared to WT images (Table 1). Furthermore, the membrane localization of Cx26WT, I30N and D50Y,

that was verified by Cx26 and Rhodamine labeled WGA co-staining suggested that a fraction of proteins in all conditions were localized to the plasma membrane (Fig. 2b, arrow heads). These suggested that mutant proteins can be found both in the Golgi apparatus and the plasma membrane.

## Altered hemichannel activities of Cx26I30N and D50Y channels

Cx26 mutations associated with KID syndrome have been shown to cause elevated hemichannel activities [19, 21, 22]. Here, we determined the ability of Cx26I30N mutation to form aberrant hemichannels in addition to D50Y mutation that was previously shown to result in elevated membrane currents in *Xenopus* oocytes by electrophysiological measurements [26] in mammalian cell lines with fluorescent dye uptake assay using Neurobiotin (NB) (Fig. 3). Under physiological calcium concentrations, Cx26WT hemichannels had 10 % increase in the mean fluorescent intensity of NB in cells compared to negative control.

**Table 1** Analysis of colocalization between Cx26 and golgin-97, the Golgi apparatus marker

|  | Cells analyzed (n) | Colocalization (n) | Ratio (%) |
|---|---|---|---|
| Cx26WT | 25 | 2 | 8[a] |
| Cx26I30N | 33 | 19 | 58[b] |
| Cx26D50Y | 26 | 13 | 50[b] |

[a,b]Different letters indicate statistically significant differences among groups

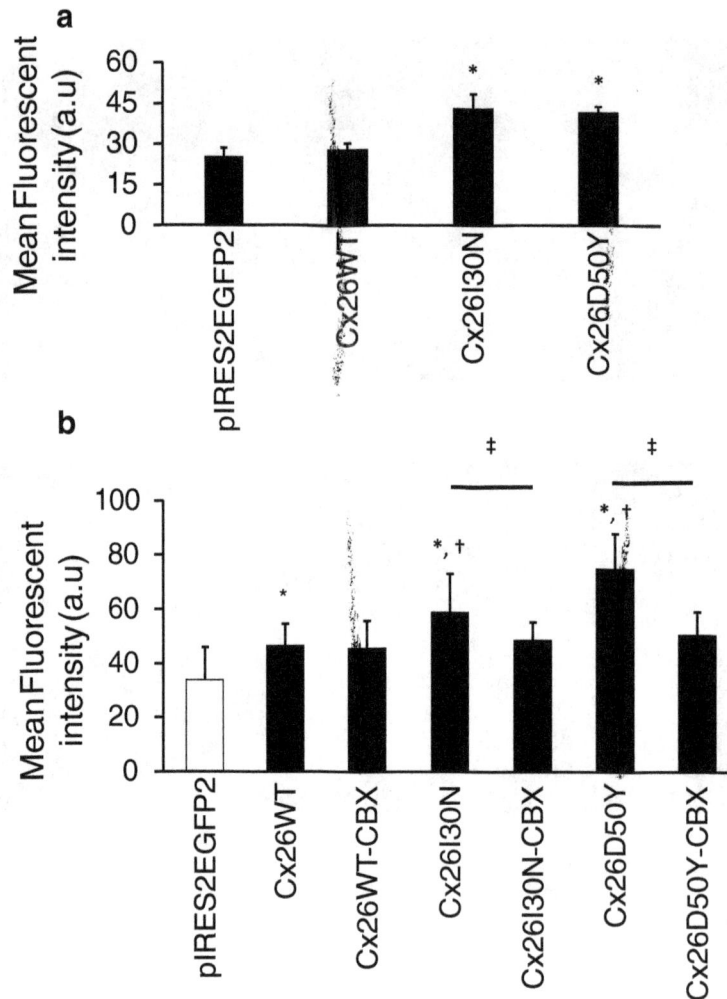

**Fig. 3** Fluorescent dye uptake in Cx26WT, I30N and D50Y transfected cells. **a** Comparison of fluorescent dye intensities in transfected cells under physiological calcium concentration. Cx26I30N and D50Y were compared with both negative control pIRES2EGFP2 and Cx26WT (*, $p < 0.01$). **b** Comparison of fluorescent dye intensities in transfected cells in divalent free medium and in the presence of 100 μM CBX, a hemichannel blocker. Cx26WT, I30N and D50Y were compared with negative control pIRES2EGFP2 (*, $p < 0.01$); Cx26I30N and D50Y were compared with Cx26WT (†, $p < 0.01$) and comparison between samples treated with or without CBX was performed (‡, $p < 0.01$)

On the other hand, Cx26I30N and D50Y expressing cells had 1.5 fold increase in NB uptake with respect to WT containing cells ($n = 15$ images, $p < 0.01$) (Fig. 3a), suggesting an aberrant hemichannel activities. To further verify the formation of abnormal hemichannels on the plasma membrane, dye uptake studies were performed under divalent free conditions and in the presence of a hemichannel blocker, carbenoxolone (CBX) (Fig. 3b). Analysis of NB fluorescent intensity in EGFP positive cells suggested that the fluorescent intensity of Cx26WT cells was 1.3 fold ($n = 50$ images, $p < 0.01$) higher than control cells and for I30N and D50Y cells the fluorescent intensities increased by 1.3 ($n = 60$ images, $p < 0.01$) and 1.6 fold ($n = 49$ images, $p < 0.01$) compared to Cx26WT cells, respectively (Fig. 3b). To determine if the uptake of NB into

the cells were mediated by Cx26I30N and D50Y abnormal hemichannels, cells were initially treated with a hemichannel blocker, carbenoxolone (CBX) before NB application. Treatment of cells with 100 μM CBX for 20 min resulted in a 17 % and 35 % (both, $p < 0.01$) reduction in the levels of mean fluorescent intensities in Cx26I30N and D50Y expressing cells compared to their CBX absent counterparts, respectively. These suggested that the increase in the uptake of NB into cells were mediated by aberrant hemichannels formed from Cx26I30N and D50Y.

### Effect of Cx26I30N and D50Y mutations on intracellular calcium signals

Calcium signals are essential for the maintenance of epidermal homeostasis [27–29]. Therefore, we next wanted to investigate whether Cx26I30N and D50Y mutations have any effect on internal calcium content by using a calcium indicator, Fluo-3 AM, and flow cytometry (Fig. 4). Comparison of Fluo-3 AM signals in pCS2+, Cx26WT, I30N and D50Y transfected cells demonstrated that intracellular calcium content in both I30N and D50Y transfected cells were elevated by 1.4 ($p < 0.01$) and 1.6 fold ($p < 0.01$) compared to cells with pCS2+ and Cx26WT clones, respectively (Fig. 4). This increase in intracellular calcium content was reduced by 34 % and 50 % (both, $p < 0.01$) with the treatment of 100 μM CBX in Cx26I30N and D50Y containing cells, respectively, suggesting the involvement of aberrant hemichannels in the elevation of intracellular calcium concentration (Fig. 4).

## Discussion

Characterization of Cx26 mutations associated with KID syndrome will improve the understanding of both Cx26 function in normal epidermal homeostasis and generation of disease phenotypes in affected individuals. Examination of several Cx26 mutations suggested that KID syndrome mutations lead to the formation of active hemichannels on the plasma membrane that might result in uncontrolled exchange of molecules between the cytosol and the extracellular environment, influencing the cellular/tissue homeostasis [19, 30, 31]. Here, we characterized two additional KID syndrome mutations, Cx26I30N and D50Y, in order to understand their effect on the protein biosynthetic pathway, hemichannel activity and

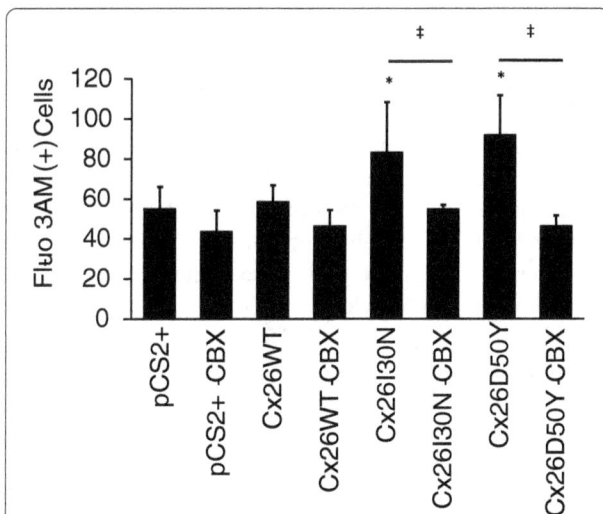

**Fig. 4** Effect of Cx26 mutations on intracellular calcium levels. Intracellular calcium concentration determined by a calcium indicator (Fluo-3 AM) using flow cytometry was compared among pCS2+, Cx26WT, I30N and D50Y containing cells (*, $p < 0.01$). Comparison between 100 μM CBX treated and non-treated samples was also performed (‡, $p < 0.01$)

intracellular calcium levels in communication deficient cell lines. We demonstrated that both mutants failed to form gap junction plaques at cell-to-cell contact sites between neighboring cells in contrast to Cx26WT proteins even though a fraction of mutant proteins was targeted to the plasma membrane. Moreover, cells with mutant constructs had increased uptake of neurobiotin fluorescent dye and elevated intracellular calcium levels compared to cells with Cx26WT which were abolished by carbenoxolone, a hemichannel blocker. These observations provided support for the contribution of aberrant hemichannels and the involvement of altered calcium signals in KID syndrome [18, 32].

Mutations in the Cx26 gene cause both non-syndromic and syndromic deafness with skin pathologies. Each mutation is unique in terms of defects they cause, that is, mutations leading to various skin disorders result in distinct molecular abnormalities within cells at different levels from protein synthesis to the formation and the function of gap junction channels [3, 12]. Mutations associated with KID syndrome, for example, were shown to mostly cause abnormal hemichannel activity leading to altered regulation of molecular exchange through the plasma membrane [19–22, 33]. Similar to many characterized KID syndrome mutations, Cx26I30N and D50Y also resulted in increased fluorescent dye uptake from the external environment, suggesting the involvement of aberrant hemichannel activities for these mutations.

Disease associated Cx26 mutations lead to different molecular abnormalities that can interfere with the cellular homeostasis [3]. Included among these are alteration in trafficking of proteins to the plasma membrane, where they could prevent the formation of gap junction plaques at cell-to-cell contact sites. Abnormal membrane trafficking might result in the accumulation of mutant proteins within the subcellular structures including the ER, the ER-Golgi intermediate compartment or the Golgi apparatus [34–37]. In this study, Cx26I30N and D50Y mutations were observed to abolish the formation of gap junction plaques between adjacent cells. Although mutant proteins were observed on the plasma membrane, they were also widely accumulated in the Golgi apparatus. How mutations in isoleucine 30 (I30) and aspartic acid at position 50 (D50) affect the localization and function of Cx26 is not known. I30 is found in the first transmembrane domain (TM1) that was shown to be important for oligomerization and channel function [1, 38]. The mutation of a nonpolar residue into a polar amino acid (asparagine) might affect the structure of the TM1 as asparagine can form hydrogen bonds with other residues that can affect the organization and stability of this domain.

The inability of I30N mutation to form gap junction channels suggest that this amino acid might influence the docking of hemichannels in the extracellular space during the formation of gap junction channels. The first extracellular loop (EC1) was shown to be involved in connexon formation, docking of connexons at the extracellular space and gap junction channel function [1, 39]. D50 in EC1 is highly conserved across species and across the connexin family [16]. The mutation of this charged amino acid might alter the organization of the EC1, affecting the interactions of connexons in the ESC as they are forming the gap junction channels [22, 30].

The maintenance of calcium homeostasis is crucial for proper functioning of keratinocytes and the epidermis. Calcium plays roles in the regulation of keratinocyte proliferation, differentiation and cellular adhesion processes [29, 40]. Furthermore, there is a calcium gradient across the epidermis with the lowest concentration in the stratum basale and the highest in the stratum granulosum. Maintenance of this gradient and intracellular calcium levels are indispensable for epidermal cells as abnormalities in these processes have been linked to a variety of skin disorders [41–43]. Here, Cx26I30N and D50Y mutations were shown to result in the elevation of intracellular calcium in transfected cells compared to Cx26WT cells. An increase in the intracellular calcium levels was also observed for other Cx26 KID syndrome associated mutations [18, 32]. Terrinoni et al. demonstrated that cells with Cx26G11E and D50N mutations have an increased amount of intracellular calcium 24 h post transfection [18]. Moreover, Garcia et al. observed a significant increase in the intracellular calcium levels in cells co-expressing Cx43 and KID syndrome associated Cx26 mutant constructs compared to Cx43 alone cells [32]. Thus, deregulation of intracellular calcium mechanisms by aberrant hemichannel activities may perturb several calcium dependent cellular processes to contribute to the development of skin phenotypes.

## Conclusion

Formation of aberrant hemichannels due to Cx26I30N and D50Y mutations provided further support for the involvement of abnormal hemichannel activities as a common mechanism for KID syndrome. Uncontrolled transport of signaling molecules such as ATP or calcium through these channels may alter keratinocyte growth, proliferation and/or differentiation that could play role in the generation of dermatological anomalies in KID patients. However, the molecular mechanisms altered by aberrant Cx26 hemichannels affecting keratinocyte proliferation, differentiation and migration remain to be elusive.

## Methods

### Site-directed mutagenesis and construction of Cx26 mutant clones

Cx26I30N and D50Y missense point mutations were generated from human wild-type Cx26 cDNA in pBlueScript II (courtesy of Prof. Dr. Thomas W. White, Stony Brook University, Stony Brook, NY, USA) by using site-directed mutagenesis. PCR products were cloned into pBlueScript II cloning vector and location and insertion of mutations were verified by sequencing (Macrogen Europe, the Netherlands). Then, Cx26WT, I30N and D50Y cDNAs were subcloned into pIRES2EGFP2 (Clontech Laboratories, USA) and pCS2+ mammalian expression vectors for immunostaining and functional studies.

### Cell culture

Gap junctional communication deficient cell lines, HeLa (human cervical cell line) and neuro-2A (N2A, mouse neuroblastoma cell line) (purchased from ATCC, USA), were maintained in Dulbecco's Modified Eagle Medium (DMEM, Thermo Scientific HyClone, USA) supplemented with 10 % fetal bovine serum (FBS) (Biological Industries, Israel) and 1 % penicillin/streptomycin (GIBCO, USA) in a humidified chamber with 5 % $CO_2$ and 37 °C. For experiments, cells were plated in 6 well plates so that they would reach 70-80 % confluence on the day of transfection with Lipofectamine 2000 reagent (Invitrogen, USA). Cells were transfected with 1:2 DNA to Lipofectamine 2000 ratio following the manufacturer's protocol and 3.2 mM $CaCl_2$ was added to prevent the cell death due to the activities of prospective hemichannels. Cells were used within 24–48 h after transfection.

### Immunofluorescent stainings

HeLa cells ($2{,}5 \times 10^5$) were grown over glass coverslips in 6 well plates for immunofluorescent staining experiments. 24 h post transfection, cells were washed with PBS and then fixed with 4 % paraformaldehyde (PFA) for 20 min, permeabilized with 0.1 % Triton-X 100 for 15 min and blocked with 3 % bovine serum albumin (BSA) for 1 h at room temperature (RT). Cells were then incubated with a 1:500 dilution of a polyclonal rabbit antibody against Cx26 (Invitrogen, USA) for 1 h at RT followed by an application of a 1:200 dilution of Alexa555-conjugated goat anti-rabbit antibody (Invitrogen, USA), 1:200 dilution of Alexa Fluor 488 conjugated Phalloidin (Invitrogen, USA) and 1 μM DAPI for 45 min at RT in dark. For co-immunofluorescent staining of golgin-97, a marker for the Golgi apparatus, and Cx26 proteins, cells were incubated with primary antibodies (rabbit anti-Cx26 antibody (1:500) and mouse anti-golgin-97 (1:1000) (Invitrogen, USA)) for 1 h at room temperature. Then, secondary antibodies (Alexa555-conjugated goat anti-rabbit (Invitrogen, USA) and Alexa488-conjugated goat anti-mouse secondary antibody (1:200) (Invitrogen, USA) and 1 µM DAPI were applied to cells for 45 min at RT in dark. After washing with PBS, coverslips were dipped in distilled water, dried and mounted on glass slides. Staining was verified under fluorescence microscope (IX83, Olympus, Japan) with x40 or x100 oil-immersion objective and images were taken with a CCD digital camera.

For co-labeling of Rhodamine labeled Wheat Germ Agglutinin (WGA) and Cx26, after washing cells with PBS twice, 1:200 dilution of Rhodamine labeled WGA (5 mg/ml, Vector Labs, USA) was applied to cells for 30 min at 4 °C. Following the wash with PBS, cells were fixed and immunostained with Cx26 as explained above. Staining was verified under fluorescent microscope (IX83, Olympus, Japan) with x100 oil objective and images were taken with a CCD digital camera.

## Dye uptake
N2A cells were transfected with pIRES2EGFP2 clones for dye uptake assays with neurobiotin (NB, 287 Da, +1 charge, Vector Labs, USA). 24 h after transfection, cells were washed with PBS and incubated with $Ca^{+2}$ free medium or medium with physiological calcium concentrations for 20 min at 37 °C. Cells were then incubated with 0.5 mg/ml NB for 20 min at 37 °C [19]. Next, cells were washed with PBS containing 3,2 mM $CaCl_2$ three times for 10 min and were fixed with 4 % PFA for 20 min at room temperature. Subsequently, fixed cells were permeabilized with 0.1 % Triton-X 100 for 10 min, blocked with 3 % BSA-0.1 % TritonX-100 for 15 min and incubated with tetra-methyl rhodamine isothiocyanate (TRIT-C) conjugated streptavidin (1:1000 dilution, Pierce, USA) for 30 min at RT in dark. After washing with PBS three times for 10 min, images were acquired with a fluorescence microscope (IX71, Olympus, Japan) using the same exposure times. For carbenoxolone (CBX) treatment, cells were initially incubated with calcium free medium containing 100 µM CBX for 20 min and then NB in the same medium was applied to the cells as explained above.

Image analysis for signal intensity determination was performed with ImageJ (NIH, USA) program. During image analysis, after subtracting the background in merged images of red and green channels, the same parameters were applied to threshold the images for the measurement of red signal intensities of fluorescence only in GFP positive cells.

## Measurement of Intracellular calcium levels
Flow cytometry was used for the measurement of internal calcium content with Fluo-3 AM $Ca^{+2}$ indicator (Invitrogen, USA) [18]. N2A cells ($5x10^5$) were plated onto 6 well plates and transfected with pCS2+ clones. Cells were washed with $Ca^{+2}$ free PBS or physiological medium twice for 5 min and incubated with 5 µM Fluo-3 AM at 37 °C for 30 min. Then, cells were washed with PBS, trypsinized and resuspended in divalent cation free Hank's Balanced Salt Solution (HBSS) [44] and analyzed with FACSCANTO (BD Biosciences, USA).

## Colocalization analysis
The determination of colocalization of Cx26 and the Golgi apparatus marker, golgin-97, was carried on by using Fiji Coloc 2 plug-in. Costes' colocalization coefficient was calculated that compares a pair of images from different fluorescent channels and quantifies the statistical significance of colocalization [45].

## Statistical analysis
All results were expressed as mean (± standard deviation). For colocalization quantification, first the pixels in the green and red channels were correlated to calculate a Pearson's coefficient (PC) and then pixels in green channel were randomized and correlated with the original red channel for 200 times. PCs from 200 perturbations were compared with the original PC of non-randomized image and if original PC is higher than 95 %, a significant colocalization was accepted for the image. Groups were compared using ANOVA followed by Tukey's HSD for all results. Colocalization data that only have positive or negative results were compared using Kruskal-Wallis rank sum test followed by Mann–Whitney test. Comparison of samples in the absence and the presence of CBX was done by using Student's $t$-test. Statistical significance was considered for $p < 0.05$.

## Additional file

Additional file 1: Raw data for the following experiments: Cx26-golgin97 colocalization, neurobiotin uptake assay under physiological condition, neurobiotin uptake assay without calcium and in the presence of carbenexolone (CBX) and intracellular calcium content experiments. (XLSX 25 kb)

Competing interests
The authors declare that they have no competing interests.

Authors' contributions
HA: Performed transfections, immunostaining experiments, intracellular calcium content determination, helped to draft the manuscript. VB: performed transfections and performed dye uptake assays. GMO: Immunostaining experiments, CBX inhibitor experiments, relevant statistical analysis, conceived, designed and coordinated the study; wrote the manuscript. All authors read and approved the final manuscript.

Acknowledgements
We thank Dr. Thomas W. White from Stony Brook University, NY, USA for kindly providing pIRES2EGFP2, pCS2+ and pBSBK-Cx26WT vectors. Expert technical help from Drs. Ozden Yalcin-Ozuysal, Engin Ozcivici and Izmir

Institute of Technology, Biotechnology and Bioengineering Research and Application Center is well appreciated. Further, we thank Drs. Ozden Yalcin-Ozuysal, Engin Ozcivici and Thomas W White for their critical comments on the manuscript.

### Grants

This work was supported by The Scientific and Technological Research Council of Turkey Career Grant (210T035) and FP7 Marie Curie Re-Integration Grant (PIRG08-GA-2010-277101).

### References

1. Yeager M, Harris AL. Gap junction channel structure in the early 21st century: facts and fantasies. Curr Opin Cell Biol. 2007;19(5):521–8.
2. Goodenough DA, Paul DL. Gap junctions. Cold Spring Harbor Perspect Biol. 2009;1(1):a002576.
3. Kelly JJ, Simek J, Laird DW. Mechanisms linking connexin mutations to human diseases. Cell Tissue Res. 2015;360(3):701–21.
4. Retamal MA, Froger N, Palacios-Prado N, Ezan P, Saez PJ, Saez JC, et al. Cx43 hemichannels and gap junction channels in astrocytes are regulated oppositely by proinflammatory cytokines released from activated microglia. J Neurosci. 2007;27(50):13781–92.
5. Laird DW. The gap junction proteome and its relationship to disease. Trends Cell Biol. 2010;20(2):92–101.
6. De Vuyst E, Decrock E, De Bock M, Yamasaki H, Naus CC, Evans WH, et al. Connexin hemichannels and gap junction channels are differentially influenced by lipopolysaccharide and basic fibroblast growth factor. Mol Biol Cell. 2007;18(1):34–46.
7. Saez JC, Schalper KA, Retamal MA, Orellana JA, Shoji KF, Bennett MV. Cell membrane permeabilization via connexin hemichannels in living and dying cells. Exp Cell Res. 2010;316(15):2377–89.
8. Wang N, De Bock M, Decrock E, Bol M, Gadicherla A, Vinken M, et al. Paracrine signaling through plasma membrane hemichannels. Biochim Biophys Acta. 2013;1828(1):35–50.
9. Duman D, Tekin M. Autosomal recessive nonsyndromic deafness genes: a review. Front Biosci. 2012;17:2213–36.
10. Mese G, Richard G, White TW. Gap junctions: basic structure and function. J Investig Dermatol. 2007;127(11):2516–24.
11. Hoang Dinh E, Ahmad S, Chang Q, Tang W, Stong B, Lin X. Diverse deafness mechanisms of connexin mutations revealed by studies using in vitro approaches and mouse models. Brain Res. 2009;1277:52–69.
12. Lee JR, White TW. Connexin-26 mutations in deafness and skin disease. Expert Rev Mol Med. 2009;11, e35.
13. Richard G. Connexin disorders of the skin. Clin Dermatol. 2005;23(1):23–32.
14. Xu J, Nicholson BJ. The role of connexins in ear and skin physiology - functional insights from disease-associated mutations. Biochim Biophys Acta. 2013;1828(1):167–78.
15. Jonard L, Feldmann D, Parsy C, Freitag S, Sinico M, Koval C, et al. A familial case of Keratitis-Ichthyosis-Deafness (KID) syndrome with the GJB2 mutation G45E. Eur J Med Genet. 2008;51(1):35–43.
16. van Steensel MA, van Geel M, Nahuys M, Smitt JH, Steijlen PM. A novel connexin 26 mutation in a patient diagnosed with keratitis-ichthyosis-deafness syndrome. J Invest Dermatol. 2002;118(4):724–7.
17. Montgomery JR, White TW, Martin BL, Turner ML, Holland SM. A novel connexin 26 gene mutation associated with features of the keratitis-ichthyosis-deafness syndrome and the follicular occlusion triad. J Am Acad Dermatol. 2004;51(3):377–82.
18. Terrinoni A, Codispoti A, Serra V, Didona B, Bruno E, Nistico R, et al. Connexin 26 (GJB2) mutations, causing KID Syndrome, are associated with cell death due to calcium gating deregulation. Biochem Biophys Res Commun. 2010;394(4):909–14.
19. Mese G, Sellitto C, Li L, Wang HZ, Valiunas V, Richard G, et al. The Cx26-G45E mutation displays increased hemichannel activity in a mouse model of the lethal form of keratitis-ichthyosis-deafness syndrome. Mol Biol Cell. 2011;22(24):4776–86.
20. Stong BC, Chang Q, Ahmad S, Lin X. A novel mechanism for connexin 26 mutation linked deafness: cell death caused by leaky gap junction hemichannels. Laryngoscope. 2006;116(12):2205–10.
21. Gerido DA, DeRosa AM, Richard G, White TW. Aberrant hemichannel properties of Cx26 mutations causing skin disease and deafness. Am J Physiol Cell Physiol. 2007;293(1):C337–45.
22. Lee JR, Derosa AM, White TW. Connexin mutations causing skin disease and deafness increase hemichannel activity and cell death when expressed in Xenopus oocytes. J Investig Dermatol. 2009;129(4):870–8.
23. Yotsumoto S, Hashiguchi T, Chen X, Ohtake N, Tomitaka A, Akamatsu H, et al. Novel mutations in GJB2 encoding connexin-26 in Japanese patients with keratitis-ichthyosis-deafness syndrome. Br J Dermatol. 2003;148(4):649–53.
24. Choung YH, Shin YR, Kim HJ, Kim YC, Ahn JH, Choi SJ, et al. Cochlear implantation and connexin expression in the child with keratitis-ichthyosis-deafness syndrome. Int J Pediatr Otorhinolaryngol. 2008;72(6):911–5.
25. Arndt S, Aschendorff A, Schild C, Beck R, Maier W, Laszig R, et al. A novel dominant and a de novo mutation in the GJB2 gene (connexin-26) cause keratitis-ichthyosis-deafness syndrome: implication for cochlear implantation. Otol Neurotol. 2010;31(2):210–5.
26. Lopez W, Gonzalez J, Liu Y, Harris AL, Contreras JE. Insights on the mechanisms of Ca(2+) regulation of connexin26 hemichannels revealed by human pathogenic mutations (D50N/Y). J Gen Physiol. 2013;142(1):23–35.
27. Fuchs E. Scratching the surface of skin development. Nature. 2007;445(7130):834–42.
28. Blanpain C, Fuchs E. Epidermal stem cells of the skin. Annu Rev Cell Dev Biol. 2006;22:339–73.
29. Martin PE, Easton JA, Hodgins MB, Wright CS. Connexins: sensors of epidermal integrity that are therapeutic targets. FEBS Lett. 2014;588(8):1304–14.
30. Mhaske PV, Levit NA, Li L, Wang HZ, Lee JR, Shuja Z, et al. The human Cx26-D50A and Cx26-A88V mutations causing keratitis-ichthyosis-deafness syndrome display increased hemichannel activity. Am J Physiol Cell Physiol. 2013;304(12):C1150–8.
31. Levit NA, Sellitto C, Wang HZ, Li L, Srinivas M, Brink PR, et al. Aberrant connexin26 hemichannels underlying keratitis-ichthyosis-deafness syndrome are potently inhibited by mefloquine. J Invest Dermatol. 2015;135(4):1033–42.
32. Garcia IE, Maripillan J, Jara O, Ceriani R, Palacios-Munoz A, Ramachandran J, et al. Keratitis-ichthyosis-deafness syndrome-associated Cx26 mutants produce nonfunctional gap junctions but hyperactive hemichannels when co-expressed with wild type Cx43. J Invest Dermatol. 2015;135(5):1338–47.
33. Sanchez HA, Mese G, Srinivas M, White TW, Verselis VK. Differentially altered Ca2+ regulation and Ca2+ permeability in Cx26 hemichannels formed by the A40V and G45E mutations that cause keratitis ichthyosis deafness syndrome. J Gen Physiol. 2010;136(1):47–62.
34. Thomas T, Telford D, Laird DW. Functional domain mapping and selective trans-dominant effects exhibited by Cx26 disease-causing mutations. J Biol Chem. 2004;279(18):19157–68.
35. Marziano NK, Casalotti SO, Portelli AE, Becker DL, Forge A. Mutations in the gene for connexin 26 (GJB2) that cause hearing loss have a dominant negative effect on connexin 30. Hum Mol Genet. 2003;12(8):805–12.
36. Su CC, Li SY, Su MC, Chen WC, Yang JJ. Mutation R184Q of connexin 26 in hearing loss patients has a dominant-negative effect on connexin 26 and connexin 30. Eur J Hum Genet. 2010;18(9):1061–4.
37. Berger AC, Kelly JJ, Lajoie P, Shao Q, Laird DW. Mutations in Cx30 that are linked to skin disease and non-syndromic hearing loss exhibit several distinct cellular pathologies. J Cell Sci. 2014;127(Pt 8):1751–64.
38. Jara O, Acuna R, Garcia IE, Maripillan J, Figueroa V, Saez JC, et al. Critical role of the first transmembrane domain of Cx26 in regulating oligomerization and function. Mol Biol Cell. 2012;23(17):3299–311.
39. Bai D, Wang AH. Extracellular domains play different roles in gap junction formation and docking compatibility. Biochem J. 2014;458(1):1–10.
40. Pani B, Singh BB. Darier's disease: a calcium-signaling perspective. Cell Mol Life Sci. 2008;65(2):205–11.
41. Bikle DD, Oda Y, Xie Z. Calcium and 1,25(OH)2D: interacting drivers of epidermal differentiation. J Steroid Biochem Mol Biol. 2004;89–90(1–5):355–60.
42. Proksch E, Brandner JM, Jensen JM. The skin: an indispensable barrier. Exp Dermatol. 2008;17(12):1063–72.
43. Lopez-Pajares V, Yan K, Zarnegar BJ, Jameson KL, Khavari PA. Genetic pathways in disorders of epidermal differentiation. Trends Genet. 2013;29(1):31–40.
44. Caro AA, Cederbaum AI. Role of calcium and calcium-activated proteases in CYP2E1-dependent toxicity in HEPG2 cells. J Biol Chem. 2002;277(1):104–13.
45. Bolte S, Cordelieres FP. A guided tour into subcellular colocalization analysis in light microscopy. J Microsc. 2006;224(Pt 3):213–32.

# Permissions

The contributors of this book come from diverse backgrounds, making this book a truly international effort. This book will bring forth new frontiers with its revolutionizing research information and detailed analysis of the nascent developments around the world.

We would like to thank all the contributing authors for lending their expertise to make the book truly unique. They have played a crucial role in the development of this book. Without their invaluable contributions this book wouldn't have been possible. They have made vital efforts to compile up to date information on the varied aspects of this subject to make this book a valuable addition to the collection of many professionals and students.

This book was conceptualized with the vision of imparting up-to-date information and advanced data in this field. To ensure the same, a matchless editorial board was set up. Every individual on the board went through rigorous rounds of assessment to prove their worth. After which they invested a large part of their time researching and compiling the most relevant data for our readers.

The editorial board has been involved in producing this book since its inception. They have spent rigorous hours researching and exploring the diverse topics which have resulted in the successful publishing of this book. They have passed on their knowledge of decades through this book. To expedite this challenging task, the publisher supported the team at every step. A small team of assistant editors was also appointed to further simplify the editing procedure and attain best results for the readers.

Apart from the editorial board, the designing team has also invested a significant amount of their time in understanding the subject and creating the most relevant covers. They scrutinized every image to scout for the most suitable representation of the subject and create an appropriate cover for the book.

The publishing team has been an ardent support to the editorial, designing and production team. Their endless efforts to recruit the best for this project, has resulted in the accomplishment of this book. They are a veteran in the field of academics and their pool of knowledge is as vast as their experience in printing. Their expertise and guidance has proved useful at every step. Their uncompromising quality standards have made this book an exceptional effort. Their encouragement from time to time has been an inspiration for everyone.

The publisher and the editorial board hope that this book will prove to be a valuable piece of knowledge for researchers, students, practitioners and scholars across the globe.

# List of Contributors

Lina Rimkutė, Vaidas Jotautis, Alina Marandykina, Renata Sveikatienė, Ieva Antanavičiūtė and Vytenis Arvydas Skeberdis
Institute of Cardiology, Lithuanian University of Health Sciences, 17 Sukilėlių Ave., 50009 Kaunas, Lithuania

Luis A. Cea and Christian Arriagada
Program of Anatomy and Developmental Biology, Faculty of Medicine, Institute of Biomedical Sciences, University of Chile, Av. Independencia #1027, Independencia, Santiago, Chile

Jorge A. Bevilacqua
Program of Anatomy and Developmental Biology, Faculty of Medicine, Institute of Biomedical Sciences, University of Chile, Av. Independencia #1027, Independencia, Santiago, Chile
Departamento de Neurología y Neurocirugía, Hospital Clínico Universidad de Chile, Universidad de Chile, Santiago, Chile

Ana María Cárdenas
Centro Interdisciplinario de Neurociencias de Valparaíso, Universidad de Valparaíso, Valparaíso, Chile

Anne Bigot and Vincent Mouly
Center for Research in Myology, Sorbonne Universités, UPMC Univ Paris 06, INSERM UMRS974, CNRS FRE3617, 47 Boulevard de l'hôpital, 75013 Paris, France

Juan C. Sáez
Centro Interdisciplinario de Neurociencias de Valparaíso, Universidad de Valparaíso, Valparaíso, Chile
Departamento de Fisiología, Facultad de Ciencias Biológicas, Pontificia Universidad Católica de Chile, Santiago, Chile

Pablo Caviedes
Programa de Farmacología Molecular y Clínica, Facultad de Medicina, Instituto de Ciencias Biomédicas, Universidad de Chile, Santiago, Chile

Qi Zhang, Man Shang, Mengxiao Zhang, Yao Wang, Yan Chen, Yanna Wu, Junqiu Song and Yanxia Liu
Department of Pharmacology, School of Basic Medical Sciences, Tianjin Medical University, No. 22, Qixiangtai Road, Heping District, Tianjin 300070, People's Republic of China

Minglin Liu
Section of Endocrinology, Department of Medicine, Temple University School of Medicine, 3500 North Broad Street, Room 480A, Philadelphia, PA 19140, USA
Department of Dermatology, Perelman School of Medicine, University of Pennsylvania, Philadelphia PA, 19104, USA

Qi Cao, Yiping Wang, Ya Wang, Vincent W. S. Lee, Padmashree Rao, David C. H. Harris, Tian Kui Tan and Guoping Zheng
Centre for Transplant and Renal Research, Westmead Institute for Medical Research, the University of Sydney, 176 Hawkesbury Road, Sydney, NSW 2145, Australia

Ye Zhao
Centre for Transplant and Renal Research, Westmead Institute for Medical Research, the University of Sydney, 176 Hawkesbury Road, Sydney, NSW 2145, Australia
The School of Biomedical Sciences, Chengdu Medical College, Chengdu 610500, PR China

Xi Qiao
Centre for Transplant and Renal Research, Westmead Institute for Medical Research, the University of Sydney, 176 Hawkesbury Road, Sydney, NSW 2145, Australia
Department of Nephrology, Second Hospital of Shanxi Medical University, Shanxi Kidney Disease Institute, WuYi Road 382, Taiyuan 030001, Shanxi, PR China

Lihua Wang
Department of Nephrology, Second Hospital of Shanxi Medical University, Shanxi Kidney Disease Institute, WuYi Road 382, Taiyuan 030001, Shanxi, PR China

**Hong Zhao and Jianlin Zhang**
Department of Biochemistry and Molecular Biology, Shanxi Medical University, Xinjian Road 56, Taiyuan 030001, Shanxi, PR China

**Yun Zhang**
Experimental Centre of Science and Research, the First Clinical Hospital of Shanxi Medical University, Xinjian Road 382, Taiyuan 030001, Shanxi, PR China

**Yuan Min Wang and Stephen I. Alexander**
Centre for Kidney Research, Children's Hospital at Westmead, 212 Hawkesbury Road, Sydney, NSW, Australia

**Pengcheng Wang**
Department of Immunology and Microbiology, Medical School of Jinan University, Guangdong, China

**Xiangming Zeng**
Department of Immunology and Microbiology, Medical School of Jinan University, Guangdong, China
Lundberg-Kienlen Lung Biology and Toxicology Laboratory, Department of Physiological Sciences, Stillwater, OK, USA

**Chaoqun Huang, Lakmini Senavirathna and Lin Liu**
Lundberg-Kienlen Lung Biology and Toxicology Laboratory, Department of Physiological Sciences, Stillwater, OK, USA
Oklahoma Center for Respiratory and Infectious Diseases, Oklahoma State University, Stillwater, OK, USA

**M. Jarad, E. A. Kuczynski, J. Morrison, A. M. Viloria-Petit and B. L. Coomber**
Department of Biomedical Sciences, Ontario Veterinary College, University of Guelph, OVC Room 3645, Guelph N1G 2W1, ON, Canada

**Amr Alraies, Rachel J. Waddington and Alastair J. Sloan**
Mineralised Tissue Group, Oral and Biomedical Sciences, School of Dentistry, College of Biomedical and Life Sciences, Cardiff University, Heath Park, CF14 4XY, Cardiff, UK
Cardiff Institute Tissue Engineering and Repair (CITER), Cardiff University, Cardiff CF14 4XY, UK

**Nadia Y. A. Alaidaroos and Ryan Moseley**
Cardiff Institute Tissue Engineering and Repair (CITER), Cardiff University, Cardiff CF14 4XY, UK
Stem Cells, Wound Repair and Regeneration, Oral and Biomedical Sciences, School of Dentistry, College of Biomedical and Life Sciences, Cardiff University, Heath Park, Cardiff CF14 4XY, UK

**Weronika Zarychta-Wiśniewska, Anna Burdzinska, Agnieszka Kulesza, Kamila Gala and Marek Sabat**
Department of Immunology, Transplantology and Internal Medicine, Transplantation Institute, Medical University of Warsaw, Nowogrodzka str. 59, 02-006 Warsaw, Poland

**Beata Kaleta**
Department of Clinical Immunology, Transplantation Institute, Medical University of Warsaw, Warsaw, Poland

**Katarzyna Zielniok**
Department of Physiological Sciences, Faculty of Veterinary Medicine, Warsaw University of Life Sciences, Warsaw, Poland

**Katarzyna Siennicka**
Department of Regenerative Medicine, Maria Sklodowska-Curie Memorial Cancer Center, Warsaw, Poland

**Leszek Paczek**
Department of Immunology, Transplantology and Internal Medicine, Transplantation Institute, Medical University of Warsaw, Nowogrodzka str. 59, 02-006 Warsaw, Poland
Department of Bioinformatics, Institute of Biochemistry and Biophysics, Polish Academy of Sciences, Warsaw, Poland

**Jung Seon Seo, Young Ha Choi, Hyeon Soo Kim and Sun-Hwa Park**
Department of Anatomy, Institute of Human Genetics, Korea University College of Medicine, 73, Inchon-ro, Seongbuk-gu, Seoul 02841, Republic of Korea

**Ji Wook Moon**
Department of Pathology, Korea University College of Medicine, 73, Inchon-ro, Seongbuk-gu, Seoul 02841, Republic of Korea

**R. Witt, A. Weigand, A. M. Boos, A. Cai, M. Hardt, A. Arkudas, R. E. Horch and J. P. Beier**
Department of Plastic and Hand Surgery and Laboratory for Tissue Engineering and Regenerative Medicine, University Hospital of Erlangen, Friedrich-Alexander University of Erlangen-Nürnberg (FAU), Krankenhausstraße 12, 91054 Erlangen, Germany

**D. Dippold**
Institute of Biomaterials, Department of Materials Science and Engineering, University of Erlangen-Nürnberg (FAU), Cauerstraße 6, 91058 Erlangen, Germany
Institute of Polymer Materials, Department of Materials Science and Engineering, University of Erlangen- Nürnberg (FAU), Martensstrasse 7, 91058 Erlangen, Germany

**A. R. Boccaccini**
Institute of Biomaterials, Department of Materials Science and Engineering, University of Erlangen-Nürnberg (FAU), Cauerstraße 6, 91058 Erlangen, Germany

**D. W. Schubert**
Institute of Polymer Materials, Department of Materials Science and Engineering, University of Erlangen- Nürnberg (FAU), Martensstrasse 7, 91058 Erlangen, Germany

**C. Lange**
Interdisciplinary Clinic for Stem Cell Transplantation, University Cancer Center Hamburg (UCCH), 20246 Hamburg, Germany

**Yuyu Li, Zhiai Hu, Chenchen Zhou, Yang Xu, Li Huang, Xin Wang and Shujuan Zou**
Department of Orthodontics, West China Hospital of Stomatology, Sichuan University, No. 14, 3rd Section, Renmin South Road, Chengdu 610041, China
State Key Laboratory of Oral Diseases, National Clinical Research Center for Oral Diseases, West China Hospital of Stomatology, Sichuan University, No. 14, 3rd Section, Renmin South Road, Chengdu 610041, China

**Johanna Huun and Liv B. Gansmo**
Section of Oncology, Department of Clinical Science, University of Bergen, 5020 Bergen, Norway

**Per E. Lønning and Stian Knappskog**
Section of Oncology, Department of Clinical Science, University of Bergen, 5020 Bergen, Norway
Department of Oncology, Haukeland University Hospital, Bergen, Norway

**Bård Mannsåker**
Section of Oncology, Department of Clinical Science, University of Bergen, 5020 Bergen, Norway
Department of Oncology, Haukeland University Hospital, Bergen, Norway
Present address: Department of Oncology and Palliative Medicine, Bodø, Norway

**Gjertrud Titlestad Iversen**
Department of Oncology, Haukeland University Hospital, Bergen, Norway

**Jan Inge Øvrebø**
Department of Biology, University of Bergen, Bergen, Norway
Present address: Huntsman Cancer Institute, University of Utah Health Care, Salt Lake City, USA

**Tünde Szatmári, Ashish Kumar-Singh, Lena Möbus and Rita Ötvös**
Department of Laboratory Medicine, Division of Pathology, Karolinska Institutet, SE-14186 Stockholm, Sweden

**Anders Hjerpe and Katalin Dobra**
Department of Laboratory Medicine, Division of Pathology, Karolinska Institutet, SE-14186 Stockholm, Sweden
Division of Clinical Pathology/Cytology, Karolinska University Laboratory, Karolinska University Hospital, SE-14186 Stockholm, Sweden

**Filip Mundt**
Division of Clinical Pathology/ Cytology, Karolinska University Laboratory, Karolinska University Hospital, SE-14186 Stockholm, Sweden

**Zhaoxing Dong, Wen Lei, Yin Wang, ZhenKun Li and Tao Zhang**
Department of Respiratory, The 2nd Affiliated Hospital of Kunming Medical University, Dianmian Road 374, Kunming, Yunnan 650101, China

**Wenlin Tai**
Department of Clinical Laboratory, Yunnan Molecular Diagnostic Center, The 2nd Affiliated Hospital of Kunming Medical University, Dianmian Road, Kunming, Yunnan, China

**Simon Boggild and Simon Molgaard**
MIND Centre, Stereology and Electron Microscopy Laboratory, Aarhus University, 8000 C Aarhus, Denmark
MIND Centre, Department of Biomedicine, Aarhus University, Ole Worms Allé 3, 8000 C Aarhus, Denmark

**Simon Glerup**
MIND Centre, Department of Biomedicine, Aarhus University, Ole Worms Allé 3, 8000 C Aarhus, Denmark

**Jens Randel Nyengaard**
MIND Centre, Stereology and Electron Microscopy Laboratory, Aarhus University, 8000 C Aarhus, Denmark
Centre for Stochastic Geometry and Advanced Bioimaging, Aarhus University, 8000 C Aarhus, Denmark

**Hande Aypek, Veysel Bay and Gülistan Meşe**
Department of Molecular Biology and Genetics, Izmir Institute of Technology, Urla, Izmir, Turkey

# Index

www.ingramcontent.com/pod-product-compliance
Lightning Source LLC
Chambersburg PA
CBHW082013190326
41458CB00010B/3170